Nonlinear Optimization and Related Topics

Applied Optimization

Volume 36

Series Editors:

Panos M. Pardalos
University of Florida, U.S.A.

Donald Hearn
University of Florida, U.S.A.

The titles published in this series are listed at the end of this volume.

Nonlinear Optimization and Related Topics

edited by

Gianni Di Pillo
University Di Roma La Sapienza

and

Franco Giannessi
University of Pisa

KLUWER ACADEMIC PUBLISHERS
DORDRECHT / BOSTON / LONDON

Library of Congress Cataloging-in-Publication Data

Nonlinear optimization and related topics / edited by Gianni Di Pillo and Franco Giannessi.
p. cm. -- (Applied optimization ; v. 36)
ISBN 0-7923-6109-1 (hc. : alk. paper)
1. Mathematical optimization--Congresses. 2. Nonlinear theories--Congresses. I. Di Pillo, G. II. Giannessi, F. III. Workshop on Nonlinear Optimization (1998 : Erice, Italy) IV. Applied optimization ; vol. 36.

QA402.5 .N55 1999

99-056953

ISBN 0-7923-6109-1

Published by Kluwer Academic Publishers,
P.O. Box 17, 3300 AA Dordrecht, The Netherlands.

Sold and distributed in North, Central and South America
by Kluwer Academic Publishers,
101 Philip Drive, Norwell, MA 02061, U.S.A.

In all other countries, sold and distributed
by Kluwer Academic Publishers,
P.O. Box 322, 3300 AH Dordrecht, The Netherlands.

Printed on acid-free paper

Printed in the Netherlands.

Contents

Preface

This volume contains the edited texts of the lectures presented at the Workshop on Nonlinear Optimization held in Erice, Sicily, at the "G. Stampacchia" School of Mathematics of the "E. Majorana" Centre for Scientific Culture, June 23 - July 2, 1998.

In the tradition of these meetings, the main purpose was to review and discuss recent advances and promising research trends concerning theory, algorithms and innovative applications in the field of Nonlinear Optimization, and of related topics such as Convex Optimization, Nonsmooth Optimization, Variational Inequalities and Complementarity Problems.

The meeting was attended by 83 people from 21 countries. Besides the lectures, several formal and informal discussions took place. The result was a wide and deep knowledge of the present research tendencies in the field.

We wish to express our appreciation for the active contribution of all the participants in the meeting. Our gratitude is due to the Ettore Majorana Centre in Erice, which offered its facilities and rewarding environment: its staff was certainly instrumental for the success of the meeting. Our gratitude is also due to Francisco Facchinei and Massimo Roma for the effort and time devoted as members of the Organising Committee. We are indebted to the Italian National Research Council, and in particular to the Group on Functional Analysis and its Applications and to the Committees on Engineering Sciences and on Information Sciences and Technologies for their financial support. Finally, we address our thanks to Kluwer Academic Publishers for having offered to publish this volume.

Gianni Di Pillo (Roma)
Franco Giannessi (Pisa)

Generalized Lagrange multipliers: regularity and boundedness [1]

Giancarlo Bigi (bigi@dm.unipi.it)
Department of Mathematics, Via F.Buonarroti, 2, 56127 Pisa, Italy.

Massimo Pappalardo (pappalar@dm.unipi.it)
Department of Mathematics, Via F.Buonarroti, 2, 56127 Pisa, Italy.

Abstract

A generalized Dini derivative for vector-valued functions is introduced in order to obtain a set of generalized Lagrange multipliers for vector minimum points. The relationships between the boundedness of this set and the regularity conditions introduced in [2] are studied.

Keywords: global Dini derivatives, total regularity, generalized Lagrange multipliers.

1 Introduction

Necessary optimality conditions for vector optimization problems can be established exploiting a vector Dini derivative, which is obtained, via Kuratowski limit, with a global operator and not with a componentwise one. These optimality conditions are formulated in terms of the impossibility of positively homogeneous systems. By linear separation arguments and adding convexlikeness assumptions on this vector Dini derivative, we obtain a set of generalized Lagrange multipliers for vector minimum points; such multipliers can be viewed as the gradient of the separating hyperplane.
The subsequent analysis is devoted to establish regularity conditions (or constraint qualifications) in order to have additional properties on these multipliers. Most of this analysis has been developed in [2, 3]. In this paper we apply the result of those papers

[1]This work was partially supported by MURST National Research Program "Metodi per l'Ottimizzazione di Sistemi e Tecnologie (MOST)".

G. Di Pillo and F. Giannessi (eds.), Nonlinear Optimization and Related Topics, 1–14.

paper is to show that the regularity conditions introduced in [2, 3] are equivalent to the boundedness of the set of multipliers and generalize classical results of the scalar case [5, 9, 11].

Let us recall some classical definitions and notations that will be used in the sequel. The sets *cl* A and *int* A are the closure and the interior of A. Let $||\cdot||$ be any norm on $\mathbb{R}^n$, $B_\delta(x)$ be the ball of center x and radius δ and $\langle \cdot , \cdot \rangle$ be the usual inner product. The function g: $\mathbb{R}^n \longrightarrow \mathbb{R}^m$ is locally Lipschitz at $\bar{x}$ iff

$$\exists\, \delta > 0, \;\; \exists\, L > 0, \;\; \forall\, x, y \in B_\delta(\bar{x}) \;\; : \;\; ||g(x) - g(y)|| \leq L||x - y||.$$

A set-valued map $D : \mathbb{R}^n \rightrightarrows \mathbb{R}^m$ is a mapping from $\mathbb{R}^n$ to the subset of $\mathbb{R}^m$; we say that D is convexlike on $X \subseteq \mathbb{R}^n$ iff

$$\forall\, x, y \in \mathrm{X}, \;\; \forall\, \lambda \in \; [0,1], \;\; \exists\, z \in \mathrm{X} \;\; : \;\; \lambda D(x) + (1-\lambda) D(y) \subseteq D(z) + \mathbb{R}^m_+.$$

Given a convex cone $\mathrm{K} \subseteq \mathbb{R}^q$ and a set $\mathrm{A} \subseteq \mathbb{R}^q$, we recall the following definition.

Definition 1.1.
(i) $\bar{x} \in A$ *is a minimum point of* A *iff there is no* $y \in A \setminus \{\bar{x}\}$ *such that* $\bar{x} - y \in K$.

(ii) $\bar{z} \in A$ *is a maximum point of* A *iff there is no* $y \in A \setminus \{\bar{z}\}$ *such that* $y - \bar{z} \in K$.

We will denote by $min_K\ A$ $(max_K\ A)$ the set of all minimum (maximum) points of A. Moreover, we also recall the following classical result (see [6, 7, 14]).

Theorem 1.2. *Let* K *be a pointed convex cone. If* A *is compact, then*

(i) $max_K\ A$ *and* $min_K\ A$ *are not empty;*

(ii) $\forall\, y \in A, \;\; \exists\, \bar{x} \in min_K\ A, \exists\, \bar{z} \in max_K\ A$ *such that* $y - \bar{x} \in K$ *and* $\bar{z} - y \in K$.

The above definitions allow to introduce the concept of solution of a vector constrained extremum problem in the following way.

Definition 1.3. *Given* $X \subseteq \mathbb{R}^n$ *and* $f : X \longrightarrow \mathbb{R}^q$, *we say that* $\bar{x} \in X$ *is a vector minimum point of the problem*

$$min_K \; \{ \; f(x) \;\; : \;\; x \in X \; \}$$

iff $f(\bar{x}) \in min_K\ f(X)$.

2 Global Dini derivatives

In this section we exploit new concepts of maximum and minimum limit of a vector-valued function and we employ them to introduce a set-valued Dini derivative of locally Lipschitz functions.

Given $f : X \longrightarrow \mathbb{R}^q$, $X \subseteq \mathbb{R}^n$, $\bar{x} \in cl\ X$, we consider the following set

$$A(f)(\bar{x}) := \bigcap_{\varepsilon>0} cl\ \{\ f(x)\ :\ x \in \tilde{B}_\varepsilon(\bar{x}) \cap X\ \}$$

where $\tilde{B}_\varepsilon(\bar{x}) := B_\varepsilon(\bar{x}) \backslash \{\bar{x}\}$. Observe that $A(f)(\bar{x})$ is always closed. Moreover, it can be considered as the set of all possible limits of $f(x)$ as x approaches to $\bar{x}$.

Proposition 2.1. *The following characterization holds:*

$$A(f)(\bar{x}) = \{\ y \in \mathbb{R}^q : \exists\ \{x_n\} \subseteq X, x_n \neq \bar{x},\ s.t.\ x_n \rightarrow \bar{x} \text{ and } f(x_n) \rightarrow y\ \}.$$

Proof. it is obvious that

$$\{\ y \in \mathbb{R}^q : \exists\ \{x_n\} \subseteq X, x_n \neq \bar{x},\ s.t.\ x_n \rightarrow \bar{x} \text{ and } f(x_n) \rightarrow y\ \} \subseteq A(f)(\bar{x}).$$

Vice versa, suppose that $y \in A(f)(\bar{x})$. Thus, for each $\varepsilon > 0$

$$\exists\{z_n(\varepsilon)\} \subseteq \tilde{B}_\varepsilon(\bar{x}) \cap X\ :\ f(z_n(\varepsilon)) \rightarrow y.$$

Let us define the sequence $\{x_k\} \subseteq X$, choosing $x_k = z_n(k^{-1})$ with n such that

$$||y - f(z_n(k^{-1}))|| < k^{-1}, \quad ||z_n(k^{-1}) - \bar{x}|| < k^{-1}.$$

Thus, we have both $x_k \rightarrow \bar{x}$ and $f(x_k) \rightarrow y$. □

Corollary 2.2. *If $f(B_\varepsilon(\bar{x}) \cap X)$ is bounded for some $\varepsilon > 0$, then $A(f)(\bar{x})$ is a nonempty compact set.*

When $f(B_\varepsilon(\bar{x}) \cap X)$ is bounded for some $\varepsilon > 0$, we will shortly say that f is bounded around $\bar{x}$.
The above result and Theorem 1.2 imply that the sets of minimum and maximum points of $A(f)(\bar{x})$ are not empty for all the functions, which are bounded around $\bar{x}$. Thus, the following definitions of maximum and minimum limit can be introduced in this framework without any ambiguity.

Definition 2.3. *Let f be bounded around $\bar{x}$.*

(i) $\limsup_{K,\, x\rightarrow\bar{x}} f(x) := \max_K A(f)(\bar{x})$.

(ii) $\liminf_{K,\, x\rightarrow\bar{x}} f(x) := \min_K A(f)(\bar{x})$.

Notice that generally these maximum and minimum limits are sets which contain more than one element. However, when $q = 1$ and $K = \mathbb{R}_+$, the above definitions collapse to the well-known concepts of lim sup and lim inf of a real-valued function. Moreover, these definitions allow to extend the characterization of continuity to vector-valued bounded functions.

Proposition 2.4. *Let f be bounded around $\bar{x}$. Then,*

f is continuous at $\bar{x} \in X$ iff $\limsup_{x \to \bar{x}} {}_K f(x) = \liminf_{x \to \bar{x}} {}_K f(x) = \{f(\bar{x})\}$.

Another possible definition is the componentwise one, that is considering the vectors whose components are the maximum and minimum limits of the single components of f. The two definitions are strictly connected, as shown by the following result [13].

Proposition 2.5. *Let f be bounded around $\bar{x}$. Then,*

(i) $\limsup_{x \to \bar{x}} {}_{\mathbb{R}^q_+} f(x) = \{(\limsup_{x \to \bar{x}} f_1(x), ..., \limsup_{x \to \bar{x}} f_q(x))\}$ *iff* $(\limsup_{x \to \bar{x}} f_1(x), ..., \limsup_{x \to \bar{x}} f_q(x)) \in A(f)(\bar{x})$.

(ii) $\liminf_{x \to \bar{x}} {}_{\mathbb{R}^q_+} f(x) = \{(\liminf_{x \to \bar{x}} f_1(x), ..., \liminf_{x \to \bar{x}} f_q(x))\}$ *iff* $(\liminf_{x \to \bar{x}} f_1(x), ..., \liminf_{x \to \bar{x}} f_q(x)) \in A(f)(\bar{x})$.

Many examples show that the componentwise maximum and minimum limits do not always belong to $A(f)(\bar{x})$ if $q > 1$. However, the following relationships between the componentwise definitions and the global ones always hold [13].

Proposition 2.6. *Let f be bounded around $\bar{x}$. Then,*

(i) $(\limsup_{x \to \bar{x}} f_1(x), ..., \limsup_{x \to \bar{x}} f_q(x)) - y \in \mathbb{R}^q_+ \quad \forall\, y \in \limsup_{x \to \bar{x}} {}_{\mathbb{R}^q_+} f(x)$;

(ii) $y - (\liminf_{x \to \bar{x}} f_1(x), ..., \liminf_{x \to \bar{x}} f_q(x)) \in \mathbb{R}^q_+ \quad \forall\, y \in \liminf_{x \to \bar{x}} {}_{\mathbb{R}^q_+} f(x)$.

The above introduced concepts of global lim sup and lim inf can be employed to introduce generalized derivatives of vector-valued functions, which do not rely on generalized derivatives of the single components.

Definition 2.7. *The incremental ratio of f is the function*

$$h_f(x, y, t) := t^{-1}[f(x + ty) - f(x)].$$

When f is locally Lipschitz at $\bar{x}$, the incremental ratio $h_f(\bar{x}, \bar{y}, \cdot)$ is bounded around $\bar{t} = 0$ for all directions $\bar{y} \in \mathbb{R}^n$. Thus, the maximum and minimum limit of this ratio can be considered in this framework.

Definition 2.8. *Let f be locally Lipschitz at $\bar{x}$. The lower Dini global derivative of f at $\bar{x}$ in the direction $\bar{y}$ is the set*

$$D^- f(\bar{x}; \bar{y}) := \liminf_{t \downarrow 0} {}_{\mathbb{R}^q_+} h_f(\bar{x}, \bar{y}, t).$$

In the same way it is possible to introduce the upper Dini global derivative, just considering the lim sup of the incremental ratio. By Proposition 2.1, this last concept coincides with the one introduced in [15].

The function $h_f(\bar{x}, \bar{y}, \cdot)$ depends on the chosen $\bar{x}$ and $\bar{y}$; moreover, the only considered limit is $t \downarrow 0$. Thus, we can denote $A(h_f(\bar{x}, \bar{y}, \cdot))(0)$ by $A_{h_f}(\bar{x}, \bar{y})$ without any possible misunderstanding. Using this notation, we observe that

$$D^- f(\bar{x}; \bar{y}) = \liminf_{t \downarrow 0}{}_{\mathbb{R}^q_+} \, h_f(\bar{x}, \bar{y}, t) = \min{}_{\mathbb{R}^q_+} A_{h_f}(\bar{x}, \bar{y}).$$

Given two vector-valued functions f and g, we can consider the pair (f, g) as an unique vector-valued function. Thus, any generalized derivative of the pair can be regarded as a joint generalized derivative of f and g. This can be useful in constrained extremum problems, where f and g are, respectively, the objective and the constraining functions.

Definition 2.9. *Let f and g be locally Lipschitz at $\bar{x}$. The joint lower Dini global derivative of f and g at $\bar{x}$ in the direction $\bar{y}$ is the set*

$$D^-(f, g)(\bar{x}; \bar{y}) := \liminf_{t \downarrow 0}{}_{\mathbb{R}^q_+} \, h_{(f,g)}(\bar{x}, \bar{y}, t),$$

where $h_{(f,g)}(\bar{x}, \bar{y}, t) := (h_f(\bar{x}, \bar{y}, t), h_g(\bar{x}, \bar{y}, t))$.

The joint generalized derivatives of f and g cannot be studied just exploiting the generalized derivatives of the single functions f and g. In fact, though, by Proposition 2.1, it is easy to prove that

$$A_{h_{(f,g)}}(\bar{x}, \bar{y}) \subseteq A_{h_f}(\bar{x}, \bar{y}) \times A_{h_g}(\bar{x}, \bar{y}),$$

we remark that it is not true that

$$D^-(f, g)(\bar{x}; \bar{y}) \subseteq D^- f(\bar{x}; \bar{y}) \times D^- g(\bar{x}; \bar{y}),$$

as shown by the following example.

Example 2.10. Let $f, g : \mathbb{R} \longrightarrow \mathbb{R}$ be the locally Lipschitz functions

$$f(x) := x \sin x^{-1}, \quad g(x) := x \cos x^{-1}, \quad x \neq 0,$$

with $f(0) = g(0) = 0$. Considering the direction $\bar{y} = 1$, easy calculations show that

$$A_{h_{(f,g)}}(0,1) = \{\ (a,b) \in \mathbb{R}^2 \ :\ a^2 + b^2 = 1\ \},$$
$$A_{h_f}(0,1) = A_{h_g}(0,1) = [-1,1].$$

Thus, taking the minimum points, we get

$$D^-(f,g)(0;1) = \{\ (a,b) \in \mathbb{R}^2_- \ :\ a^2 + b^2 = 1\ \},$$
$$D^- f(0;1) = D^- g(0;1) = \{-1\}.$$

Further studies on the structure of $D^-(f,g)(\bar{x};\bar{y})$ can be carried on. Since they go beyond the aim of the present paper, they will be the subject of future investigations.

3 Vector Optimality

We apply the concepts exploited in the previous section to obtain a new first order optimality condition for vector constrained extremum problems. Let

$$\text{(P)} \quad \min_{int\ \mathbb{R}^q_+} \{\ f(x) \ :\ g(x) \in \mathbb{R}^m_-,\ x \in X\ \},$$

where $X \subseteq \mathbb{R}^n$, $f : X \longrightarrow \mathbb{R}^q$ and $g : X \longrightarrow \mathbb{R}^m$.
Taking into account the relationships between generalized derivatives and tangent cones [4], we also consider the feasible (or radial) tangent cone:

$$F(X,\bar{x}) := \{\ y \in \mathbb{R}^n : \exists\ \lambda > 0,\ \forall\ t \in]0,\lambda[,\ \ \bar{x} + ty \in X\ \}.$$

The main idea is to use the joint Dini global derivatives of both the objective and the constraining functions, instead of the single derivatives. Such a joint derivative can be somehow considered as a "derivative of the whole problem".

Theorem 3.1. *Suppose that f and g are locally Lipschitz at $\bar{x}$. If $\bar{x}$ is a vector minimum point of (P), then*

$$D^-(f,g)(\bar{x};\bar{y}) \cap int\ \mathbb{R}^{q+m}_- = \emptyset, \qquad \forall\ \bar{y} \in F(X,\bar{x}). \tag{1}$$

Proof. Ab absurdo, suppose there exist $\bar{y} \in F(X,\bar{x})$ and $(u,v) \in D^-(f,g)(\bar{x};\bar{y})$ with $u \in int\ \mathbb{R}^q_-$ and $v \in int\ \mathbb{R}^m_-$. By Proposition 2.1,

$$\exists\ t_n \downarrow 0 \ :\ h_f(\bar{x},\bar{y},t_n) \to u,\ \ h_g(\bar{x},\bar{y},t_n) \to v.$$

Let $\varepsilon > 0$ be such that $u + B_\varepsilon(0) \subseteq int\ \mathbb{R}^q_-$ and $v + B_\varepsilon(0) \subseteq int\ \mathbb{R}^m_-$. Thus, for n

large enough we have that

$$t_n^{-1}[f(\bar{x}+t_n\bar{y})-f(\bar{x})]\in u+B_\varepsilon(0),\quad t_n^{-1}[g(\bar{x}+t_n\bar{y})-g(\bar{x})]\in v+B_\varepsilon(0).$$

Thus,

$$f(\bar{x}+t_n\bar{y})-f(\bar{x})\in t_n[u+B_\varepsilon(0)]\subseteq int\ \mathbb{R}^q_-,$$

$$g(\bar{x}+t_n\bar{y})\in g(\bar{x})+t_n[u+B_\varepsilon(0)]\subseteq \mathbb{R}^m_- + int\ \mathbb{R}^m_- = int\ \mathbb{R}^m_-.$$

Since $\bar{y}\in F(X,\bar{x})$, $\bar{x}+t_n\bar{y}\in X$ for all n large enough. Hence, choosing a suitable n, the feasible point $\bar{x}+t_n\bar{y}$ contradicts the optimality of $\bar{x}$. □

Remark 3.2. By Theorem 1.2, condition (1) is equivalent to the following one:

$$A_{h_{(f,g)}}(\bar{x},\bar{y})\cap int\ \mathbb{R}^{q+m}_- = \emptyset,\qquad \forall\, \bar{y}\in F(X,\bar{x}).$$

The use of the joint derivative is an essential tool in Theorem 3.1. In fact, generally it is not true that

$$\forall\, \bar{y}\in F(X,\bar{x})\ :\ \left\{\begin{array}{l} D^-f(\bar{x};\bar{y})\ \cap\ int\ \mathbb{R}^q_- = \emptyset \\ D^-g(\bar{x};\bar{y})\ \cap\ int\ \mathbb{R}^m_- = \emptyset \end{array}\right. \tag{2}$$

is a necessary optimality condition, as shown by the following example.

Example 3.3. Let $X=\mathbb{R}_+$, $f,g : X \longrightarrow \mathbb{R}$ be the locally Lipschitz functions

$$f(x) := -x\sin x^{-1},\quad g(x) := x\sin x^{-1},\quad x\neq 0,$$

with $f(0)=g(0)=0$. It is clear that $\bar{x}=0$ is a vector minimum point of (P). Moreover, given any $\bar{y}\in F(X,\bar{x})=\mathbb{R}_+$, we have

$$A_{h_f}(0,\bar{y}) = A_{h_g}(0,\bar{y}) = [-\bar{y},\bar{y}]$$

and thus

$$D^-f(0;\bar{y}) = D^-g(0;\bar{y}) = \{-\bar{y}\},$$

so that condition (2) does not hold. On the contrary, condition (1) holds, as Theorem 3.1 requires, since

$$D^-(f,g)(0;\bar{y}) = A_{h_{(f,g)}}(0,\bar{y}) = \{\ (-a,a)\in\mathbb{R}^2\ :\ a\in[0,\bar{y}]\ \}.$$

If we adapt a result of [13] to our framework, it is easy to prove the following necessary optimality condition, which does not involve joint derivatives.

Theorem 3.4. *Suppose that f and g are locally Lipschitz at $\bar{x}$. If $\bar{x}$ is a vector* *minimum point of (P), then it is impossible to find $\bar{y} \in F(X, \bar{x})$ such that*

$$\begin{cases} A_{h_f}(\bar{x}, \bar{y}) \cap int\ \mathbb{R}^q_- \neq \emptyset \\ A_{h_g}(\bar{x}, \bar{y}) \ \subseteq int\ \mathbb{R}^m_-. \end{cases} \tag{3}$$

However, Theorem 3.1 gives an optimality condition, which is stronger than the above one. In fact, we have the following result.

Proposition 3.5. *Suppose that g is locally Lipschitz at $\bar{x}$. If condition (1) holds, then the system (3) has no solution $\bar{y} \in F(X, \bar{x})$.*

Proof. Ab absurdo, suppose that $\bar{y} \in F(X, \bar{x})$ satisfies (3). Then, there exists $u \in A_{h_f}(\bar{x}, \bar{y})$ with $u \in int\ \mathbb{R}^q_-$. By Proposition 2.1

$$\exists\ t_n \downarrow 0\ :\ h_f(\bar{x}, \bar{y}, t_n) \to u.$$

Let us define

$$a_n := h_g(\bar{x}, \bar{y}, t_n).$$

Since g is locally Lipschitz at $\bar{x}$, the sequence $\{a_n\}$ is bounded. Taking the suitable subsequence, we can suppose that $a_n \to v$. Hence, $v \in A_{h_g}(\bar{x}, \bar{y}) \subseteq int\ \mathbb{R}^m_-$. Moreover, $(u, v) \in A_{h_{(f,g)}}(\bar{x}, \bar{y})$. By Remark 3.2, this contradicts (1). □

The vice versa does not hold, as Example 3.3 shows.

4 Regularity

Under suitable convexity assumptions, the optimality condition given in Theorem 3.1 can be turned into a generalized Karush-Kuhn-Tucker condition. Thus, the regularity scheme introduced in [2] can be considered.

¿From now onwards, we will denote all the elements of $D^-(f, g)(\bar{x}; \bar{y})$ as pairs $(u, v) \in \mathbb{R}^q \times \mathbb{R}^m$ in order to distinguish the roles of f and g in the joint derivative.

Theorem 4.1. *Suppose that f and g are locally Lipschitz at $\bar{x}$. Let the set-valued map $D^-(f, g)(\bar{x};\ \cdot)$ be convexlike on $F(X, \bar{x})$. If $\bar{x}$ is a vector minimum point of (P), then*

$$\begin{array}{l} \exists\ (\theta, \lambda) \in (\mathbb{R}^q_+ \times \mathbb{R}^m_+) \backslash \{0\}\ \text{such that}\ \forall\ \bar{y} \in F(X, \bar{x}) \\ \langle \theta\ ,\ u \rangle + \langle \lambda\ ,\ v \rangle \geq 0, \quad \forall\ (u, v) \in D^-(f, g)(\bar{x}; \bar{y}). \end{array} \tag{4}$$

Proof. To simplify notations, let $\mathcal{K} = \bigcup D^-(f, g)(\bar{x}; \bar{y})$ where the union is made over all $\bar{y} \in F(X, \bar{x})$. By Theorem 3.1, we have

$$\mathcal{K} \cap int\ \mathbb{R}_-^{q+m} = \emptyset.$$

Suppose that

$$\exists\ k \in \mathcal{K},\ r \in \mathbb{R}_+^{q+m}\ :\ k + r \in \mathbb{R}_-^{q+m}.$$

Thus,

$$k \in -r + int\ \mathbb{R}_-^{q+m} \subseteq \mathbb{R}_-^{q+m}$$

which is a contradiction. Hence, we also have

$$\mathcal{K} + \mathbb{R}_+^{q+m} \cap int\ \mathbb{R}_-^{q+m} = \emptyset.$$

Since $D^-(f,g)(\bar{x};\ \cdot)$ is convexlike on $F(X,\bar{x})$, the set $\mathcal{K}+\mathbb{R}_+^{q+m}$ is convex. Thus, (4) follows immediately from the Separation Theorem. □

The nonzero vectors (θ, λ), which satisfy (4), will be called generalized multipliers and we will denote by $\Gamma(\bar{x})$ the set of all generalized multipliers. We remark that $\Gamma(\bar{x})$ is a convex cone.

When $\theta = 0$, the objective function f does not play any role in the optimality condition (4); when $\theta_i = 0$, the same thing happens to the corresponding component of f. Therefore, the following classification has been proposed in [2].

Definition 4.2.

(i) *(P) is weak-regular at $\bar{x}$ iff* $\exists\ (\theta, \lambda) \in \Gamma(\bar{x})$ *s.t.* $\theta \neq 0$.

(ii) *(P) is regular at $\bar{x}$ iff* $\exists\ (\theta, \lambda) \in \Gamma(\bar{x})$ *s.t.* $\forall\ i = 1, ..., q\ :\ \theta_i > 0$.

(iii) *(P) is totally weak-regular at $\bar{x}$ iff* $\forall\ (\theta, \lambda) \in \Gamma(\bar{x})\ :\ \theta \neq 0$.

(iv) *(P) is totally regular at $\bar{x}$ iff* $\forall\ (\theta, \lambda) \in \Gamma(\bar{x}),\ \forall\ i = 1, ..., q\ :\ \theta_i > 0$.

Observe that, when f is real-valued, weak-regularity coincide with regularity and total weak-regularity with total regularity.
It is clear that (iv) implies (iii) and (ii), which both imply (i). Many examples show that, in general, no relationship between (iii) and (ii) holds.
As far as we know, (iii) and (iv) have never been considered before, even if they guarantee that the undesirable situations, we mentioned above, can never happen.
Since we are dealing with necessary optimality conditions, we may suppose that $\bar{x}$ is any feasible solution of (P) such that $\Gamma(\bar{x}) \neq \emptyset$. This means exactly that $\bar{x}$ is just a stationary point of (P).
Arguing as in [2], we can prove the following technical results.

Theorem 4.3. *Let $\bar{x}$ be any feasible solution of (P) such that $\Gamma(\bar{x}) \neq \emptyset$. Suppose that $D^-(f,g)(\bar{x};\ \cdot)$ is convexlike on $F(X,\bar{x})$. Then, (P) is totally weak-regular at $\bar{x}$*

iff

$$\exists\, \bar{y} \in F(X,\bar{x}),\ \exists\, (u,v) \in D^-(f,g)(\bar{x};\bar{y}) \ : v \in int\ \mathbb{R}^m_- . \tag{5}$$

Theorem 4.4. *Let $\bar{x}$ be any feasible solution of (P) such that $\Gamma(\bar{x}) \neq \emptyset$. Suppose that $D^-(f,g)(\bar{x};\ \cdot)$ is convexlike on $F(X,\bar{x})$. Then, (P) is totally regular at $\bar{x}$ iff*

$$\begin{array}{c}\forall\, i = 1,...,q,\ \exists\, \bar{y}^i \in F(X,\bar{x}),\ \exists\, (u^i,v^i) \in D^-(f,g)(\bar{x};\bar{y}^i) \text{ such that} \\ u^i_j < 0, \quad \forall\, j \neq i,\ v\ \in int\ \mathbb{R}^m_- .\end{array} \tag{6}$$

Both (5) and (6) are Mangasarian-Fromovitz type conditions; in the following section they will be employed to guarantee the boundedness of suitable sets of generalized multipliers.

5 Boundedness of generalized multipliers

In this section we want to study the relationships between total weak-regularity, total regularity and the boundedness of generalized multipliers. Besides its intrinsic interest, this property seems to be crucial to obtain stability results for vector parametric optimization and to estabilish duality results.

Since the set of generalized multipliers is a (convex) cone, the study of boundedness must be devoted to subsets of $\Gamma(\bar{x})$. In the scalar case the approach consists in considering the multipliers (θ,λ) such that $\theta = 1$. At least two different extensions can be considered in the vector case; in fact, since θ is not a number but a vector, both

$$\Gamma_+(\bar{x}) := \Gamma(\bar{x}) \cap \{\ (\theta,\lambda)\ :\ ||\ \theta\ || = 1\ \}$$

and

$$\Gamma_*(\bar{x}) := \Gamma(\bar{x})\ \cap\ \{(\theta,\lambda) : \exists\, i \text{ such that } \theta_i = 1\ \}$$

can be considered. We show that the boundedness of the former set is equivalent to total weak-regularity and the boundedness of the latter is equivalent to total regularity. Moreover, we give estimates of the diameters of these sets, which represent bounds for the multipliers' size.

Theorem 5.1. *Let $\bar{x}$ be any feasible solution of (P) such that $\Gamma_+(\bar{x}) \neq \emptyset$. Suppose that $D^-(f,g)(\bar{x};\ \cdot)$ is convexlike on $F(X,\bar{x})$. Then,*

$$\exists\, k \geq 0, \quad \forall\, (\theta,\lambda) \in \Gamma_+(\bar{x})\ :\ ||(\theta,\lambda)|| \leq k.$$

iff the generalized Mangasarian-Fromovitiz condition (5) holds.

Proof. $\Longleftarrow$) Ab absurdo, suppose that there exists $\{(\theta_k,\lambda_k)\} \subseteq \Gamma_+(\bar{x})$ such that $||(\theta_k,\lambda_k)|| \to +\infty$. Since $||\ \theta_k\ || = 1$ for each k, we must have $||\lambda_k|| \to +\infty$. Taking the suitable subsequences, we can suppose that $\{||\lambda_k||^{-1}\lambda_k\}$ converges to a nonzero vector $\bar{\lambda} \in \mathbb{R}^m_+$ and $\{\theta_k\}$ converges to a vector $\bar{\theta}$. Thanks to the assumption, we have

$$||\lambda_k||^{-1}(\langle\theta_k\ ,\ u\rangle + \langle\lambda_k\ ,\ v\rangle) \geq 0, \qquad \forall\ (u,v) \in D^-(f,g)(\bar{x};\bar{y}),\ \ \forall\ \bar{y} \in F(X,\bar{x}).$$

Taking the limit k$\rightarrow +\infty$, we get

$$\langle 0\ ,\ u\rangle + \langle\bar{\lambda}\ ,\ v\rangle \geq 0, \qquad \forall\ (u,v) \in D^-(f,g)(\bar{x};\bar{y}),\ \ \forall\ \bar{y} \in F(X,\bar{x}),$$

so that $(0,\bar{\lambda}) \in \Gamma(\bar{x})$. By Theorem 4.3, this contradicts the assumption.
$\Longrightarrow$) Ab absurdo, suppose that (5) does not hold. By Theorem 4.3, there exists $(0,\lambda_0) \in \Gamma(\bar{x})$. Moreover, there exists $(\theta,\lambda) \in \Gamma_+(\bar{x})$. In particular, we have $(\theta, t\lambda_0+\lambda) \in \Gamma_+(\bar{x})$ for each $t > 0$. But $||(\theta, t\lambda_0+\lambda)|| \rightarrow +\infty$ as $t \rightarrow +\infty$ in contradiction with the assumption. □

Moreover, the generalized Mangasarian-Fromovitz condition (5) can also be employed to give an estimate of the bound in the following way.

Theorem 5.2. *Suppose that $D^-(f,g)(\bar{x};\ \cdot)$ is convexlike on $F(X,\bar{x})$. If (P) is totally weak-regular at $\bar{x}$ and $\bar{y} \in F(X,\bar{x})$, $(u,v) \in D^-(f,g)(\bar{x};\bar{y})$ satisfy (5), then the number*

$$max\ \{\ ||(\theta,\lambda)||\ :\ (\theta,\lambda) \in \Gamma_+(\bar{x})\ \textit{and}\ \langle\theta\ ,\ u\rangle + \langle\lambda\ ,\ v\rangle = 0\ \}$$

is a bound for $\{\ ||(\theta,\lambda)||\ :\ (\theta,\lambda) \in \Gamma_+(\bar{x})\ \}$.

Proof. If $(\theta,\lambda) \in \Gamma(\bar{x})$, we observe that $\langle\theta\ ,\ u\rangle + \langle\lambda\ ,\ v\rangle \geq 0$ and thus

$$\sup\ \{\ ||(\theta,\lambda)||\ :\ (\theta,\lambda) \in \Gamma_+(\bar{x})\ \text{and}\ \langle\theta\ ,\ u\rangle + \langle\lambda\ ,\ v\rangle \geq 0\ \} \tag{7}$$

is a bound. The feasible region of (7) is bounded: if ab absurdo it were not, there would exist a sequence $\{(\theta_k,\lambda_k)\} \subseteq \Gamma_+(\bar{x})$ such that $||\lambda_k|| \longrightarrow +\infty$ and

$$||\lambda_k||^{-1}(\langle\theta_k\ ,\ u\rangle + \langle\lambda_k\ ,\ v\rangle) \geq 0. \tag{8}$$

Taking suitable subsequences, we can suppose that $\{||\lambda_k||^{-1}\lambda_k\}$ converges to a nonzero vector $\bar{\lambda} \in \mathbb{R}^m_+$ and $\{\theta_k\}$ converges to a vector $\bar{\theta} \in B_1(0)$. Taking the limit in (8) we get $\langle\bar{\lambda}\ ,\ v\rangle \geq 0$, which is impossible since $\bar{\lambda} \in \mathbb{R}^m_+\backslash\{0\}$ and $v\ \in$ *int* $\mathbb{R}^m_-$. Moreover, it is easy to see that the feasible region is also closed, hence it is a compact set and the maximum in (7) is obtained. To conclude the proof, we must show that the maximum is obtained by a vector $(\bar{\theta},\bar{\lambda})$ such that $\langle\bar{\theta}\ ,\ u\rangle + \langle\bar{\lambda}\ ,\ v\rangle = 0$. In fact, if it were not, we would have $\langle\bar{\theta}\ ,\ u\rangle > -\langle\bar{\lambda}\ ,\ v\rangle$, but, since $(\bar{\theta},\bar{\lambda})$ is feasible, we also have $\langle\bar{\theta}\ ,\ u\rangle \geq 0$. Thus, we could find t > 1 such that $\langle\bar{\theta}\ ,\ u\rangle \geq -\langle t\bar{\lambda}\ ,\ v\rangle$ and therefore $(\bar{\theta}, t\bar{\lambda})$ would be feasible and moreover $||(\bar{\theta}, t\bar{\lambda})|| > ||(\bar{\theta},\bar{\lambda})||$. □

Theorem 5.3. *Let $\bar{x}$ be any feasible solution of (P) such that $\Gamma_*(\bar{x}) \neq \emptyset$. Suppose* *that* $D^-(f,g)(\bar{x};\ \cdot)$ *is convexlike on* $F(X,\bar{x})$. *Then,*

$$\exists\ k \geq 0,\quad \forall\ (\theta,\lambda) \in \Gamma_*(\bar{x})\ :\ ||(\theta,\lambda)|| \leq k$$

iff the generalized Mangasarian-Fromovitz condition (6) holds.

Proof. Fixed $i \in \{\ 1,...,q\ \}$, let

$$\Gamma_i(\bar{x}) := \{\ (\theta,\lambda) \in \Gamma(\bar{x}) : \theta_i = 0\ \},\qquad \Gamma_*^i(\bar{x}) := \{\ (\theta,\lambda) \in \Gamma_*(\bar{x}) : \theta_i = 1\ \}.$$

Arguing as in Theorem 5.1, we can prove that

$$\Gamma_i(\bar{x}) = \emptyset \Longleftrightarrow \exists\ k_i \geq 0,\quad \forall\ (\theta,\lambda) \in \Gamma_*^i(\bar{x})\ :\ ||(\theta,\lambda)|| \leq k_i.$$

Considering all indexes, we get that $\Gamma_*(\bar{x})$ is bounded iff (P) is totally regular at $\bar{x}$. Thus, the theorem follows immediately from Theorem 4.4. □

Moreover, the generalized Mangasarian-Fromovitz condition (6) provides the following estimate of the bound.

Theorem 5.4. *Suppose that* $D^-(f,g)(\bar{x};\ \cdot)$ *is convexlike on* $F(X,\bar{x})$. *If (P) is totally regular at* $\bar{x}$ *and* $\bar{y}^i \in F(X,\bar{x})$, $(u^i,v^i) \in D^-(f,g)(\bar{x};\bar{y}^i)$, $i = 1,..,q$, *satisfy (6), then the number*

$$max\ \{k_i\ :\ i = 1,...,q\},$$

where

$$k_i := max\{\ ||(\theta,\lambda)||\ :\ (\theta,\lambda) \in \Gamma_*^i(\bar{x})\ and\ \langle\theta\ ,\ u^i\rangle + \langle\lambda\ ,\ v^i\rangle = 0\},$$

is a bound for $\{\ ||(\theta,\lambda)||\ :\ (\theta,\lambda) \in \Gamma_*(\bar{x})\ \}$.

Proof. Arguing as in Theorem 5.2, we can prove that k_i is a bound for $\Gamma_*^i(\bar{x})$ and thus the thesis follows. □

When f is real-valued, $\Gamma_+(\bar{x})$ and $\Gamma_*(\bar{x})$ are the same set and the results of this section follow the same line of the ones of [3]. The following example shows that, when f is vector-valued, the distinction between $\Gamma_+(\bar{x})$ and $\Gamma_*(\bar{x})$ is required.

Example 5.5. Let $X = \mathbb{R}^2$, $f : \mathbb{R}^2 \longrightarrow \mathbb{R}^2$ and $g : \mathbb{R}^2 \longrightarrow \mathbb{R}$ be defined by

$$f(x_1,x_2) = (x_1^2 + x_2, x_2),\qquad g(x_1,x_2) = x_1^2 - x_2.$$

It is easy to check that $\bar{x} = (0,0)$ is a vector minimum point of (P) and that

$$\Gamma(\bar{x}) = \{ \ (a,b,a+b) \ : \ a,b \geq 0, \ a+b > 0 \ \}.$$

Since $(t,1,t+1) \in \Gamma_*(\bar{x})$ for all $t \geq 0$, $\Gamma_*(\bar{x})$ is unbounded. On the contrary,

$$\Gamma_+(\bar{x}) = \{ \ (a,b,a+b) \ : \ a,b \geq 0, \ ||(a,b)|| = 1 \ \}$$

is obviously bounded.

References

[1] Aubin, J.P. and Frankowska, H. (1990), *Set-valued Analysis*, Birkhäuser, Boston.

[2] Bigi, G. and Pappalardo, M. (1999), "Regularity Conditions in Vector Optimization", *forthcoming on Journal of Optimization Theory and Applications*, Vol. 102.

[3] Castellani, M., Mastroeni, G., and Pappalardo, M. (1997), "Separation of Sets, Lagrange Multipliers and Totally Regular Extremum Problems", *Journal of Optimization Theory and Applications*, Vol. 92, pp. 249-261.

[4] Elster, K.H. and Thierfelder, J. (1989), "On Cone Approximations and Generalized Directional Derivatives", in *Nonsmooth Optimization and Related Topics*, Clarke, F.H., Dem'yamov, V.F. and Giannessi, F. (eds), Plenum, New York, pp. 133-154.

[5] Gauvin, J. (1977), "A Necessary and Sufficient Regularity Condition to Have Bounded Multipliers in Nonconvex Programming", *Mathematical Programming*, Vol. 12, pp. 136-138.

[6] Jahn, J. (1986), *Mathematical Vector Optimization in Partially Ordered Linear Spaces*, Peter Lang, Frankfurt.

[7] Luc, D.T. (1989), *Theory of Vector Optimization*, Springer, Berlin.

[8] Mangasarian, O.L. (1969), *Nonlinear Programming*, Mc Graw-Hill, New York.

[9] Mangasarian, O.L. (1985), "Computable Numerical Bounds for Lagrange Multipliers of Stationary Points of Nonconvex Differentiable Nonlinear Programs", *Operations Research Letters*, Vol. 4, pp. 47-48.

[10] Mangasarian, O.L., and Fromovitz, S. (1967), "The Fritz John Necessary Optimality Conditions in Presence of Equality and Inequality Constraints", *Journal of Mathematical Analysis and Applications*, Vol. 17, pp. 37-47.

[11] Nguyen, V.H., Strodiot, J.J. and Mifflin, R. (1980), "On Conditions to Have Bounded Multipliers in Locally Lipschitz Programming", *Mathematical Programming*, Vol. 18, pp. 100-106.

[12] Pappalardo, M. (1992), "Error Bounds for Generalized Lagrange Multipliers in Locally Lipschitz Programming", *Journal of Optimization Theory and Applications*, Vol. 73, pp. 205-210.

[13] Pappalardo, M. and Stöcklin, W. (1998), "Necessary Optimality Conditions in Nondifferentiable Vector Optimization", *submitted to Optimization.*

[14] Sawaragi, Y., Nakayama, H. and Tanino, T. (1985), *Theory of Multiobjective Optimization*, Academic, Orlando.

[15] Yang, X.Q. (1998), "A Generalized Upper Dini-Directional Derivative in Vector Optimization", *Optimization*, Vol. 43, pp. 339-351.

A primal-dual algorithm for minimizing a non-convex function subject to bound and linear equality constraints

Andrew R. Conn (arconn@watson.ibm.com)
IBM T.J. Watson Research Center, P.O.Box 218, Yorktown Heights, NY 10598, USA

Nicholas I. M. Gould (n.gould@rl.ac.uk)
Computational Science and Engineering Department,
Rutherford Appleton Laboratory, Chilton, Oxfordshire OX11 0QX, England

Philippe L. Toint (philippe.toint@fundp.ac.be)
Department of Mathematics, Facultés Universitaires ND de la Paix,
61, rue de Bruxelles, B-5000 Namur, Belgium, EU

Abstract

A new primal-dual algorithm is proposed for the minimization of non-convex objective functions subject to simple bounds and linear equality constraints. The method alternates between a classical primal-dual step and a Newton-like modified barrier step in order to ensure descent on a suitable merit function. Convergence of a well-defined subsequence of iterates is proved from arbitrary starting points. Preliminary numerical results are presented.

Keywords: Primal-dual algorithms, non-convex optimization, linear constraints.

G. Di Pillo and F. Giannessi (eds.), Nonlinear Optimization and Related Topics, 15–49.

1 Introduction: the problem and the algorithm

1.1 The problem

In this paper, we consider algorithms for solving general (ie, non-convex), linearly constrained, differentiable optimization problems. We shall distinguish between simple bounds and general linear constraints, and find it convenient to reformulate inequalities as equalities via slack variables. We thus consider the problem

$$\begin{array}{ll} \text{minimize} & f(x) \\ & Ax = b \\ \text{s.t.} & x \geq 0 \end{array} \tag{1.1}$$

where $f(\cdot)$ is a real valued function on $\mathbf{R}^n$, x is a vector in $\mathbf{R}^n$, A is an $m \times n$ matrix and b is a vector of $\mathbf{R}^m$.

In part, we are motivated to consider the above problem because of our experiences with the general large-scale nonlinear programming package LANCELOT (Conn, Gould and Toint, 1992). In this package, simple bounds are treated explicitly and all other constraints are converted to equations and incorporated into an augmented Lagrangian merit function. While this proves to be a robust approach (Conn, Gould and Toint, 1996*a*), it has a number of obvious drawbacks. One of these is that augmentation may not be the ideal way to treat linear constraints, and a more attractive approach is to handle all linear constraints explicitly (Conn, Gould, Sartenaer and Toint, 1996*b*). We note that there has been a relatively long history of methods that use linearly constrained subproblems at their heart. References.include the methods of Rosen and Kreuser (1972), Robinson (1972), and Murtagh and Saunders (1978), the latter being the basis of the well-known large-scale nonlinear programming package MINOS.

Another drawback with the LANCELOT approach is the use of the simple bounds that are active at the generalized Cauchy point to predict those which will be active at the solution (see the trust region based kernel algorithm SBMIN, Conn, Gould and Toint, 1988). Unfortunately this approach does not appear to be very effective when the problem is either degenerate or close to degenerate. On the other hand interior point methods, particularly primal-dual approaches, have enjoyed much success in linear programming and it is generally accepted that any state-of-the-art library for linear programming should include both interior point and simplex methods (for example OSL of Corporation (1990) and CPLEX, 4.0 (1995)). It is usually acknowledged that interior point methods are less sensitive to degeneracy than active set methods, see for example Shanno (1994). Thus we were motivated to consider an interior point method in which linear constraints $Ax = b$ are handled explicitly and simple bounds are handled via a logarithmic barrier term. For the record, we still expect to handle general nonlinear constraints using the augmented Lagrangian. However, we do want to retain the flexibility of not necessarily satisfying the linear constraints during the earlier iterations.

In addition, since the linear programming problem is a convex linear problem, it is the case that the first order conditions are sufficient to characterize a solution and thus it is possible to dispense with a merit function entirely. In the non-convex case, the merit function is an essential ingredient of any successful algorithm and the choice of merit function was a considerable concern in the present paper.

However noble one may believe these goals, there are some significant difficulties in an interior point approach. Besides those already mentioned there is an additional discussion in the conclusions of this paper. Although we are not successful in addressing all these issues, and indeed some of the most important practical issues will depend upon much more extensive testing, what we do hope we have achieved in the present paper is a consistent method with a single merit function and a guaranteed descent direction that either is the primal dual direction or a very modified barrier step. In addition, linear equalities are treated explicitly without requiring primal feasibility initially.

Considering the vast literature on primal-dual methods for convex problems, there has been remarkably little work on extending these methods to the non-convex case. This may be because dual variables are not globally meaningful for non-convex problems, but one is tempted to believe that in the neighbourhood of a minimizer some sort of local convexity may be amenable to a primal-dual approach. Indeed, Simantiraki and Shanno (1995) analyse such a local method. Globally, of course, one would expect to require a merit function to force convergence, and Forsgren and Gill (1996) attempt to provide such a function for primal-dual methods. A complete analysis of an interior-point algorithm for non-convex linearly constrained optimization is provided by Bonnans and Pola (1993), but this algorithm appears to require both a strictly interior starting point and a convex model of the objective.

Although the emphasis here is on theoretical issues, we do include preliminary results on a non-trivial set of general quadratic programming problems from the the CUTE test set (see, Bongartz, Conn, Gould and Toint, 1995) which we compare with a state-of-the-art active set method designed for solving quadratic programs. Before going into further details of the proposed algorithm we include some additional notation and our assumptions.

If we denote the Euclidean inner product by $\langle \cdot, \cdot \rangle$ and let e be the vector of all ones, we assume that

AS1. $f(\cdot)$ is a twice continuously differentiable,

AS2. the iterates of our algorithm remain in a convex bounded subset $\mathcal{D}$ of the positive orthant,

AS3. A has full rank, and

AS4. there exists a point $x_\odot$ strictly interior to the positive orthant such that $Ax_\odot = b$.

Note that (AS1) and (AS2) together imply that the function $f(x) - \mu\langle \log(x), e\rangle$ is bounded below on $\mathcal{D}$ for every $\mu > 0$. Also note that AS2 automatically holds if, as is frequently the case, the feasible domain of problem (1.1) is bounded.

1.2 The primal-dual search direction

The first order criticality conditions for problem (1.1) may be written as

$$\begin{aligned} g(x) + A^T y - z &= 0 \\ Ax &= b \\ XZe &= 0, \\ (x,z) &\geq 0, \end{aligned} \tag{1.2}$$

where z is a vector in $\mathbf{R}^n$, y a vector in $\mathbf{R}^m$, $g(x) \stackrel{\text{def}}{=} \nabla_x f(x)$ and

$$X = \text{diag}(x_1, \ldots, x_n) \text{ and } Z = \text{diag}(z_1, \ldots, z_n).$$

In order to build our algorithm, we consider a perturbed version of this system of equations given by

$$\begin{aligned} g(x) + A^T y - z &= 0 \\ Ax &= b \\ XZe &= \mu e, \\ (x,z) &\geq 0, \end{aligned} \tag{1.3}$$

where

$$\mu = \sigma \frac{\langle x, z\rangle}{n},$$

for some given $\sigma \in (0,1)$. Our algorithm moves from the current estimate $(x_k, z_k) > 0$ of the (x,z) components of the solution of (1.1) to a new estimate $(x_{k+1}, z_{k+1}) > 0$ given by

$$x_{k+1} = x_k + \alpha_k^{(x)} \Delta x_k \text{ and } z_{k+1} = z_k + \alpha_k^{(z)} \Delta z_k, \tag{1.4}$$

for some scalar stepsizes $\alpha_k^{(x)}, \alpha_k^{(z)} \in (0,1]$, where Δx_k and Δz_k may, for instance, be chosen as Δx_k^{PD} and Δz_k^{PD} which solve the system

$$\begin{aligned} H_k \Delta x_k^{\text{PD}} + A^T y_{k+1}^{\text{PD}} - \Delta z_k^{\text{PD}} &= -g_k + z_k, \\ A \Delta x_k^{\text{PD}} &= b - Ax_k \\ Z_k \Delta x_k^{\text{PD}} + X_k \Delta z_k^{\text{PD}} &= \mu_k e - X_k Z_k e, \end{aligned} \tag{1.5}$$

where $H_k \stackrel{\text{def}}{=} H(x_k) \stackrel{\text{def}}{=} \nabla_{xx} f(x_k)$ and where $g_k \stackrel{\text{def}}{=} g(x_k)$. This system is a linearization, at (x_k, z_k), of (1.3), in which y_{k+1} is the new estimate of the Lagrange multiplier associated with the constraint $Ax = b$. Eliminating Δz_k^{PD}, and defining

$$r_k = Ax_k - b, \tag{1.6}$$

we obtain that

$$\begin{pmatrix} H_k + X_k^{-1}Z_k & A^T \\ A & 0 \end{pmatrix} \begin{pmatrix} \Delta x_k^{\text{PD}} \\ y_{k+1}^{\text{PD}} \end{pmatrix} = - \begin{pmatrix} g_k - \mu_k X_k^{-1} e \\ r_k \end{pmatrix} \tag{1.7}$$

and

$$\Delta z_k^{\text{PD}} = -z_k - X_k^{-1} Z_k \Delta x_k^{\text{PD}} + \mu_k X_k^{-1} e. \tag{1.8}$$

Note that (1.7) fully defines Δx_k^{PD}, and y_{k+1}^{PD} provided the matrix $H_k + X_k^{-1} Z_k \stackrel{\text{def}}{=} G_k$ is nonsingular in the nullspace of A. This is obviously the case if $f(x)$ is strictly convex, but may not be true in general. We discuss below how G_k might be modified or how Δx_k^{PD} may be defined in more general situations. Observe also that, if this quantity is well defined, Δz_k^{PD} is in turn well defined by (1.8). The strict positivity of x_{k+1} and z_{k+1} is ensured by suitably restricting the stepsizes $\alpha_k^{(x)}$ and $\alpha_k^{(z)}$, as is detailed below. Thus, the zero components of x_* or z_* at the solution can only be attained in the limit.

Observe that we may now choose to represent the infeasibility with respect to the linear constraints by introducing an artificial variable ξ in the system (1.7), which is defined by

$$Ax - b = \xi r_0, \tag{1.9}$$

which is possible for a scalar variable because of the second equation of (1.5). If $r_0 \neq 0$, the initial value ξ_0 of the artificial variable is set to one; at each iteration, we have that

$$\xi_k r_0 = r_k, \tag{1.10}$$

and we may augment the primal-dual step with the correction

$$\Delta \xi_k^{\text{PD}} = -\xi_k \tag{1.11}$$

to ξ_k. Thus if a unit step is ever taken, the linear equality constraints will be satisfied exactly from then on. We will use the notation $v = (x, \xi)$ to denote points in the (x, ξ)-space.

1.3 An alternative search direction

When $\xi > 0$, we may now consider the alternative problem of minimizing the shifted penalty function

$$f(x) + \frac{1}{2}\rho(\xi + 1)^2$$

subject to the constraints (1.9) and

$$x \geq 0.$$

In this formulation, the shifted penalty terms drives the variable ξ below zero for sufficiently large ρ. We then intend to stop the minimization prematurely as soon as ξ attains the value zero. Writing the first order optimality conditions for this modified problem, we obtain that

$$\begin{aligned} g(x) + A^T y - z &= 0, \\ -\langle r_0, y\rangle + \rho(\xi + 1) &= 0, \\ Ax - \xi r_0 &= b, \\ XZe &= 0, \\ (x, z) &\geq 0. \end{aligned}$$

We perturb the system in the same manner as above and write the corresponding Newton's iteration, which yields that

$$\begin{aligned} H_k \Delta x_k^{\text{MB}} + A^T y_{k+1}^{\text{MB}} - \Delta z_k^{\text{MB}} &= -g_k + z_k, \\ A\Delta x_k^{\text{MB}} - \Delta\xi_k^{\text{MB}} r_0 &= 0, \\ -\langle r_0, y_{k+1}^{\text{MB}}\rangle + \rho_k \Delta\xi_k^{\text{MB}} &= -\rho_k(\xi_k + 1), \\ Z_k \Delta x_k^{\text{MB}} + X_k \Delta z_k^{\text{MB}} &= \mu_k e - X_k Z_k e. \end{aligned} \tag{1.12}$$

As before, we may eliminate Δz_k^{MB}, and obtain that

$$\begin{pmatrix} H_k + X_k^{-1} Z_k & A^T & 0 \\ A & 0 & -r_0 \\ 0 & -r_0^T & \rho_k \end{pmatrix} \begin{pmatrix} \Delta x_k^{\text{MB}} \\ y_{k+1}^{\text{MB}} \\ \Delta\xi_k^{\text{MB}} \end{pmatrix} = - \begin{pmatrix} g_k - \mu_k X_k^{-1} e \\ 0 \\ \rho_k(\xi_k + 1) \end{pmatrix} \tag{1.13}$$

and

$$\Delta z_k^{\text{MB}} = -z_k - X_k^{-1} Z_k \Delta x_k^{\text{MB}} + \mu_k X_k^{-1} e. \tag{1.14}$$

Observe that the system (1.13) has a bordered form obtained from (1.7).

1.4 The merit function

We now introduce, for given $\mu, \rho > 0$, the logarithmic penalty function defined by

$$\phi(v, \mu, \rho) = f(x) + \frac{1}{2}\rho(\xi + 1)^2 - \mu\langle \log(x), e\rangle.$$

Examining now the derivative of this function, we find that

$$\nabla_x \phi(v, \mu, \rho) = g(x) - \mu X^{-1} e \ \text{ and } \ \nabla_\xi \phi(v, \mu, \rho) = \rho(\xi + 1). \tag{1.15}$$

We first consider the slope of this function at a given iterate v_k^T along the step

$$\Delta v_k^{\text{MB}} = ((\Delta x_k^{\text{MB}})^T, \Delta\xi_k^{\text{MB}})$$

defined by (1.12) (or, equivalently, (1.13) and (1.14)), and we obtain from (1.15) that

$$\langle \nabla_v \phi_k, \Delta v_k^{\text{MB}}\rangle = -\langle \Delta x_k^{\text{MB}}, G_k \Delta x_k^{\text{MB}}\rangle - \rho_k(\Delta\xi_k^{\text{MB}})^2, \tag{1.16}$$

where we have defined $\phi_k = \phi(v_k, \mu_k, \rho_k)$.

On the other hand, the direction

$$(\Delta v_k^{\rm PD})^T = ((\Delta x_k^{\rm PD})^T, \Delta\xi_k^{\rm PD}),$$

defined by (1.5) (or, equivalently, (1.7) and (1.8)) and (1.11), yields the slope

$$\begin{aligned} \langle \nabla_v \phi_k, \Delta v_k^{\rm PD}\rangle &= \langle \Delta x_k^{\rm PD}, g_k - \mu_k X_k^{-1} e\rangle - \rho_k \xi_k(\xi_k + 1) \\ &= -\langle \Delta x_k^{\rm PD}, G_k \Delta x_k^{\rm PD}\rangle - \langle \Delta x_k^{\rm PD}, A^T y_{k+1}^{\rm PD}\rangle - \rho_k \xi_k(\xi_k + 1) \\ &= -\langle \Delta x_k^{\rm PD}, G_k \Delta x_k^{\rm PD}\rangle - \rho_k(\Delta\xi_k^{\rm PD})^2 - \xi_k(\rho_k - \langle r_0, y_{k+1}^{\rm PD}\rangle), \end{aligned} \tag{1.17}$$

where we have used (1.5), the definition of ξ and (1.11).

We now examine under which conditions the slopes given by (1.16) and (1.17) are negative. To this aim, we introduce the following definition: the matrix G is said to be *second-order sufficient* with respect to A if and only if the augmented matrix

$$K = \begin{pmatrix} G & A^T \\ A & 0 \end{pmatrix} \tag{1.18}$$

is nonsingular and has precisely m negative eigenvalues. This is equivalent to requiring that $\langle y, Gy\rangle > 0$ for all nonzero y satisfying $Ay = 0$, or to the reduced matrix $N^T G N$ being positive definite, where the columns of N span the nullspace of A (see, for instance, Gould, 1985). The matrix is *second-order necessary* if we drop the requirement that K be nonsingular; this is then equivalent to requiring that $\langle y, Gy\rangle \geq 0$ for all y satisfying $Ay = 0$ or to the reduced matrix $N^T G N$ being positive semidefinite.

If $\xi_k = 0$ then the identity (1.10) and (1.7) gives that $A\Delta x_k = 0$. Thus, if the matrix G_k is second-order sufficient with respect to A, we may deduce that

$$\langle \nabla_v \phi_k, \Delta v_k^{\rm PD}\rangle = -\langle \Delta x_k^{\rm PD}, G_k \Delta x_k^{\rm PD}\rangle < 0. \tag{1.19}$$

Similarly,

$$\langle \nabla_v \phi_k, \Delta v_k^{\rm MB}\rangle = -\langle \Delta x_k^{\rm MB}, G_k \Delta x_k^{\rm MB}\rangle < 0. \tag{1.20}$$

If we now consider the case where G_k is second-order sufficient with respect to A but $\xi_k \neq 0$, it turns out that we can still show that the slopes (1.16) and (1.17) are negative provided we choose ρ_k large enough. This result from the two following lemmas.

Lemma 1 *Assume that the matrix G is second-order sufficient with respect to A and that the columns of N form an orthonormal basis of the nullspace of A. Then the smallest eigenvalue of $N^T G N$ is at least equal to the smallest positive eigenvalue of K, where K is defined by (1.18).*

Proof. The proof can be found in Appendix. □

Lemma 2 [1] *Assume that the matrix G is second-order sufficient with respect to A, and that the smallest strictly positive eigenvalue of K is $\lambda > 0$. Then, if one chooses an arbitrary m-dimensional vector r and if*

$$\rho \geq \lambda + \frac{2}{\lambda}\|r\|^2, \tag{1.21}$$

the matrix

$$\bar{G} = \begin{pmatrix} G & 0 \\ 0 & \rho \end{pmatrix}$$

is second-order sufficient with respect to $(A\ r)$ and

$$\langle v, \bar{G}v \rangle \geq \frac{1}{2}\lambda\|v\|^2 \tag{1.22}$$

for $v = (x, \xi)$ in the nullspace of $(A\ r)$.

Proof. The proof can be found in appendix. □

Returning to the sign of the slopes of (1.16) and (1.17) in the case where G_k is second-order sufficient with respect to A and $\xi_k \neq 0$, we see immediately, from (1.16), Lemma 2 and the second equation of (1.12), that

$$\langle \nabla_v \phi_k, \Delta v_k^{\text{MB}} \rangle \leq -\frac{1}{2}\lambda(\|\Delta x_k^{\text{MB}}\|^2 + (\Delta \xi_k^{\text{MB}})^2). \tag{1.23}$$

so long as

$$\rho_k \geq \lambda + \frac{2}{\lambda}\|r_0\|^2,$$

[1]It is interesting to note that Lemma 2 does not hold if second-order sufficiency is replaced by second-order necessity. For, suppose that

$$G = \begin{pmatrix} 0 & 0 & 1 \\ 0 & 1 & 1 \\ 1 & 1 & 1 \end{pmatrix}, \quad A = \begin{pmatrix} 0 & 0 & 1 \end{pmatrix} \text{ and } r = -1.$$

Then the columns of the matrix

$$N = \begin{pmatrix} 1 & 0 & 0 & 0 \\ 0 & 1 & 0 & 0 \\ 0 & 0 & 1 & 1 \end{pmatrix}$$

form a basis of the nullspace of $(A\ r)$, and the resulting "reduced matrix" is

$$N^T \begin{pmatrix} G & 0 \\ 0 & \rho \end{pmatrix} N = \begin{pmatrix} 0 & 0 & 1 \\ 0 & 1 & 1 \\ 1 & 1 & 1+\rho \end{pmatrix}.$$

Unfortunately, this latter matrix has a negative eigenvalue for all ρ.

where λ is the smallest eigenvalue of $N^T G_k N$. We also see that the second equation of (1.7) can be rewritten as

$$A\Delta x_k^{\text{PD}} - \Delta\xi_k r_0 = 0, \tag{1.24}$$

and thus may deduce from (1.17), Lemma 2, (1.11), and (1.24) that

$$\langle \nabla_v \phi_k, \Delta v_k^{\text{PD}} \rangle \leq -\frac{1}{2}\lambda(\|\Delta x_k^{\text{PD}}\|^2 + (\Delta\xi_k^{\text{PD}})^2), \tag{1.25}$$

whenever

$$\rho_k \geq \max\left[\lambda + \frac{2}{\lambda}\|r_0\|^2, \langle r_0, y_{k+1}^{\text{PD}} \rangle\right]. \tag{1.26}$$

Observe that condition (1.26) depends on the size of $\langle r_0, y_{k+1}^{\text{PD}} \rangle$. The penalty parameter ρ_k may thus become too large because of this latter quantity, in which case we might prefer to use the alternative formulation using the shifted quadratic penalty term for which descent is always obtained (see (1.23)) if G_k is second order sufficient with respect to A (see (1.16)). Our algorithm takes advantage of this observation.

1.5 Modifications

If G_k is not second-order sufficient with respect to A, we may add a positive semidefinite modification ΔG_k to G_k, so that $G_k + \Delta G_k$ is uniformly second-order sufficient with respect to A, meaning that the minimum eigenvalue of $N^T(G_k + \Delta G_k)N$ is larger than some $\lambda > 0$ independent of k. This in turn yields well defined Δx_k^{PD} and y_{k+1}^{PD}, and ensures (1.19). The smallest such modification may need to be as large as $\|N^T G_k N\| + \lambda$, but here we merely require that

$$\|\Delta G_k\| \leq \kappa_2 \|G_k\| + \lambda \tag{1.27}$$

for some $\kappa_2 \geq 1$. The modification ΔG_k to make $N^T G_k N$ positive definite may be much smaller than that required to make G_k itself positive definite.

The technique of ensuring the second-order sufficiency of G_k with respect to A is not the only one which can be considered to make the slope (1.17) negative. One could also modify Δx_k^{PD} to include a sufficient contribution of a *direction of negative curvature*, provided the second equation of (1.7) remains satisfied. This then leads to a trust-region like method, which is the object of current research (Conn, Gould, Orban and Toint, 1999).

The fact that the directional derivative (1.19) is negative ensures that the (possibly modified) primal-dual step Δv_k^{PD} is a descent direction for ϕ_k, when v_k is not a minimizer. We may thus consider this function as a "merit function" associated with this step, that is with the linearization of conditions (1.3).

The viability of such approaches using modifications to G_k are discussed further, with additional references in Forsgren and Murray (1993), Gould (1995) and Higham and Cheng (1998). Of course, we must estimate the value of λ obtained when using such techniques, as it appears in condition (1.21).

1.6 The step

We now turn to the question of determining the stepsizes in (1.4). A first and crucial constraint on the stepsizes is induced by our decision to maintain both x_{k+1} and z_{k+1} strictly positive. We thus have to specify some bounds on $\alpha_k^{(x)}$ and $\alpha_k^{(z)}$ that will guarantee that the iterates remains "sufficiently" inside the positive orthant of the (x,z)-space. When both stepsizes are chosen equal (i.e. $\alpha_k^{(x)} = \alpha_k^{(z)}$), a set of suitable conditions (see Simantiraki and Shanno (1995) or Zhang and Zhang (1994)) on the (unique) stepsize is given by the inequalities

$$X_{k+1}Z_{k+1}e \geq \gamma \frac{\langle x_{k+1}, z_{k+1} \rangle}{n} e \tag{1.28}$$

and

$$\langle x_{k+1}, z_{k+1} \rangle \geq \gamma \| g_{k+1} + A^T y_{k+1} - z_{k+1} \|, \tag{1.29}$$

where $\gamma \in (0,1)$. We observe that conditions (1.28) and (1.29) clearly ensure that x_{k+1} and z_{k+1} both have all components strictly positive so long as the conditions

$$z_{k+1} = g_{k+1} + A^T y_{k+1} \text{ and } \langle x_{k+1}, z_{k+1} \rangle = 0$$

are violated. On the other hand, condition (1.28) and (1.29) appear to be somewhat restrictive in practice because (1.28) often restricts the step in x more than necessary. We might thus prefer to keep independent stepsizes in x and z and require, instead of (1.28), that

$$x_{k+1} \geq \omega(\mu_k) x_k \tag{1.30}$$

where $\omega(\mu_k) \in (0,1)$ is a small parameter possibly dependent on the value of μ_k. Note that the largest stepsize ensuring (1.30) is given by

$$\bar{\alpha}_k^{(x)} = \min \left[1, (1 - \omega(\mu_k)) \min_{[\Delta x_k]_i < 0} \frac{-[x_k]_i}{[\Delta x_k]_i} \right], \tag{1.31}$$

where $[w]_i$ denotes the i-th component of the vector w. However, if this maximum stepsize is adequate for the primal-dual step Δv_k^{PD} in that (1.11) ensures that

$$\xi_k + \bar{\alpha}_k^{(x)} \Delta \xi_k^{\text{PD}} \geq 0,$$

this may not be the case for the modified barrier step Δv_k^{MB} because $\Delta \xi_k^{\text{MB}}$ is now defined from the solution of (1.12). Indeed, for ρ_k large enough, we would expect ξ_k to tend to -1. We thus have to limit the stepsize to maintain ξ_{k+1} non-negative: the largest stepsize in ξ is now given by

$$\bar{\alpha}_k^{(\xi)} = \begin{cases} \min[1, -\xi_k / \Delta \xi_k^{\text{MB}}] & \text{if } \Delta \xi_k^{\text{MB}} < 0, \\ 1 & \text{otherwise.} \end{cases} \tag{1.32}$$

(Note that a zero value of ξ_k is desirable, as it implies primal feasibility of the iterates.) Combining these bounds, we obtain that the maximum stepsize in the $v = (x, \xi)$ space is given by

$$\bar{\alpha}_k^{(v)} = \begin{cases} \bar{\alpha}_k^{(x)} & \text{if } \Delta v_k = \Delta v_k^{\text{PD}}, \\ \min[\bar{\alpha}_k^{(x)}, \bar{\alpha}_k^{(\xi)}] & \text{if } \Delta v_k = \Delta v_k^{\text{MB}}. \end{cases} \tag{1.33}$$

We may then calculate the actual stepsize

$$\alpha_k^{(v)} = \beta^{j_k} \bar{\alpha}_k^{(v)}, \tag{1.34}$$

by a classical Armijo linesearch procedure, that is by determining the smallest nonnegative integer j_k such that, for some $\beta \in (0,1)$ and some $\eta \in (0, \frac{1}{2})$,

$$\phi(v_k + \beta^{j_k} \bar{\alpha}_k^{(v)} \Delta v_k, \mu_k, \rho_k) \leq \phi_k + \eta \beta^{j_k} \bar{\alpha}_k^{(v)} \langle \nabla_v \phi_k, \Delta v_k \rangle. \tag{1.35}$$

1.7 The algorithm

We are now in position to formally state our algorithm.

ALGORITHM:

Step 0: Set $k = 0$. The starting iterate $(x_0, 1, z_0)$ is given, such that $x_0, z_0 > 0$, as well as the initial barrier parameter $\mu_0 > 0$ and the constants $0 < \beta, \lambda, \eta, \nu_1, \sigma, \bar{\omega} < 1$, $\theta^{\text{DF}}, \theta^{\text{PF}} > 0$, $\delta, \nu_2 > 1$, $\theta^{\text{C}} \in (1, 1/\sigma)$, and $\rho_0 \geq \rho_{\min} \stackrel{\text{def}}{=} \lambda + 2\|r_0\|/\lambda$. Define $\xi_0 = 1$ and set $\omega(\mu_0) \in (0, \bar{\omega}]$.

Step 1: Compute the primal-dual step Δv_k^{PD} and y_{k+1}^{PD} from (1.7) and (1.11), modifying G_k if necessary to ensure that it is uniformly second-order sufficient with respect to A (with constant λ).

Step 2: If either $\xi_k = 0$ or (1.25) holds, define $\Delta v_k = \Delta v_k^{\text{PD}}$ and $y_{k+1} = y_{k+1}^{\text{PD}}$. Otherwise, compute the modified barrier step Δv_k^{MB} and y_{k+1}^{MB} from (1.13) and set $\Delta v_k = \Delta v_k^{\text{MB}}$ and $y_{k+1} = y_{k+1}^{\text{MB}}$.

Step 3: Compute $\alpha_k^{(v)}$ from (1.34) and (1.35). Then set

$$x_{k+1} = x_k + \alpha_k^{(v)} \Delta x_k \text{ and } \xi_{k+1} = \xi_k + \alpha_k^{(v)} \Delta \xi_k.$$

Step 4: Define

$$\Delta z_k = -z_k - X_k^{-1} Z_k \Delta x_k + \mu_k X_k^{-1} e. \tag{1.36}$$

If $z_k + \Delta z_k$ lies (componentwise) in the interval

$$\left[\nu_1 \min(e, z_k, \mu_k X_{k+1}^{-1} e),\ \max(\nu_2 e, z_k, \nu_2 \mu_k^{-1} e, \nu_2 \mu_k X_{k+1}^{-1} e)\right], \tag{1.37}$$

then set $z_{k+1} = z_k + \Delta z_k$; otherwise choose any z_{k+1} in the interval (1.37).

Step 5: Set $\rho_{k+1} = \rho_k$. If

$$\|g_{k+1} - A^T y_{k+1} - z_{k+1}\| \leq \theta^{\text{DF}} \mu_k \tag{1.38}$$

and

$$\langle x_{k+1}, z_{k+1} \rangle \leq n\theta^{\text{C}} \mu_k, \tag{1.39}$$

then test whether

$$\xi_{k+1} \leq \theta^{\text{PF}} \mu_k. \tag{1.40}$$

If all of these inequalities hold, define

$$\mu_{k+1} = \sigma \langle x_{k+1}, z_{k+1} \rangle / n \tag{1.41}$$

and possibly redefine $\rho_{k+1} \geq \rho_{\min}$, $\omega(\mu_{k+1}) \in (0, \bar{\omega}]$.

If (1.38) and (1.39) hold, but (1.40) fails, set $\mu_{k+1} = \mu_k$, and redefine $\rho_{k+1} = \delta \rho_k$ if

$$\Delta v_k = \Delta v_k^{\text{MB}} \text{ and } \alpha_k^{(v)} \geq \frac{\xi_k}{1 + \xi_k}. \tag{1.42}$$

If either of (1.38) or (1.39) fails, set $\mu_{k+1} = \mu_k$.

In all cases, increment k by one and go back to Step 1.

1.8 Comments on the algorithm

This algorithm suggests a few comments.

1. The requirement that z_{k+1} belongs to the interval (1.37) appears somewhat complex, but it is designed for maximum flexibility in the choice in z_{k+1}. The theory below only requires that the components of z_{k+1} are bounded above and away from zero while μ_k is not updated, and that the choice $z_{k+1} = \mu_k X_{k+1}^{-1} e$ is asymptotically acceptable when Δx_k tends to zero. This is similar to the conditions of Gill, Murray, Ponceléon and Saunders (1995), where these bounds are fixed a priori. Note that $z_{k+1} = z_k$ is always a

feasible choice when $z_k + \Delta z_k$ does not belong to the interval (1.37), and that then the nonnegativity of z_{k+1} is always guaranteed.

There are many algorithmic possibilities for computing a suitable z_{k+1} when $z_k + \Delta z_k$ does not belong to the interval (1.37). One could, for instance, use a backtracking strategy starting from $z_k + \Delta z_k$, or choose z_{k+1} to minimize $\|X_{k+1}Z_{k+1}e - \mu_k e\|$ subject to being in the desired interval.

Also note that the condition that $z_k + \Delta z_k$ must belong to the interval (1.37) does not restrict the step in x.

2. The tests of Step 5 aim to allow for frequent updating of μ_k, and hence for the rapid progress of the algorithm. We will say that iteration k is *μ-critical* whenever conditions (1.38), (1.39) and (1.40) hold. Condition (1.38) may be viewed as ensuring sufficient dual feasibility (hence θ^{DF}), (1.40) as ensuring sufficient primal feasibility (hence θ^{PF}) and (1.39) as ensuring a sufficient decrease in the value of the complementarity (hence θ^{C}). This latter condition is inspired by the literature on primal-dual algorithms (see Simantiraki and Shanno (1995), Zhang and Zhang (1994)) or Carpenter, Lustig, Mulvey and Shanno (1993), for instance).

   The conditions (1.42) are intended to allow ρ_k to increase when the value of this latter penalty parameter is not large enough to ensure primal feasibility, that is to ensure that the minimum of the merit function lies sufficiently close to the line $\xi = 0$. This is of concern only when a modified barrier step is used, as the primal-dual step always ensure improved primal feasibility. Hence the first condition. The second guarantees that a significant contribution to the minimization of the merit function is derived from the change in ξ.

3. The role of the modified barrier step Δv_k^{MB} and y_k^{MB} (possibly computed in Step 2) is to ensure adequate progress when the primal-dual step is uphill. In the numerical tests, it seems that it is effective in bringing the iterates back in a region where the primal-dual step is again acceptable. It may thus be viewed as an implicit centering device.

4. The dependence of the parameters $\omega(\mu_k)$ on μ_k is introduced with the aim of ensuring that, if μ_k is decreasing rapidly because of (1.41), the linesearch bound (1.30) should not prevent fast convergence by unduly restricting the stepsize. The threshold $\omega(\mu_k)$ may thus be adapted to avoid this effect. For instance, one might want to choose $\omega(\mu_k)$ to be of the order of μ_k, but the design of a truly efficient strategy will require much more detailed numerical experiments.

5. Suitable values for the constants might be, for instance,

$$\eta = 0.0001, \quad \sigma = \nu_1 = \bar{\omega} = \omega(\mu_k) = 0.01, \tag{1.43}$$

$$\theta^{\text{DF}} = \theta^{\text{PF}} = 1, \quad \beta = 0.5, \quad \delta = 10 \quad \theta^{\text{C}} = 99 \text{ and } \nu_2 = 100, \qquad (1.44)$$

but this remains to be confirmed by numerical experiences.

6. Observe that the algorithm does not update the value of y_k from iteration to iteration. This is possible because (1.7) and (1.13) directly compute y_{k+1}^{PD} and y_{k+1}^{MB}. Thus, although we expect y_{k+1} to converge to the Lagrange multipliers at the solution, these multipliers are recomputed afresh at each iteration.

   The fact that y_k is not recurred explicitly has the further advantage that we may modify Δx_k when G_k is not second-order sufficient with respect to A without considering any implied change in y_k.

7. If primal feasibility is obtained during the course of the calculation, that is if $\xi_k = 0$ for some k, the algorithm reduces to a purely (feasible) primal-dual framework.

8. The modified barrier step Δv_k^{MB} can be obtained at low cost from the factorization used to compute Δv_k^{PD}.

9. As the iterates approach a constrained minimum, we may expect G_k to become second-order necessary with respect to A, which implies that no modification of the primal-dual step should be necessary asymptotically, if the threshold value λ is chosen small enough. (This is expected because the problem becomes convex in a neighbourhood of such a minimum.) This property would *not* hold if we had chosen to make G_k positive definite, instead of $N^T G_k N$, possibly resulting in slower asymptotic convergence.

10. Observe that the penalty parameter ρ_k may be updated whenever the barrier parameter μ_k is reduced. This update may be an increase or a decrease. It provides the possibility of dynamically adapting ρ_k as the algorithm proceeds, without restricting the sequence of penalty parameters to be monotonically increasing.

2 Global convergence

We now intend to prove that our algorithm is globally convergent. More precisely, we wish to show that all limit points of a well-defined subsequence[2] of iterates are critical points for problem (1.1), for every choice of the starting iterate $(x_0, 1, z_0)$ for which (x_0, z_0) is strictly interior to the positive orthant in the (x, z)-space, as expressed in the following theorem.

[2]They are, in fact, the "major" iterations of the algorithm, if expressed as a two-level procedure.

Theorem 3 *Let* $\{(x_k, \xi_k, z_k)\}$ *be a sequence of iterates generated by the algorithm and define*

$$\mathcal{K} = \{k \mid \mu_k < \mu_{k-1}\}.$$

Then, K is infinite and we have that

$$\lim_{\substack{k\to\infty \\ k\in\mathcal{K}}} X_k Z_k = 0, \tag{2.1}$$

$$\lim_{\substack{k\to\infty \\ k\in\mathcal{K}}} \|g_k + A^T y_k - z_k\| = 0 \tag{2.2}$$

and

$$\lim_{\substack{k\to\infty \\ k\in\mathcal{K}}} \|Ax_k - b\| = 0. \tag{2.3}$$

The proof, which is detailed at the end of the section, uses an argument by contradiction. We will assume that convergence does not occur in that the barrier parameter μ_k stays bounded away from zero, and distinguish two cases. In the first, we assume that the penalty parameter ρ_k stays bounded; we will then show that a μ-critical iterate is eventually found if the primal-dual step is used, while (1.40) may not be obtained if the modified barrier step is used. In this latter case, we show that the penalty parameter has to increase. If, on the other hand, ρ_k tends to infinity, this can only happen for modified barrier steps, in which case we will prove that μ-criticality again eventually hold. This then implies that the barrier parameter is reduced contradicting our initial assumption, and convergence is thus obtained.

First note that assumptions (AS1) and (AS2) imply that there exists a constant $\kappa_8 > 0$ such that, for all k,

$$\|g_k\| \le \kappa_8 \text{ and } \|H_k\| \le \kappa_8. \tag{2.4}$$

Before proceeding further, we state some useful properties of the algorithm.

Lemma 4 *Let* $\{(x_k, \xi_k, z_k)\}$ *be a sequence of iterates generated by the algorithm. Then,*

(i) the sequence $\{\mu_k\}$ *is non-increasing and*

$$\mu_{k+1} \ne \mu_k \textit{ implies that } \mu_{k+1} \le \sigma\theta^{\mathrm{C}}\mu_k, \tag{2.5}$$

(ii) one has that, for all k,

$$\rho_k \ge \rho_{\min}, \quad A\Delta x_k = \Delta\xi_k r_0, \tag{2.6}$$

and

$$\langle \nabla_v \phi_k, \Delta v_k \rangle \le -\frac{1}{2}\lambda(\|\Delta v_k\|^2 + (\Delta\xi_k)^2). \tag{2.7}$$

Furthermore, if $\mu_k = \bar{\mu}$ and $\rho_k = \bar{\rho}$ for some $\bar{\mu}, \bar{\rho} > 0$ and all $k \geq 0$, then there exists a constant $\kappa_1 > 0$ such that

$$0 \leq \xi_k \leq \kappa_1 \tag{2.8}$$

for all k.

Proof. The non-increasing nature of the sequence $\{\mu_k\}$ and (2.5) immediately follow from (1.41), condition (1.39) and the inequality $\sigma\theta^c < 1$. The first bound of (2.6) results from the initial value $\rho_0 \geq \rho_{\min}$ and the fact that $\rho_k \geq \rho_{\min}$ for all k, because of the mechanism of Step 5. The second equation of (2.6) is a consequence of the mechanism of Steps 2 and 3, (1.24) and the second equation of (1.12). The inequality (2.7) then follows from (1.16), Lemma 2, the first bound of (2.6) and (1.25).

We conclude our proof by showing that, if μ_k and ρ_k are fixed at $\bar{\mu}$ and $\bar{\rho}$, respectively, then ξ_k remains bounded. First notice that the mechanism of Step 2 and Step 3 imposes that, for all k,

$$\phi(v_k, \bar{\mu}, \bar{\rho}) \leq \phi(v_{k-1}, \bar{\mu}, \bar{\rho}) \leq \phi(v_0, \bar{\mu}, \bar{\rho})$$

and thus that

$$(\xi_k + 1)^2 \leq (\xi_0 + 1)^2 + \frac{2}{\bar{\rho}}[f(x_0) - \bar{\mu}\langle \log(x_0), e\rangle - (f(x_k) - \bar{\mu}\langle \log(x_k), e\rangle)]. \tag{2.9}$$

Now, if

$$(\xi_k + 1)^2 \leq (\xi_0 + 1)^2, \tag{2.10}$$

then one obtains that

$$\xi_k \leq \xi_0. \tag{2.11}$$

On the other hand, if (2.10) does not hold, then the expression within brackets in the right-hand side of (2.9) is positive, and thus

$$(\xi_k + 1)^2 \leq (\xi_0 + 1)^2 + \frac{2}{\bar{\rho}}[f(x_0) - \bar{\mu}\langle \log(x_0), e\rangle - \kappa_3], \tag{2.12}$$

where $\kappa_3 = \min_{x \in \mathcal{D}}[f(x) - \bar{\mu}\langle \log(x), e\rangle]$ is finite because of (AS1) and (AS2). The bounds (2.11) and (2.12) and the fact that $\xi_k \geq 0$ because of (1.32) then yield (2.8), completing the proof □

We next prove a technical result showing under what conditions the primal-dual and modified barrier steps are bounded when μ and ρ are fixed.

Lemma 5 *Assume that $\mu_k = \bar{\mu}$ and $\rho_k = \bar{\rho}$ for some $\bar{\mu}, \bar{\rho} > 0$ and all $k \geq 0$. Assume furthermore that there exists a $\kappa_4 > 0$ such that, for all k,*

$$\bar{\rho} \leq \kappa_4, \quad \|g_k - \mu_k X_k^{-1} e\| \leq \kappa_4, \quad \textit{and} \quad \|G_k\| \leq \kappa_4. \tag{2.13}$$

Then, there exists a positive constant $\kappa_5 > 0$ such that, for all k,

$$\|\Delta x_k^{\rm PD}\| \le \kappa_5, \quad |\Delta \xi_k^{\rm PD}| \le \kappa_5 \text{ and } \|y_{k+1}^{\rm PD}\| \le \kappa_5, \tag{2.14}$$

and

$$\|\Delta x_k^{\rm MB}\| \le \kappa_5, \quad |\Delta \xi_k^{\rm MB}| \le \kappa_5 \text{ and } \|y_{k+1}^{\rm MB}\| \le \kappa_5. \tag{2.15}$$

Proof. Consider the primal-dual step first. Writing

$$\Delta x_k^{\rm PD} = A^T \Delta x_k^{(a)} + N \Delta x_k^{(n)}, \tag{2.16}$$

we obtain from the second equation of (1.7) that

$$AA^T \Delta x_k^{(a)} = -r_k$$

which implies, since A has full rank (AS3), that

$$\|A^T \Delta x_k^{(a)}\| = \|A^T(AA^T)^{-1} r_k\| \le \kappa_1 \|r_0\| \, \|A\| \, \|(AA^T)^{-1}\| \stackrel{\rm def}{=} \kappa_6, \tag{2.17}$$

where we have used (2.8) and (1.10) to deduce the last inequality. On the other hand, the first equation of (1.7) gives that

$$N^T G_k N \Delta x_k^{(n)} = -N^T(g_k - \mu_k X_k^{-1} e) - N^T G_k A^T \Delta x_k^{(a)} \tag{2.18}$$

The second-order sufficiency of G_k (possibly modified) with respect to A, (2.13), (2.17) and (2.18) then ensure that

$$\lambda \|\Delta x_k^{(n)}\| \le \kappa_4 (1 + \kappa_6) \|N\|, \tag{2.19}$$

where λ is the smallest eigenvalue of the (possibly modified) G_k restricted to the nullspace of A. Combining (2.16), (2.17) and (2.19), we deduce that

$$\|\Delta x_k^{\rm PD}\| \le \kappa_6 + \frac{2\kappa_4}{\lambda}(1 + \kappa_6)\|N\|^2 \stackrel{\rm def}{=} \kappa_7. \tag{2.20}$$

Similarly, we obtain from the first equation of (1.7) that

$$AA^T y_{k+1}^{\rm PD} = -A(g_k - \mu_k X_k^{-1} e) - AG_k \Delta x_k^{\rm PD},$$

which yields, using (2.13) and (2.20), that

$$\|y_{k+1}^{\rm PD}\| \le \kappa_4 (1 + \kappa_7) \|A\| \, \|(AA^T)^{-1}\|. \tag{2.21}$$

Finally, from (1.11) and (2.8),

$$|\Delta \xi_k^{\rm PD}| = |\xi_k| \le \kappa_1. \tag{2.22}$$

Together, the bounds (2.20), (2.21) and (2.22) prove (2.14) with

$$\kappa_5 = \kappa_5^{\mathrm{PD}} \stackrel{\mathrm{def}}{=} \max\left[\kappa_7, \kappa_4(1+\kappa_7)\|A\|\,\|(AA^T)^{-1}\|, \kappa_1\right].$$

Consider now the modified barrier step. Premultiplying the first equation of (1.13) by Δx_k^{MB}, we obtain, using successively the second and third equations of the same system, that

$$\begin{aligned} -\langle \Delta x_k^{\mathrm{MB}}, g_k - \mu_k X_k^{-1} e\rangle &= \langle \Delta x_k^{\mathrm{MB}}, G_k \Delta x_k^{\mathrm{MB}}\rangle + \langle \Delta x_k^{\mathrm{MB}}, A^T y_{k+1}^{\mathrm{MB}}\rangle \\ &= \langle \Delta x_k^{\mathrm{MB}}, G_k \Delta x_k^{\mathrm{MB}}\rangle + \Delta \xi_k^{\mathrm{MB}} \langle r_0, y_{k+1}^{\mathrm{MB}}\rangle \\ &= \langle \Delta x_k^{\mathrm{MB}}, G_k \Delta x_k^{\mathrm{MB}}\rangle + \Delta \xi_k^{\mathrm{MB}}(\rho_k \Delta \xi_k^{\mathrm{MB}} + \rho_k(\xi_k + 1)). \end{aligned}$$

Now, using the second-order sufficiency of G_k with respect to A and Lemma 2, we have that

$$\begin{aligned} \tfrac{1}{2}\lambda \|\Delta v_k^{\mathrm{MB}}\|^2 &\le \langle \Delta x_k^{\mathrm{MB}}, G_k \Delta x_k^{\mathrm{MB}}\rangle + \rho_k (\Delta \xi_k^{\mathrm{MB}})^2 \\ &= -\langle \Delta x_k^{\mathrm{MB}}, g_k - \mu_k X_k^{-1} e\rangle - \rho_k \Delta \xi_k^{\mathrm{MB}} (\xi_k + 1) \\ &\le \|\Delta v_k^{\mathrm{MB}}\| (\|g_k - \mu_k X_k^{-1} e\| + \rho_k(\xi_k + 1)). \end{aligned}$$

We therefore obtain, using (2.13), (2.8) and the first part of (2.13), that

$$\frac{1}{2}\lambda \|\Delta v_k^{\mathrm{MB}}\| \le \|g_k - \mu_k X_k^{-1} e\| + \rho_k(|\xi_k| + 1) \le \kappa_4 + \kappa_4(\kappa_1 + 1). \tag{2.23}$$

We also obtain from the first equation of (1.13) that

$$AA^T y_{k+1}^{\mathrm{MB}} = -A(g_k - \mu_k X_k^{-1} e) - AG_k \Delta x_k^{\mathrm{MB}},$$

and thus, using again (2.13), (2.23) and the inequality $\|\Delta x_k^{\mathrm{MB}}\| \le \|\Delta v_k^{\mathrm{MB}}\|$, that

$$\|y_{k+1}^{\mathrm{MB}}\| \le \kappa_4 \|A\| \|(AA^T)^{-1}\| (1 + \frac{2\kappa_4}{\lambda}(2 + \kappa_1)). \tag{2.24}$$

Combining (2.23), (2.24) and $|\Delta \xi_k^{\mathrm{MB}}| \le \|\Delta v_k^{\mathrm{MB}}\|$, we obtain (2.15) with

$$\kappa_5 = \kappa_5^{\mathrm{MB}} \stackrel{\mathrm{def}}{=} \max\left[\frac{2\kappa_4}{\lambda}(2+\kappa_1), \kappa_4 \|A\| \|(AA^T)^{-1}\| \left(1 + \frac{2\kappa_4}{\lambda}(2+\kappa_1)\right)\right].$$

The complete result then follows by taking $\kappa_5 = \max[\kappa_5^{\mathrm{PD}}, \kappa_5^{\mathrm{MB}}]$. □

We next examine the behaviour of a sequence of iterates for fixed μ and ρ.

Lemma 6 *Let $\{(x_k, \xi_k, z_k)\}$ be a sequence of iterates generated by the algorithm and assume that*

$$\mu_k = \bar{\mu} \text{ and } \rho_k = \bar{\rho} \tag{2.25}$$

for all k. Then, we have that

$$\lim_{k\to\infty} \|\Delta x_k\| = 0, \tag{2.26}$$

$$\lim_{k\to\infty} \Delta \xi_k = 0, \tag{2.27}$$

$$\lim_{k\to\infty} X_{k+1} Z_{k+1} e = \bar{\mu} e, \tag{2.28}$$

$$\lim_{k\to\infty} \|g_{k+1} + A^T y_{k+1} - \bar{\mu} X_{k+1}^{-1} e\| = 0. \tag{2.29}$$

Proof. We start our proof by noting that, for fixed $\bar{\mu}$ and $\bar{\rho}$, the iteration then reduces to approximately minimizing the function $\phi(v, \bar{\mu}, \bar{\rho})$. Moreover, as a consequence of (2.25), and because the level set

$$L_0 = \{(x, \xi) \in \mathcal{D} \times [0, \infty] \mid \phi(x, \xi, \bar{\mu}, \bar{\rho}) \leq \phi(x_0, \xi_0, \bar{\mu}, \bar{\rho})\}$$

is bounded away from the boundary of the positive orthant in x, we may deduce that, for all k,

$$X_k e \geq \kappa_{12} e \tag{2.30}$$

for some $\kappa_{12} \in (0, 1)$. On the other hand, (1.37) and (2.30) imply that

$$\|z_k\| \leq \max\left(\nu_2\sqrt{n}, \|z_0\|, \nu_2\frac{\sqrt{n}}{\bar{\mu}}, \nu_2\frac{\bar{\mu}\sqrt{n}}{\kappa_{12}}\right). \tag{2.31}$$

for all k. Combining now this last bound with (2.30) and the second bound of (2.4), we then deduce from the definition of G_k that there exists a $\kappa_{13} > 0$ such that

$$\|G_k\| \leq \kappa_{13}.$$

Furthermore, we obtain from (1.27) that we may choose, for each k, a ΔG_k ensuring that $G_k + \Delta G_k$ is second-order sufficient with respect to A (with constant λ), such that, using (1.27),

$$\|G_k + \Delta G_k\| \leq \|G_k\| + \|\Delta G_k\| \leq (1 + \kappa_2)\kappa_{13} + \lambda \tag{2.32}$$

and the minimum eigenvalue of $G_k + \Delta G_k$ in the nullspace of A is at least λ. If we now examine the gradient of the merit function with respect to x, we verify that

$$\|g_k - \bar{\mu} X_k^{-1} e\| \leq \|g_k\| + \bar{\mu}\|X_k^{-1} e\| \leq \kappa_8 + \bar{\mu}\frac{\sqrt{n}}{\kappa_{12}}, \tag{2.33}$$

where we have used the first bound of (2.4) and (2.30). Combining (2.25), (2.32) and (2.33), we see that all the conditions of Lemma 5 are satisfied for

$$\kappa_4 = \max\left[2\kappa_{13} + \lambda, \kappa_8 + \bar{\mu}\frac{\sqrt{n}}{\kappa_{12}}, \bar{\rho}\right].$$

We may thus deduce from this lemma that (2.14) and (2.15) hold, which gives that,

$$\|\Delta x_k\| \leq \|\Delta v_k\| \leq \kappa_{14} \tag{2.34}$$

for some $\kappa_{14} > 0$.

We now show that we can deduce a contradiction if the minimization of $\phi(v, \bar{\mu}, \bar{\rho})$ is not successful. To this aim, we make the additional assumption that

$$\|\Delta x_k\| \geq \kappa_{15} \tag{2.35}$$

for all $k \in J$, where J is the index set of a subsequence, and for some $\kappa_{15} \in (0, \kappa_{14})$. We then deduce from (2.7) that, for $k \in J$,

$$\langle \nabla_v \phi_k, \Delta v_k \rangle \leq -\frac{1}{2}\lambda \kappa_{15}^2 \stackrel{\text{def}}{=} -\kappa_{16} < 0. \tag{2.36}$$

We now observe that (1.31), (2.30) and (2.34) give that

$$\bar{\alpha}_k^{(x)} \geq \min\left[1, (1-\omega(\bar{\mu}))\frac{\kappa_{12}}{\kappa_{14}}\right] \tag{2.37}$$

for all k. Furthermore, we note that the mechanism of the algorithm implies that the situation where

$$\alpha_k^{(v)} = \bar{\alpha}_k^{(\xi)} < 1 \tag{2.38}$$

can only happen for a unique k, ξ_k being identically zero (and thus $\bar{\alpha}_k^{(\xi)}$ being identically one) for all subsequent iterations. Hence, if k_ξ is the index of the iteration where (2.38) holds (defining $k_\xi = \infty$ if (2.38) never holds), we see that

$$\alpha_k^{(v)} < \bar{\alpha}_k^{(\xi)}, \text{ if } k < k_\xi, \qquad \text{or} \qquad \alpha_k^{(v)} \leq 1, \text{ if } k > k_\xi.$$

We therefore conclude that, for k sufficiently large, the inequality $\alpha_k^{(v)} \leq \bar{\alpha}_k^{(\xi)}$ does not limit the stepsize in the linesearch procedure (1.35) to a value strictly below one. Moreover, combining (1.33), (2.37) and the definition of k_ξ, we have that

$$\bar{\alpha}_k^{(v)} \geq \alpha^{(v)} \stackrel{\text{def}}{=} \min\left[1, (1-\omega(\bar{\mu}))\frac{\kappa_{12}}{\kappa_{14}}, \bar{\alpha}_{k_\xi}^{(\xi)}\right]$$

for all $k \in J$.

We next consider iteration $k \in J$ and distinguish two cases. The first is when

$$\langle \nabla_v \phi(v_k + \alpha \Delta v_k, \bar{\mu}, \bar{\rho}), \Delta v_k \rangle < \eta \langle \nabla_v \phi_k, \Delta v_k \rangle \tag{2.39}$$

for all $\alpha \in (0, \alpha^{(v)}]$. In the second case, we assume that there exists a (smallest) $\bar{\alpha} \in (0, \alpha^{(v)}]$ such that

$$\langle \nabla_v \phi(v_k + \bar{\alpha} \Delta v_k, \bar{\mu}, \bar{\rho}), \Delta v_k \rangle = \eta \langle \nabla_v \phi_k, \Delta v_k \rangle. \tag{2.40}$$

But (2.36), (2.40) and the mean-value theorem then give that

$$\begin{aligned} 0 &< \langle \nabla_v \phi(v_k + \bar{\alpha} \Delta v_k, \bar{\mu}, \bar{\rho}), \Delta v_k \rangle - \langle \nabla_v \phi_k, \Delta v_k \rangle \\ &= \bar{\alpha} \langle \nabla_{vv}^2 \phi(v_k + \zeta_1 \Delta v_k, \bar{\mu}, \bar{\rho}) \Delta v_k, \Delta v_k \rangle \end{aligned}$$

for some $\zeta_1 \in (0, \alpha^{(v)})$. Hence, recalling (2.40), we obtain that

$$\bar{\alpha} = \frac{-(1-\eta)\langle \nabla_v \phi_k, \Delta v_k \rangle}{\langle \nabla_{vv}^2 \phi(v_k + \zeta_1 \Delta v_k, \bar{\mu}, \bar{\rho}) \Delta v_k, \Delta v_k \rangle}. \tag{2.41}$$

Observe now that

$$\nabla^2_{vv}\phi(v_k + \zeta_1 \Delta v_k, \bar{\mu}, \bar{\rho}) = \begin{pmatrix} H(x_k + \zeta_1 \Delta x_k) + \bar{\mu}(X_k + \zeta_1 \Delta X_k)^{-2} & 0 \\ 0 & \bar{\rho} \end{pmatrix}. \quad (2.42)$$

Observe also that (AS1), (AS2), (1.31), (1.33), (1.34) and the fact that $\zeta_1 \leq \alpha^{(v)}$ ensure that

$$\|H(x_k + \zeta_1 \Delta x_k)\| \leq \kappa_{17} \quad (2.43)$$

for some $\kappa_{17} > 0$. We also deduce from (2.30), (1.31), (1.33), (1.34) and the fact that $\zeta_1 \leq \alpha^{(v)}$ that the components of $x_k + \zeta_1 \Delta x_k$ are bounded below by a positive constant. This fact, (2.43) and (2.42) then imply that

$$\|\nabla^2_{vv}\phi(v_k + \zeta_1 \Delta v_k), \bar{\mu}, \bar{\rho})\| \leq \kappa_{18}$$

for some $\kappa_{18} > 0$. Substituting this bound, (2.34) and (2.36) in (2.41) and using the Cauchy-Schwarz inequality, we obtain that

$$\bar{\alpha} \geq \frac{(1-\eta)\kappa_{16}}{\kappa_{18}\kappa_{14}^2}.$$

Thus, gathering the two cases, we conclude that, for all $k \in J$, (2.39) holds for every α between 0 and

$$\alpha^* \stackrel{\text{def}}{=} \min\left(\frac{(1-\eta)\kappa_{16}}{\kappa_{18}\kappa_{14}^2}, \alpha^{(v)}\right).$$

Returning to the function $\phi(v, \bar{\mu}, \bar{\rho})$ itself, we therefore obtain that, for each $\alpha \in [0, \alpha^*]$ and all $k \in J$,

$$\begin{aligned} \phi(v_k + \alpha \Delta v_k, \bar{\mu}, \bar{\rho}) &= \phi_k + \alpha \langle \nabla_v \phi(v_k + \zeta_2 \Delta v_k, \bar{\mu}, \bar{\rho}), \Delta v_k \rangle \\ &\leq \phi_k + \eta\alpha \langle \nabla_v \phi_k, \Delta v_k \rangle, \end{aligned}$$

where $\zeta_2 \in (0, \alpha)$. As a consequence, the stepsize determined by (1.35) must satisfy

$$\alpha_k^{(v)} \geq \min\left[\beta\alpha^*, (1-\omega(\bar{\mu}))\frac{\kappa_{12}}{\kappa_{14}}\right] \stackrel{\text{def}}{=} \kappa_{19}$$

finally yielding, together with (2.36), that

$$\phi_{k+1} = \phi(v_k + \alpha_k^{(v)} \Delta v_k, \bar{\mu}, \bar{\rho}) \leq \phi_k - \eta\kappa_{16}\kappa_{19}, \quad (2.44)$$

for all $k \in J$ sufficiently large. But (2.44) implies that the sequence $\{\phi_k\}$ tends to minus infinity, which is impossible because (AS1) and (AS.2) imply that ϕ is bounded below in $\mathcal{D}$. Hence our assumption (2.35) must be false and we obtain that (2.26) holds. This limit, (2.30) and the inequality $\alpha_k^{(v)} \leq 1$ in turn imply that

$$\lim_{k\to\infty} \|X_k^{-1} - X_{k+1}^{-1}\| = \lim_{k\to\infty} \max_{i=1,\ldots,n} \left[\frac{\alpha_k^{(v)}|[\Delta x_k]_i|}{|[x_k]_i[x_{k+1}]_i|}\right] \leq \lim_{k\to\infty} \frac{\|\Delta x_k\|}{\kappa_{12}^2} = 0. \quad (2.45)$$

But, since

$$\begin{aligned}\|z_k + \Delta z_k - \bar{\mu}X_{k+1}^{-1}e\| &\leq \|z_k + \Delta z_k - \bar{\mu}X_k^{-1}e\| + \bar{\mu}\|(X_k^{-1} - X_{k+1}^{-1})e\| \\ &\leq \|X_k^{-1}Z_k\|\,\|\Delta x_k\| + \bar{\mu}\sqrt{n}\|X_k^{-1} - X_{k+1}^{-1}\|,\end{aligned}$$

where we have used (1.36), we also obtain from (2.26), (2.30), (2.31) and (2.45) that

$$\lim_{k\to\infty} \|z_k + \Delta z_k - \bar{\mu}X_{k+1}^{-1}e\| = 0.$$

But this limit and the inequalities $\nu_1 < 1$ and $\nu_2 > 1$ give that

$$\nu_1\bar{\mu}X_{k+1}^{-1}e \leq z_k + \Delta z_k \leq \nu_2\bar{\mu}X_{k+1}^{-1}e$$

for k sufficiently large. Hence, from the definition of Step 4 of the algorithm, $z_{k+1} = z_k + \Delta z_k$ for sufficiently large k. Thus (1.36) yields that

$$X_{k+1}Z_{k+1}e = X_{k+1}X_k^{-1}(-Z_k\Delta x_k + \bar{\mu}e). \quad (2.46)$$

On the other hand, since

$$\frac{[x_{k+1}]_i}{[x_k]_i} = \frac{[x_k + \alpha_k^{(v)}\Delta x_k]_i}{[x_k]_i} = 1 + \alpha_k^{(v)}\frac{[\Delta x_k]_i}{[x_k]_i},$$

we deduce from (2.26), (2.30) and $\alpha_k^{(v)} \in (0, 1]$ that

$$\lim_{k\to\infty} X_{k+1}X_k^{-1} = I_n, \quad (2.47)$$

where I_n is the identity matrix of dimension n. The limit (2.28) then follows from combining (2.46), (2.47), (2.26) and (2.31).

We also note that (1.7), (1.13), (2.30), (2.31) and (2.26) give that

$$\lim_{k\to\infty} \|g_k + A^Ty_{k+1} - \bar{\mu}X_k^{-1}e\| = 0.$$

We then use the continuity of the gradient, (2.26) and (2.45) to obtain (2.29). Finally, (2.27) follows from (2.26) and the second part of (2.6). □

The next stage in our theory is to analyze the situation where the penalty parameter ρ_k tends to infinity, and show that infeasibilities with respect to the linear equality constraints must then decrease.

Lemma 7 *Let $\{(x_k, \xi_k, z_k)\}$ be a sequence of iterates generated by the algorithm and define I to be the index set of all iterations such that ρ_k is increased at Step 5. Assume furthermore that*

$$\mu_k = \bar{\mu}$$

for all k and that the subsequence indexed by I is infinite. Then, there exists an infinite subsequence indexed by $J \subseteq I$ such that, for $k \in J$,

$$\xi_{k+1} \leq \theta^{\mathrm{PF}} \bar{\mu}. \tag{2.48}$$

Proof. Note that the first part of (1.42) implies that $\Delta v_k = \Delta v_k^{\mathrm{MB}}$ for all $k \in I$. Observe that, for $k \in I$, some components of y_{k+1} could be bounded in norm. Let us denote by $y_{k+1}^{\mathcal{B}}$ the vector whose entries are those of y_{k+1} in this (possibly empty) components' set and zero elsewhere, and by $\kappa_{20} > 0$ the associated upper bound. We thus obtain that, for $k \in I$,

$$y_{k+1} = y_{k+1}^{\mathcal{U}} + y_{k+1}^{\mathcal{B}} \text{ where } \|y_{k+1}^{\mathcal{B}}\| \leq \kappa_{20},$$

thus defining $y_{k+1}^{\mathcal{U}}$.

We now consider two cases. The first is when the set of bounded components of y_{k+1} is a proper subset of $\{1, \ldots, m\}$. In this case, there must be an infinite subsequence of I indexed by J and some subset $\mathcal{Z} \subseteq \{1, \ldots, n\}$ such that

$$\lim_{\substack{k\to\infty \\ k\in J}} |[A^T y_{k+1}]_i| = \lim_{\substack{k\to\infty \\ k\in J}} |[A^T y_{k+1}^{\mathcal{U}}]_i| = \infty \tag{2.49}$$

for $i \in \mathcal{Z}$, where we used the fact that A has full rank (AS3), while

$$|[A^T y_{k+1}]_i| \leq \kappa_{21} \tag{2.50}$$

for some $\kappa_{21} > 0$ and for $i \notin \mathcal{Z}$. Now, (1.38) holds for $k \in J \subseteq I$, which, with (2.4) and (2.49), implies, since (2.4) has no dependence on J, that

$$\lim_{\substack{k\to\infty \\ k\in J}} [A^T y_{k+1}]_i = \lim_{\substack{k\to\infty \\ k\in J}} [z_{k+1}]_i \quad (i \in \mathcal{Z}). \tag{2.51}$$

Since $z_{k+1} > 0$, we immediately deduce from (2.49) and (2.51) that

$$\lim_{\substack{k\to\infty \\ k\in J}} [A^T y_{k+1}]_i = \infty \quad (i \in \mathcal{Z}). \tag{2.52}$$

Furthermore, (2.51) and (2.52) yield that

$$\lim_{\substack{k\to\infty \\ k\in J}} [z_{k+1}]_i = \infty \quad (i \in \mathcal{Z}).$$

But this last limit is only possible if the last term in the upper bound of (1.37) tends itself to infinity, that is if

$$\lim_{\substack{k\to\infty \\ k\in J}} [x_{k+1}]_i = 0 \quad (i \in \mathcal{Z}). \tag{2.53}$$

Now, for $k \in J$,

$$\xi_{k+1}\langle r_0, y_{k+1}^{\mathcal{U}}\rangle = \langle r_{k+1}, y_{k+1}^{\mathcal{U}}\rangle = \langle A(x_{k+1} - x_\odot), y_{k+1}^{\mathcal{U}}\rangle = \langle x_{k+1} - x_\odot, A^T y_{k+1}^{\mathcal{U}}\rangle, \tag{2.54}$$

where $x_\odot$ is given in (AS4) and where we have used (1.10), (1.6) and the identity $Ax_\odot = b$. Clearly, (2.53) gives that

$$\lim_{\substack{k\to\infty \\ k\in J}} [x_{k+1} - x_\odot]_i = -[x_\odot]_i,$$

for $i \in \mathcal{Z}$, which, together with (2.52), (2.50), (2.54) and the inequality $x_\odot > 0$, yields that

$$\lim_{\substack{k\to\infty \\ k\in J}} \xi_{k+1}\langle r_0, y_{k+1}^{\mathcal{U}}\rangle = -\infty. \tag{2.55}$$

If we now turn to the second case, that is when all components of y_{k+1} are bounded by κ_{20}, we then have that $y_{k=1}^{\mathcal{U}}$ is identically zero and we define $J = I$.

In both cases, we obtain from the Cauchy-Schwarz inequality and $\xi_{k+1} \geq 0$ that

$$\langle r_0, y_{k+1}\rangle = \langle r_0, y_{k+1}^{\mathcal{B}}\rangle + \langle r_0, y_{k+1}^{\mathcal{U}}\rangle \leq \kappa_{20}\|r_0\| \tag{2.56}$$

for $k \in J$ sufficiently large, where we used (2.55) to obtain the last inequality if $y_{k+1}^{\mathcal{U}}$ is nonzero. Now the third equation of (1.12), the bound $\alpha_k^{(v)} \leq 1$ and the second part of (1.42) then give that, for $k \in J$,

$$\xi_{k+1} = \xi_k + \alpha_k^{(v)}\Delta\xi_k = \frac{\alpha_k^{(v)}}{\rho_k}\langle r_0, y_{k+1}\rangle + \xi_k - \alpha_k^{(v)}(\xi_k + 1) \leq \frac{1}{\rho_k}\langle r_0, y_{k+1}\rangle. \tag{2.57}$$

Substituting (2.56) in (2.57) then gives that, for $k \in J$ sufficiently large,

$$\xi_{k+1} \leq \frac{\kappa_{20}\|r_0\|}{\rho_k}.$$

This and the fact that ρ_k tends to infinity ensures that (2.48) holds for $k \in J$ sufficiently large, as required. □

We are now ready to prove our main convergence result.

Proof of Theorem 3. In order to prove our main convergence result, we will now consider the behaviour of the algorithm if convergence never occurs, and later deduce that this behaviour is impossible. Assume therefore, for the purpose of establishing a contradiction, that, for all k,

$$\mu_k \geq \bar{\mu} > 0. \tag{2.58}$$

Because Lemma 4 ensures that the sequence $\{\mu_k\}$ is non-increasing, (2.58) implies that the update (1.41) is never performed for k sufficiently large and we may thus assume, without loss of generality, that

$$\mu_k = \bar{\mu} \tag{2.59}$$

for all $k \geq 0$.

Assume first that

$$\Delta v_k = \Delta v_k^{\text{MB}} \tag{2.60}$$

and

$$\rho_k = \bar{\rho} \tag{2.61}$$

hold for all k sufficiently large. Because of equalities (2.59) and (2.61), we may then apply Lemma 6 and deduce that (2.26), (2.27), (2.28) and (2.29) hold. But these limits imply that conditions (1.38) and (1.39) are satisfied for k sufficiently large. Furthermore (1.31), (2.26) and (2.30) ensure that $\alpha_k^{(x)} = 1$ for all k sufficiently large. Moreover, as (2.59) guarantees that (1.40) cannot be true, (1.32) ensures that $\alpha_k^{(\xi)} = 1$ for all k sufficiently large. Hence $\alpha_k^{(v)} = 1$, and (1.42) are satisfied for all k sufficiently large. Since ρ_k remains constant, the mechanism of Step 5 then ensures that (1.40) must also be satisfied for such k. As a consequence, μ_k is eventually reduced according to (1.41), which contradicts (2.59). Hence, if μ_k remains constant and (2.60) holds for all sufficiently large k, ρ_k must tend to infinity and is increased in Step 5, for some infinite subsequence I. We may then apply Lemma 7 and deduce (2.48) for some subsequence J for which conditions (1.38), (1.39) and (1.40) hold. As above, this in turn implies that μ_k is reduced according to (1.41), again contradicting (2.59). We therefore deduce that (2.60) cannot hold for all sufficiently large k if (2.59) holds. As a consequence, if this last relation holds, there must exist an infinite subsequence indexed by L such that

$$\Delta v_k = \Delta v_k^{\text{PD}}$$

for $k \in L$. Applying now Lemma 7 as above, we also conclude that, if ρ_k is increased infinitely often in Step 5, then μ_k must be reduced, which is impossible because of (2.59). As a consequence, we therefore deduce that ρ_k remains constant (and equal to some $\bar{\rho}$) for sufficiently large k. We may then apply Lemma 6 again, and deduce that (2.26), (2.28) and (2.29) hold for sufficiently large k. But the first of these limits, the third part of (2.6) and the second block of (1.7) together then imply that

$$\|r_0\| \lim_{k\to\infty} \xi_k = \lim_{k\to\infty} \|r_k\| \leq \|A\| \lim_{k\to\infty} \|\Delta x_k^{\text{PD}}\| = 0 \tag{2.62}$$

for $k \in L$. Once more, we see that, for $k \in L$ sufficiently large, μ_k must then be reduced using (1.41), since (2.28), (2.29) and (2.62) ensure that a μ-critical iteration must occur eventually. This again contradicts (2.59), finally proving that this last assumption is impossible.

Hence μ_k is not bounded away from zero. But (1.41) implies that $\mu_k > 0$ for all k, and thus that the subsequence indexed by $\mathcal{K}$ is infinite, and we deduce from condition (2.5) of Lemma 4 and the inequality $\sigma\theta^c < 1$ that

$$\lim_{k\to\infty} \mu_k = 0. \tag{2.63}$$

Recalling now the definition of $\mathcal{K}$, the index set of all iterations immediately following an update of μ_k using (1.41), we then see that (1.39) implies that

$$X_k Z_k e \leq \langle x_k, z_k \rangle e = \frac{n}{\sigma}\mu_k e$$

for $k \in \mathcal{K}$. But this inequality and (2.63) together yield the limit (2.1). Combining (2.1), (1.41), (1.38) and (1.40), one obtains (2.2) and (2.3), which concludes the proof. □

We conclude this section by noting we may consider relaxing even further the conditions of Step 4 on the update of the dual variables z. In the algorithm, we have enforced the choice $z_{k+1} = z_k + \Delta z_k$ whenever this vector falls in the interval (1.37). This can be relaxed somewhat, in that our theory still holds if we only require that z_{k+1} is any vector satisfying the bounds given by (1.37) with the property that

$$\lim_{k\to\infty} \Delta x_k = 0 \text{ implies } \lim_{k\to\infty} X_{k+1} Z_{k+1} = \mu_k e.$$

This implication is indeed all we need to obtain the limit (2.28) from (2.26) at the end of the proof of Lemma 6. The main interest of this slight extension is that it now covers the case where

$$z_k = \mu_k X_k^{-1} e. \tag{2.64}$$

If the choice (2.64) is made, the algorithm reduces, for iteration k, to a pure primal method in that z_k is entirely eliminated from the computation: Δz_k need not be computed and (1.36) may thus be skipped altogether. We then obtain that

$$G_k = H_k + \mu_k X_k^{-2},$$

which is exactly the Hessian of the merit function $\phi(v_k, \mu_k, \rho_k)$ in the x-space. This may be attractive if one wishes to exploit directions of negative curvature for the merit function, as they then correspond to linear combinations of eigenvectors of G_k associated with negative eigenvalues. Again, the detail of these considerations is beyond the scope of the present paper and we postpone their presentation for future work.

3 Preliminary numerical tests

In order to investigate the effectiveness of the method discussed in this paper, we have written a prototype fortran 90 implementation of the algorithm proposed in Section 1.7 to solve *quadratic programs*, that is for problems for which $f(x)$ is a quadratic function. In this implementation, z_{k+1} is simply chosen as $z_k + \alpha_k^{(z)} \Delta z_k$, where $\alpha_k^{(z)}$ is the minimum of 1 and the largest stepsize such that $z_k + \alpha_k^{(z)} \Delta z_k$ remains in the interval (1.37).

The solution of linear systems of the form

$$\begin{pmatrix} G & A^T \\ A & 0 \end{pmatrix} \begin{pmatrix} \Delta x \\ y \end{pmatrix} = - \begin{pmatrix} g \\ r \end{pmatrix}$$

lie at the heart of the algorithm. For convenience, we have used the Harwell Subroutine Library (1995) package MA27 to solve such systems. The multifrontal scheme used (see, Duff and Reid, 1983) has the additional advantages of being able to cope with large, sparse systems and of reporting the inertia of the relevant coefficient matrix, K. Although some authors (for instance, Gill, Murray, Saunders and Wright, 1990) have reported that such an approach is handicapped by the severe indefiniteness of K, we followed the advice of Gill, Murray, Ponceléon and Saunders (1991) and use a very small pivot threshold (10^{-6}) together with iterative refinement as an effective means of solution. If MA27 reports that G is not second-order sufficient, we use the naive expedient of replacing G by $G + ||G||I$. More sophisticated strategies are being considered (see Gould, 1995), but are beyond the scope of this paper.

The actual algorithm implemented is the obvious generalization of the algorithm described above designed to cope with simple bounds of the form $l \leq x \leq u$ rather than nonnegativities alone. All fixed variables are removed automatically and the minimization performed with respect to the remaining variables. A given starting point x is adjusted so that each component lies at least a distance ten on the feasible side of its nearest bound; if this is impossible the mid point between the two bounds is chosen. Similarly, the dual variable associated with each simple bound is supplied by the user (we used zero in our tests) and adjusted so that it is at least a distance 10 to the feasible side of its relevant dual bound. We use $\mu_0 = \langle x_0, z_0 \rangle / n$ and the parameter values suggested in (1.43)(1.44). The algorithm is halted as soon as the norm of the residual of (1.2) is smaller than 10^{-4}, or more than 1000 iterations have been performed. Other experiments where this limit was raised to 10000 gave essentially the same results.

To test our algorithm, we have selected all of the larger quadratic programs in the CUTE test set (see, Bongartz et al., 1995). Although it is desirable in practice to preprocess the problems (for instance, to remove redundant constraints and scale the problem, see Andersen, Gondzio, Mészáros and Xu, 1996), we have not done so.

In Table 1, we give the results of our preliminary tests. They were performed in double precision on an IBM RISC System/6000 3BT workstation with 64 Megabytes of RAM, using the xlf90 compiler and optimization level -O3.

Name	n	m	type	its	MB	mods	time	VE09-its	VE09-time
AUG2DCQP	3280	1600	C	21	5	0	4.86	3112	133.15
AUG2DQP	3280	1600	C	21	3	0	4.72	3019	127.71
AUG3DCQP	3873	1000	C	16	0	0	5.00	3056	106.58
AUG3DQP	3873	1000	C	16	0	0	6.26	2097	71.44
BLOCKQP1	2006	1001	NC	26	0	18	6.34	1006	28.76
BLOCKQP2	2006	1001	NC	10	0	3	2.92	1006	40.42
BLOCKQP3	2006	1001	NC	> 1000				1006	28.80
BLOWEYA	2002	1002	C	9	1	0	1.60	1597	68.96
BLOWEYB	2002	1002	C	7	0	0	1.29	1497	67.86
BLOWEYC	2002	1002	C	10	1	0	1.72	1697	53.97
CVXQP1	1000	500	C	30	0	0	44.44	861	70.92
CVXQP2	1000	250	C	32	0	0	19.15	370	13.50
CVXQP3	1000	750	C	31	0	0	132.40	1389	107.11
DUALC1	223	215	C	44	0	0	1.17	12	0.23
DUALC2	235	229	C	37	0	0	0.83	14	0.25
DUALC5	285	278	C	12	0	0	0.37	10	0.37
DUALC8	510	503	C	20	0	0	1.05	11	0.94
GOULDQP2	699	349	C	4	0	0	0.19	251	1.84
GOULDQP3	699	349	C	7	0	0	0.35	463	2.52
KSIP	1021	1001	C	30	0	0	6.38	1388	36.42
MOSARQP1	1500	600	C	16	0	0	2.01	5859	91.63
MOSARQP2	1500	600	C	13	0	0	1.60	1679	27.73
NCVXQP1	1000	500	NC	956	297	927	3762.58	1561	51.04
NCVXQP2	1000	500	NC	> 1000				1840	61.50
NCVXQP3	1000	500	NC	481	0	470	1913.23	too ill-cond. basis	
NCVXQP4	1000	250	NC	> 1000				649	2.97
NCVXQP5	1000	250	NC	> 1000				565	2.78
NCVXQP6	1000	250	NC	332	0	319	459.49	532	3.53
NCVXQP8	1000	750	NC	> 1000				1901	141.70
NCVXQP7	1000	750	NC	> 1000				1567	120.56
NCVXQP9	1000	750	NC	322	0	288	2862.79	too ill-cond. basis	
PRIMALC1	239	9	C	83	40	0	2.62	20	0.17
PRIMALC2	238	7	C	61	4	0	1.69	4	0.14
PRIMALC5	295	8	C	16	1	0	0.43	14	0.23
PRIMALC8	528	8	C	16	1	0	0.76	20	0.71
PRIMAL1	410	85	C	17	0	0	3.13	361	4.12
PRIMAL2	745	96	C	11	0	0	3.02	677	12.25
PRIMAL3	856	111	C	13	0	0	14.22	798	35.49
PRIMAL4	1564	75	C	11	0	0	6.51	1515	40.16
QPCBOEI1	726	351	C	113	11	0	8.45	823	6.41
QPCBOEI2	305	166	C	109	4	0	3.38	303	1.14
QPCSTAIR	614	356	C	174	8	0	13.26	987	16.05

Table 1: Preliminary numerical results.

Name	n	m	type	its	MB	mods	time	VE09-its	VE09-time
QPNBOEI1	726	351	NC	> 1000				736	5.66
QPNBOEI2	305	166	NC	652	1	639	61.88	299	1.14
QPNSTAIR	614	356	NC	226	13	207	30.39	993	15.61
SOSQP1	2000	1001	SOS	5	0	0	0.73	996	14.49
STCQP1	4097	2052	NC	A rank deficient				2845	67.81
STCQP2	4097	2052	NC	11	0	4	51.09	2040	98.22
STNQP1	4097	2052	NC	A rank deficient				3158	68.01
STNQP2	4097	2052	NC	24	0	15	201.12	1408	39.11
UBH1	909	600	C	5	1	0	0.29	315	5.12
YAO	1002	500	C	847	7	0	38.34	3	2.06

Table 1: Preliminary numerical results (continued).

For each example, we report its name along with its dimensions (n is the number of variables, m the number of constraints), the problem type (C for convex, SOS for second-order sufficient and NC for non-convex and not second-order sufficient), the number of iterations performed (its), the number of these which were modified barrier (1.13) iterations (MB) and the number for which G was modified (mods), and the time taken in seconds (time). For comparison, the tables also show the number of iterations and time taken by a fortran-90 version of VE09, a quadratic programming subroutine from the Harwell Subroutine Library (VE09-its and VE09-time, respectively). This latter algorithm is designed to handle non-convex problems and is of the active-set type, each of its iterations corresponding to a pivoting operation. Thus an iteration of VE09 is much cheaper that an iteration of the new algorithm–the former corresponds to a factorization update, while the latter is a refactorization.The reader is referred to Gould (1991) for further details on this method. We also ran tests using MINOS of Murtagh and Saunders (1993) which we do not report here because they are quantitatively similar to those obtained with VE09.

We immediately note that the primal-dual algorithm performs well on convex problems (C and SOS), with the possible exception of YAO. On the other hand, its performance on the non-convex (NC) ones is somewhat disappointing. A closer examination of these runs indicates that our naive matrix modification technique is really too naive; when the Hessian involves many negative eigenvalues, these appear to be removed one at a time, resulting in a large number of iterations before second-order sufficiency is achieved. A more sophisticated way of treating negative curvature directions is therefore highly desirable. Generally however, given the crude nature of the present preliminary implementation, the new primal-dual method definitely shows some potential.

4 Conclusion

We have presented a primal-dual algorithmic framework for minimization under linear equality constraints and non-negativity constraints on the variables, whose merit function is adapted to problems with twice differentiable non-convex objective functions. We proved global convergence for this framework under the assumptions that the iterates remain in a bounded subset of the positive orthant, that the matrix associated with the linear equality constraints is full rank and there exists a point which is strictly feasible. Some preliminary numerical results have been given and discussed, indicating a clear potential for further research.

In particular, the use of negative curvature directions appears to require more sophistication. Although the current method works in its current naive form, it converges slowly for problems involving massive indefiniteness. Less naive strategies are thus needed and are the object of current investigations.

Unsurprisingly there are disadvantages to the approach we have taken. The primary numerical linear algebraic computation is essentially a calculation involving the Karush-Kuhn-Tucker matrix

$$\begin{pmatrix} H_k + X_k^{-1} Z_k & A^T \\ A & 0 \end{pmatrix} \tag{4.1}$$

which is inherently non-trivial to handle because one expects small components in x_k without corresponding small components in z_k. The analogous matrix in the case of linear programming is the matrix

$$\begin{pmatrix} X_k^{-1} Z_k & A^T \\ A & 0 \end{pmatrix}. \tag{4.2}$$

The fact that in this case the upper left-hand block is diagonal (and for non-degenerate problems this matrix is asymptotically non-singular) makes this form of ill-conditioning easier to handle, (see for example, Wright (1992)). However, Ponceleón (1990) and Forsgren, Gill and Shinnerl (1996) show how one can treat the general case.

A more direct concern is that it is inappropriate to use the normal equations when considering (4.1) instead of (4.2). Many authors have suggested using a direct factorization of (4.1)/(4.2) (see for example Duff, Gould, Reid, Scott and Turner (1991), Fourr and Mehrotra (1993), Vanderbei and Carpenter (1993) and Andersen et al. (1996)) which can be very successful. Other issues we would like to consider in future include trying to justify why a primal-dual approach should be more successful *globally* even for non-convex problems than a primal approach, and trying to explain why the central path appears to be so important for numerical efficiency. Since it is also generally recommended that, at least in the case of interior point approaches to the linear programming problem, one makes use of predictor-corrector techniques to enhance performance, we remark that we wish to extend the methods considered here to include such improvements.

5 Acknowledgements

The authors are pleased to acknowledge the useful comments made by M. H. Wright.

Appendix

Proof of Lemma 1

Proof. If G is second-order sufficient with respect to A and N is orthogonal, we have that the minimum eigenvalue of N^TGN, which we denote by $\epsilon > 0$, is the solution of the minimization problem

$$\min_s \{\langle s, Gs\rangle \mid As = 0 \text{ and } \|s\| = 1\}.$$

(Here and below, the symbol $\|\cdot\|$ denotes the Euclidean norm.) The minimizer of this problem satisfies the first-order optimality conditions

$$\begin{aligned} Gs + A^Tu &= \epsilon s \\ As &= 0 \end{aligned} \tag{A.1}$$

and $\epsilon = \langle s, Gs\rangle$. Adding ϵu on both sides of the second equation, we see that (A.1) yields that

$$\begin{pmatrix} G & A^T \\ A & \epsilon I \end{pmatrix} \begin{pmatrix} s \\ u \end{pmatrix} = \epsilon \begin{pmatrix} s \\ u \end{pmatrix},$$

and thus ϵ is an eigenvalue of the matrix

$$\begin{pmatrix} G & A^T \\ A & \epsilon I \end{pmatrix}. \tag{A.2}$$

Now, we can view the matrix K defined in (1.18) as a symmetric perturbation of (A.2), and deduce from Wilkinson (1965, Section 44, p. 101), that K has an eigenvalue in the range $[0, \epsilon]$. Since K is nonsingular, this eigenvalue must be in the interval $(0, \epsilon]$, which proves the result. □

Proof of Lemma 2

Proof. Consider the matrix

$$\bar{K}_1 = \begin{pmatrix} G & 0 & A^T \\ 0 & \rho - \lambda & r^T \\ A & r & 0 \end{pmatrix}$$

Pivoting on the 2-2 block and using Sylvester's law of inertia, we obtain that

$$In(\bar{K}_1) = (1,0,0) + In(K(\rho)), \quad \text{where } K(\rho) \stackrel{\text{def}}{=} \begin{pmatrix} G & A^T \\ A & -rr^T/(\rho-\lambda) \end{pmatrix}. \tag{A.3}$$

As, by assumption, K is nonsingular and has exactly m negative eigenvalues, Wilkinson (1965, Section 40, p. 97) implies that the smallest positive eigenvalue of $K(\rho)$ is at least $\frac{1}{2}\lambda$ provided that

$$\left\| \frac{rr^T}{\rho-\lambda} \right\| = \frac{\|r\|^2}{\rho-\lambda} \le \frac{1}{2}\lambda,$$

i.e., provided ρ satisfies (1.21). The continuity of the eigenvalues of $K(\rho)$ then implies that both $K(\rho)$, and, in view of (A.3), $\bar{K}_1$, also have precisely m negative eigenvalues for all ρ satisfying (1.21). Thus $\bar{N}^T\bar{G}_1\bar{N}$ is positive definite, where $\bar{G}_1 = \text{diag}(G, \rho-\lambda)$ and where the columns of $\bar{N}$ span the nullspace of $(A\ r)$. As a consequence,

$$\bar{N}^T\bar{G}\bar{N} = \bar{N}^T(\bar{G}_1 + \text{diag}(0,\lambda))\bar{N} \tag{A.4}$$

is also positive definite, which proves the first part of the lemma.

To prove the second part, observe that

$$\bar{K} = \begin{pmatrix} G & 0 & A^T \\ 0 & \lambda & 0 \\ A & 0 & -rr^T/(\rho-\lambda) \end{pmatrix} + \frac{1}{\rho-\lambda} \begin{pmatrix} 0 \\ \rho-\lambda \\ r \end{pmatrix} \begin{pmatrix} 0 & \rho-\lambda & r^T \end{pmatrix} \stackrel{\text{def}}{=} K_1 + K_2.$$

But the eigenvalues of K_1 are λ and those of $K(\rho)$: the smallest positive eigenvalue of K_1 is thus at least $\frac{1}{2}\lambda$ so long as (1.21) holds. Moreover, K_2 is a positive rank-one term, which implies that the eigenvalues of $\bar{K}$ are not smaller than those of K_1. Recalling that $\bar{K}$ has exactly m negative eigenvalues if (1.21) holds, we see that its smallest positive eigenvalue is at least $\frac{1}{2}\lambda$. Applying now Lemma 1 gives (1.22). □

References

CPLEX 4.0. *High-performance linear, integer and quadratic programming software.* CPLEX Optimization, Inc., Incline Village, Nevada, USA, 1995.

E. D. Andersen, J. Gondzio, C. Mészáros, and X. Xu. Implementation of interior point methods for large scale linear programming. *In* T. Terlaky, ed., 'Interior Point Methods in Mathematical Programming', pp. 189–252, Dordrecht, The Netherlands, 1996. Kluwer Academic Publishers.

I. Bongartz, A. R. Conn, N.I.M. Gould, and Ph. L. Toint. CUTE: Constrained and Unconstrained Testing Environment. *ACM Transactions on Mathematical Software*, **21**(1), 123–160, 1995.

J. F. Bonnans and C Pola. A trust-region interior point algorithm for linearly constrained optimization. Technical Report 1948, INRIA, Roquencourt, France, 1993.

T. M. Carpenter, I. J. Lustig, J. M. Mulvey, and D. F. Shanno. Higher-order predictor-corrector interior point methods with application to quadratic objectives. *SIAM Journal on Optimization*, **3**(4), 696–733, 1993.

A. R. Conn, N. I. M. Gould, and Ph. L. Toint. Global convergence of a class of trust region algorithms for optimization with simple bounds. *SIAM Journal on Numerical Analysis*, **25**, 433–460, 1988. See also same journal 26:764–767, 1989.

A. R. Conn, N. I. M. Gould, and Ph. L. Toint. LANCELOT*: a Fortran package for large-scale nonlinear optimization (Release A)*. Number 17 *in* 'Springer Series in Computational Mathematics'. Springer Verlag, Heidelberg, Berlin, New York, 1992.

A. R. Conn, N. I. M. Gould, and Ph. L. Toint. Numerical experiments with the lancelot package (Release A) for large-scale nonlinear optimization. *Mathematical Programming, Series A*, **73**(1), 73–110, 1996*a*.

A. R. Conn, N. I. M. Gould, D. Orban, and Ph. L. Toint. A primal-dual trust-region algorithm for minimizing a non-convex function subject to bound and linear equality constraints. Technical Report (in preparation), Department of Mathematics, FUNDP, 1999.

A. R. Conn, Nick Gould, A. Sartenaer, and Ph. L. Toint. Convergence properties of an augmented Lagrangian algorithm for optimization with a combination of general equality and linear constraints. *SIAM Journal on Optimization*, **6**(3), 674–703, 1996*b*.

International Business Machine Corporation. *Optimization Subroutine Library: Guide and Reference*, second edn, 1990.

I. S. Duff and J. K. Reid. The multifrontal solution of indefinite sparse symmetric linear equations. *ACM Transactions on Mathematical Software*, **9**(3), 302–325, 1983.

I. S. Duff, N. I. M. Gould, J. K. Reid, J. A. Scott, and K. Turner. The factorization of sparse symmetric indefinite matrices. *IMA Journal of Numerical Analysis*, **11**, 181–204, 1991.

A. Forsgren, P. E. Gill, and J. R. Shinnerl. Stability of symmetric ill-conditioned systems arising in interior point methods for constrained optimization. *SIAM Journal on Matrix Analysis and Applications*, **17**(1), 187–211, 1996.

A. L. Forsgren and P. E. Gill. Primal-dual interior methods for nonconvex nonlinear programming. Technical Report TRITA-MAT-1996-OS4, Royal Institute of Technology, Stockholm, Sweden, 1996.

A. L. Forsgren and W. Murray. Newton methods for large-scale linear equality-constrained minimization. *SIAM Journal on Matrix Analysis and Applications*, **14**(2), 560–587, 1993.

R. Fourr and S. Mehrotra. Solving symmetrical indefinite systems for an interior-point method for linear programming. *Mathematical Programming, Series A*, **62**(1), 15–39, 1993.

P. E. Gill, W. Murray, D. B. Ponceléon, and M. A. Saunders. Solving reduced KKT systems in barrier methods for linear and quadratic programming. Technical Report SOL 91-7, Dept. of Operations Research Stanford University, Stanford, USA, 1991.

P. E. Gill, W. Murray, D. B. Ponceléon, and M. A. Saunders. Primal-dual methods for linear programming. *Mathematical Programming, Series A*, **70**(3), 251–278, 1995.

P. E. Gill, W. Murray, M. A. Saunders, and M. H. Wright. A Schur complement method for sparse quadratic programming. *In* M. G. Cox and S. Hammarling, eds, 'Reliable Numerical Computation', pp. 113–138, Oxford, 1990. Clarendon Press.

N. I. M. Gould. On practical conditions for the existence and uniqueness of solutions to the general equality quadratic programming problem. *Mathematical Programming*, **32**, 90–99, 1985.

N. I. M. Gould. An algorithm for large-scale quadratic programming. *IMA Journal of Numerical Analysis*, **11**(3), 299–324, 1991.

N. I. M. Gould. Constructing appropriate models for large-scale, linearly-constrained, nonconvex, nonlinear, optimization algorithms. Technical Report RAL-TR 95-037, Rutherford Appleton Laboratory, Chilton, England, 1995.

Harwell Subroutine Library. *A catalogue of subroutines (release 12)*. Advanced Computing Department, Harwell Laboratory, Harwell, UK, 1995.

N. J. Higham and S. Cheng. Modifying the inertia of matrices arising in optimization. *Linear Algebra and its Applications*, **275–276**, 261–279, 1998.

B. A. Murtagh and M. A. Saunders. Large-scale linearly constrained optimization. *Mathematical Programming*, **14**, 41–72, 1978.

B. A. Murtagh and M. A. Saunders. MINOS 5.4 USER'S GUIDE. Technical Report SOL 83-20R, Department of Operations Research, Stanford University, Stanford, USA, 1993. Revised 1995.

D. B. Ponceleón. *Barrier methods for large-scale quadratic programming*. PhD thesis, Department of Computer Science, Stanford University, Stanford, California 94305, USA, 1990.

S. M. Robinson. A quadratically convergent algorithm for general nonlinear programming problems. *Mathematical Programming*, **3**, 145–156, 1972.

J. B. Rosen and J. Kreuser. A gradient projection algorithm for nonlinear constraints. *In* F. Lootsma, ed., 'Numerical Methods for Nonlinear Optimization', London, 1972. Academic Press.

D. F. Shanno. Algorithms for linear programming. *In* E. Spedicato, ed., 'Algorithms for continuous optimization: the state of the art', Dordrecht, The Netherlands, 1994. Kluwer Academic Publishers.

E. M. Simantiraki and D. F. Shanno. An infeasible-interior-point method for linear complementarity problems. Technical Report RRR 7-95, RUTCOR, Rutgers University, New Jersey, USA, 1995.

R. J. Vanderbei and T. J. Carpenter. Symmetrical indefinite systems for interior point methods. *Mathematical Programming*, **58**(1), 1–32, 1993.

J. H. Wilkinson. *The Algebraic Eigenvalue Problem*. Clarendon Press, Oxford, UK, 1965.

S. J. Wright. An interior point algorithm for linearly constrained optimization. *SIAM Journal on Optimization*, **2**(3), 450–473, 1992.

Y. Zhang and D. Zhang. Superlinear convergence of infeasible-interior-point methods for linear programming. *Mathematical Programming*, **66**(3), 361–377, 1994.

Minimal Convexificators of a positively Homogeneous Function and a Characterization of Its Convexity and Concavity [1]

Vladimir F. Demyanov (vladimir.demyanov@pobox.spbu.ru)
Applied Mathematics Department
St. Petersburg State University 198904, Staryi Peterhof, St.Petersburg, RUSSIA

Abstract

The notions of convexificator and minimal convexificator of a posotively homogeneous (p.h.) function $h : \mathbb{R}^n \to \mathbb{R}$ are used to get a characterization of the convexity and concavity of h. It is shown that the uniqueness of a minimal (by inclusion) convexificator is a necessary and sufficient condition for a p.h. function to be convex or concave. Representations of a p.h. function in terms of its minimal convexificators are also derived.

Keywords: positively homogeneous function, convexificator, minimal convexificator, convexity, concavity.

1 Introduction

In Optimization Theory an important role is played by positively homogeneous (p.h.) functions: necessary and/or sufficient optimality conditions are often formulated in terms of directional derivatives [1, 3] or (if the function under study is not directionally differentiable) different generalizations of directional derivatives (such as the Hadamard and Dini upper and lower directional derivatives, the Clarke and Michel-Penot derivatives – see, e.g., [1, 4, 5]). All the derivatives mentioned are positively homogeneous functions (as functions of direction).

[1]This research was supported by the Russian Foundation for Fundamental Studies under Grant RFFI No. 97-01-00499

G. Di Pillo and F. Giannessi (eds.), Nonlinear Optimization and Related Topics, 51–65.

B.N.Pschenichnyi [3] introduced the notions of upper convex approximation and lower concave approximation for a p.h. function. Exhaustive families of upper convex approximations and lower concave approximations were introduced and studied by A.M.Rubinov in 1982 (see [1] for related references). The notion of convexificator of a p.h. function was introduced in [6] (see also [1]). A convexificator of a p.h. function h generates two functions: one is an upper convex approximation of h, another – a lower concave approximation of h. The notion of convexificator has proved to be a useful tool to study different problems of Nonsmooth Analysis [7, 9]. For example, a mean-value theorem may be formulated in terms of convexificators [7], an implicit function theorem for systems of nonsmooth equations can be established by making use of convexificators [10].

In the present paper a characterization of convexity and concavity of a p.h. function in the finite dimensional case is stated in terms of minimal convexificators: it turns out that for a p.h. function h to be convex or concave it is necessary and sufficient that there exist a unique minimal convexificator of the function h (Section 2). In Section 3 some representations of a p.h. function in terms of the family of its minimal convexificators are stated.

2 A characterization of convexity and concavity of a p.h. function

Let $h : \mathbb{R}^n \to \mathbb{R}$ be a positively homogeneous (p.h) (of degree one) function, i.e.

$$h(\lambda g) = \lambda h(g) \quad \forall \lambda \geq 0,\ \forall g \in \mathbb{R}^n.$$

B.N.Pschenichnyi [3] introduced the notions of upper convex and lower concave approximations of a p.h. function h : A p.h. convex function $\overline{h} : \mathbb{R}^n \to \mathbb{R}$ is called an upper convex approximation of h if

$$h(g) \leq \overline{h}(g) \quad \forall g \in \mathbb{R}^n. \tag{1}$$

A p.h. function $\underline{h} : \mathbb{R}^n \to \mathbb{R}$ is called a lower concave approximation of h if

$$\underline{h}(g) \leq h(g) \quad \forall g \in \mathbb{R}^n. \tag{2}$$

Every convex p.h. function $\overline{h}$ can be uniquely represented in the form (see [2])

$$\overline{h}(g) = \max_{v \in \overline{C}} (v, g)$$

where $\overline{C} \subset \mathbb{R}^n$ is a convex compact set (called the subdifferential of $\overline{h}$.) Every concave p.h. function $\underline{h}$ can be uniquely represented in the form

$$\underline{h}(g) = \min_{w \in \underline{C}} (w, g)$$

with $\underline{C} \subset \mathbb{R}^n$ being a convex compact set (called the superdifferential of $\underline{h}$). We say that a convex compact set $C \subset \mathbb{R}^n$ is a convexificator (CF) of h if

$$\min_{w \in C}(w, g) \leq h(g) \leq \max_{v \in C}(v, g) \quad \forall g \in \mathbb{R}^n. \tag{3}$$

It follows from the definitions (see (1)-(3)) that the function $\underline{h}(g) = \min_{w \in C}(w, g)$ is a lower concave approximation of h while the function $\overline{h}(g) = \max_{v \in C}(v, g)$ is an upper convex approximation of h.

If h is bounded on $S_1 = \{g \in \mathbb{R}^n \mid \|g\| = 1\}$ then there is no problem of existence of a convexificator: any ball centered at the origin with the radius sufficiently large can be taken as a convexificator of the function h. Note that the functions $\underline{h}$ and $\overline{h}$ above are constructed by means of the same set C. Let $\mathcal{C}(h)$ be the family of all CF's of h. We say that $C^* \in \mathcal{C}(h)$ is a minimal convexificator (MCF) of h if there exists no other $C \in \mathcal{C}(h)$ such that $C \subset C^*$, $C \neq C^*$.

It is clear that if $C^* \in \mathcal{C}(h)$ and

$$C^* \subset C \quad \forall C \in \mathcal{C}(h) \tag{4}$$

then C^* is a unique MCF. Condition (4) is necessary and sufficient for C^* to be a unique MCF.

Theorem 2.1 *Let $C^* \in \mathcal{C}(h)$ be an MCF of a continuous p.h. (of degree one) function h. For C^* to be a unique MCF of h it is necessary and sufficient that for every $g \in \mathbb{R}^n$ be either*

$$\max_{v \in C^*}(v, g) = h(g), \tag{5}$$

or

$$\min_{v \in C^*}(v, -g) = h(-g). \tag{6}$$

Proof. Necessity. Let $C^* \in \mathcal{C}(h)$ be a unique MCF of h. It is necessary to prove that for every $g \in \mathbb{R}^n$ one of the relations (5) and (6) holds. Assume the contrary. Then there exist vectors $v_0 \in C^*, g_0 \in \mathbb{R}^n, \|g_0\| = 1$, and a real $\epsilon > 0$ such that

$$h(g_0) \leq \max_{v \in C^*}(v, g_0) - 2\epsilon = (v_0, g_0) - 2\epsilon,$$

$$h(-g_0) \geq \min_{v \in C^*}(v, -g_0) + 2\epsilon = (v_0, -g_0) + 2\epsilon.$$

By the continuity of h there exists a $\delta > 0$ such that

$$h(g) \leq \max_{v \in C^*}(v, g) - \epsilon,$$

$$h(-g) \geq \min_{v \in C^*}(v, -g) + \epsilon$$

for all $g \in S_1$, $\|g - g_0\| \leq \delta$. Here $S_1 = \{g \in \mathbb{R}^n \mid \|g\| = 1\}$.

Let (see Fig. 1)

$$\alpha = \arcsin \delta/2, \; d = \max_{v \in C^*}(v, g_0),$$

$$a = \max_{v \in C^*, \|g-g_0\| \le 1, \|g\|=\|g_0\|=1} (v, g),$$

$$r = \left| \frac{a - (d-\epsilon)\cos\alpha}{\sin\alpha} \right|,$$

$$A = (d-\epsilon)g_0, \; L = \{x \in \mathbb{R}^n \mid (x - A, g_0) = 0\},$$

$$B = \{x \in \mathbb{R}^n \mid (x, g_0) \le d - \epsilon\}.$$

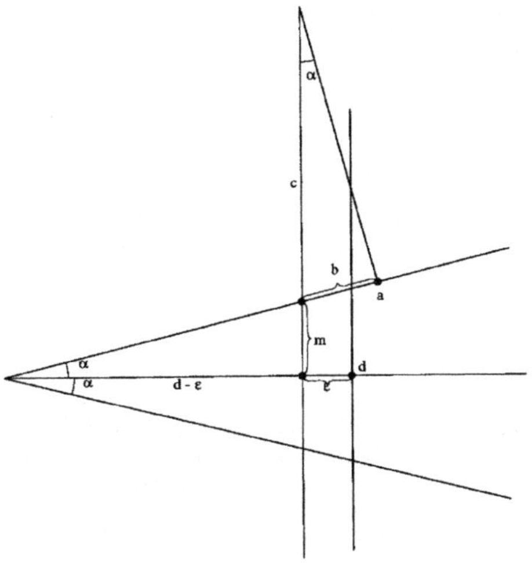

Figure 1: The case where (5) and (6) don't hold

In Fig. 1

$$l = \frac{d-\epsilon}{\cos\alpha}, \quad m = (d-\epsilon)\text{tg } \alpha, \quad b = a - \frac{d-\epsilon}{\cos\alpha},$$

$$c = \frac{b}{\sin\alpha} = \frac{a\cos\alpha - d + \epsilon}{\cos\alpha \, \sin\alpha},$$

$$r = c + m = \frac{a - (d - \epsilon)\cos\alpha}{\sin\alpha}.$$

The set $C = \text{co}\,\{C^* \bigcap B,\ S_r(A) \bigcap L\}$ is a CF and $C^* \not\subset C$ (i.e., there exists a $v_0 \in C^*$ such that $v_0 \notin C$. Since condition (4) doesn't hold then the set C^* is not a unique MCF. This contradiction proves the necessity.

Sufficiency. Let for every $g \in \mathbb{R}^n$ one of the conditions (5) and (6) hold. we must prove that C^* is a unique MCF. Assume the contrary. Then (see (4)) a $C \in \mathcal{C}(h)$ exists such that $C^* \not\subset C$. Find

$$\max_{v \in C^*} \min_{w \in C} ||v - w|| = ||v^* - w^*|| = a^* > 0.$$

Put

$$g^* = (v^* - w^*)/||v^* - w^*|| = (v^* - w^*)/a^*.$$

Then

$$(v^* - w^*, g^*) = (v^*, g^*) - (w^*, g^*) = a^*,$$

$$\max_{v \in C^*}(v, g^*) = (v^*, g^*), \quad \min_{v \in C^*}(v, -g^*) = (v^*, -g^*),$$

$$\max_{w \in C}(w, g^*) = (w^*, g^*), \quad \min_{w \in C}(w, -g^*) = (w^*, -g^*).$$

We have

$$h(g^*) \leq \max_{w \in C}(w, g^*) = (w^*, g^*) = (v^*, g^*) - a^* =$$

$$= \max_{v \in C^*}(v, g^*) - a^* < \max_{v \in C^*}(v, g^*),$$

$$h(-g^*) \geq \min_{w \in C}(w, -g^*) = (w^*, -g^*) = (v^*, -g^*) + a^* =$$

$$= \min_{v \in C^*}(v, -g^*) + a^* > \min_{v \in C^*}(v, -g^*).$$

Hence, for the direction g^* neither condition (5) nor condition (6) is satisfied. This contradiction proves the sufficiency.

Corollary 2.1 *Let a $C^* \in \mathcal{C}(h)$ be a unique MCF of h. Then for every $g \in \mathbb{R}^n$ either*

$$h(g) = \max_{v \in C^*}(v, g), \quad h(-g) = \max_{v \in C^*}(v, -g), \tag{7}$$

or

$$h(g) = \min_{v \in C^*}(v, g), \quad h(-g) = \min_{v \in C^*}(v, -g). \tag{8}$$

Proof. In the case

$$\max_{v \in C^*}(v, g) = \min_{v \in C^*}(v, g)$$

the corollary is transparent. Assume that

$$\max_{v \in C^*}(v, g) > \min_{v \in C^*}(v, g). \tag{9}$$

By Theorem 1 one of two relations (5) and (6) holds. Let

$$h(g) = \max_{v \in C^*}(v, g) = (v_0, g). \tag{10}$$

For the direction $g_1 = -g$ one of the relations

$$h(g_1) = h(-g) = \max_{v \in C^*}(v, -g) = (v_1, -g) \tag{11}$$

and

$$h(-g_1) = h(g) = \min_{v \in C^*}(v, g) = (v_1, g) \tag{12}$$

should also hold.

Relations (9) and (10) imply that (12) is impossible, hence, (11) does hold. Thus, if (10) holds, then the second relation in (7) takes place. Analogously, in the case $h(g) = \min_{v \in C^*}(v, g)$ the second relation in (8) holds.

Theorem 2.2 *Let a function $h : \mathbb{R}^n \to \mathbb{R}$ be continuous and p.h. of degree one. For the function h to be convex or concave on $\mathbb{R}^n$ it is necessary and sufficient that there exist a unique MCF of h.*

Proof. Necessity is clear since any convex function h has a unique MCF coinciding with its subdifferential at the point 0_n, and any concave function has a unique MCF coinciding with its superdifferential at 0_n. Sufficiency. Let h have a unique MCF C^*. If C^* is a singleton: $C^* = \{v\}$ then $h(g) = (v, g)$, and the statement is evident.

Consider now the case where C^* contains more than one point. Let $v_1 \in C^*$, $v_2 \in C^*$, $v_1 \neq v_2$. Put $g_0' = (v_2 - v_1)/||v_2 - v_1||$. By Corollary 1 for g_0' either (7) or (8) holds. Let, e.g.,

$$h(g_0') = \max_{v \in C^*}(v, g_0'). \tag{13}$$

Then (see (7)),

$$h(-g_0') = \max_{v \in C^*}(v, -g_0').$$

Let us show that

$$h(g) = \max_{v \in C^*}(v, g) \quad \forall g \in \mathbb{R}^n, \quad ||g|| = 1. \tag{14}$$

First of all consider the case $(g_0', g) \neq 0$. Assume that (14) doesn't hold, let

$$h(g) = \min_{v \in C^*}(v, g) < \max_{v \in C^*}(v, g).$$

Without loss of generality we may assume that $(g_0', g) = a_0 > 0$ (if it is not true, take $-g_0'$ instead of g_0'). It is clear that

$$\max_{v \in C^*}(v, g) - \min_{v \in C^*}(v, g) \geq (g_0', g)||v_2 - v_1|| = a_0||v_2 - v_1|| = a > 0$$

Put $g_0'' = g$. Note that $g_0'' \neq -g_0'$ (since $(g_0', g_0'') = (g_0', g) = a_0 > 0$).

Take

$$\overline{g}_0 = \frac{g_0' + g_0''}{||g_0' + g_0''||}.$$

If $h(\overline{g}_0) = \max_{v\in C^*}(v, \overline{g}_0)$ then put $g_1' = \overline{g}_0$, $g_1'' = g_0''$. If $h(\overline{g}_0) = \min_{v\in C^*}(v, \overline{g}_0)$ then put $g_1' = g_0'$, $g_1'' = \overline{g}_0$. Clearly,

$$(g_1', g_1'') > (g_0', g_0'') = a_0 > 0.$$

Continue in the same manner. Let g_k' and g_k'' be found such that

$$(g_k', g_k'') > a_0, \quad h(g_k') = \max_{v\in C^*}(v, g_k'), \quad h(g_k'') = \min_{v\in C^*}(v, g_k''). \tag{15}$$

Take

$$\overline{g}_k = \frac{g_k' + g_k''}{||g_k' + g_k''||}.$$

We have

$$\max_{v\in C^*}(v, \overline{g}_k) - \min_{v\in C^*}(v, \overline{g}_k) \geq \ ||v_2 - v_1||(g_0', \overline{g}_k) \geq a > 0.$$

If $h(\overline{g}_k) = \max_{v\in C^*}(v, \overline{g}_k)$ then put $g_{k+1}' = \overline{g}^k$, $g_{k+1}'' = g_k''$. If $h(\overline{g}_k) = \min_{v\in C^*}(v, \overline{g}_k)$ then take $g_{k+1}' = g_k'$, $g_{k+1}'' = g_k$. Note that

$$(g_{k+1}', g_{k+1}'') > (g_k', g_k'') > a_0.$$

As a result, the sequences $\{g_k'\}$ and $\{g_k''\}$ are constructed such that

$$||g_k' - g_k''|| \to 0, \ ||g_k'|| = ||g_k''|| = 1,$$

$$\max_{v\in C^*}(v, g_k') - \min_{v\in C^*}(v, g_k') \geq a > 0. \tag{16}$$

There exists a unique point $\overline{g}$ such that

$$||\overline{g}|| = 1, \ g_k' \to \overline{g}, \ g_k'' \to \overline{g}. \tag{17}$$

Since h is continuous then (15) and (17) imply

$$h(\overline{g}) = \max_{v\in C^*}(v, \overline{g}) = \min_{v\in C^*}(v, \overline{g}) \tag{18}$$

which contradicts (16). Thus, if $(g_0', g) \neq 0$ then (14) holds. Now let us consider the case $(g_0', g) = 0$. Choose a sequence $\{g_k\}$ such that

$$g_k \in \mathbb{R}^n, \ g_k \to g, \ ||g_k|| = 1, \ (g_0', g_k) \neq 0.$$

It has already been proved that $h(g_k) = \max_{v\in C^*}(v, g_k)$. Passing to the limit as $k \to \infty$ we get

$$h(g) = \max_{v\in C^*}(v, g). \tag{19}$$

It was assumed that $||g|| = 1$. Due to the positive homogeneity of the function h, the relation (14) is valid for all $g \in \mathbb{R}^n$.

In the case $h(g_0') = \min_{v\in C^*}(v, g_0')$ it is analogously demonstrated that

$$h(g) = \min_{v\in C^*}(v, g) \ \ \forall g \in \mathbb{R}^n. \tag{20}$$

3 A relationship between a p.h. function and its minimal convexificators

Let $h : \mathbb{R}^n \to \mathbb{R}$ be a p.h. function which is Lipschitz on $B_1 = \{g \in \mathbb{R}^n | \; ||g|| \leq 1\}$.

Theorem 3.1 *For every $g \in \mathbb{R}^n, g \neq 0_n$, there exists a minimal convexificator $C_g \in \mathcal{C}(h)$ such that either*

$$\max_{v \in C_g}(v, g) = h(g) \tag{21}$$

or

$$\min_{v \in C_g}(v, -g) = h(-g). \tag{22}$$

Proof. Put $b = \max_{g \in S_1} |h(g)|$ where $S_1 = \{g \in \mathbb{R}^n \mid \; ||g|| = 1\}$. If $b = 0$ then

$$h(g) = 0 \quad \forall g \in \mathbb{R}^n$$

and the set $C^* = 0_n$ is a unique MCF of h.

Let $b > 0$. First consider the case

$$\max\{h(g), h(-g)\} = a(g) > 0. \tag{23}$$

Without loss of generality we may assume that $||g|| = 1$. It is clear that $b \geq a(g)$. If $b = a(g)$ then let us take

$$C = B_b = \{v \in \mathbb{R}^n \mid \; ||v|| \leq b\}.$$

If $b > a(g)$ then (see Fig. 2 where $c = x \sin \alpha, x = c/\sin \alpha = L/\cos(\alpha/2)$)

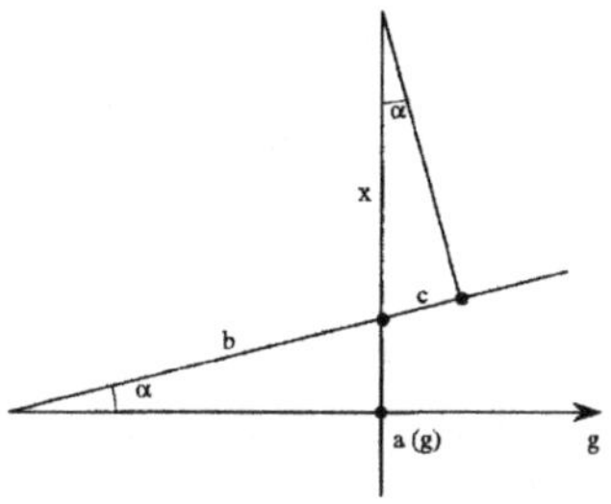

Figure 2: The case $b > a(g)$

we put $\alpha = \arccos\,(a(g)/b)$. Observe that $0 < \alpha < \pi/2$. Since h is Lipschitz on B_1 then

$$|h(g') - h(g)| \leq L||g' - g|| \tag{24}$$

where $L < \infty$ is a Lipschitz constant of h. For

$$g' \in S_{1\alpha} = \{g' \in S_1 \mid (g', g) \geq \cos\alpha\}$$

we have (see Fig. 3) $||g' - g|| = 2\sin(\beta/2)$ where $\beta = (g', g)$.

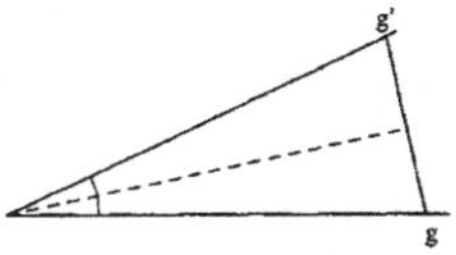

Figure 3: The case $\alpha > 0$

Therefore (24) implies

$$|h(g') - h(g)| \leq 2L\sin(\beta/2) \leq 2L\sin(\alpha/2) \quad \forall g' \in S_{1\alpha}.$$

Put

$$x = \frac{2L\sin\frac{\alpha}{2}}{\sin\alpha} = \frac{L}{\cos\frac{\alpha}{2}},$$

$$R = x + a(g)\text{tg}\ \alpha = \frac{L}{\cos\frac{\alpha}{2}} + a(g)\text{tg}\ \alpha, \tag{25}$$

$$A = a(g)g, \quad \Omega = \{v \in calR^n \mid (v - A, A) = 0, \quad ||v - A|| \leq R\},$$

$$\mathcal{L}_A = \{v \in \mathbb{R}^n \mid (v - A, A) \leq 0\},$$

$$C = \text{co}\ (B_b \cup \Omega) \cap \mathcal{L}_A.$$

Since

$$\min_{w\in\Omega}(w, g') \leq h(g') \leq \max_{v\in\Omega}(v, g') \quad \forall g' \in S_{1\alpha},$$

$$\min_{w\in B_b}(w, g') \leq h(g') \leq \max_{v\in B_b}(v, g') \quad \forall g' \in S_1, g' \notin S_{1\alpha}$$

then

$$\min_{w\in C}(w, g') \leq h(g') \leq \max_{v\in C}(v, g') \quad \forall g' \in S_1$$

and, due to the homogeneity of h, C is a convexificator. It is also clear that either

$$\max_{v\in C}(v, g) = h(g)$$

or

$$\min_{v\in C}(v,-g) = h(-g).$$

Now let us find an MCF C_g contained in C. For C_g one of the relations (21) or (22) holds.

The case $a(g) < 0$ is treated analogously, it is required to take in (25)

$$R = L/\cos(\alpha/2) - a(g)\text{tg }\alpha.$$

Finally, consider the case $a(g) = 0$. We may assume (see (23)) that $h(g) = 0$ (if $h(g) < 0$ then $h(-g) = 0$ and we take $-g$ instead of g).Then

$$|h(g') - h(g)| = |h(g')| \le 2L\sin\frac{\pi}{4} = L\sqrt{2} \quad \forall g' \in S_1, (g',g) \ge 0.$$

Take

$$\Omega = \{v \in \mathbb{R}^n \mid (v,g) = 0, \ \|v\| \le L\sqrt{2}\},$$

$$\mathcal{L} = \{v \in \mathbb{R}^n \mid (v,g) \le 0\}, \quad C = \text{co }\{B_b \cup \Omega\} \cap \mathcal{L}.$$

The set C is a convexificator of h.

Remark 3.1 *Arguing in a similar way for the direction $g_1 = -g$ one constructs the point $A_1 = -a(g)g$ and the sets*

$$\mathcal{L}_{1A_1} = \{v \in \mathbb{R}^n \mid (v - A_1, A_1) \le 0\},$$

$$C_1 = \text{co }(B_b \cup \Omega \cup \Omega_1) \cup \mathcal{L}_A \cup \mathcal{L}_{1A_1}.$$

Now find an MCF C_g^* which is contained in C_1. For this MCF we have either

$$\max_{v\in C_g^*}(v,g) = h(g) \tag{26}$$

or

$$\min_{v\in C_g^*}(v,-g) = h(-g) \tag{27}$$

and simultaneously either

$$\max_{v\in C_g^*}(v,-g) = h(-g) \tag{28}$$

or

$$\min_{v\in C_g^*}(v,g) = h(g). \tag{29}$$

If both relations (26) and (27) hold then

$$h(g) = \max_{v\in C_g^*}(v,g) = \min_{v\in C_g^*}(v,g). \tag{30}$$

Since C_g^* is a CF then

$$\min_{v\in C_g^*}(v,-g) \le h(-g) \le \max_{v\in C_g^*}(v,-g).$$

Therefore (26) and (27) yield

$$h(-g) = -\max_{v \in C_g^*}(v, g) = -\min_{v \in C_g^*}(v, g). \tag{31}$$

It follows from (30) and (31) that $h(-g) = -h(g)$. If

$$\max_{v \in C_g^*}(v, g) > \min_{v \in C_g^*}(v, g)$$

then either

$$h(g) = \max_{v \in C_g^*}(v, g), \quad h(-g) = \max_{v \in C_g^*}(v, -g) \tag{32}$$

or

$$h(g) = \min_{v \in C_g^*}(v, g), \quad h(-g) = \min_{v \in C_g^*}(v, -g). \tag{33}$$

In the case $h(-g) = -h(g)$ the set C_g^* belongs to the hyperplane $(v - A, g) = 0$ where $A = h(g)g$.

Corollary 3.1 *For every $g \in \mathbb{R}^n$ one of the following relations holds: either*

$$\min_{C \in H^*} \max_{v \in C}(v, g) = h(g) \tag{34}$$

or

$$\max_{C \in H^*} \min_{v \in C}(v, g) = h(g) \tag{35}$$

where $H*$ is the family of all minimal convexificators of the function h.

Remark 3.2 *Remark 3.1 implies that for every $g \in \mathbb{R}^n$ either*

$$h(g) = \min_{C \in H^*} \max_{v \in C}(v, g), \quad h(-g) = \min_{C \in H^*} \max_{v \in C}(v, -g)$$

or

$$h(g) = \max_{C \in H^*} \min_{v \in C}(v, g), \quad h(-g) = \max_{C \in H^*} \min_{v \in C}(v, -g).$$

Remark 3.3 *It follows also from Remark 3.2 that for every $g \in \mathbb{R}^n, g \neq 0_n$, there exists an MCF such that either (32) holds, or (33) takes place. If, in addition, $h(-g) = -h(g)$ then C_g^* is contained in the hyperplane $(v - A, g) = 0$ where $A = h(g)g$, i.e. $(v, g) = h(g)||g||^2$.*

Let H^* be the set of all MCF's of a p.h. function h. By Corollary 3.1 for every $g \in \mathbb{R}^n$ one of the relations (34) and (35) holds. Remark 3.2 says that for the direction $g_1 = -g$ one of the relations (34) and (35) also holds.

Lemma 3.1 *Let h be a p.h. Lipschitz function. If $\mathcal{K} \subset \mathbb{R}^n$ is a convex cone such that*

$$h(-g) \neq -h(g) \quad \forall g \in \mathcal{K} \tag{36}$$

then either

$$h(g) = \min_{C \in H^*} \max_{v \in C} (v, g) \quad \forall g \in \mathcal{K} \tag{37}$$

or

$$h(g) = \max_{C \in H^*} \min_{v \in C} (v, g) \quad \forall g \in \mathcal{K}. \tag{38}$$

Proof. Take any $g'_0 \in \mathcal{K}$, $||g'_0|| = 1$. Assume, for example, that

$$h(g'_0) = \min_{C \in H^*} \max_{v \in C} (v, g'_0) = \max_{v \in C'_0} (v, g'_0).$$

Here $C'_0 = C^*_{g_0}$. Let us prove that (37) is valid.

Arguing by contradiction, we find a $g''_0 \in \mathcal{K}$, $||g''_0|| = 1$, such that

$$h(g''_0) = \max_{C \in H^*} \min_{v \in C} (v, g''_0) = \min_{v \in C''_0} (v, g''_0).$$

It follows from (36) that $g''_0 \neq -g''_0$. Put $\overline{g}_0 = (g'_0 + g''_0)/||g'_0 + g''_0||$. Due to the convexity of $\mathcal{K}$ we get $\overline{g}_0 \in \mathcal{K}$. If

$$h(\overline{g}_0) = \min_{C \in H^*} \max_{v \in C} (v, \overline{g}_0)$$

then we take $g'_1 = \overline{g}_0$, $g''_1 = g''_0$. In the case

$$h(\overline{g}_0) = \max_{C \in H^*} \min_{v \in C} (v, \overline{g}_0)$$

we put $g'_1 = g'_0$, $g''_1 = \overline{g}_0$. Continuing in the same manner we construct the sequences of vectors $\{g'_k\}$ and $\{g''_k\}$ such that

$$g'_k \to g_0, \; g''_k \to g_0, \; ||g_0|| = 1, \quad g_0 \in \mathcal{K},$$

$$h(g'_k) = \max_{v \in C'_k} (v, g'_k), \quad h(g''_k) = \min_{v \in C''_k} (v, g''_k).$$

The continuity of h yields

$$h(g'_k) - h(g''_k) = \max_{v \in C'_k} (v, g'_k) - \min_{v \in C''_k} (v, g''_k) \;\to 0. \tag{39}$$

Employing (32) and (33) (see Remark 3.1) we get

$$h(-g'_k) - h(-g''_k) = \max_{v \in C'_k} (v, -g'_k) - \min_{v \in C''_k} (v, -g''_k) =$$

$$= \max_{v \in C''_k} (v, g''_k) - \min_{v \in C'_k} (v, g'_k) \;\to 0. \tag{40}$$

Summing up (39) and (40) we have $a_k + b_k \longrightarrow 0$ where

$$a_k = \max_{v \in C'_k}(v, g'_k) - \min_{v \in C'_k}(v, g'_k) \geq 0,$$

$$b_k = \max_{v \in C''_k}(v, g''_k) - \min_{v \in C''_k}(v, g''_k) \geq 0.$$

Hence,

$$a_k \longrightarrow 0, \quad b_k \longrightarrow 0. \tag{41}$$

Since C'_k are convexificators (even minimal ones) then

$$\min_{v \in C'_k}(v, g'_k) \leq h(g'_k) \leq \max_{v \in C'_k}(v, g'_k), \tag{42}$$

$$\min_{v \in C'_k}(v, -g'_k) \leq h(-g'_k) \leq \max_{v \in C'_k}(v, -g'_k). \tag{43}$$

Summing (42) and (43) yields

$$\min_{v \in C'_k}(v, g'_k) - \max_{v \in C'_k}(v, g'_k) \leq h(g'_k) + h(-g'_k) \leq$$

$$\leq \max_{v \in C'_k}(v, g'_k) - \min_{v \in C'_k}(v, g'_k)$$

or

$$-a_k \leq h(g'_k) + h(-g'_k) \leq a_k.$$

It follows from (41) that $h(g_0) + h(-g_0) = 0$ which contradicts (36).

Lemma 3.2 *Let $h : \mathbb{R}^n \rightarrow \mathbb{R}$ be a p.h. Lipschitz function. If there exists only a finite number of directions $g_1, ..., g_N$ such that*

$$||g_k|| = 1, \ h(-g_k) = -h(g_k) \quad \forall k \in 1 : N,$$

then either

$$h(g) = \min_{C \in H^*} \max_{v \in C}(v, g) \quad \forall g \in \mathbb{R}^n$$

or

$$h(g) = \max_{C \in H^*} \min_{v \in C}(v, g) \quad \forall g \in \mathbb{R}^n.$$

Proof. Let

$$\sigma = \{g \mid ||g|| = 1, \quad h(-g) = -h(g)\} = \{g_k \mid k \in 1 : N\}.$$

Take any g'_0 such that

$$h(-g'_0) \neq -h(g'_0). \tag{44}$$

For definiteness, assume that

$$h(g'_0) = \max_{v \in C'_0}(v, g'_0) > \min_{v \in C'_0}(v, g'_0). \tag{45}$$

By the continuity of h, the inequality (45) holds in some sphere $S_{r_1}(g_0') = \{g \mid \|g - g_0'\| \leq r_1\}$ where $r_1 > 0$. By Lemma 3.1, (45) guarantees that (37) holds for all $g \in S_{r_1}(g_0')$. Assume that there exists a g_0'' such that $\|g_0''\| = 1$,

$$h(g_0'') = \min_{v \in C_0''}(v, g_0'') < \max_{v \in C_0''}(v, g_0''). \tag{46}$$

(47) implies

$$h(-g_0'') \neq -h(g_0''). \tag{47}$$

By the continuity of h, (47) holds in some sphere $S_{r_2}(g_0'')$ with $r_2 > 0$. By Lemma 3.1 we conclude that (38) holds for all $g \in S_{r_2}(g_0'')$. Since σ contains a finite number of elements then there exist $g_1' \in S_{r_1}$ and $g_1'' \in S_{r_2}$ such that

$$\mathcal{P} \cap \sigma \neq \emptyset \tag{48}$$

where

$$\mathcal{P} = \{g_\alpha \mid g_\alpha = (\alpha g_1' + (1-\alpha)g_1'')/\|\alpha g_1' + (1-\alpha)g_1''\|, \quad \alpha \in [0,1]\}.$$

Arguing as in the proof of Lemma 3.1, we find a $g_0 \in \mathcal{P}$ such that $h(-g_0) = -h(g_0)$, i.e. $g_0 \in \sigma$ which contradicts (48).

Corollary 3.2 *If all minimal convexificators of h are solid (i.e. int $C \neq \emptyset$ $\forall C \in H^*$) then either (34) holds for every $g \in \mathbb{R}^n$ or (35) is valid for every $g \in \mathbb{R}^n$. In such a case there exist only two p.h. functions having the same family of MCF's:*

$$h_1(g) = \min_{C \in H^*} \max_{v \in C}(v, g), \quad h_2(g) = \max_{C \in H^*} \min_{v \in C}(v, g).$$

This property (as follows from Lemma 3.2) holds also in the case where there exists a finite number of unit directions g such that $h(-g) = -h(g)$. It is clear that $h_1(-g) = -h_2(g)$.

Aknowledgement. The author thanks Dr. Amos Uderzo for useful remarks and discussions during his stay at St. Petersburg State University.

References

[1] Demyanov, V.F., Rubinov, A.M. (1995), *Constructive Nonsmooth Analysis*, Verlag Peter Lang, Franfurt a/M.

[2] Rockafellar, R.T. (1970), *Convex Analysis*, Princeton University Press, Princeton N.J.

[3] Pschenichnyi, B.N.(1980), *Convex Analysis and Extremal Problems*, Nauka, Moscow.

[4] Clarke, F.M. (1983), *Optimization and Nonsmooth Analysis*, Wiley Intersciencem New York.

[5] Michel, P., Penot, J.-P. (1984), Calcus sous-differential pour les fonctions lipschitzienness et non-lipschitziennes, *C.R. Acad. Sc. Paris, Ser. I* 298, pp. 269-272.

[6] Demyanov, V.F. (1994) , Convexification and concavification of a positively homogeneous function by the same family of linear functions, Universita di Pisa , Report 3,208,802.

[7] Demyanov, V.F., Jeyakumar, V. (1997), Hunting for a smaller convex subdifferential, *J. of Global Optimization*, Vol, 10, No. 3, pp.305-326.

[8] Jeyakumar, V., Demyanov, V.F. (1996), A mean-value theorem and a characterization of convexity using convexificators, Applied Math. Research Report AMR 96/13, Univ. of New South Wales, Sydney, Australia.

[9] Jeyakumar, V., Luc, D.T., Schaible, S. (1998), Characterizations of generalized monotone nonsmooth continuous maps using Approximate Jacobians, *J. of Convex Analysis,* Vol. 5, No. 1, pp. 119-132.

[10] 10. Demyanov, V.F., Murzabekova, G.E. (1999), Convexificators and Implicit functions in Nonsmooth Systems, Forthcoming in *J. of Comput. Math. and Math. Physics.*

Optimal Control Problems and Penalization[1]

Vladimir F. Demyanov (vladimir.demyanov@pobox.spbu.ru)
Department of Applied Mathematics, St. Petersburg State University Staryi Peterhof, 198904, Russia

Franco Giannessi (gianness@dm.unipi.it)
Department of Mathematics, Università di Pisa, Via Buonarroti 2, 56127 Pisa, Italy

Vladimir V. Karelin (vladimir.karelin@pobox.spbu.ru)
Department of Applied Mathematics, St. Petersburg State University Staryi Peterhof, 198904, Russia

Abstract

The Exact Penalization Technique is applied to treat optimal control problems in a system described by ordinary differential equations. The resulting functional is essentially nonsmooth but directionally differentiable (even subdifferentiable). Differential equations are viewed as constraints and are "removed" by introducing an exact penalty function. The aim of the paper is to illustrate that well-known optimality conditions can be derived via Exact Penalty approach.

Key words: Optimal Control, Exact Penalization, Directional differentiability, Necessary Optimality Conditions, Nonsmooth Analysis.

[1]The research of the 1st and 3rd Authors was supported by the Russian Foundation for Fundamental Studies under Grant No. 97–01–00499

G. Di Pillo and F. Giannessi (eds.), Nonlinear Optimization and Related Topics, 67–78.

1 Introduction

In our previous paper [1] Exact Penalization Technique [2-5] was applied to treat one class of optimal control problems where "controls" where just some parameters of the system of differential equations describing the behaviour of some controlled object. In the present paper we consider a general optimal control problem. The system of differential equations is viewed as "constraints" and it is shown how these constraints can be "deleted" by introducing a proper Exact Penalty function. The resulting new functional is essentially nonsmooth, however, it has some nice differential properties and modern methods of Nondifferentiable Optimization allow to attack this problem numerically.

In Section 2 differential properties of the function replacing differential equations are studied. In Section 3 an exact penalty function is introduced and its directional differentiability is established. The case of unconstrained controls is discussed in Section 4. A necessary condition for a minimum is formulated. the well-known classical results are corollaries of this necessary condition.

2 The statement of the Problem

Let $P_m[0,T]$ be the family of all m-dimensional vector-functions defined and piecewise continuous on $[0,T]$ with $T>0$ being fixed and let

$$U = \{u \in P_m[0,T] | u(t) \in V \quad \forall\, t \in [0,T]\},$$

where $V \subset \mathbb{R}^m$ is a given compact set. A vector - function $u \in U$ will be referred to as a control function or just a control, U will be called the set of controls. Let a function $f(x,u,t) : \mathbb{R}^n \times \mathbb{R}^m \times \mathbb{R} \to \mathbb{R}^n$ be differentiable with respect to x. The functions f and $\frac{\partial f}{\partial x}$ are assumed to be continuous on $\mathbb{R}^n \times \mathbb{R}^m \times \mathbb{R}$.

For any given $u \in U$, consider the following system of differential equations:

$$\dot{x} = f(x,u,t), \tag{1}$$

$$x(0) = x_0. \tag{2}$$

If $t \in [0,T]$ is a discontinuity point of u then $\dot{x}(t)$ is the right - hand side derivative at t. Under the conditions stated for every $u \in U$ there exists a unique piecewise differentiable function $x(t)$ defined on $[0,T]$ and satisfying (1) - (2). Consider the set

$$\Omega_u := \{z \in P_n[0,T], \quad x(t) \in P_n[0,T], \quad \varphi_u(z) = 0\}, \tag{3}$$

where

$$\varphi_u(z) := \left[\int_0^T \left(z(t) - f(x_0 + \int_0^t z(\tau)d\tau, u(t), t)\right)^2 dt\right]^{1/2}. \tag{4}$$

Note that $\varphi_u(z) \geq 0 \ \ \forall z \in P_n[0,T],\ \forall u \in U$. If $z \in \Omega_u$, then obviously the function $x(t) = x_0 + \int_0^t z(\tau)d\tau$ satisfies (1), (2), and vice versa, if $x(t)$ is a solution of (1), (2) then $z \in \Omega_u$ (with $z(t) = f(x(t), u(t), t)$). Thus, the problem of finding a solution of (1), (2) for some fixed $u \in U$ is equivalent to finding a $z \in P_n[0,T]$ such that $\varphi_u(z) = 0$.

Now let us study the differentiability properties of the function φ_u. Firstly, consider the case:

$$\varphi_u(z) > 0. \tag{5}$$

Take $v \in P_n[0,T]$ and find the directional derivative of φ_u at some point z satisfying (5) in a direction v. By definition

$$\varphi_u'(z;v) = \lim_{\alpha \downarrow 0} \frac{1}{\alpha}[\varphi(z+\alpha v) - \varphi_u(z)]. \tag{6}$$

We shall prove that this limit does exist and find its value.

Lemma 2.1 *If $\varphi_u(z) > 0$ (i.e., $z \notin \Omega_u$), then the function φ_u is Gâteaux differentiable at z.*

Proof. We have

$$\varphi_u(z+\alpha v) = \left[\int_0^T \left(z(t) + \alpha v(t) - f(x_0 + \int_0^t (z+\alpha v)(\tau))d\tau,\ u(t), t)\right)^2 dt\right]^{1/2} =$$

$$\left\{\int_0^T \left[z(t) - f(x_0 + \int_0^t z(\tau)d\tau, u(t), t) + \right.\right.$$

$$\left.\left. +\alpha[v(t) - \frac{\partial f(t)}{\partial x}\int_0^t v(\tau)d\tau] + o(\alpha)\right]^2 dt\right\}^{1/2} \tag{7}$$

Here $\frac{\partial f(t)}{\partial x} = \frac{\partial f(x(t),u(t),t)}{\partial x}$. Since $h(\alpha) = [(a+\alpha b)^2]^{1/2} = (a^2 + 2\alpha(a,b) + \alpha^2 b^2)^{1/2}$, then $h'(0) = \frac{(a,b)}{\|a\|}$. Therefore (7) implies

$$\varphi_u'(z;v) = \int_0^T \left(w(t)\ ,\ v(t) - \frac{\partial f(t)}{\partial x}\int_0^t v(\tau)d\tau\right) dt, \tag{8}$$

where

$$w(t) := \frac{1}{\varphi_u(z)}\left(z(t) - f(x_0 + \int_0^t z(\tau)d\tau, u(t), t)\right). \tag{9}$$

It is clear that

$$\|w\| := \left[\int_0^T (w(t))^2 dt\right]^{1/2} = 1. \tag{10}$$

In (8) us set an "*" denotes transposition. Let us set

$$B := \int_0^T \left(\frac{\partial f(t)}{\partial x} \int_0^t v(\tau)d\tau, w(t) \right) dt =$$

$$= \int_0^T \left(\int_0^t v(\tau)d\tau, \left(\frac{\partial f(t)}{\partial x} \right)^* w(t) \right) dt. \tag{11}$$

Integrating (11) by parts, we have

$$B = \int_0^T \left(v(t), \int_t^T \left(\frac{\partial f(\tau)}{\partial x} \right)^* w(\tau)d\tau \right) dt. \tag{12}$$

Substituting (12) in (8) we get

$$\varphi_u'(z;v) = \int_0^T \left(v(t), w(t) - \int_t^T \left(\frac{\partial f(\tau)}{\partial x} \right)^* w(\tau)d\tau \right) dt = (\nabla\varphi_u, v), \tag{13}$$

where

$$\nabla\varphi_u = w(t) - \int_t^T \left(\frac{\partial f(\tau)}{\partial x} \right)^* w(\tau)d\tau \tag{14}$$

and $(\nabla\varphi_u, v)$ will be referred to as "the scalar product" of $\nabla\varphi_u$ and v. Since (13) is linear in v and the complement of Ω_u is open, we conclude that φ_u is Gâteaux differentiable at z with "the gradient" $\nabla\varphi_u$ (in the space $P_n[0,T]$).

Remark 1. Under conditions imposed on f and U one may assume that

$P_n[0,T]$ is an open set of piecewise continuous on $[0,T]$ functions which are pointwise bounded on $[0,T]$ by some constant $K < \infty$.

Lemma 2.2 *There exists an $a > 0$ such that*

$$\min_{\|v\|=1} (\nabla\varphi_u, v) \leq -a < 0 \qquad \forall z \in P_n[0,T] : u \in U, \quad z \notin \Omega. \tag{15}$$

Proof. Let us prove, first of all, that

$$\varphi_u \neq \mathbf{0}. \tag{16}$$

Here 0 is the zero element of the space $P_n[0,T]$. Assuming the opposite, we have:

$$w(t) - \int_t^T \left(\frac{\partial f(\tau)}{\partial x} \right)^* w(\tau)d\tau = 0_n \quad \forall t \in [0,T]. \tag{17}$$

Then (17) implies $w(t) = 0_n$ $\forall t \in [0,T]$ which contradicts (10). Thus (16) holds. Suppose now that (15) is invalid. Then, there exists a sequence $[z_k, u_k]$ such that

$$z \in P_n[0,T], u_k \in U, z_k \notin \Omega, \quad \nabla\varphi_k \to 0, \tag{18}$$

where

$$\nabla\varphi_k = w_k(t) - \int_t^T \left(\frac{\partial f_k(\tau)}{\partial x}\right)^* w_k(\tau)d\tau, \quad \frac{\partial f_k(t)}{\partial x} = \frac{\partial f(x_k(t), u_k, t)}{\partial x},$$

$$x_k(t) = x_0 + \int_0^t z_k(\tau)d\tau,$$

$$w_k(t) = \frac{1}{\varphi_{u_k}(z_k)}\left(z_k(t) - f(x_0 + \int_0^t z_k(\tau)d\tau, u_k, t\right).$$

Note that

$$\|w_k\| = 1. \tag{19}$$

Relations (18) imply

$$\|h_k\| \to 0, \tag{20}$$

where

$$h_k(t) = w_k(t) - \int_t^T \left(\frac{\partial f_k(\tau)}{\partial x}\right)^* w_k(\tau)d\tau. \tag{21}$$

Relations (20) and (21) yield (due to the continuous dependence of the solutions of integral equations on the right-hand sides) $\|w_k\| \to 0$ which contradicts (19). This completes the proof. ◻

Now consider the case where $\varphi_k(z) = 0$. Note that

$$\varphi_u(z) = \max_{\|\bar{v}\|=1} \int_0^T \left(z(t) - f(x_0 + \int_0^t z(\tau)d\tau, u(t), t), \bar{v}(t)\right) dt. \tag{22}$$

If $\varphi_u(z) = 0$, then $h(t) = z(t) - f(x_0 + \int_0^t z(\tau)d\tau, u(t), t) = 0 \ \forall t \in [0, T]$.
Since

$$z(t) + \alpha v(t) - f(x_0 + \int_0^t (z(\tau) + \alpha v(\tau))d\tau, u(t), t) =$$

$$= h(t) + \alpha\left[v(t) - \frac{\partial f(t)}{\partial x}\int_0^t v(\tau)d\tau\right] + o(\alpha),$$

then (see (22)):

$$\varphi_u'(z, v) = \max_{\|\bar{v}\|=1} \int_0^T \left(\bar{v}(t), v(t) - \frac{\partial f(t)}{\partial x}\int_0^t v(\tau)d\tau\right) dt. \tag{23}$$

Using the same procedure as in (11)–(12), from (23) we get

$$\varphi_u'(z, v) = \max_{\|\bar{v}\|=1} \left[\int_0^T \left(v(t), \bar{v}(t) - \int_t^T \left(\frac{\partial f(t)}{\partial x}\right)^* \bar{v}(\tau)d\tau\right) dt\right]. \tag{24}$$

(23) and (24) show that the following proposition holds.

Lemma 2.3 *If $\varphi_u(z) = 0$, then the function φ is directionally differentiable at z; it is even subdifferentiable, i.e.*

$$\varphi_u'(z, v) = \max_{v^* \in \partial\varphi_u(z)} (v^*, v), \tag{25}$$

where

$$\partial\varphi_u(z) = \Big\{ v^* \in P_n[0,T], |v^*(t) = \bar{v}(t) -$$

$$- \int_t^T \left(\frac{\partial f(\tau)}{\partial x}\right)^* \bar{v}(\tau)d\tau, \ , \|\bar{v}\| \leq 1 \Big\}. \tag{26}$$

3 Optimal Control Problems

Let us consider the following problem: find

$$\min_{u \in U} \mathcal{I}(u)$$

where

$$\mathcal{I}(u) = \int_0^T f_0(x(t,u), u(t), t)dt, \tag{27}$$

$x(t,u)$ is a solution of (1), (2) with $u \in U$, and $f_0 : \mathbb{R}^n \times \mathbb{R}^m \times \mathbb{R} \to \mathbb{R}^n$ is differentiable with respect to x. The functions f_0 and $\frac{\partial f_0}{\partial x}$ are assumed to be continuous on $\mathbb{R}^n \times \mathbb{R}^m \times \mathbb{R}$.

It follows from Section 2 that the above problem is equivalent to the following one: Find

$$\min_{z \in P[0,T], u \in U, \phi_u(z)=0} \Phi(z, u)$$

where

$$\Phi(z; u) = \int_0^T f_0(x_0 + \int_0^t z(\tau)d\tau, u(t), t)dt. \tag{28}$$

Let us compute the derivative of the function $\Phi_u(z) = \Phi(z; u)$ in a direction $v \in P_n[0,T]$

$$\Phi_u'(z; v) = \lim_{\alpha \downarrow 0} \frac{1}{\alpha} [\Phi(z + \alpha, u) - \Phi(z; u)] = \int_0^T (v(t), \nabla\Phi_u(z))dt \tag{29}$$

where

$$\nabla\Phi_u(z) = \int_0^T \frac{\partial f_0(\tau)}{\partial x} d\tau, \ f_0(t) = f_0(x_0 + \int_0^t z(\tau)d\tau, u(t), t) \tag{30}$$

It follows from (29) and (30) that Φ_u is Gâteaux differentiable with the "gradient" $\nabla\Phi_u(z)$ given by (30) where

$$\dot{\psi} = -\frac{\partial f_0(x(t), u(t), t)}{\partial x} d\tau, \tag{31}$$

$$\psi_u(T) = 0. \tag{32}$$

Arguing as in [6] we are able to prove the following.

Theorem 3.1 *If the function f_0 is Lipschitz with respect to x and u then there exists a $\lambda^* \geq 0$ such that for any $\lambda \geq \lambda^*$ the set of minimizers of $J(u)$ on the set U coincides with the set of minimizers of the function*

$$F(z,u) = \int_0^T f_0(x_0 + \int_0^t z(\tau)d\tau, u(t), t)dt +$$

$$\lambda \left\{ \int_0^T [z(t) - f(x_0 + \int_0^t z(\tau)d\tau, u(t), t)]^2 dt \right\}^{1/2} \tag{33}$$

on the entire space $P_n[0,T] \times \mathbb{R}^m$.

Remark 2. Observe that in (27) additional constraints are present: $x(t,u)$ is the solution of (1),(2). In (33) $z(t)$ is an arbitrary piecewise continuous n - function. Hence, in (33) no differential equations are involved. The function $F_u = F(z,u)$ is directionally differentiable (as a function of z) for any fixed $u \in U$: if $\varphi_u(z) > 0$ then by Lemma 1 the function F_u is even Gâteaux differentiable with the "gradient"

$$\psi(t) + \lambda \left[w(t) - \int_t^T \left(\frac{\partial f(\tau)}{\partial x} \right)^* w(\tau)d\tau \right] \tag{34}$$

where $\psi(t)$ satisfies (31) – (32) and $\omega(t)$ – (9).

If $\varphi_u(z) = 0$ (*i.e.*, $z \in \Omega_u$) then by Lemma 3 the function $F_u(t)$ is subdifferentiable with the subdifferential

$$\partial F_u(z) = \psi + \lambda \partial \varphi_u(z) \tag{35}$$

where $\partial \varphi_u(z)$ is given by (26).

4 The case of unconstrained controls

In some optimal control problem the function u in (1) is of the form $u(t) = K(t)x(t)$ where $K(t)$ is a matrix – valued continuous function. In such a case the initial problem can be reformulated and stated in the following form. (K being a new control function denoted again by u.)

Let $C_m[0,T]$ be the family of all m – dimensional vector – functions defined and continuous on $[0,T]$ with $T > 0$ being fixed. Let a function $f(x,u,t) : \mathbb{R}^n \times \mathbb{R}^m \times \mathbb{R} \longrightarrow \mathbb{R}^n$ be differentiable with respect to x and u. The functions f, $\frac{\partial f}{\partial x}$ and $\frac{\partial f}{\partial u}$ are assumed to be continuous on $\mathbb{R}^n \times \mathbb{R}^m \times \mathbb{R}$.

For any given $u \in C_m[0,T]$, consider the system (1), (2). Under the conditions stated for every $u \in C_m[0,T]$ there exists a unique continuously differentiable function $x(t) \in C_n[0,T]$ satisfying (1)–(2). Consider the set

$$\Omega := \left\{ [z,u] \middle| z \in C_n[0,T], \quad u \in C_m[0,T], \quad \varphi(z,u) = 0 \right\}, \tag{36}$$

where

$$\varphi(z,u) := \left[\int_0^T \left(z(t) - f(x_0 + \int_0^t z(\tau)d\tau, u(t), t)\right)^2 dt\right]^{1/2}. \tag{37}$$

Note that $\varphi(u,z) \geq 0 \ \forall z \in C_n[0,T], \ \forall u \in C_m[0,T]$. If $[z,u] \in \Omega$, then obviously the function $x(t) = x_0 + \int_0^t z(\tau)d\tau$ satisfies (1), (2), and vice versa, if $x(t)$ is a solution of (1), (2) then $[z,u] \in \Omega$ (with $z(t) = f(x(t),u(t),t)$). Thus, the problem of finding a solution of (1), (2) for some $u \in C_m[0,T]$ is equivalent to finding a $z \in P_n[0,T]$ such that $\varphi(z,u) = 0$.

Let $g := [v,q]$ where $v \in C_n[0,T], q \in C_m[0,T]$. Put $\|g\| := \max\{\|v\|, \|q\|\}$ where

$$\|v\| := \left[\int_0^T (v(t))^2\right]^{\frac{1}{2}}, \|q\| := \left[\int_0^T (q(t))^2\right]^{\frac{1}{2}}$$

The pair g will be called a direction (in the space $C_n[0,T] \times C_m[0,T]$.) If there exists the finite limit

$$\varphi'(z,u;g) = \lim_{\alpha \downarrow 0} \frac{1}{\alpha}[\varphi(z + \alpha v, u + \alpha q) - \varphi(z,u)].$$

the φ is called differentiable at $[z,u]$ in the direction g. Arguing as in Section 2 (see also [1]) we get the following propositions.

Lemma 4.1 *If $\varphi(u,z) > 0$ (i.e., $[z,u] \notin \Omega_u$), then the function φ is Gâteaux differentiable at $[z,u]$ and*

$$\varphi'(z,u,g) = \int_0^T \left(w(t) - \int_t^T \left(\frac{\partial f(\tau)}{\partial x}\right)^* w(\tau)d\tau, v(t)\right) dt -$$

$$- \int_t^T \left(\left(\frac{\partial f(t)}{\partial u}\right)^* w(t), q(t)\right) dt = (\nabla\varphi, g), \tag{38}$$

where

$$\nabla\varphi(t) = \left[w(t) - \int_t^T \left(\frac{\partial f(\tau)}{\partial x}\right)^* w(\tau)d\tau, -\left(\frac{\partial f(t)}{\partial u}\right)^* w(t)\right] \tag{39}$$

$$w(t) := \frac{1}{\varphi(z,u)}\left(z(t) - f(x_0 + \int_0^t z(\tau)d\tau, u(t), t)\right). \tag{40}$$

Here, as usual, $\frac{\partial f(t)}{\partial x} = \frac{\partial f(x(t),u(t),t)}{\partial x}, \frac{\partial f(t)}{\partial x} = \frac{\partial f(x(t),u(t),t)}{\partial u}$.

Note that $\|w(t)\| = 1$. The pair of vector functions $\nabla\varphi(t)$ will be referred to as the "gradient" of φ at $[z,u]$.

Lemma 4.2 *If $\varphi(z,u) = 0$, then the function φ is directionally differentiable at z,u; it is even subdifferentiable, i.e.*

$$\varphi'(z,u,g) = \max_{\|\bar{v}\|=1}\left[\int_0^T \left(\bar{v}(t) - \int_t^T \left(\frac{\partial f(t)}{\partial x}\right)^* \bar{v}(\tau)d\tau, v(t)\right) dt - \right.$$

$$\left. - \int_0^T \left(\left(\frac{\partial f(t)}{\partial u}\right)^* \bar{v}(t), q(t)\right) dt\right]. \tag{41}$$

Remark 3. (41) means that if $\varphi(z,u) = 0$, then the function φ is subdifferentiable at $[z,u]$:

$$\varphi'(z,u;g) = \max_{G\in\partial\varphi(z,u)} (G,g), \tag{42}$$

where

$$\partial\varphi(z,u) = \left\{G = [v^*,q^*]\middle| v^* \in C_n[0,T], q^* \in C_m[0,T], \right.$$

$$v^*(t) = \bar{v}(t) - \int_t^T \left(\frac{\partial f(\tau)}{\partial x}\right)^* \bar{v}(\tau)d\tau,$$

$$\left. q^* = -\left(\frac{\partial f(\tau)}{\partial u}\right)^* \bar{v}(t), \quad \bar{v} \in C_n[0,T], \|\bar{v}\| \le 1\right\}. \tag{43}$$

Lemma 4.3 *There exists an $a > 0$ such that*

$$\min_{\|g\|=1} (\nabla\varphi, g) \le -a < 0 \qquad \forall [z,u] \notin \Omega. \tag{44}$$

Now consider the problem of minimizing the functional

$$x\mathcal{I}(u) = \int_0^T f_0(x(t,u),u(t),t)dt, \tag{45}$$

where $x(t,u)$ is the solution of (1), (2) with $u \in C_m[0,T]$ and $f_0 : \mathbf{R}^n \times \mathbf{R}^m \times \mathbf{R} \to \mathbf{R}$ is differentiable in x and u. The functions f_0, $\frac{\partial f_0}{\partial x}$ and $\frac{\partial f_0}{\partial u}$ are assumed to be continuous on $\mathbf{R}^n \times \mathbf{R}^m \times \mathbf{R}$. The above problem is equivalent to the following one: Find

$$\min_{z\in C_m[0,T], u\in C_m[0,T], \varphi(z,u)=0} \Phi(z,u) \tag{46}$$

where

$$\Phi(z;u) = \int_0^T f_0\left(x_0 + \int_0^t z(\tau)d\tau, u(t), t\right)dt.$$

Let us compute the derivative of $\Phi(z;u)$ in a direction $g = [v,q]$ (where $v \in C_n[0,T]$, $q \in C_n[0,T]$) :

$$\Phi'(z,u;g) = \lim_{\alpha\downarrow 0} \frac{1}{\alpha}\left[\Phi(z+\alpha, u+\alpha q) - \Phi(z;u)\right] =$$

$$= \int_0^T \left[\left(\int_t^T \frac{\partial f_0(\tau)}{\partial x} d\tau, v(t)\right) + \left(\frac{\partial f_0(t)}{\partial u}, q(t)\right)\right] dt = \left(\nabla\Phi, g\right) \tag{47}$$

where

$$\nabla\Phi(t) = \left[\int_0^T \frac{\partial f_0(\tau)}{\partial x}d\tau,\ \frac{\partial f_0(t)}{\partial u}\right]$$

(47) shows that Φ is Gâteaux differentiable and $\nabla\Phi$ is the "gradient" of Φ at $[z, u]$. Arguing as in [1] we get the following.

Theorem 4.1 *If Φ is Lipschitz on $C_n[0,T]\times C_m[0,T]$ then there exists a $\lambda_0 \geq 0$ such that for any $\lambda \geq \lambda_0$ the set of minimisers of Φ on the set $\Omega = \{[z,u] \mid \ \varphi(z,u) = 0\}$ coincides with the set of minimisers of the function*

$$F(z,u) = \Phi(z,u) + \lambda\varphi(z,u) = \int_0^T f_0\Big(x_0 + \int_0^t z(\tau)d\tau, u(t), t\Big)dt +$$

$$+\lambda\Big\{\int_0^T \Big[z(t) - f(x_0 + \int_0^t z(\tau)d\tau, u(t), t)\Big]^2 dt\Big\}^{1/2} \tag{48}$$

on the entire space $C_n[0,T]\times C_m[0,T]$.

Thus , the constrained optimization problem (46) is reduced to the unconstrained problem of minimizing (48).

Since the function Φ is differentiable and the function φ is directionally differentiable (with the directional derivatives defined by (38) or (42)) then one can employ numerical methods available for unconstrained optimization.

Let $[z^*, u^*]$ be a minimiser of (46). Then $\varphi(z^*, u^*) = 0$.

The functional $F(z,u)$ is subdifferentiable and its subdifferential at $[z^*, u^*]$ is

$$\partial F(z^*, u^*) = \nabla\Phi + \lambda\partial\varphi, \tag{49}$$

where $\nabla\Phi$ is given by (47) and $\partial\varphi$ – by (43) (both calculated at $[z^*, u^*]$).

By the necessary condition for a minimum of a subdifferentiable function (see [7]).

$$\mathbf{0} \in \partial F(z^*, u^*). \tag{50}$$

Here $\mathbf{0}$ is the zero element of $C_n[0,T]\times C_m[0,T]$. It follows from (43), (47), (49) and (50) that there exists a $\bar v \in C_n[0,T]$, $\|\bar v\| \leq 1$, such that

$$\int_t^T \frac{\partial f_0(x^*(\tau), u^*(\tau), \tau)d\tau}{\partial x} +$$

$$+\lambda\left[\bar v(t) - \int_t^T \left(\frac{\partial f(x^*(\tau), u^*(\tau), \tau)}{\partial x}\right)^* \bar v(\tau)d\tau\right] = 0 \quad \forall t \in [0,T], \tag{51}$$

$$\frac{\partial f_0(x^*(t), u^*(t), t)}{\partial u} - \lambda\left(\frac{\partial f(x^*(t), u^*(t), t)}{\partial u}\right)^* \bar v(t) = 0. \tag{52}$$

Denoting $\lambda\bar{v}(t)$ by $\psi(t)$ and differentiating (51) with respect to t, we get

$$\dot{\psi}(t) = -\left(\frac{\partial f(x^*(t), u^*(t), t)}{\partial x}\right)^* \psi(t) + \frac{\partial f_0(x^*(t), u^*(t), t)}{\partial x}, \tag{53}$$

$$\psi(T) = 0_n, \tag{54}$$

$$\left(\frac{\partial f(x^*(t), u^*(t), t)}{\partial u}\right)^* \psi(t) - \frac{\partial f_0(x^*(t), u^*(t), t)}{\partial u} = 0. \tag{55}$$

Finally, we got the following necessary condition.

Theorem 4.2 *For a function $u^* \in C_m[0,T]$ to be a minimiser of functional $J(u)$ it is necessary that there exist a solution $\psi(t) \in C_m[0,T]$ of the differential equation (53)-(54) satisfying the condition (55).*

Remark 4. Let us introduce a new variable x_0, put $\bar{x} = (x_0, x_1, ..., x_n)$ and consider the following differential equation

$$\dot{\bar{x}} = \bar{f}(\bar{x}, u, t), \tag{56}$$

$$\bar{x}(0) = \bar{x}_0. \tag{57}$$

where

$$\bar{f} = (f_0, ..., f_n), f_i(\bar{x}, u, t) = f_i(x_1, ..., x_n, u, t),$$

$\bar{x} = (0, x_1, ..., x_n) \in \mathbf{R}^{n+1}$ (the functions f_i do not depend explicitly on x_0.)

Put $\bar{\psi}(t) = (\psi_0(t), \psi_1(t), ..., \psi_n(t))$. Then the equation (53)-(54) and the condition (55) can be rewritten in a more symmetric form

$$\dot{\bar{\psi}}(t) = -\left(\frac{\partial \bar{f}(\bar{x}^*(t), u^*(t), t)}{\partial \bar{x}}\right)^* \bar{\psi}(t),$$

$$\bar{\psi}(T) = (-1, 0, ..., 0),$$

$$\left(\frac{\partial \bar{f}(\bar{x}^*(t), u^*(t), t)}{\partial u}\right)^* \bar{\psi}(t) = 0.$$

Remark 5. The result above is just a "linearized" maximim principle. A similar exact penalty approach in the case of absence of the differentiability of f in u will produce the full-scale ("nonlinearized") Pontryagin maximum principle.

References

[1] V.F.Demyanov, F.Giannessi and V.V.Karelin, "Optimal Control Problems via Exact Penalty Functions", Jornal of Global Optimization, Vol. 12, N3, pp.215–223, 1998.

[2] Eremin I.I. "The method of penalties in convex programming", Doklady Acad. Nauk SSSR, Vol.173, pp. 748 - 751, 1967.

[3] Zangwill W.L., "Nonlinear programming via penalty functions", Management Science, Vol. 13, pp. 344–358, 1967.

[4] Fletcher R., "Penalty functions". In "Mathematical programming: the state of the art." (Eds. A. Bachen, M. Grötschel, B. Korte), Springer–Verlag, Berlin, pp. 87–114, 1983.

[5] Di Pillo G., Grippo L., "On the Exactness of a class of Nondifferentiable Penalty functions". J. Optim. Theory Appl. Vol. 57, pp. 397–408, 1988.

[6] Demyanov V.F., Di Pillo G., Facchinei F., "Exact penalization via Dini and Hadamard constraint derivatives", Optimization Methods and Software, Vol. 9, pp. 19-36, 1998. forthcoming.

[7] Demyanov V.F., Rubinov A.M., "Constructive Nonsmooth Analysis". Peter Lang Verlag, Frankfurt a/M., 1995.

A truncated Newton method for constrained optimization [1]

Gianni Di Pillo (dipillo@dis.uniroma1.it)

Stefano Lucidi (lucidi@dis.uniroma1.it)

Laura Palagi (palagi@dis.uniroma1.it)

Dipartimento di Informatica e Sistemistica, Università di Roma "La Sapienza", via Buonarroti, 12, 00185 Roma, Italy.

Abstract

In this paper we consider the extension of the unconstrained truncated Newton method approach to the constrained optimization problem $\{min f(x) \; | g(x) \leq 0\}$. To this aim, we first consider a primal-dual system of nonlinear equations $F(x, \lambda) = 0$ with the property that $F(x, \lambda) = 0 \iff (x, \lambda)$ is a *KKT point* for the constrained problem. This system is derived on the basis of an efficient strategy for the identification of the constraints active at the solution. Then, we adopt a truncated Newton scheme for the solution of the system, based on a QR factorization of the gradient of the estimated active constraints and on conjugate gradient methods in the tangent space of these constraints. This approach is particularly advantageous when the number of variables is much larger than the number of constraints.

Under mild assumption and without requiring the strict complementarity, we prove local convergence with superlinear convergence rate of the iterates. Then, we show how to globalize the convergence of the iterates, making use of a line search technique applied to a primal-dual merit function in the class of exact augmented Lagrangian functions.

Preliminary numerical results are reported.

Keywords: constrained optimization, nonlinear programming, nonlinear programming algorithms, QR factorization, truncated Newton methods, augmented Lagrangian functions.

---

[1]This work was partially supported by MURST National Research Program "Metodi per l'Ottimizzazione di Sistemi e Tecnologie (MOST)".

G. Di Pillo and F. Giannessi (eds.), Nonlinear Optimization and Related Topics, 79–103.

1 Introduction

Truncated Newton methods have been introduced in unconstrained optimization wit the aim to cope with large dimensional problems. In fact, in large scale optimizatioı solving the Newton equation at each iteration can be expensive and may not b justified when far from a solution. Instead, an inaccurate solution to the Newto equation can be computed using a conjugate gradient method [2, 9].

In this paper we consider the extension of the unconstrained truncated Newto method approach to a constrained optimization problem of the type:

$$\begin{aligned} \min \ & f(x) \\ & g(x) \leq 0 \end{aligned} \qquad (1$$

where $x \in \mathbb{R}^n$, $f: \mathbb{R}^n \to \mathbb{R}$, $g: \mathbb{R}^n \to \mathbb{R}^m$ are twice continuously differentiabl functions.

To this aim, we first consider a primal-dual system of nonlinear equations with th property that its solutions are *KKT points* for the constrained problem. This systeı is derived on the basis of an efficient strategy for the identification of the constraint active at the solution. Then, we adopt a truncated scheme for the computation of a approximate Newton direction for the solution of the system. The truncated schem is based on a QR factorization of the gradient of the estimated active constraints an on conjugate gradient methods in the tangent space of these constraints. In this wa we can tackle problems in which the number of variables is large with respect to th number of constraints, similarly to the unconstrained case. Under mild assumptio and without requiring the strict complementarity, we prove local convergence wit superlinear convergence rate of the iterates. Then, we show how to globalize th convergence of the iterates, making use of a line search technique applied to a prima dual merit function in the class of exact augmented Lagrangian functions.

In the sequel we denote by $\mathcal{F}$ the feasible set of Problem (1).
We denote by $L(x,\lambda) = f(x) + \lambda' g(x)$, the Lagrangian function for Problem (1 where $\lambda \in \mathbb{R}^m$ is the Karush-Kuhn-Tucker (KKT) multiplier.
We denote by $(\bar{x},\bar{\lambda}) \in \mathbb{R}^{n+m}$ a KKT point for Problem (1), namely a point such tha $\bar{x}$ is feasible and:

$$\nabla_x L(\bar{x},\bar{\lambda}) = 0, \quad \bar{\lambda}' g(\bar{x}) = 0, \quad \bar{\lambda} \geq 0.$$

At a given point $x \in \mathbb{R}^n$, not necessarily feasible, we associate the index sets:

$$\mathcal{A}_0(x) = \{j : g_j(x) = 0\}, \quad \mathcal{N}_0(x) = \{j : g_j(x) < 0\}.$$

Moreover, in correspondence of a KKT point $(\bar{x},\bar{\lambda})$ we consider also the index set:

$$\mathcal{A}_+(\bar{x},\bar{\lambda}) = \{j \in \mathcal{A}_0(\bar{x}) : \bar{\lambda}_j > 0\}. \qquad (2$$

The *strict complementarity* condition holds at $(\bar{x},\bar{\lambda})$ if $\mathcal{A}_+(\bar{x},\bar{\lambda}) = \mathcal{A}_0(\bar{x})$.

We say that the *linear independence constraints qualification* (LICQ) holds at x if the gradients $\nabla g_j(x)$ with $j \in \mathcal{A}_0(x)$ are linearly independent.
At a KKT point $(\bar{x}, \bar{\lambda})$, we say that the *strong second order sufficient condition* (SSOSC) holds if

$$z' \nabla_x^2 L(\bar{x}, \bar{\lambda}) z > 0,$$

for all vector $z \neq 0$ such that

$$\nabla g_j(\bar{x})' z = 0, \quad j \in \mathcal{A}_+(\bar{x}, \bar{\lambda}).$$

We conclude this section by introducing some notation.
Given a vector $v \in \mathbb{R}^p$, we indicate by the uppercase V the diagonal matrix $V = \text{diag}_{1 \leq i \leq p}\{v_i\}$.
Given two vectors $v, w \in \mathbb{R}^p$, the operation $\max\{v, w\}$ is intended component-wise, namely $\max\{v, w\}$ denotes the vector with components $\max\{v_i, w_i\}$.
Let $\mathcal{K} \subseteq \{1, \ldots, m\}$ be an index subset, v be an m vector and Q be a $n \times m$ matrix. We denote by $v_\mathcal{K}$ the subvector of v with components v_i such that $i \in \mathcal{K}$ and by $Q_\mathcal{K}$ the submatrix of Q made up of the columns q_i of Q, with $i \in \mathcal{K}$. For the transpose of $Q_\mathcal{K}$, $(Q_\mathcal{K})'$, we adopt the short notation $Q'_\mathcal{K}$.
In partitioned matrices, 0 and I denote respectively a null submatrix and an identity submatrix of the appropriate dimensions.

2 A local truncated Newton algorithm

In order to define the truncated Newton algorithm we consider preliminarily a primal-dual local algorithm described by the iteration:

$$\begin{aligned} x^{k+1} &= x^k + s_x^k \\ \lambda^{k+1} &= \lambda^k + s_\lambda^k, \end{aligned} \tag{3}$$

where $s^k = (s_x^k, s_\lambda^k) \in \mathbb{R}^n \times \mathbb{R}^m$ is the search direction.
The search direction s^k is defined on the basis of an estimate of active and nonactive constraints at a KKT point. At a given point (x, λ), we estimate the index sets of the constraints that are respectively active and nonactive at a KKT point by means of the sets $\mathcal{A} = \mathcal{A}(x, \lambda)$ and $\mathcal{N} = \mathcal{N}(x, \lambda)$ given by:

$$\begin{aligned} \mathcal{A}(x, \lambda) &= \{i : g_i(x) \geq -\zeta \lambda_i\} \\ \mathcal{N}(x, \lambda) &= \{i : g_i(x) < -\zeta \lambda_i\}, \end{aligned} \tag{4}$$

where $\zeta > 0$. We note that $\mathcal{A}, \mathcal{N}$ constitute a partition of the set $\{1, \ldots, m\}$.
In [8] it has been shown that there exists a neighborhood $\mathcal{B}$ of a KKT point $(\bar{x}, \bar{\lambda})$ such that for all $(x, \lambda) \in \mathcal{B}$ the sets $\mathcal{A}, \mathcal{N}$ satisfy:

$$\begin{aligned} &\mathcal{A}_+(\bar{x}, \bar{\lambda}) \subseteq \mathcal{A}(x, \lambda) \subseteq \mathcal{A}_0(\bar{x}) \\ &\mathcal{N}_0(\bar{x}) \subseteq \mathcal{N}(x, \lambda) \subseteq \{i : \bar{\lambda}_i = 0\}. \end{aligned} \tag{5}$$

Moreover, in [5] it has been shown that $(\bar{x}, \bar{\lambda})$ is a KKT point for Problem (1) if and only if it is a solution of the nonlinear system of equations:

$$F(x,\lambda) = \begin{bmatrix} \nabla L(x,\lambda) \\ g_{\mathcal{A}(x,\lambda)}(x) \\ \lambda_{\mathcal{N}(x,\lambda)} \end{bmatrix} = 0. \tag{6}$$

On these basis, the search direction $s^k = (s_x^k, s_\lambda^k)$ is defined as the solution of the linear system :

$$\begin{bmatrix} \nabla_x^2 L(x^k,\lambda^k) & \nabla g_{\mathcal{A}^k}(x^k) & \nabla g_{\mathcal{N}^k}(x^k) \\ \nabla g_{\mathcal{A}^k}(x^k)' & 0 & 0 \\ 0 & 0 & I \end{bmatrix} \begin{bmatrix} s_x^k \\ s_{\lambda_{\mathcal{A}^k}}^k \\ s_{\lambda_{\mathcal{N}^k}}^k \end{bmatrix} = - \begin{bmatrix} \nabla_x L(x^k,\lambda^k) \\ g_{\mathcal{A}^k}(x^k) \\ \lambda_{\mathcal{N}^k}^k \end{bmatrix}, \tag{7}$$

where $\mathcal{A}^k = \mathcal{A}(x^k, \lambda^k)$, $\mathcal{N}^k = \mathcal{N}(x^k, \lambda^k)$; if the sets $\mathcal{A}^k, \mathcal{N}^k$ are constant in a neighborhood of (x^k, λ^k), (7) turns out to be the Newton system at (x^k, λ^k) for the nonlinear system (6) [5, 8].

In [8] the following proposition has been proved:

Proposition 2.1 *Let f and g_i, $i = 1,\dots,m$, be twice continuously differentiable functions with Lipschitz continuous Hessian matrices. Let $(\bar{x}, \bar{\lambda})$ be a* KKT *point for Problem (1) which satisfies the SSOSC and assume that in $\bar{x}$ LICQ holds. Then a neighborhood $\mathcal{B}$ of $(\bar{x}, \bar{\lambda})$ exists such that if $(x^0, \lambda^0) \in \mathcal{B}$, the system (7) is well defined and the sequence $\{(x^k, \lambda^k)\}$ produced by (3), with s^k given by (7), satisfies $(x^k, \lambda^k) \in \mathcal{B}$ for all k, converges to $(\bar{x}, \bar{\lambda})$ and the rate of convergence is quadratic (superlinear for the sequence $\{x^k\}$).*

2.1 The truncated Newton-type direction

It is easy to verify that the solution of system (7) can be equivalently obtained by solving w.r.t. s_x^k and $w^k \in \mathbb{R}^{m_a}$, with $m_a = |\mathcal{A}^k|$, the smaller linear system:

$$\begin{bmatrix} \nabla_x^2 L(x^k,\lambda^k) & \nabla g_{\mathcal{A}^k}(x^k) \\ \nabla g_{\mathcal{A}^k}(x^k)' & 0 \end{bmatrix} \begin{bmatrix} s_x^k \\ w^k \end{bmatrix} = - \begin{bmatrix} \nabla f(x^k) \\ g_{\mathcal{A}^k}(x^k) \end{bmatrix}, \tag{8}$$

and then by letting:

$$s_{\lambda_{\mathcal{A}^k}}^k = w^k - \lambda_{\mathcal{A}^k}^k, \tag{9}$$

$$s_{\lambda_{\mathcal{N}^k}}^k = -\lambda_{\mathcal{N}^k}^k. \tag{10}$$

System (8) can be solved by using several techniques. Here we consider an approach based on the QR factorization of the matrix $\nabla g_{\mathcal{A}^k}(x^k)$ [7, 11]. For sake of simplicity, in the sequel of this section we make the following assumption:

Assumption 1 *For every k the matrix $\nabla g_{\mathcal{A}^k}(x^k)$ has full rank.*

We recall that the QR factorization of $\nabla g_{\mathcal{A}^k}(x^k)$ produces an $n \times m_a$ matrix Y^k, an $n \times (n - m_a)$ matrix Z^k and a $m_a \times m_a$ upper triangular matrix R^k such that:

$$\nabla g_{\mathcal{A}^k}(x^k) = (Y^k \quad Z^k) \begin{pmatrix} R^k \\ 0 \end{pmatrix}, \tag{11}$$

where Y^k and Z^k have orthonormal columns, the columns of Y^k form an orthonormal basis for $Range(\nabla g_{\mathcal{A}^k}(x^k))$ and the columns of Z^k form an orthonormal basis for $Null(\nabla g_{\mathcal{A}^k}(x^k)')$; moreover, under Assumption 1, R^k is nonsingular.

Making use of the matrices Y^k and Z^k we can write s_x^k in terms of components $s_Y^k \in \mathbb{R}^{m_a}$ and $s_Z^k \in \mathbb{R}^{n-m_a}$ in the following way:

$$s_x^k = Y^k s_Y^k + Z^k s_Z^k.$$

It is well known that, by the properties of the QR factorization, system (8) is equivalent to the system of equations in the unknowns s_Y^k, s_Z^k, w^k:

$$(R^k)' s_Y^k = -g_{\mathcal{A}^k}(x^k), \tag{12}$$

$$(Z^k)' \nabla_x^2 L(x^k, \lambda^k) Z^k s_Z^k = -(Z^k)' \left[\nabla f(x^k) + \nabla_x^2 L(x^k, \lambda^k) Y^k s_Y^k\right], \tag{13}$$

$$R^k w^k = -(Y^k)' \left[\nabla f(x^k) + \nabla_x^2 L(x^k, \lambda^k) s_x^k\right]. \tag{14}$$

¿From (12)-(14), it appears that the computation of s_Y^k, s_Z^k, w^k is decomposed into three stages. In particular the component s_Y^k guarantees that the direction s_x^k satisfies the linearized constraints $\nabla g_{\mathcal{A}^k}(x^k)' s_x^k + g_{\mathcal{A}^k} = 0$; the component s_Z^k satisfies the first equation of system (8) in the tangent space of the active constraints $Null(\nabla g_{\mathcal{A}^k}(x^k)')$, namely it solves $Z'(\nabla_x^2 L(x^k, \lambda^k) s_x^k + \nabla g_{\mathcal{A}^k}(x^k) w^k + \nabla f(x^k)) = 0$; as regard w^k, it satisfies the first equation of system (8) in the orthogonal space $Range(\nabla g_{\mathcal{A}^k}(x^k)')$, namely it solves $Y'(\nabla_x^2 L(x^k, \lambda^k) s_x^k + \nabla g_{\mathcal{A}^k}(x^k) w^k + \nabla f(x^k)) = 0$.
Systems (12) and (14) are triangular and hence s_Y^k and w^k are easy to compute. As regard system (13), its solution can be a heavy computational task when $n - m_a$ is large. However, as in unconstrained optimization, it can be solved by using a truncated scheme that yields a suitable approximate solution d_Z^k for the component s_Z^k that guarantees superlinear and global convergence properties. By using the short notation

$$B^k = (Z^k)' \nabla_x^2 L(x^k, \lambda^k) Z^k, \tag{15}$$

$$h^k = (Z^k)' \left[\nabla f(x^k) + \nabla_x^2 L(x^k, \lambda^k) Y^k s_Y^k\right], \tag{16}$$

system (13) can be written as

$$B^k s_Z^k = -h^k;$$

then d_Z^k is the solution of the perturbed system

$$B^k d_Z^k = -h^k + \rho^k$$

where $\rho^k \in \mathbb{R}^{n-m_a}$ is the residual.

When $\rho^k \neq 0$, we obtain an approximation $(d_x^k,\ d_\lambda^k)$ of the direction $(s_x^k,\ s_\lambda^k)$ that satisfies the following equations:

$$\begin{aligned}
d_x^k &= Y^k d_Y^k + Z^k d_Z^k, &(17)\\
(R^k)' d_Y^k &= -g_{\mathcal{A}^k}(x^k), &(18)\\
B^k d_Z^k &= -h^k + \rho^k &(19)\\
R^k \tilde{w}^k &= -(Y^k)' \left[\nabla f(x^k) + \nabla_x^2 L(x^k, \lambda^k) d_x^k\right] &(20)\\
d_{\lambda_{\mathcal{A}^k}}^k &= \tilde{w}^k - \lambda_{\mathcal{A}^k}^k, &(21)\\
d_{\lambda_{\mathcal{N}^k}}^k &= -\lambda_{\mathcal{N}^k}^k. &(22)
\end{aligned}$$

In conclusion the local algorithm based on the truncated scheme is defined by the iteration

$$\begin{aligned}
x^{k+1} &= x^k + d_x^k\\
\lambda^{k+1} &= \lambda^k + d_\lambda^k,
\end{aligned} \tag{23}$$

where (d_x^k, d_λ^k) is given by (17)-(22).

In order to prove convergence properties of the truncated algorithm (23), we assume that it satisfies the following assumption:

Assumption 2

- *the component d_Z^k satisfies the conditions:*

$$\begin{aligned}
(d_Z^k)' h^k &\leq -K_1 \|h^k\|^2, &(24)\\
\|d_Z^k\| &\leq K_2 \|h^k\|, &(25)
\end{aligned}$$

with K_1, K_2 positive scalars;

- *moreover, whenever B^k is positive definite, the component d_Z^k satisfies:*

$$\|B^k d_Z^k + h^k\| = \|\rho^k\| \leq r^k, \tag{26}$$

where r^k is such that

$$\lim_{k\to\infty} \frac{r^k}{\|h^k\|} = 0. \tag{27}$$

Examples of truncated algorithms which fulfil the preceding requirements can be defined by following the approaches proposed in [2, 9]. These algorithms are based on conjugate gradient methods that employ only matrix-vector products and hence do not require matrix storage and manipulation.

We can state the following result on the convergence of the truncated scheme (23).

Proposition 2.2 *Let $(\bar{x}, \bar{\lambda})$ be a KKT point for Problem (1) which satisfies LICQ and SSOSC. Assume that the Hessian matrices $\nabla^2 f$ and $\nabla^2 g_j$ for $j = 1, ...m$, are Lipschitz continuous and that (27) holds. Then there exists a neighborhood $\mathcal{B}$ of $(\bar{x}, \bar{\lambda})$ such that, if $(x^0, \lambda^0) \in \mathcal{B}$, the sequence $\{(x^k, \lambda^k)\}$ produced by the iteration (23) converges to $(\bar{x}, \bar{\lambda})$ with superlinear rate of convergence.*

The proof easily follows from Proposition 2.1 and following the same approach of inexact Newton methods for unconstrained minimization.

2.2 Some properties of the direction d^k

In this section, we preliminary develop some equalities and inequalities involving the direction d^k that will be used in this and in the next sections. Then, we establish a result on the relationships between limit points of the sequence $\{(x^k, \lambda^k)\}$ and limit points of the sequence $\{d^k\}$.

By (24), (25) and substituting the value of h^k given by (16), we get

$$\begin{aligned}
(d_Z^k)'h^k &\leq -\frac{K_1}{K_2^2}\|d_Z^k\|^2, \\
(d_Z^k)'(Z^k)'\nabla f(x^k) &\leq -\frac{K_1}{K_2^2}\|d_Z^k\|^2 - (d_Z^k)'(Z^k)'\nabla_x^2 L(x^k, \lambda^k)Y^k d_Y^k,
\end{aligned}$$

and, recalling that by the properties of the matrix Z^k it results $(Z^k)'\nabla g_{\mathcal{A}^k} = 0$, we can write:

$$(d_Z^k)'(Z^k)'(\nabla f(x^k) + \nabla g_{\mathcal{A}^k}^k \lambda_{\mathcal{A}^k}^k) \leq -\frac{K_1}{K_2^2}\|d_Z^k\|^2 - (d_Z^k)'(Z^k)'\nabla_x^2 L(x^k, \lambda^k)Y^k d_Y^k. \quad (28)$$

By using again (24) we have: $K_1\|h^k\|^2 \leq \|d_Z^k\| \, \|h^k\|$, that is:

$$\|h^k\| \leq \frac{1}{K_1}\|d_Z^k\|. \quad (29)$$

By using again (25), the properties of the matrix Z^k ($(Z^k)'\nabla f = (Z^k)'\nabla_x L - (Z^k)'\nabla g_{\mathcal{N}^k}\lambda_{\mathcal{N}^k}$) and (22) we obtain:

$$\|d_Z^k\| \leq K_2\|(Z^k)'(\nabla_x L(x^k, \lambda^k) + \nabla g_{\mathcal{N}^k}(x^k)d_{\lambda_{\mathcal{N}^k}}^k + \nabla_x^2 L(x^k, \lambda^k)Y^k d_Y^k)\|. \quad (30)$$

By using (18) we have:

$$\nabla g_{\mathcal{A}^k}' d_x^k = \begin{pmatrix} (R^k)' & 0 \end{pmatrix} \begin{pmatrix} (Y^k)' \\ (Z^k)' \end{pmatrix} (Y^k d_Y^k + Z^k d_Z^k) = (R^k)' d_Y^k = -g_{\mathcal{A}^k}(x^k). \quad (31)$$

By using (20) and (21) we have:

$$\begin{aligned}(Y^k)'\nabla g_{A^k} d_{\lambda_{A^k}} &= (Y^k)'[(Y^k \quad Z^k)\begin{pmatrix} R^k \\ 0 \end{pmatrix}\tilde{w}^k - \nabla g_{A^k}\lambda_{A^k}] \\ &= R^k\tilde{w}^k - (Y^k)'\nabla g_{A^k}\lambda_{A^k} \\ &= -(Y^k)'(\nabla f(x^k) + \nabla g_{A^k}\lambda_{A^k}) - (Y^k)'\nabla_x^2 L(x^k, \lambda^k) d_x^k,\end{aligned} \tag{32}$$

from which we get:

$$-(Y^k)'(\nabla f(x^k) + \nabla g_{A^k}\lambda_{A^k}) = (Y^k)'(\nabla_x^2 L(x^k, \lambda^k) d_x^k + \nabla g_{A^k} d_{\lambda_{A^k}}), \tag{33}$$

that can be written also as:

$$-(Y^k)'\nabla_x L(x^k, \lambda^k) = (Y^k)'(\nabla_x^2 L(x^k, \lambda^k) d_x^k + \nabla g d_\lambda). \tag{34}$$

Finally by recalling the properties of the matrices Z^k and Y^k, and by using (22), (29) and (34), we have:

$$\begin{aligned}\|\nabla_x L(x^k, \lambda^k)\| &= \|(Y^k)'\nabla_x L(x^k, \lambda^k)\| + \|(Z^k)'\nabla_x L(x^k, \lambda^k)\| \\ &\leq \|(Y^k)'\nabla_x L(x^k, \lambda^k)\| + \|(Z^k)'(\nabla_x f(x^k) + \nabla_x^2 L(x^k, \lambda^k) Y^k d_Y^k)\| \\ &\quad + \|(Z^k)'(\nabla_x^2 L(x^k, \lambda^k) Y^k d_Y^k + \nabla g_{\mathcal{N}^k}(x^k) d_{\lambda_{\mathcal{N}^k}}^k)\| \\ &\leq \|(Y^k)'(\nabla_x^2 L(x^k, \lambda^k) d_x^k + \nabla g(x^k) d_\lambda^k)\| \\ &\quad + \|(Z^k)'(\nabla_x^2 L(x^k, \lambda^k) Y^k d_Y^k + \nabla g_{\mathcal{N}^k}(x^k) d_{\lambda_{\mathcal{N}^k}}^k)\| + \frac{1}{K_1}\|d_Z^k\|.\end{aligned} \tag{35}$$

Now we can state the following proposition.

Proposition 2.3 *Let $\{(x^k, \lambda^k)\}$ be a bounded sequence.*

(a) *Suppose that d^k exists. If*

$$\lim_{k\to\infty} \|d^k\| = 0,$$

then every accumulation point $(\bar{x}, \bar{\lambda})$ of $\{(x^k, \lambda^k)\}$ is a KKT point.

(b) *Suppose that LICQ holds at any KKT point. If every accumulation point $(\bar{x}, \bar{\lambda})$ of $\{(x^k, \lambda^k)\}$ is a KKT point, then eventually the matrices $\nabla g_{A^k}(x^k)$ are full rank, the direction d^k is defined and it satisfies*

$$\lim_{k\to\infty} \|d^k\| = 0.$$

Proof. (a) If $(\bar{x}, \bar{\lambda})$ is an accumulation point of the sequence $\{(x^k, \lambda^k)\}$ and if we take into account that the number of subsets of $\{1, \ldots, m\}$ is finite and, hence, also the number of possible different estimates $\mathcal{A}^k$ and $\mathcal{N}^k$ is finite, we can extract a subsequence (that without of loss of generality we recall $\{(x^k, \lambda^k)\}$) such that:

$$\begin{array}{lll} (x^k, \lambda^k) & \to & (\bar{x}, \bar{\lambda}) \\ \mathcal{A}^k = \bar{\mathcal{A}}, & \mathcal{N}^k = \bar{\mathcal{N}}, & \text{for all } k. \end{array}$$

Now, passing to the limit in (22), (18) and (35)

$$\bar{\lambda}_{\bar{\mathcal{N}}} = 0, \tag{36}$$

$$g_{\bar{\mathcal{A}}}(\bar{x}) = 0, \tag{37}$$

$$\nabla_x L(\bar{x}, \bar{\lambda}) = 0. \tag{38}$$

Recalling the definitions of $\mathcal{A}^k$ and $\mathcal{N}^k$ we have:

$$g_{\bar{\mathcal{N}}}(\bar{x}) < 0, \tag{39}$$

$$\bar{\lambda}_{\bar{\mathcal{A}}} \geq 0. \tag{40}$$

Hence (36)-(40) imply that $(\bar{x}, \bar{\lambda})$ is a KKT point.
(b) We prove this part by contradiction. If point (b) does not hold we can extract subsequences, that we recall again $\{(x^k, \lambda^k)\}$, $\{d^k\}$, such that

$$\lim_{k\to\infty} (x^k, \lambda^k) = (\bar{x}, \bar{\lambda}),$$

$$\lim_{k\to\infty} \|d^k\| > 0, \tag{41}$$

where $(\bar{x}, \bar{\lambda})$ is a KKT point.
Since the number of possible different estimates $\mathcal{A}^k$ and $\mathcal{N}^k$ is finite, also in this case we assume, without loss of generality, that $\mathcal{A}^k = \bar{\mathcal{A}}$ and $\mathcal{N}^k = \bar{\mathcal{N}}$ for every k. Furthermore, for k sufficiently large by (5) we have:

$$\bar{\mathcal{A}} \subseteq \{i : g_i(\bar{x}) = 0\}, \tag{42}$$

$$\bar{\mathcal{N}} \subseteq \{i : \bar{\lambda}_i = 0\}. \tag{43}$$

The assumption that the gradients of the active constraints are linearly independent at $\bar{x}$ ensures that, for sufficiently large k, the gradients $\nabla g_i(x^k)$, for $i \in \bar{\mathcal{A}}$, are linearly independent, so that the matrix R^k is nonsingular. For k sufficiently large, we get from (42) $g_{\bar{\mathcal{A}}}(x^k) = 0$, and from (18) we get $d_Y^k = 0$; hence $\lim_{k\to\infty} d_Y^k = \bar{d}_Y = 0$. ¿From (43) we get, for sufficiently large k, $\lambda_{\bar{\mathcal{N}}}^k = 0$, and, by (22), $d_{\lambda_{\bar{\mathcal{N}}}}^k = 0$; hence, $\lim_{k\to\infty} d_{\lambda_{\bar{\mathcal{N}}}}^k = \bar{d}_{\lambda_{\bar{\mathcal{N}}}} = 0$.
Then, the term $(Z^k)'(\nabla_x L^k + \nabla g_{\bar{\mathcal{N}}} d_{\lambda_{\bar{\mathcal{N}}}})$ tends to zero due to the fact that, by assumption, $\{(x^k, \lambda^k)\} \to (\bar{x}, \bar{\lambda})$; hence, using (30), we get also $\lim_{k\to\infty} d_Z^k = \bar{d}_Z = 0$. Therefore, we have $\lim_{k\to\infty} d_x^k = \bar{d}_x = 0$.
Now, taking the limit of the following equation:

$$(Y^k)'(\nabla f^k + \nabla g_{\bar{\mathcal{A}}}^k \lambda_{\bar{\mathcal{A}}}^k) = (Y^k)' \nabla_x L^k$$

and using (32) we obtain $\lim_{k\to\infty} d^k_{\lambda_{\bar{A}}} = \bar{d}_{\lambda_{\bar{A}}} = 0$. Hence it results $\lim_{k\to\infty} d^k = \bar{d} = 0$ that contradicts (41). □

As a consequence of Proposition 2.3 we have that either a $\bar{k}$ exists such that $d^{\bar{k}} = 0$ and $(x^{\bar{k}}, \lambda^{\bar{k}})$ is a KKT point or that $d^k \neq 0$ for any finite value of k.

3 The globalization scheme

The iteration (23) is only locally convergent. To force the global convergence we adopt a linesearch procedure of the type

$$\begin{pmatrix} x^{k+1} \\ \lambda^{k+1} \end{pmatrix} = \begin{pmatrix} x^k \\ \lambda^k \end{pmatrix} + \mu^k \begin{pmatrix} d^k_x \\ d^k_\lambda \end{pmatrix}, \tag{44}$$

where the stepsize μ^k is determined making use of a merit function in the primal and dual variables (x, λ).

As usual, the merit function performs the tasks of measuring the progress of the iterates towards KKT point of Problem (1), and of providing alternative descent directions when the direction d^k is not suitable; in addition, the particular merit function used here ensures the boundedness of the sequence $\{(x^k, \lambda^k)\}$.

We use here the *exact* augmented Lagrangian function $L_a(x, \lambda; \epsilon)$ introduced in [4]. We first introduce shortly the function; then we prove a descent property of the direction d^k with respect to L_a in a neighborhood of a KKT point.

3.1 The augmented Lagrangian function

Let $\alpha, s \in \mathbb{R}$ be given scalars such that $\alpha > 0$ and $s \geq 2$, and let:

$$a(x) = \alpha - \sum_{i=1}^{m} \max\{g_i(x), 0\}^s; \tag{45}$$

it turns out that the set

$$\mathcal{P} = \{x \in \mathbb{R}^n : \; a(x) > 0\}$$

is an open perturbation of the feasible set $\mathcal{F}$.

Let us now consider the function:

$$p(x, \lambda) = \frac{a(x)}{1 + \|\lambda\|^2};$$

then we can define on the set $\mathcal{P} \times \mathbb{R}^m$ the augmented Lagrangian function:

$$\begin{aligned} L_a(x, \lambda; \epsilon) \;&=\; f(x) + \lambda' \max\{g(x), -\epsilon p(x, \lambda)\lambda\} \qquad (46) \\ &+ \frac{1}{2\epsilon p(x, \lambda)} \|\max\{g(x), -\epsilon p(x, \lambda)\lambda\}\|^2 + \|\nabla g(x)' \nabla_x L(x, \lambda) + G(x)^2 \lambda\|^2, \end{aligned}$$

where ϵ is a penalty parameter such that $\epsilon \in (0, E]$, with E an arbitrary positive constant. Note that the term $1/p(x,\lambda)$ is a barrier term that goes to $+\infty$ when (x,λ) approaches the boundary of the set $\mathcal{P} \times \mathbb{R}^m$, where the function $L_a(x,\lambda;\epsilon)$ is defined.

The function L_a has been extensively studied in [4]. It has been shown there that L_a is an *exact augmented Lagrangian function* in the sense that for sufficiently small values of ϵ, there exists a one-to-one correspondence between local (global) solutions of Problem (1) and local (global) minimizers of L_a, as well as between KKT points of Problem (1) and stationary points of L_a.

The exactness properties of the function L_a can be stated under the following assumptions:

Assumption 3 *One of the two following conditions is satisfied:*

(a) *$\mathcal{P}$ is a bounded set;*

(b) *a point $\hat{x} \in \mathcal{F}$ is known and $f(x)$ is such that for any $\{x_k\} \subseteq \mathcal{P}$ with $\|x_k\| \to \infty$ it results $f(x_k) \to \infty$.*

Assumption 4 *For every $x \in \mathcal{F}$, LICQ holds.*

Assumption 5 *One of the two following conditions is satisfied:*

(a) *at every point $x \in \mathcal{P}$, if:*

$$\sum_{i=1}^{m} c_i(x)\nabla g_i(x) = 0,$$

with

$$c_i(x) = \left[1 + \frac{s}{2}\frac{\|\max\{g(x),0\}\|^2 \max\{g_i(x),0\}^{(s-2)}}{a(x)}\right]\max\{g_i(x),0\},$$

then $x \in \mathcal{F}$;

(b) *a point $\hat{x} \in \mathcal{F}$ is known.*

These assumptions have been discussed in [4]. In the sequel, we assume that Assumptions 3, 4 and 5 hold.

From the computational point of view, we are interested in the level set

$$\Omega(x_0,\lambda_0;\epsilon) = \{(x,\lambda) \in \mathcal{P} \times \mathbb{R}^m : \; L_a(x,\lambda;\epsilon) \le L_a(x_0,\lambda_0;\epsilon)\}.$$

where $(x_0,\lambda_0) \in \mathcal{P} \times \mathbb{R}^m$, and, in particular, $(x_0,\lambda_0) \in \mathcal{F} \times \mathbb{R}^m$ if a feasible point x_0 is known.

The following propositions state the properties of the function L_a that are of interest in the sequel.

Proposition 3.1

(a) *For every* $\epsilon > 0$ *the level set* $\Omega(x_0, \lambda_0; \epsilon)$ *is compact.*

(b) *There exists a compact set* $\Omega_x \subset \mathcal{P}$ *such that* $\Omega(x_0, \lambda_0; \epsilon) \subseteq \Omega_x \times \mathrm{IR}^m$ *for all* $\epsilon \in (0, E]$.

Proposition 3.2 *There exists an* $\bar{\epsilon}$ *such that for all* $\epsilon \in (0, \bar{\epsilon}]$ *and all* $(x, \lambda) \in \Omega(x_0, \lambda_0; \epsilon)$ *we have:*

$$\|\nabla L_a(x, \lambda; \epsilon)\| \geq \|\max\{g(x), -\epsilon p(x, \lambda)\lambda\}\|.$$

Proposition 3.3

(a) *Let* $(\bar{x}, \bar{\lambda}) \in \mathcal{P} \times \mathrm{IR}^m$, *be a stationary point for* $L_a(x, \lambda; \epsilon)$, *and assume that* $\max\{g(\bar{x}), -\epsilon p(\bar{x}, \bar{\lambda})\bar{\lambda}\} = 0$. *Then,* $(\bar{x}, \bar{\lambda})$ *is a KKT pair of Problem (1).*

(b) *There exists an* $\bar{\epsilon} > 0$ *such that for all* $\epsilon \in (0, \bar{\epsilon}]$, *if* $(\bar{x}, \bar{\lambda}) \in \Omega(x_0, \lambda_0; \epsilon)$ *is a stationary point of* $L_a(x, \lambda; \epsilon)$, *the point* $(\bar{x}, \bar{\lambda})$ *is a KKT point for Problem (1).*

In order to establish a connection between the direction d and the function L_a, we assume, from now on, that in the estimates of the sets of active and nonactive constraints given by (4) the coefficient ζ is specified as $\epsilon p(x, \lambda)$; that is we assume, from now on, that the sets $\mathcal{A}$ and $\mathcal{N}$ are given by:

$$\begin{aligned} \mathcal{A}(x, \lambda; \epsilon) &= \{i : g_i(x) \geq -\epsilon p(x, \lambda)\lambda_i\} \\ \mathcal{N}(x, \lambda; \epsilon) &= \{i : g_i(x) < -\epsilon p(x, \lambda)\lambda_i\}. \end{aligned} \tag{47}$$

3.2 The descent property of the direction d^k w.r.t. L_a

¿From the definition and the differentiability assumptions on f and g, it follows that the function $L_a(x, \lambda; \epsilon)$ is an SC^1 function for all $(x, \lambda) \in \mathcal{P} \times \mathrm{IR}^m$, that is a continuously differentiable function with a semismooth gradient ∇L_a (see [13]). The gradient of L_a is obtained from (46) as:

$$\begin{aligned} \nabla_x L_a(x, \lambda; \epsilon) &= \nabla_x L(x, \lambda) + \frac{1}{\epsilon p(x, \lambda)} \nabla g(x) \max\{g(x), -\epsilon p(x, \lambda)\lambda\} \qquad (48) \\ &\quad + \frac{s}{2\epsilon a(x) p(x, \lambda)} \|\max\{g(x), -\epsilon p(x, \lambda)\lambda\}\|^2 \sum_{i:\ g_i > 0} \nabla g_i(x) g_i(x)^{s-1} \\ &\quad + 2\left[\nabla_x^2 L(x, \lambda)\nabla g(x) + \sum_{i=1}^m \nabla_x^2 g_i(x) \nabla_x L(x, \lambda) e_i' + 2\nabla g(x) G(x) \Lambda\right] \\ &\quad \times \left[\nabla g(x)' \nabla_x L(x, \lambda) + G(x)^2 \lambda\right], \\ \nabla_\lambda L_a(x, \lambda; \epsilon) &= \max\{g(x), -\epsilon p(x, \lambda)\lambda\} + \frac{1}{\epsilon a(x)} \|\max\{g(x), -\epsilon p(x, \lambda)\lambda\}\|^2 \lambda \qquad (49) \\ &\quad + 2\left[\nabla g(x)' \nabla g(x) + G^2(x)\right]\left[\nabla g(x)' \nabla_x L(x, \lambda) + G(x)^2 \lambda\right], \end{aligned}$$

where e_i denotes the $i-th$ column of the $m \times m$ identity matrix.

In order to establish the descent property of truncated direction d^k, we introduce the following function $\phi : \mathbb{R}^{2n+m} \to \mathbb{R}$:

$$\begin{aligned}\phi(x,\lambda,d;\epsilon) &= d_x'(\nabla f + \nabla g_A \lambda_A + \frac{1}{\epsilon p}\nabla g_A g_A - \nabla g_A d_{\lambda_A}) \\ &\quad + \epsilon p \|d_{\lambda_N}\|^2 - 2\|\nabla g' \nabla_x^2 L d_x + (\nabla g' \nabla g + G^2) d_\lambda\|^2,\end{aligned} \tag{50}$$

and we preliminarily prove the following technical result:

Proposition 3.4 *Suppose that Assumption 4 holds. Let $\{(x^k,\lambda^k)\}$ be a bounded sequence and assume that the direction d^k obtained by (17)-(22) exists and that the sequence $\{d^k\}$ is bounded. Then $\bar{\epsilon} > 0$ and $\eta > 0$ exist, such that for all $\epsilon \in (0,\bar{\epsilon}]$, for all (x^k,λ^k) and d^k it results:*

$$\phi(x^k,\lambda^k,d^k;\epsilon) \le -\eta \|d^k\|^2. \tag{51}$$

Proof. The proof is by contradiction. Assume that there exist sequences $\{\epsilon^j\}$, $\{\eta^j\}$ $\{(x^j,\lambda^j)\} \subseteq \{(x^k,\lambda^k)\}$, and $\{d^j\} \subseteq \{d^k\}$ such that

$$\epsilon^j \to 0, \qquad \eta^j \to 0, \qquad (x^j,\lambda^j) \to (\hat{x},\hat{\lambda}), \qquad p^j \to \hat{p}, \tag{52}$$

$$d^j \neq 0, \qquad \frac{d^j}{\|d^j\|} \to \hat{d} \neq 0, \qquad \|\hat{d}\| = 1, \tag{53}$$

$$\phi(x^j,\lambda^j,d^j;\epsilon^j) > -\eta^j \|d^j\|^2. \tag{54}$$

Now recalling (17),(28), (31), and (33) we have:

$$\begin{aligned}&(d_x^j)'(\nabla f^j + \nabla g_{A^j}^j \lambda_{A^j}^j + \frac{1}{\epsilon^j p^j}\nabla g_{A^j}^j g_{A^j}^j) - (d_{\lambda_{A^j}}^j)'\nabla g_{A^j}^j{}' d_x^j \\ &= \left(Z^j d_Z^j + Y^j d_Y^j\right)'(\nabla f^j + \nabla g_{A^j}^j \lambda_{A^j}^j) + \frac{1}{\epsilon^j p^j}(d_x^j)'\nabla g_{A^j}^j g_{A^j}^j \\ &\quad -(d_{\lambda_{A^j}}^j)'\nabla g_{A^j}^j{}' Y^j d_Y^j \\ &\le -\frac{K_1}{K_2^2}\|d_Z^j\|^2 - (Z^j d_Z^j)'\nabla_x^2 L(x^j,\lambda^j) Y^j d_Y^j \\ &\quad -(Y^j d_Y^j)'(\nabla_x^2 L(x^j,\lambda^j) d_x^j + \nabla g_{A^j} d_{\lambda_{A^j}}^j) - \frac{1}{\epsilon^j p^j}\|g_{A^j}^j\|^2 - (d_{\lambda_{A^j}}^j)'\nabla g_{A^j}^j{}' Y^j d_Y^j \\ &= -\frac{K_1}{K_2^2}\|d_Z^j\|^2 - 2(Z^j d_Z^j)'\nabla_x^2 L(x^j,\lambda^j) Y^j d_Y^j - (Y^j d_Y^j)'\nabla_x^2 L(x^j,\lambda^j) Y^j d_Y^j \\ &\quad -2(Y^j d_Y^j)'\nabla g_{A^j} d_{\lambda_{A^j}}^j - \frac{1}{\epsilon^j p^j}\|g_{A^j}^j\|^2.\end{aligned} \tag{55}$$

Therefore, by using (54), the definition of ϕ, and (55), we have:

$$\frac{K_1}{K_2^2}\|d_Z^j\|^2 + 2(Z^j d_Z^j)'\nabla_x^2 L(x^j,\lambda^j) Y^j d_Y^j + (Y^j d_Y^j)'\nabla_x^2 L(x^j,\lambda^j) Y^j d_Y^j$$

$$+2(Y^j d_Y^j)'\nabla g_{\mathcal{A}^j}^j d_{\lambda_{\mathcal{A}^j}}^j + \frac{1}{\epsilon^j p^j}\|g_{\mathcal{A}^j}^j\|^2 \tag{56}$$
$$+2\|\nabla g_{\mathcal{A}^j}^j{}'\nabla_x^2 L(x^j,\lambda^j)d_x^j + (\nabla g^{j\prime}\nabla g^j + G^{j2})d_\lambda^j\|^2 - \epsilon^j p^j\|d_{\lambda_{\mathcal{N}^j}}^j\|^2 - \eta^j\|d^j\|^2 < 0.$$

Then, dividing (56) by $\|d^j\|^2$ and denoting $\hat{d}^j = d^j/\|d^j\|$, we have:

$$\frac{K_1}{K_2^2}\|\hat{d}_Z^j\|^2 + 2(Z^j\hat{d}_Z^j)'\nabla_x^2 L(x^j,\lambda^j)Y^j\hat{d}_Y^j + (Y^j\hat{d}_Y^j)'\nabla_x^2 L(x^j,\lambda^j)Y^j\hat{d}_Y^j$$
$$+2(Y^j\hat{d}_Y^j)'\nabla g_{\mathcal{A}^j}^j\hat{d}_{\lambda_{\mathcal{A}^j}}^j + 2\|\nabla g^{j\prime}\nabla_x^2 L(x^j,\lambda^j)\hat{d}_x^j + (\nabla g^{j\prime}\nabla g^j + G^{j2})\hat{d}_\lambda^j\|^2$$
$$+\frac{1}{\epsilon p^j}\left(\frac{\|g_{\mathcal{A}^j}^j\|}{\|d^j\|}\right)^2 - \epsilon^j p^j\|\hat{d}_{\lambda_{\mathcal{N}^j}}^j\|^2 - \eta^j < 0. \tag{57}$$

Without loss of generality, we assume now that $\mathcal{A}^j$ and $\mathcal{N}^j$ are constant so that we can write:

$$\mathcal{A}^j = \hat{\mathcal{A}}, \qquad \mathcal{N}^j = \hat{\mathcal{N}}.$$

By the continuity assumptions, by the fact that the sequences $\{(x^j,\lambda^j)\}$, $\{p^j\}$, $\{Z^j\}$, $\{Y^j\}$ and $\{\hat{d}^j\}$ are bounded, we obtain from (52) and (57):

$$\lim_{j\to\infty}\frac{\|g_{\hat{\mathcal{A}}}(x^j)\|}{\|d^j\|} = 0, \tag{58}$$

which, using the assumption that the sequence $\{d^j\}$ is bounded, yields

$$g_{\hat{\mathcal{A}}}(\hat{x}) = 0. \tag{59}$$

This limit shows that the index set $\hat{\mathcal{A}}$ is such that:

$$\hat{\mathcal{A}} \subseteq \mathcal{A}_0(\hat{x}). \tag{60}$$

Recalling the definition (47) of the index set $\mathcal{N}(x,\lambda;\epsilon)$, the boundedness of the sequences $\{(x^j,\lambda^j)\}$ and $\{p^j\}$, and (52), we obtain:

$$\lim_{k\to\infty} g_{\hat{\mathcal{N}}}(x^j) = g_{\hat{\mathcal{N}}}(\hat{x}) < 0; \tag{61}$$

then (59) and (61) imply that

$$\hat{x} \in \mathcal{F}. \tag{62}$$

Now (60), (62), Assumption 4 and the properties of the QR factorization imply that, for sufficiently large indeces j, we have:

$$Y^j d_Y^j = -\nabla g_{\hat{\mathcal{A}}}^j(\nabla g_{\hat{\mathcal{A}}}^j{}'\nabla g_{\hat{\mathcal{A}}}^j)^{-1}\frac{g_{\hat{\mathcal{A}}}^j}{\|d^j\|},$$

which, recalling (58) implies

$$\lim_{j\to\infty} Y^j d_Y^j = 0. \tag{63}$$

For $j \to \infty$, (53),(56) and (63) yield:

$$\lim_{j\to\infty} \hat{d}_Z^j = 0, \tag{64}$$

$$\left\|\nabla g(\hat{x})'\nabla_x^2 L(\hat{x},\hat{\lambda})\hat{d}_x + \left(\nabla g(\hat{x})'\nabla g(\hat{x}) + \begin{bmatrix} 0 & 0 \\ 0 & G_{\hat{\mathcal{N}}}(\hat{x})^2 \end{bmatrix}\right)\hat{d}_\lambda\right\| = 0. \tag{65}$$

Now, (17), (63) and (64) yield:

$$\hat{d}_x = 0. \tag{66}$$

By (65) and (66) we obtain:

$$\left(\nabla g(\hat{x})'\nabla g(\hat{x}) + \begin{bmatrix} 0 & 0 \\ 0 & G_{\hat{\mathcal{N}}}(\hat{x})^2 \end{bmatrix}\right)\hat{d}_\lambda = 0. \tag{67}$$

Since $\hat{x} \in \mathcal{F}$, by Assumption 4 it turns out that the coefficient matrix in (67) is non singular [3]; therefore

$$\hat{d}_\lambda = 0. \tag{68}$$

Finally (66), (68) contradict the fact that $\|\hat{d}\| = 1$. □

Now we can prove that, by controlling the magnitude of the residual in the truncated Newton scheme, it is possible to ensure that the direction d^k is a descent direction for L_a in a neighborhood of a KKT point of Problem (1).

Proposition 3.5 *Suppose that Assumption 4 holds. Let $\{(x^k, \lambda^k)\}$ be a bounded sequence and assume that the direction d^k obtained by (17)-(22) exists and that the sequence $\{d^k\}$ is bounded. Furthermore assume that d^k satisfies Assumption 2 and that*

$$r^k \leq \theta(x^k, \lambda^k)\|h^k\|, \tag{69}$$

where θ is a continuous function such that $\theta(\bar{x}, \bar{\lambda}) = 0$ for every KKT point $(\bar{x}, \bar{\lambda})$. Let $\bar{\epsilon}$ and η be values for which (51) of Proposition 3.4 holds. Then for every $\epsilon \in (0, \bar{\epsilon}]$ and for every KKT point of Problem (1) which satisfies the SSOSC there exists a neighborhood $\mathcal{B}_\epsilon$ of $(\bar{x}, \bar{\lambda})$ such that:

$$\nabla L_a(x^k, \lambda^k; \epsilon)'d^k \leq -\frac{\eta}{2}\|d^k\|^2, \tag{70}$$

for all $(x^k, \lambda^k) \in \mathcal{B}_\epsilon$.

Proof. Preliminarily we observe that, by the definition (47)of the sets $\mathcal{A}(x^k, \lambda^k; \epsilon)$ and $\mathcal{N}(x^k, \lambda^k; \epsilon)$, the term $\max\{g(x), -\epsilon p(x,\lambda)\lambda\}$ that appears in the gradient formulas (48),(49) can be partitioned as:

$$\max\{g(x^k), -\epsilon p(x^k,\lambda^k)\lambda^k\} = \begin{pmatrix} g_{A^k}(x^k) \\ -\epsilon p(x^k,\lambda^k)\lambda^k_{\mathcal{N}} \end{pmatrix}.$$

Recalling also (22) and (31) we can write, for the directional derivative $\nabla L_a(x^k, \lambda^k; \epsilon)'d^k$ (we omit the superscript k):

$$\begin{aligned} d'\nabla L_a &= d_x'\nabla_x L_a + d'_{\lambda_{\mathcal{A}}}\nabla_{\lambda_{\mathcal{A}}} L_a + d'_{\lambda_{\mathcal{N}}}\nabla_{\lambda_{\mathcal{N}}} L_a \\ &= d_x'(\nabla_x f + \nabla g_{\mathcal{A}}\lambda_{\mathcal{A}}) - \frac{1}{\epsilon p}\|\nabla g_{\mathcal{A}}' d_x\|^2 \\ &+ \frac{s}{2\epsilon a p}(g_{\mathcal{A}}'\nabla g_{\mathcal{A}}' d_x + \epsilon^2 p^2 \lambda_{\mathcal{N}}' d_{\lambda_{\mathcal{N}}}) \sum_{i:\ g_i>0} g_i^{s-1}\nabla g_i' d_x \\ &+ 2d_x'\Big[\nabla_x^2 L \nabla g + \sum_{i=1}^m \nabla_x^2 g_i \nabla_x L e_i' + 2\nabla g G\Lambda\Big]\Big[\nabla g'\nabla_x L + G^2\lambda\Big], \\ &- d'_{\lambda_{\mathcal{A}}}\nabla g_{\mathcal{A}}' d_x - \frac{1}{\epsilon a}\Big[d'_{\lambda_{\mathcal{A}}}\lambda_{\mathcal{A}} g_{\mathcal{A}}'\nabla g_{\mathcal{A}}' d_x + \epsilon^2 p^2 d'_{\lambda_{\mathcal{A}}}\lambda_{\mathcal{A}}\lambda_{\mathcal{N}}' d_{\lambda_{\mathcal{N}}}\Big]. \\ &+ 2d'_{\lambda_{\mathcal{A}}}\Big[\nabla g_{\mathcal{A}}'\nabla g + (\,G_{\mathcal{A}}^2 \quad 0\,)\Big]\Big[\nabla g'\nabla_x L + G^2\lambda\Big] \\ &+ 2d'_{\lambda_{\mathcal{N}}}\Big[\nabla g_{\mathcal{N}}'\nabla g + (\,0 \quad G_{\mathcal{N}}^2\,)\Big]\Big[\nabla g'\nabla_x L + G^2\lambda\Big] \\ &+ \epsilon p\|d_{\lambda_{\mathcal{N}}}\|^2 - \frac{1}{\epsilon a}\Big[\|g_{\mathcal{A}}\|^2 + \epsilon^2 p^2\|\lambda_{\mathcal{N}}\|^2\Big]\|d_{\lambda_{\mathcal{N}}}\|^2, \end{aligned}$$

which can be rewritten as:

$$\begin{aligned} d'\nabla L_a &= d_x'(\nabla_x f + \nabla g_{\mathcal{A}}\lambda_{\mathcal{A}}) - \frac{1}{\epsilon p}\|\nabla g_{\mathcal{A}}' d_x\|^2 - d'_{\lambda_{\mathcal{A}}}\nabla g_{\mathcal{A}}' d_x + \epsilon p\|d_{\lambda_{\mathcal{N}}}\|^2 \\ &+ \frac{s}{2\epsilon a p}(g_{\mathcal{A}}'\nabla g_{\mathcal{A}}' d_x + \epsilon^2 p^2 \lambda_{\mathcal{N}}' d_{\lambda_{\mathcal{N}}}) \sum_{i:\ g_i>0} g_i^{s-1}\nabla g_i' d_x \\ &+ 2\Big[\nabla g'\nabla_x^2 L d_x + (\nabla g'\nabla g + G^2)d_\lambda\Big]'\Big[\nabla g'\nabla_x L + G^2\lambda\Big] \\ &+ 2d_x'\Big[\sum_{i=1}^m \nabla_x^2 g_i \nabla_x L e_i' + 2\nabla g G\Lambda\Big]\Big[\nabla g'\nabla_x L + G^2\lambda\Big] \\ &- \frac{1}{\epsilon a}\Big[d'_{\lambda_{\mathcal{A}}}\lambda_{\mathcal{A}} g_{\mathcal{A}}'\nabla g_{\mathcal{A}}' d_x + \epsilon^2 p^2 d'_{\lambda_{\mathcal{A}}}\lambda_{\mathcal{A}}\lambda_{\mathcal{N}}' d_{\lambda_{\mathcal{N}}}\Big] \\ &- \frac{1}{\epsilon a}\Big[\|g_{\mathcal{A}}\|^2 + \epsilon^2 p^2\|\lambda_{\mathcal{N}}\|^2\Big]\|d_{\lambda_{\mathcal{N}}}\|^2. \end{aligned}$$

Recalling the definition of ϕ, given by (50), and using again (22) and (31) we have:

$$\begin{aligned} d'\nabla L_a &= \phi(x,\lambda,d;\epsilon) \\ &+ 2\Big[\nabla g'\nabla_x^2 L d_x + (\nabla g'\nabla g + G^2)d_\lambda\Big]\Big[\nabla g'(\nabla_x L + \nabla_x^2 L d_x + \nabla g d_\lambda)\Big] \end{aligned} \tag{71}$$

$$
\begin{aligned}
&+2\Big[\nabla g'\nabla_x^2 L d_x + (\nabla g'\nabla g + G^2)d_\lambda\Big]\Big[G^2 d_\lambda - \begin{pmatrix} G_{\mathcal{A}}\Lambda_{\mathcal{A}}\nabla g'_{\mathcal{A}} d_x \\ G_{\mathcal{N}}^2 d_{\lambda_{\mathcal{N}}} \end{pmatrix}\Big] \\
&+2d'_x\Big[\sum_{i=1}^{m}\nabla_x^2 g_i \nabla_x L e'_i + 2\nabla g G\Lambda\Big]\Big[-\nabla g'(\nabla_x^2 L d_x + \nabla g d_\lambda) - \begin{pmatrix} G_{\mathcal{A}}\Lambda_{\mathcal{A}}\nabla g'_{\mathcal{A}} d_x \\ G_{\mathcal{N}}^2 d_{\lambda_{\mathcal{N}}} \end{pmatrix}\Big] \\
&+2d'_x\Big[\sum_{i=1}^{m}\nabla_x^2 g_i \nabla_x L e'_i + 2\nabla g G\Lambda\Big]\Big[\nabla g'(\nabla_x L + \nabla_x^2 L d_x + \nabla g d_\lambda)\Big] \\
&-\frac{1}{\epsilon a}\Big[d'_{\lambda_{\mathcal{A}}}\lambda_{\mathcal{A}} g'_{\mathcal{A}}\nabla g'_{\mathcal{A}} d_x + \epsilon^2 p^2 d'_{\lambda_{\mathcal{A}}}\lambda_{\mathcal{A}}\lambda'_{\mathcal{N}} d_{\lambda_{\mathcal{N}}}\Big] - \frac{1}{\epsilon a}\Big[\|g_{\mathcal{A}}\|^2 + \epsilon^2 p^2\|\lambda_{\mathcal{N}}\|^2\Big]\|d_{\lambda_{\mathcal{N}}}\|^2 \\
&+\frac{s}{2\epsilon a p}(g'_{\mathcal{A}}\nabla g'_{\mathcal{A}} d_x + \epsilon^2 p^2 \lambda'_{\mathcal{N}} d_{\lambda_{\mathcal{N}}}) \sum_{i:\ g_i>0} g_i^{s-1}\nabla g'_i d_x.
\end{aligned}
$$

Now, by the properties of the QR decomposition, taking into account (22) and (34), it results:

$$
\begin{aligned}
&\|\nabla_x L(x,\lambda) + \nabla_x^2 L(x,\lambda) d_x + \nabla g(x) d_\lambda\| \\
&= \left\| \begin{pmatrix} Y'(\nabla_x L(x,\lambda) + \nabla_x^2 L(x,\lambda) d_x + \nabla g(x) d_\lambda) \\ Z'(\nabla_x L(x,\lambda) + \nabla_x^2 L(x,\lambda) d_x + \nabla g(x) d_\lambda) \end{pmatrix} \right\| \\
&= \|B d_Z + h\|.
\end{aligned} \tag{72}
$$

Since by SSOSC and (5) there exists a neighborhood of $(\bar{x}, \bar{\lambda})$ such that the matrix B given in (15) is positive definite, by (26) we have $\|Bd_Z + h\| \leq r$. Now, by using (29) and (69) we get:

$$
\|\nabla_x L + \nabla_x^2 L d_x + \nabla g d_\lambda\| \leq \frac{\theta(x,\lambda)}{K_1}\|d_Z\|.
$$

In conclusion, by Proposition 3.4, we get from (71) for the directional derivative:

$$
d^{k\prime}\nabla L_a(x^k, \lambda^k; \epsilon) \leq -\eta\|d^k\|^2 + \Phi(x^k, \lambda^k; \epsilon^k)\|d^k\|^2
$$

As $(x^k, \lambda^k) \to (\bar{x}, \bar{\lambda})$, we have that $g_{\mathcal{A}^k}(x^k) \to 0$, $\lambda^k_{\mathcal{N}^k} \to 0$, $\nabla_x L(x^k, \lambda^k) \to 0$, $G(x^k)\Lambda^k \to 0$ and hence $\Phi(x^k, \lambda^k; \epsilon^k) \to 0$. This concludes the proof. □

4 The algorithm model

In this section we describe an algorithm model, which employs the truncated Newton-type direction d^k introduced in Section 2 and the exact augmented Lagrangian function L_a considered in Section 3 so as to ensure global convergence, with superlinear convergence rate, towards KKT points of Problem (1). In particular, in order to guarantee global convergence, the algorithm is provided with an automatic adjustment rule for the penalty parameter ϵ.

Algorithm Model Erice98

Data: $(\tilde{x}, \tilde{\lambda}) \in \mathbb{R}^n \times \mathbb{R}^m$, $\bar{\epsilon} > 0$, $\delta > 0$; *choose* α *such that* $\tilde{x} \in \mathcal{P}$.

Step 1. (Initialization) *Set* $k = 0$, $(x^0, \lambda^0) = (\tilde{x}, \tilde{\lambda})$, $\epsilon = \bar{\epsilon}$.

Step 2. (Test for convergence) *If* (x^k, λ^k) *is a KKT point stop.*

Step 3. (Truncated Newton-type direction) *If the direction* d^k *can be computed by the scheme (17)-(22) then:*

- *if*

$$\nabla L_a(x^k, \lambda^k; \epsilon)' d^k \le -\frac{\epsilon}{2}\|d^k\|^2, \tag{73}$$

then go to Step 6;

- *else:*
  - *if* $\phi(x^k, \lambda^k, p^k; \epsilon) > -\epsilon\|d^k\|^2$, *and* $\|d^k\| \le \delta$ *and* $\|\nabla g(x^k)'(\nabla_x L(x^k, \lambda^k) + G(x^k)^2\lambda)\| \le \delta$, *then go to Step 5.*

Step 4. (Gradient related direction) *Compute a direction* d^k *such that*

$$\nabla L_a(x^k, \lambda^k; \epsilon)' d^k \le -c_1\|\nabla L_a(x^k, \lambda^k; \epsilon)\|^2, \qquad \|d^k\| \le c_2\|\nabla L_a(x^k, \lambda^k; \epsilon)\|, \tag{74}$$

with $c_1 > 0$ *and* $c_2 > 0$;

- *if* $\nabla L_a(x^k, \lambda^k; \epsilon)' d^k \le -\epsilon\|\max\{g(x^k), -\epsilon p(x^k, \lambda^k)\lambda^k\}\|^2$ *then go to Step 6.*

Step 5. (Update the penalty parameter) *Choose* $\epsilon \in (0, \rho\epsilon)$ *with* $\rho \in (0, 1)$;

- *if* $L_a(\tilde{x}, \tilde{\lambda}; \epsilon) \le L_a(x^k, \lambda^k; \epsilon)$ *then set* $(x^{k+1}, \lambda^{k+1}) = (\tilde{x}, \tilde{\lambda})$,
- *else set* $(x^{k+1}, \lambda^{k+1}) = (x^k, \lambda^k)$.

Set $k = k + 1$ *and go Step 3.*

Step 6. (Line search) *Let* $\sigma \in (0, 1)$ *and* $\gamma \in (0, 1/2)$. *Compute the smallest integer* $j = 0, 1 \ldots$ *such that*

$$x^k + \sigma^j d_x^k \in \mathcal{P},$$
$$L_a(x^k + \sigma^j d_x^k, \lambda^k + \sigma^j d_\lambda^k; \epsilon) \le L_a(x^k, \lambda^k; \epsilon) + \gamma\sigma^j \nabla L_a(x^k, \lambda^k; \epsilon)' d^k,$$

set $\mu^k = \sigma^j$, $(x^{k+1}, \lambda^{k+1}) = (x^k + \mu^k d_x^k, \lambda^k + \mu^k d_\lambda^k)$, $k = k + 1$ *and go to Step 2.*

The following proposition establishes the convergence properties of the algorithm.

Proposition 4.1 *Suppose that Assumptions 3, 4, 5 hold. Assume that the direction d^k obtained by (17)-(22) satisfies Assumption 2 and (69). Let $\{(x^k, \lambda^k)\}$, $\{d^k\}$ be sequences produced by the algorithm* Erice98, *then:*

(a) *after a finite number of times the penalty parameter ϵ is no longer updated;*

(b) *the sequence $\{(x^k, \lambda^k)\}$ is bounded;*

(c) *every accumulation point of the sequence $\{(x^k, \lambda^k)\}$ is a KKT point;*

(d) *if $(\bar{x}, \bar{\lambda})$ is an accumulation point of the sequence $\{(x^k, \lambda^k)\}$ which is an isolated KKT point, then the whole sequence $\{(x^k, \lambda^k)\}$ converges to $(\bar{x}, \bar{\lambda})$;*

(e) *if the sequence $\{(x^k, \lambda^k)\}$ converges to a KKT point $(\bar{x}, \bar{\lambda})$ satisfying the SSOSC then eventually the truncated Newton direction is taken as search direction;*

(f) *if the sequence $\{(x^k, \lambda^k)\}$ converges to a KKT point $(\bar{x}, \bar{\lambda})$ satisfying the SSOSC then eventually $\mu^k = 1$ and the rate of convergence is superlinear.*

Proof. (a) The proof is by contradiction. Suppose that the assertion is false. Then the penalty parameter is updated an infinite number of times either by the instructions at Step 3 or by the ones at Step 4. If the test at Step 4 is violated an infinite number of times, there exist sequences $\{\epsilon^j\}$, $\{(x^j, \lambda^j)\}$, $\{d^j\}$ such that

$$\begin{array}{c} \epsilon^j \to 0, \\ \nabla L_a(x^j, \lambda^j; \epsilon^j)'d^j > -\epsilon^j \| \max\{g(x^j), -\epsilon^j p(x^j, \lambda^j)\lambda^j\}\|^2. \end{array} \tag{75}$$

Now, recalling the first of (74) and Proposition 3.2, we have that for sufficiently small values of ϵ^j:

$$\nabla L_a(x^j, \lambda^j; \epsilon^j)'d^j \leq -c_1 \|\nabla L_a(x^j, \lambda^j; \epsilon^j)\|^2 \leq -c_1 \| \max\{g(x^j), -\epsilon^j p(x^j, \lambda^j)\lambda^j\}\|^2,$$

which for sufficiently large j contradicts (75).
If the penalty parameter is updated an infinite number of times by the instructions at Step 3, there exist sequences $\{\epsilon^j\}$, $\{(x^j, \lambda^j)\}$, $\{d^j\}$ such that each d^j satisfies (73) and:

$$\epsilon^j \to 0, \tag{76}$$
$$\phi(x^j, \lambda^j, p^j; \epsilon^j) > -\epsilon^j \|d^j\|^2, \tag{77}$$
$$\|d^j\| \leq \delta, \tag{78}$$
$$\|\nabla g(x^j)'(\nabla_x L(x^j, \lambda^j) + G(x^j)^2 \lambda^j)\| \leq \delta. \tag{79}$$

Now, since $\{(x^j, \lambda^j)\} \in \Omega(x^0, \lambda^0; \epsilon^j)$, (b) of Proposition 3.1 and (79) imply that $\{(x^j, \lambda^j)\}$ is bounded and (78) shows that $\{d^j\}$ is bounded. Therefore the assumptions of Proposition 3.4 are satisfied so that we get a contradiction between the assertion of Proposition 3.4 and (76)-(77); this completes the proof of (a).

(b) The proof of this point follows directly from the instructions of Steps 5 and 6, point (a) and by (a) of Proposition 3.1.

(c) Let $\{(x^k, \lambda^k)\}_K$ be a subsequence of $\{(x^k, \lambda^k)\}$ converging to $(\bar{x}, \bar{\lambda})$. As usual, we can identify the subsequence with the whole sequence. We show that $(\bar{x}, \bar{\lambda})$ is a KKT point for Problem (1). By point (a) we can assume, without loss of generality, that $\epsilon = \bar{\epsilon}$. Since, by point (b), the sequence $\{(x^k, \lambda^k)\}$ is bounded we have that the Armijo line search technique used in Step 6 implies

$$\lim_{k \to \infty,} \mu^k \nabla L_a(x^k, \lambda^k; \bar{\epsilon})' d^k = 0. \tag{80}$$

Following a standard procedure, it is easy to show that

$$\lim_{k \to \infty} \nabla L_a(x^k, \lambda^k; \bar{\epsilon})' d^k = 0. \tag{81}$$

Indeed if (81) is not true we can assume that μ^k goes to 0 and that

$$\lim_{k \to \infty} \nabla L_a(x^k, \lambda^k; \bar{\epsilon})' d^k = \nu < 0.$$

By (b) of Proposition 3.1 and the rules of Step 6, $\mu^k \to 0$ implies, for k sufficiently large:

$$\begin{array}{rcl} x^k + \dfrac{\mu^k}{\sigma} d_x^k & \in & \mathcal{P} \\ L_a\left(x^k + \dfrac{\mu^k}{\sigma} d_x^k, \lambda^k + \dfrac{\mu^k}{\sigma} d_\lambda^k; \bar{\epsilon}\right) - L_a(x^k, \lambda^k; \bar{\epsilon}) & > & \gamma \dfrac{\mu^k}{\sigma} \nabla L_a(x^k, \lambda^k; \bar{\epsilon})' d^k. \end{array} \tag{82}$$

Now, if we divide (82) by μ^k/σ and take the limit for $k \to \infty$, we obtain $\nu > \gamma\nu$, that is a contradiction to (82) since $\gamma \in (0, 1/2)$. Hence (81) holds.

Now if, for an infinite number of k, d^k is the truncated Newton-type direction that satisfies (73), then (73), (81), and point (a) of Proposition 2.3 ensure that $(\bar{x}, \bar{\lambda})$ is a KKT point.

If, for a infinite number of k, d^k is a gradient related direction which satisfies (74), then the first of (74), the test at Step 4 and (81) imply:

$$\nabla L_a(\bar{x}, \bar{\lambda}; \epsilon) = 0, \qquad \max\{g(\bar{x}), -\epsilon p(\bar{x}, \bar{\lambda})\bar{\lambda}\} = 0,$$

which, recalling (a) of Proposition 3.3, show that $(\bar{x}, \bar{\lambda})$ is a KKT point.

(d) From (b) of Proposition 3.3 we have that $(\bar{x}, \bar{\lambda})$ is a stationary point of L_a. By the instructions of Step 3, Step 4 and Step 6, and by point (b), we have that

$$\lim_{k \to \infty} \mu^k d^k = 0,$$

from which we get:

$$\lim_{k\to\infty} \|x^{k+1} - x^k\| = 0. \tag{83}$$

Now point (d) follows from (83) and from well known results (see for example Section 14.1 of [12]).

(e) Preliminarily observe that, by (c) and by point (b) of Proposition 2.3, for sufficiently large k, the truncated Newton direction can be computed at Step 3. Assume now by contradiction that the assertion is false. Then there exists a convergent subsequence $\{(x^k, \lambda^k)\}_K$ such that d^k is computed as at Step 4. Point (c) yields that the limit $(\bar{x}, \bar{\lambda})$ of $\{(x^k, \lambda^k)\}_K$ is a KKT point. Then Proposition 3.5 guarantee that, for sufficiently large k, with $k \in K$, the test (73) of Step 3 is satisfied. Therefore we get the contradiction that the algorithm would set d^k obtained by (17)-(22) for such values of k.

(f) By points (a) and (e) we have that the test (73) is eventually satisfied. Recalling that $L_a(x, \lambda; \epsilon)$ is SC^1, Proposition 3.3 of [6] ensures that eventually $\mu^k = 1$. Then the superlinear rate of convergence follows from Proposition 2.2. □

Remark 4.2 *By the proof, it is evident that, at Step 6, the Armjo-type technique can be substituted by every line search technique that guarantees*

$$\lim_{k\to\infty} \nabla L_a(x^k, \lambda^k; \epsilon)' d^k = 0.$$

5 Numerical results

In order to verify the viability of the algorithm Erice98, it has been implemented in a very preliminary form in double precision Fortran90. In this section we report the results on a small set of test problems in the CUTE collection [1], of an algorithm obtained by specializing the general scheme Erice98 considered in the previous section. The following choices were made.

- At Step 4 we use a Newton type direction for $L_a(x, \lambda; \epsilon)$ that requires only the second order derivatives of f, g [4].

- According to Remark 4.2, for the choice of the stepsize μ^k at Step 6 we implemented a nonmonotone line search scheme that follows the scheme proposed in [10].

- The starting point (x^0, λ^0) is choosen as follows. x^0 is the point given by CUTE, whereas λ^0 is taken to satisfy the system:

$$\left[\nabla g(x^0)' \nabla g(x^0) + G(x^0)^2\right] \lambda = -\nabla g(x^0)' \nabla f(x^0).$$

- In (45), the scalar $s = 3$; as concern α we adopt a safeguard rule that ensure that $x^0 \in \mathcal{P}$: if x^0 is feasible $\alpha = 1$ else

$$\alpha = \|\max\{g(x), 0\}\|_3^3 + \max\{1, 10^{-3}\|\max\{g(x), 0\}\|_3^3\}.$$

- The stopping criterion of the conjugate gradient scheme for the solution of system (13) (of dimension $n - m_a$) has been chosen as:

$$\|r^k\| \leq \left\{10^{-3}, \min\left\{\frac{10^{-4}(n - m_a)}{\max\{k, 1\}}, \|h^k\|\right\}\|h^k\|\right\}.$$

- The algorithm terminates with success if

$$\|\max\{g(x), 0\}\|_\infty \leq 10^{-5}\max\{1, \|x\|\},$$
$$\|\nabla L\|_2 + \|\max\{-\lambda, 0\}\|_\infty + \|G\lambda\|_\infty \leq 10^{-5}\max\{1, \|x\|\},$$

namely if the KKT conditions are satisfied within a given tolerance.

name	n	m	its	$\#f$	$\#g$	$\# \nabla$	cg its.	time
BDEXP3	1000	1000	9	2	22	12	67	1.75
CVXBQP1	100	200	11	18	45	32	492	61.06
MCCORMCK3	100	200	5	6	153	12	140	.32
NONSCOMP	100	200	76	166	875	281	8796	112.69
CB2	3	3	10	2	24	13	5	.09
CB3	3	3	7	2	18	10	0	.10
CHACONN1	3	3	8	4	23	13	6	.10
CHACONN2	3	3	7	2	18	10	0	.09
DIPIGRI	7	4	18	65	468	94	105	.16
HIMMELP3	2	6	6	9	25	16	8	.09
HIMMELP4	2	7	16	74	178	91	27	.12
HIMMELP5	2	7	6	10	56	17	7	.09
HIMMELP6	2	9	35	66	321	104	64	.14
PENTAGON	6	15	41	34	97	61	81	.16
POWELL20A	10	10	10	9	35	20	32	.13
SIMPLLPA	2	4	9	10	34	20	10	.10
SIMPLLPB	2	5	8	48	198	57	21	.11
TWOBARS	2	6	5	3	15	9	4	.10

Table 1: Problems from the CUTE collection

The algorithm has been tested on a selection of problems in the CUTE collection with neither equality constraints nor fixed variables. This selection includes many problems from the Hoch-Scittkowsky collection. The results are reported in Tables 1-2. For each problem we report the number of variables (n), the number of constraints (m), the number of iterations (its), the number of objective function evaluations ($\#f$), the number of gradient evaluations of the objective function and of the constraints ($\#\nabla$), the number of constraints evaluations ($\#g$), the inner iterations of the conjugate gradient (cg its) and the cpu time (time).

We point out that the problems solved are all of small dimension, except for the first ones in Table 1 that have only bound constraints on the variables. From the tables it appears that the approach is viable, but no final conclusion can be drawn on the robustness and efficiency of the algorithm, and wider computational experimentantion is required.

name	n	m	its	$\#f$	$\#g$	$\#\nabla$	cg its.	time
HS1	2	1	18	24	71	45	37	.09
HS2	2	1	11	7	34	19	15	.09
HS3	2	1	4	31	220	36	7	.09
HS4	2	2	7	18	40	30	6	.09
HS5	2	4	4	2	12	7	6	.00
HS11	2	1	5	2	14	8	5	.09
HS12	2	1	7	4	49	12	9	.10
HS15	2	3	8	4	23	13	5	.10
HS16	2	5	10	10	35	21	9	.10
HS17	2	5	9	7	29	17	8	.10
HS22	2	2	5	2	14	8	1	.08
HS23	2	9	16	32	77	53	20	.12
HS24	2	5	9	19	45	29	10	.10
HS25	3	6	1	2	6	5	2	.10
HS29	3	1	10	29	56	40	25	.11
HS30	3	7	3	4	12	8	5	.09
HS31	3	7	4	3	13	8	6	.09
HS33	3	6	7	35	55	43	5	.10
HS34	3	8	94	701	1001	800	94	.32
HS35	3	4	2	3	9	6	5	.07
HS36	3	7	15	63	105	81	12	.11
HS37	3	8	6	3	17	10	6	.10
HS38	4	8	34	20	99	55	137	.02
HS44	4	10	18	44	100	73	25	.12
HS45	5	10	7	8	27	16	6	.09
HS57	2	3	2	2	8	5	3	.10
HS65	3	7	59	155	599	251	65	.19
HS66	3	8	12	43	120	60	20	.11
HS67	3	20	22	112	1558	151	89	.58
HS70	4	9	12	28	375	41	55	.75
HS76	4	7	28	87	176	136	84	.13
HS93	6	8	99	930	1289	1048	544	.56
HS95	6	16	8	30	207	41	27	.10
HS96	6	16	11	50	343	66	30	.13
HS100	7	4	18	65	468	94	105	.15
HS108	9	14	119	300	1580	452	243	.65
HS105	8	17	34	103	220	145	320	5.79

Table 2: HS problems from the CUTE collection

References

[1] I. Bongartz, A. R. Conn, N. I. M. Gould and Ph. L. Toint (1995), "CUTE: Constrained and Unconstrained Testing Environment," *ACM Transaction on Mathematical Software*, 21:123-160.

[2] R.S. Dembo and T. Steihaug (1983), "Truncated-Newton algorithms for large scale unconstrained optimization," *Mathematical Programming*, 26:190-212.

[3] G. Di Pillo and L. Grippo (1985),"A continuously diffentiable exact penalty function for nonlinear programming problems with inequality constraints," *SIAM J. on Control and Optimization*, 23:72-84.

[4] G. Di Pillo and S. Lucidi (1999),"An augmented Lagrangian function with improved exactness properties (revised version)", TR DIS 16-99.

[5] G. Di Pillo, S. Lucidi and L. Palagi (1999),"A superlinearly convergent primal-dual algorithm for constrained optimization problems with bounded variables," TR DIS 02-99.

[6] F. Facchinei (1995), "Minimization of SC^1 functions and the Maratos effect," *Operations Research Letters*, 17:131–137.

[7] F. Facchinei and S. Lucidi (1995), "Newton-type algorithm for the solution of inequality constrained minimization problems," *Operations Reasearch Proceedings 1994*, Springer-Verlag, 33–38.

[8] F. Facchinei and S. Lucidi (1995), "Quadratically and superlinearly convergent algorithms for the solution of inequality constrained minimization problems," *Journal of Optimization Theory and Applications*, 85:265–289.

[9] L. Grippo, F. Lampariello and S. Lucidi (1989), "A truncated Newton method with nonmonotone line search for unconstrained optimization," *J. of Optimization Theory and Application*, 60:401-419.

[10] L. Grippo, F. Lampariello and S. Lucidi (1991), "A class of nonmonotone stabilization methods in unconstrained optimization," *Numerische Mathematik* 59:779–805.

[11] J. Nocedal and M. L. Overton (1985), "Projected Hessian updating algorithms for nonlinearly constrained optimization," *SIAM J. on Numerical Analysis*, 22: 821-850.

[12] J. M. Orthega and W. C. Rheinboldt (1970), *Iterative Solution of Nonlinear Equations in Several Variables*, Academic Press.

[13] L. Qi and J. Sun (1993), "A nonsmooth version of Newton's method," *Mathematical Programming*, 58:353–367.

Fixed and virtual stability center methods for convex nonsmooth minimization

Antonio Fuduli (fuduli@unical.it)
Istituto per la Sistemistica e l'Informatica del C.N.R. (ISI-CNR), c/o Dip. di Elettronica Informatica e Sistemistica, Universitá della Calabria, 87036 Rende (CS), Italia

Manlio Gaudioso (gaudioso@unical.it)
Dipartimento di Elettronica Informatica e Sistemistica, Universitá della Calabria, 87036 Rende (CS), Italia

Abstract

We tackle the problem of minimizing a convex nondifferentiable function, for which we present two new bundle type algorithms. The novelty of these methods involves a different approach for updating the stability center, that classically is chosen as the best current point (in terms of the objective function). Convergence to a minimum point for both methods, which are related to the concept of proximal trajectory, is proved under routine assumptions. Finally numerical results are reported.

Keywords: Nondifferentiable optimization, bundle methods, proximal trajectory.

1 Introduction

Nonsmooth optimization problems often arise from real-life decision making models. Moreover, several techniques, aimed at solving differentiable, but difficult, problems (e.g. large scale ones) are based on solving auxiliary nonsmooth problems; there exist also some problems that are analytically, but not numerically, differentiable, since the gradient of the objective function varies rapidly.

G. Di Pillo and F. Giannessi (eds.), Nonlinear Optimization and Related Topics, 105–122.

Historically, the oldest methods for solving convex nonsmooth minimization problems are the subgradient methods [32], devised in Soviet Union at the beginning of the 60's. Successively, stemming from the classical methods due to Lemaréchal [16, 17] and Wolfe [33], bundle methods have been developed: they can be interpreted as stabilizing methods for the cutting plane algorithm, proposed independently by Cheney and Goldstein in [4] and Kelley in [11], and have been object of several works [6, 8, 9, 13, 14, 15, 18, 24, 31] during the last twenty years.

Recently research has focused on the study of the Moreau-Yosida regularization, a powerful link between differentiable and nondifferentiable optimization, in the context of the proximal point algorithm [30]. Given a convex nondifferentiable function, if we regularize it by means of the Moreau-Yosida approach, we obtain a convex differentiable function, which has the same set of minimizers as the original function. Thus, much effort has been expended in order to minimize the Moreau-Yosida regularization, studying its second order differentiability properties [20, 28] and applying [2, 3, 7, 21, 25, 27, 29] Newton or quasi-Newton type methods. But the main difficulty of such approaches comes from the fact that the evaluation of the Moreau-Yosida function is not immediate, because it requires the solution of a nondifferentiable problem. Rockafeller in [30] proposed an approximate evaluation of the Moreau-Yosida function on the basis of some theoretical, but not implementable criteria; to circumvent this difficulty, several authors [6, 21, 25, 27] have used a bundle approach substituting the original function with a polyhedral convex function.

Finally in the last two years some authors have studied minimization techniques based on the $\mathcal{VU}$ space decomposition: near a minimizer of a convex nonsmooth function, they distinguish between the subspace $\mathcal{V}$, where the nonsmoothness of the function is concentrated and on which its graph is "$\mathcal{V}$-shaped", and its orthogonal complement $\mathcal{U}$, where the function appears to be differentiable and on which its graph is "$\mathcal{U}$-shaped". Some references are [19, 20, 25, 26].

The methods presented in this paper are related to the bundle philosophy. Thus we briefly describe how a bundle method works.

We consider the following minimization problem:

$$\min_{x \in \mathbb{R}^n} f(x), \tag{1}$$

where $f : \mathbb{R}^n \mapsto \mathbb{R} \cup \{+\infty\}$ is a closed, proper and convex function, with a nonempty set X^* of minimizing points. At any iteration of a classical bundle [10] type algorithm we maintain a bundle of k points

$$B_k = \{x_1, x_2, \ldots, x_k\}$$

and at each point x_j, $j = 1, \ldots, k$, a subgradient g_j of f is supposed to be known.

We obtain the next iterate by solving the following problem (throughout the paper, $\|\cdot\|$ indicates the Euclidean norm):

$$\min_{x \in \mathbb{R}^n} \alpha_k f_k(x) + \frac{1}{2}\|x - y_k\|^2, \tag{2}$$

where

$$f_k(x) \triangleq \max_{1 \le j \le k} f(x_j) + g_j^T(x - x_j), \tag{3}$$

is the *cutting plane function.* The point y_k is the k-th *stability center* and α_k is a positive weighting parameter whose update plays an important role [10] for the overall convergence.

Now we recall the concept of proximal trajectory [5], which is used in our methods.

Given $\alpha > 0$ and a stability center y, let $x(\alpha)$ be the minimizer over $\mathbb{R}^n$ of the function

$$\varphi_\alpha(x) = \alpha f(x) + \frac{1}{2}\|x - y\|^2, \tag{4}$$

the mapping $x(\alpha)$, $\alpha \in [0, +\infty[$, defines the *proximal trajectory* of the function f with respect to the stability center y. The proximal trajectory is a continuous curve, which emanates (for $\alpha = 0$) from the stability center and terminates (for $\alpha = \infty$) at a minimizing point of f. Moreover, it coincides [1] with the set of the weakly efficient solutions to the following two-objective program:

$$\begin{array}{l} \min_{x \in \mathbb{R}^n} f(x) \\ \min_{x \in \mathbb{R}^n} \|x - y\|. \end{array} \tag{5}$$

We observe that $x(\alpha)$ satisfies the condition

$$x(\alpha) = -\alpha g + y \quad \text{for some } g \in \partial f(x(\alpha)). \tag{6}$$

When f is a polyhedral function, its proximal trajectory is made up of line segments (with possibly a half line whenever the polyhedral function is unbounded from below) and may be constructed entirely by solving one or more structured linear programs [5].

The core of our approach is to define a minimization technique for f based on the construction of the proximal trajectories of improving polyhedral approximations of f of the cutting plane type.

We use the following notation. We indicate by $t(f_k, y)$ the entire proximal trajectory of the k-th polyhedral function f_k corresponding to the stability center y and by $x(\alpha, f_k, y)$ the solution to the following quadratic program:

$$\min_{x \in \mathbb{R}^n} \alpha f_k(x) + \frac{1}{2}\|x - y\|^2. \tag{7}$$

The paper is organized as follows. In the next two sections we describe the two new bundle algorithms, named the *fixed stability center method* and the *virtual stability center method*, respectively. Finally, we present numerical results.

2 The fixed stability center method

We describe a new bundle method, characterized by the fact that the stability center is kept fixed to the starting point and never updated. The idea involves obtaining better and better approximations $t(f_k, x_1)$ to the proximal trajectory $t(f, x_1)$, that, for $\alpha \to \infty$, provides a minimizing point of f. Thus, we proceed by increasing both k to improve the polyhedral approximation and α in order to get closer and closer to a minimizing point of f.

The algorithm, whose input parameters are

- x_1: starting point;
- $\alpha_1 > 0$: initial value of α;
- $\Delta\alpha > 0$;
- $\delta > 0$;
- $\bar{\alpha} > 0$,

may be stated as follows.

Algorithm FSC

1. $k := 1$, $B_k = \{x_1\}$;
2. calculate $x_{k+1} = x(\alpha_k, f_k, x_1)$;
3. if $f(x_{k+1}) - f_k(x_{k+1}) > \delta$ then set $\alpha_{k+1} := \alpha_k$ and go to step 5;
4. if $\alpha_k \geq \bar{\alpha}$, STOP with x_{k+1} as an estimate of an optimal solution, otherwise set $\alpha_{k+1} := \alpha_k + \Delta\alpha$;
5. set $B_{k+1} := B_k \cup \{x_{k+1}\}$, $k := k + 1$ and go to step 2.

Now we explain in detail the meaning of steps 3 and 4. In step 3 we test the "quality" of the model f_k at the point x_{k+1}. For a given $\delta > 0$, if

$$f(x_{k+1}) - f_k(x_{k+1}) > \delta, \tag{8}$$

we assume that f_k is a poor approximation to f near x_{k+1} and consequently we leave α unchanged and we enrich the bundle in order to improve the approximation. On the other hand, whenever

$$f(x_{k+1}) - f_k(x_{k+1}) \leq \delta, \tag{9}$$

we perform first the stopping test $\alpha_k \geq \bar{\alpha}$. If it fails, as the polyhedral function is a locally good model for f, we increase α, pushing forward the construction of the proximal trajectory, and at the same time we enrich the bundle.

The stopping criterion is justified by the fact that, when α_k is large, $f_k(x_{k+1})$ cannot be too much larger than f^*, the minimum value of f. Moreover, since (9) is satisfied, f_k is a good approximation of f. We report the following lemma proved in [5].

Lemma 2.1 *Let f_k be the polyhedral function defined by (3) and let x_{k+1} be the next point that will enter the bundle. If, for a given $\delta > 0$*

$$f(x_{k+1}) - f_k(x_{k+1}) > \delta, \tag{10}$$

then

$$\|x_{k+1} - x_j\| > \frac{\delta}{2L} \quad j = 1, \ldots, k,$$

where $L > 0$ is the Lipschitz constant of f on any ball containing the points

$$x_1, x_2, \ldots, x_{k+1}.$$

Proof. By definition of f_k, for all $x \in \mathbb{R}^n$ we have

$$f_k(x) \geq f(x_j) + g_j^T(x - x_j), \quad j = 1, \ldots, k,$$

Then, taking into account (10) and since

$$f(x) \geq f_k(x) \quad \text{for all } x \in \mathbb{R}^n$$

for all $j = 1 \ldots, k$ we have

$$\begin{aligned} f(x_{k+1}) - \delta > f_k(x_{k+1}) &\geq f(x_j) + g_j^T(x_{k+1} - x_j) \geq \\ &\geq f_{k+1}(x_j) + g_j^T(x_{k+1} - x_j) \geq \\ &\geq f(x_{k+1}) + g_{k+1}^T(x_j - x_{k+1}) + g_j^T(x_{k+1} - x_j), \end{aligned} \tag{11}$$

Thus

$$\delta < (g_{k+1} - g_j)^T(x_{k+1} - x_j), \tag{12}$$

from which

$$\delta < \|g_{k+1} - g_j\| \|x_{k+1} - x_j\|. \tag{13}$$

By (13) and taking into account that

$$\|g_j\| \leq L \quad \text{for all } j = 1, \ldots, k+1,$$

we obtain

$$\delta < 2L\|x_{k+1} - x_j\|.$$

□

Now we prove some properties useful to ensure the convergence of the algorithm.

Lemma 2.2 *Let $\{\alpha_k\}$ be a monotone nondecreasing sequence with $\alpha_k \to \infty$ and let*

$$f^* \triangleq \min_{x \in \mathbb{R}^n} f(x).$$

Then for any $\epsilon > 0$ there exists $\bar{k}$ such that, for $k \geq \bar{k}$,

$$f_k(x_{k+1}) < f^* + \epsilon,$$

where

$$x_{k+1} \triangleq x(\alpha_k, f_k, x_1).$$

Proof. For a given ϑ, $0 < \vartheta < \epsilon$, let

$$\sigma \triangleq \epsilon - \vartheta > 0$$

and let $\bar{x}$ be a point such that

$$f(\bar{x}) \leq f^* + \vartheta.$$

Assume that for some k

$$f_k(x_{k+1}) \geq f^* + \epsilon. \tag{14}$$

Thus, because f_k is a lower approximation of f, we have

$$f_k(x_{k+1}) \geq f^* + \epsilon \geq f(\bar{x}) - \vartheta + \epsilon \geq f_k(\bar{x}) + \sigma,$$

that is

$$f_k(x_{k+1}) \geq f_k(\bar{x}) + \sigma. \tag{15}$$

From the definition of x_{k+1} it follows that

$$\alpha_k f_k(x_{k+1}) + \frac{1}{2}\|x_{k+1} - x_1\|^2 \leq \alpha_k f_k(\bar{x}) + \frac{1}{2}\|\bar{x} - x_1\|^2,$$

so

$$\alpha_k(f_k(x_{k+1}) - f_k(\bar{x})) \leq \frac{1}{2}\|\bar{x} - x_1\|^2. \tag{16}$$

Combining (15) and (16) one obtains

$$\alpha_k \sigma \leq \frac{1}{2}\|\bar{x} - x_1\|^2,$$

i.e.

$$\alpha_k \leq \frac{1}{2\sigma}\|\bar{x} - x_1\|^2.$$

Thus, letting $\bar{k}$ be the smallest value of the index k such that

$$\alpha_k > \frac{1}{2\sigma}\|\bar{x} - x_1\|^2,$$

we conclude that the condition $f_k(x_{k+1}) < f^* + \epsilon$ is satisfied for all $k \geq \bar{k}$. □

Remark 2.1. Since ϵ may be taken arbitrarily small, the proof of the above lemma shows that for any $\epsilon > 0$ we obtain satisfaction of the condition $f_k(x_{k+1}) < f^* + \epsilon$ provided that α_k is sufficiently large, e.g. $\alpha_k > \frac{D^2}{2\epsilon}$ where D is any upper bound on the diameter of a ball around x_1 containing a minimizer of f.

Lemma 2.3 *Given $\alpha > 0$, let $\{x_{k+1}\}$ be the sequence of points*

$$x_{k+1} = x(\alpha, f_k, x_1). \tag{17}$$

Then x_{k+1} satisfies the following inequality:

$$\|x_{k+1} - x_1\| \leq 2\alpha L_1,$$

where $L_1 > 0$ is the Lipschitz constant of f on any ball around x_1.

Proof. We have

$$\alpha f_k(x_{k+1}) + \frac{1}{2}\|x_{k+1} - x_1\|^2 \leq \alpha f_k(x_1) \leq \alpha f(x_1). \tag{18}$$

Moreover, since x_1 is a point of the bundle, for all $x \in \mathbb{R}^n$ one has

$$f_k(x) \geq f(x_1) + g_1^T(x - x_1) \geq f(x_1) - \|g_1\|\|x - x_1\|,$$

where $g_1 \in \partial f(x_1)$. Then

$$\alpha f_k(x) + \frac{1}{2}\|x - x_1\|^2 \geq \alpha(f(x_1) - \|g_1\|\|x - x_1\|) + \frac{1}{2}\|x - x_1\|^2. \tag{19}$$

Combining (18) and (19) we obtain

$$-\alpha\|g_1\|\|x_{k+1} - x_1\| + \frac{1}{2}\|x_{k+1} - x_1\|^2 \leq 0,$$

from which, taking into account that $\|g_1\| \leq L_1$,

$$\|x_{k+1} - x_1\| \leq 2\alpha\|g_1\| \leq 2\alpha L_1.$$

□

Taking into account lemma 2.1 and 2.3, the following lemma holds.

Lemma 2.4 *For a given $\alpha > 0$, let $\{x_{k+1}\}$ be the sequence of points calculated according to (17). Then, for any $\delta > 0$ there exists k such that*

$$f(x_{k+1}) - f_k(x_{k+1}) \leq \delta. \tag{20}$$

The following theorem proves the convergence of the FSC algorithm.

Theorem 2.5 *Let $\eta > 0$, then there exists k such that*

$$f(x_{k+1}) - f^* < \eta,$$

where $\{x_{k+1}\}$ is the sequence generated by the FSC algorithm, provided that the input parameters δ and $\bar{\alpha}$ satisfy the conditions $\delta < \eta$ and $\bar{\alpha} > \frac{D^2}{2(\eta-\delta)}$ where D is any upper bound on the diameter of a ball around x_1 containing a minimizer of f.

Proof. By lemma 2.2 (see Remark 2.1) there exists $\bar{k}$ such that, for any $k \geq \bar{k}$,

$$f_k(x_{k+1}) < f^* + \eta - \delta. \tag{21}$$

Moreover, by lemma 2.4 there exists $k \geq \bar{k}$ such that

$$f(x_{k+1}) - f_k(x_{k+1}) \leq \delta. \tag{22}$$

(Note that in FSC α_k for $k > \bar{k}$ remains unchanged until the above condition is satisfied).

Combining (21) and (22), we have

$$f(x_{k+1}) \leq \delta + f_k(x_{k+1}) < \delta + f^* + \eta - \delta$$

and thus we obtain

$$f(x_{k+1}) - f^* < \eta.$$

□

3 The virtual stability center method

Bundle methods were conceived as stabilized versions of the cutting plane method.

The stability center was introduced as the point close to which we wish to confine the next iterate. In general it is updated to the current point, whenever a certain *sufficient decrease* condition is satisfied.

In section 2 we have presented a method in which the stability center is kept fixed to the value of the starting point throughout the algorithm.

On the other hand in [18] the authors presented, under the hypothesis of compactness of a set containing the iterates and a solution [1], a level–set type bundle algorithm where the stability center is updated at each iteration to the value of the current iterate. They observe that "...the level-sets of the model[2] are rather stable, so that extremely simple rules can be used for updating the level. This property allows us to forget about the concepts of stability center and null step: the stability center may be systematically set to the last iterate..." [18].

The algorithm that we present in this section benefits from the fixed stability center idea. In that case the idea is to remain anchored to the starting point x_1, while going towards a minimum point of the polyhedral function, and to increase the value of the weight α, whenever the polyhedral function exhibits a good agreement with the function f. Here we adopt a similar logic: we improve the quality of the model by enriching the bundle at each step and at the same time we move towards a minimum point of the model, extending the trajectory by increasing α.

[1] This hypothesis is present also in [15].

[2] The model is the polyhedral function.

The method is strictly related to the proximal trajectory theory, where problem (7) is solved by a finite sequence of linear programs. The number of such programs depends on the value of α: the greater α, the larger the number of linear programs to solve. The basic idea is the following. In general the point

$$x_k \triangleq x(\alpha_{k-1}, f_{k-1}, y),$$

of the trajectory $t(f_{k-1}, y)$ does not belong to the new proximal trajectory $t(f_k, y)$. Thus, we are expected to reconstruct the proximal trajectory $t(f_k, y)$ up to the value α_{k-1}, by solving a finite sequence of linear programs starting from the stability center (that is for $\alpha = 0$) until we get to the point of the new trajectory corresponding to $\alpha = \alpha_{k-1}$. This is expensive from the computational point of view, because it leads to solving several linear programs and the trade-off with direct solution to the quadratic program (7) is uncertain.

Thus, we propose the following variation. Since we improve the model f_{k-1} by adding a new linearization, we can vary the stability center in a way that the point x_k belongs to the new proximal trajectory for the current value α_{k-1}.

That is, given the current point

$$x_k \triangleq x(\alpha_{k-1}, f_{k-1}, y_{k-1}),$$

the idea consists in calculating, for a certain $\alpha_k > \alpha_{k-1}$, the next point in the following way:

$$x_{k+1} := x(\alpha_k, f_k, y_k),$$

where y_k is a new stability center such that

$$x_k = x(\alpha_{k-1}, f_k, y_k). \tag{23}$$

A direct way to do that is to select as the stability center the point

$$y_k := x_k + \alpha_{k-1} g_k, \tag{24}$$

with $g_k \in \partial f(x_k)$. This ensures that (23) is satisfied, because x_k satisfies the optimality condition (6) of problem (7) for $\alpha = \alpha_{k-1}$ and $y = y_k$: in fact the k-th linearization is active at x_k.

By adopting this device we avoid the necessity of solving several linear programs to reconstruct the trajectory up to x_k and we are able to find x_{k+1} from x_k along the proximal trajectory $t(f_k, y_k)$.

The considerations above justify the name of *virtual stability center* method.

3.1 The algorithm

Now we describe the algorithm, whose convergence is proved under the hypothesis that the sequence of points $\{x_k\}$ is calculated within a compact set $Q \subseteq \mathbb{R}^n$ of diameter D, such that

$$x^* \in int\ Q,$$

for some point x^* belonging to the set X^* of minimizers of f; thus we indicate by $x_Q(\alpha, f_k, y)$ the minimizer of the problem

$$\min_{x \in Q} \alpha f_k(x) + \frac{1}{2}\|x - y\|^2. \tag{25}$$

Also, we indicate by f_k^* the minimum value of f_k over Q, and by x_k^* the corresponding minimizer, so that $f_k^* = f_k(x_k^*)$. We have chosen the following stopping criterion. When at the new point x_{k+1} the polyhedral function f_k is a good model of the function f, that is

$$f(x_{k+1}) - f_k(x_{k+1}) \leq \delta, \tag{26}$$

for some positive and sufficiently small δ, we calculate the (nonnegative) difference $f_k(x_{k+1}) - f_k^*$ (observe that this requires minimization of f_k over Q). If

$$f_k(x_{k+1}) - f_k^* < \epsilon \tag{27}$$

for a given positive and sufficiently small ϵ, we stop the algorithm since an approximate optimal solution has been achieved.

Given $x_1 \in \mathbb{R}^n$, $\alpha_1 > 0$, $\delta > 0$ and $\epsilon > 0$, the algorithm may be stated as follows.

Algorithm VSC

1. $k := 1$, $B_k := \{x_1\}$, $y_k := x_1$;
2. calculate $x_{k+1} = x_Q(\alpha_k, f_k, y_k)$ and set $B_k := B_k \cup \{x_{k+1}\}$;
3. if $f(x_{k+1}) - f_k(x_{k+1}) > \delta$, go to step 4, otherwise calculate f_k^*. If $f_k(x_{k+1}) - f_k^* < \epsilon$, STOP with x_{k+1} an estimate of x^*.
4. set $y_{k+1} := x_{k+1} + \alpha_k g_{k+1}$, with $g_{k+1} \in \partial f(x_{k+1})$;
5. set $\alpha_{k+1} := \alpha_k + \alpha_k \log(k+1)$, $k := k+1$ and go to step 2.

Letting

$$x_{k+1} \triangleq x_Q(\alpha_k, f_k, y_k) \tag{28}$$

and noting that

$$x_k = x_Q(\alpha_{k-1}, f_k, y_k),$$

since

$$y_k = x_k + \alpha_{k-1} g_k \tag{29}$$

and

$$g_k \in \partial f(x_k)$$

and using

$$\alpha_k > \alpha_{k-1},$$

the following properties hold [5].

Lemma 3.1 *Let f_k be the polyhedral function defined by (3), with*

$$B_k = \{x_1, \dots, x_k\} \subset Q$$

and let $x_{k+1} \in Q$ be the next point that will enter the bundle. If for a certain $\delta > 0$

$$f(x_{k+1}) - f_k(x_{k+1}) > \delta, \tag{30}$$

then

$$\|x_{k+1} - x_j\| > \frac{\delta}{2L_Q} \quad j = 1, \dots, k,$$

where $L_Q > 0$ is the Lipschitz constant of f on Q.

From the compactness of Q and lemma 3.1 the following corollary results.

Corollary 3.2 *For any $\delta > 0$, there exists an index k such that*

$$f(x_{k+1}) - f_k(x_{k+1}) \leq \delta.$$

Lemma 3.3 *For any $\epsilon > 0$, if $\alpha_{k-1} \to \infty$ and*

$$\lim_{k\to\infty} \frac{\alpha_k}{\alpha_{k-1}} \to \infty,$$

then there exists an index $\bar{k}$ such that

$$f_k(x_{k+1}) < f_k^* + \epsilon \tag{31}$$

for all $k \geq \bar{k}$.

Proof. Suppose that, for any $\epsilon > 0$ and for some $\alpha_k > 0$, we have

$$f_k(x_{k+1}) \geq f_k(x_k^*) + \epsilon. \tag{32}$$

From definition of x_{k+1} it follows

$$\alpha_k f_k(x_{k+1}) + \frac{1}{2}\|x_{k+1} - y_k\|^2 \leq \alpha_k f_k(x_k^*) + \frac{1}{2}\|x_k^* - y_k\|^2. \tag{33}$$

Combining (32) and (33) we obtain

$$\begin{aligned}
\alpha_k\epsilon &\leq \frac{1}{2}\|x_k^* - y_k\|^2 - \frac{1}{2}\|x_{k+1} - y_k\|^2 = \\
&= \frac{1}{2}\|x_k^* - x_k - \alpha_{k-1}g_k\|^2 - \frac{1}{2}\|x_{k+1} - x_k - \alpha_{k-1}g_k\|^2 = \\
&= \frac{1}{2}\|x_k^* - x_k\|^2 + \frac{1}{2}\alpha_{k-1}^2\|g_k\|^2 - \alpha_{k-1}g_k^T(x_k^* - x_k) - \\
&- \frac{1}{2}\|x_{k+1} - x_k\|^2 - \frac{1}{2}\alpha_{k-1}^2\|g_k\|^2 + \alpha_{k-1}g_k^T(x_{k+1} - x_k) = \\
&= \frac{1}{2}\|x_k^* - x_k\|^2 - \frac{1}{2}\|x_{k+1} - x_k\|^2 + \alpha_{k-1}g_k^T(x_{k+1} - x_k^*) \leq \\
&\leq \frac{1}{2}D^2 + \alpha_{k-1}L_Q D,
\end{aligned}$$

where D is the diameter of Q and L_Q is the Lipschitz constant of f on Q. The above inequality can be rewritten as

$$\alpha_k \leq \frac{1}{\epsilon}(\frac{D^2}{2} + \alpha_{k-1}L_Q D). \tag{34}$$

The hypothesis

$$\frac{\alpha_k}{\alpha_{k-1}} \to \infty,$$

with $\alpha_{k-1} \to \infty$ ensures that there exists $\bar{k}$ such that condition (34) does not hold for all $k \geq \bar{k}$. □

The above lemma justifies the rule for increasing α_k at step 5 of the algorithm.

The following theorem proves the convergence of the VSC algorithm.

Theorem 3.4 *For any $\eta > 0$ there exists k such that*

$$f(x_{k+1}) - f^* < \eta. \tag{35}$$

provided that the input parameters δ and ϵ of VSC satisfy the condition

$$\delta + \epsilon \leq \eta \tag{36}$$

Proof. From the definition of f_k and x_k^*, since $f_k(x_k^*) \leq f^*$, we have that whenever

$$f_k(x_{k+1}) - f_k(x_k^*) < \epsilon. \tag{37}$$

we have also

$$f_k(x_{k+1}) - f^* < \epsilon. \tag{38}$$

Let $\bar{k}$ be an index such that, for any $k \geq \bar{k}$, we have

$$\alpha_k > \frac{1}{\epsilon}(\frac{D^2}{2} + \alpha_{k-1} L_Q D). \tag{39}$$

Then, by lemma 3.3, for any $k \geq \bar{k}$ we have that (38) holds.

Moreover, by corollary 3.2, there exists $k \geq \bar{k}$ such that

$$f(x_{k+1}) - f_k(x_{k+1}) \leq \delta. \tag{40}$$

Combining (38) and (40) and taking into account (36), the thesis follows.

□

4 Numerical results

Both the FSC and VSC algorithms have been implemented in Fortran 90 (double precision), using the *Microsoft Fortran PowerStation - Version 4.0* for Windows 95, provided by *Microsoft Corporation.*

In particular we focus on the VSC method, which has appeared more promising in terms of numerical results.

Although the convergence of the VSC method has been proved under the hypothesis of compactness of the search space Q, in practice it has not been necessary to take this into account explicitly in order to get convergence. Solving quadratic program (7) has been treated in terms of proximal trajectory theory, where moving from the point

$$x_k = x(\alpha_{k-1}, f_k, y_k) = x(\alpha_{k-1}, f_{k-1}, y_{k-1})$$

to the point

$$x_{k+1} = x(\alpha_k, f_k, y_k),$$

along the proximal trajectory

$$t(f_k, y_k),$$

is done by solving a finite sequence of structured linear programs. The number of such programs depends on how large α_k is, in the sense that, in general, the greater the difference $\alpha_{k+1} - \alpha_k$, the greater the number of linear programs that we must solve.

Since the subdifferential at each new point x_k is usually a singleton, continuation of the proximal trajectory, for small values of the increase of α_k, may be obtained by solving a trivial linear program.

Function	n	f^*	k_{max}	f	f/g	LP
CB2	2	1.952225	5	1.952225	30	20
CB3	2	2	5	2.000000	14	9
DEM	2	-3	5	-3.000000	3	2
QL	2	7.2	5	7.200000	30	23
LQ	2	-1.414213	5	-1.414213	16	10
Mifflin1	2	-1	10	-1.000000	8	5
Mifflin2	2	-1	10	-0.999999	30	15
Rosen	4	-44	10	-43.999998	50	49
Shor	5	22.60016	10	22.60016	42	38
Maxquad	10	-0.8414084	10	-0.841407	66	61
Maxq	20	0	200	0.000000	168	0
Maxl	20	0	200	0.000000	27	24
Goffin	50	0	200	0.000000	63	59

Table 1: numerical results of the VSC algorithm.

In table 1 the numerical results relative to classical test functions [23] for non-smooth minimization are reported. In particular, for each function, we report the following parameters:

- n: dimension of the problem;
- f^*: optimum objective function value;
- k_{max}: maximum dimension of the bundle;
- f: the best function value, calculated by the VSC algorithm;
- f/g: number of function/subgradient evaluations (coinciding with the total number of iterations);
- LP: total number of nontrivial linear programs solved by the VSC algorithm.

To manage the bundle we have adopted Kiwiel's aggregation technique [12], using the FIFO strategy.

The values of k_{max} have been chosen equal to those ones defined in [22] and [31], in order to make significant the comparison of the results.

A very particular case has occurred for function Maxq, for which at each iteration the VSC algorithm has solved only the trivial linear program, that is no real linear program has ever been solved.

In table 2 we compare our results with those obtained by Schramm and Zowe in [31]. They use a bundle method (named Bundle-Trust-Region - for short BT), based on the solution, at each iteration, of a quadratic program. The total number of quadratic programs that the BT algorithm solves has been indicated by QP.

			VSC			BT		
Function	n	f^*	f	f/g	LP	f	f/g	QP
CB2	2	1.952225	1.952225	30	20	1.952225	16	13
CB3	2	2	2.000000	14	9	2.000000	21	13
DEM	2	-3	-3.000000	3	2	-3.000000	13	9
QL	2	7.2	7.200000	30	23	7.200009	17	12
LQ	2	-1.414213	-1.414213	16	10	-1.414214	11	10
Mifflin1	2	-1	-1.000000	8	5	-1.000000	74	49
Mifflin2	2	-1	-0.999999	30	15	-1.000000	13	6
Rosen	4	-44	-43.999998	50	49	-43.99998	32	22
Shor	5	22.60016	22.60016	42	38	22.60016	30	29
Maxquad	10	-0.8414084	-0.841407	66	61	-0.8414083	56	45
Maxq	20	0	0.000000	168	0	0.0	128	125
Maxl	20	0	0.000000	27	24	0.0	84	74
Goffin	50	0	0.000000	63	59	0.0	53	51

Table 2: comparison with the BT algorithm in [31].

			VSC			PB		
Function	n	f^*	f	f/g	LP	f	f/g	QP
CB2	2	1.952225	1.952225	30	20	1.952225	14	13
CB3	2	2	2.000000	14	9	2.000000	16	15
DEM	2	-3	-3.000000	3	2	-3.000000	8	7
QL	2	7.2	7.200000	30	23	7.200001	18	17
LQ	2	-1.414213	-1.414213	16	10	-1.414213	16	10
Mifflin1	2	-1	-1.000000	8	5	-0.999998	22	21
Mifflin2	2	-1	-0.999999	30	15	-0.999999	25	18
Rosen	4	-44	-43.999998	50	49	-44	36	35
Shor	5	22.60016	22.60016	42	38	22.60016	27	26
Maxquad	10	-0.8414084	-0.841407	66	61	-0.841408	45	44
Maxq	20	0	0.000000	168	0	0.000000	191	174
Maxl	20	0	0.000000	27	24	0.000000	35	34
Goffin	50	0	0.000000	63	59	0.000000	52	51

Table 3: comparison with the PB algorithm in [22].

Our method has turned out effective for relatively large scale problems (Maxquad, Maxq, Maxl, Goffin), but it has had also a good behaviour for functions CB3, DEM, LQ and Mifflin1.

Similar observations may be done (table 3) comparing our results with those reported by Mäkelä in [22], relative to the Proximal-Bundle algorithm (for short PB), in which the calculation of the search direction is done by solving a quadratic program.

Acknowledgements. The authors are greatly indebted to an anonymous referee whose accurate comments have provided a substantial help to improve the paper.

References

[1] A. Astorino, A. Fuduli and M. Gaudioso (1997), "Analysis of regularization techniques in convex nondifferentiable optimization," In U. Zimmermann, U. Derigs, W. Gaul, R.H. Möhring and K.P. Shusters editors, *Operations Research Proceedings 1996*, Springer-Verlag, Berlin, Heidelberg, 20-25.

[2] J.F. Bonnans, J.C. Gilbert, C. Lemaréchal and C.A. Sagastizábal (1995), "A family of variable metric proximal methods," *Mathematical Programming*, 68, 15-47.

[3] X. Chen and M. Fukushima (1998), "Proximal quasi-Newton methods for nondifferentiable convex optimization," *To appear in Mathematical Programming.*

[4] E.W. Cheney and A.A. Goldstein (1959), "Newton's method for convex programming and Tchebycheff approximation," *Numerische Matematik*, 68, 15-47.

[5] A. Fuduli and M. Gaudioso (1998), "The proximal trajectory algorithm for convex minimization," *Submitted.*

[6] M. Fukushima (1984), "A descent algorithm for nonsmooth convex optimization," *Mathematical Programming*, 30, 163-175.

[7] M. Fukushima and L. Qi (1996), "A globally and superlinearly convergent algorithm for nonsmooth convex minimization," *SIAM J. on Opt.*, 6, 1106-1120.

[8] M. Gaudioso and M.F. Monaco (1982), "A bundle type approach to the unconstrained minimization of convex nonsmooth functions," *Mathematical Programming*, 23, 216-223.

[9] M. Gaudioso and M.F. Monaco (1982), "Variants to the cutting plane approach for convex nondifferentiable optimization," *Optimization*, 25, 65-72.

[10] J.-P. Hiriart-Urruty and C. Lemaréchal. (1993), *Convex analysis and minimization algorithms, Vol. I-II*, Springer-Verlag, Berlin, Heidelberg, New York.

[11] J.E. Kelley (1960), "The cutting plane method for solving convex programs," *Journal of the SIAM*, 8, 703-712.

[12] K.C. Kiwiel. (1985), *Methods of descent for nondifferentiable optimization*, Springer-Verlag, Berlin, Heidelberg.

[13] K.C. Kiwiel (1990), "Proximity control in bundle methods for convex nondifferentiable minimization," *Mathematical Programming*, 46, 105-122.

[14] K.C. Kiwiel (1991), "A tilted cutting plane proximal bundle method for convex nondifferentiable optimization," *Operations Research Letters*, 10, 75-81.

[15] K.C. Kiwiel (1995), "Proximal level bundle methods for convex nondifferentiable optimization, saddle point problems and variational inequalities," *Mathematical Programming*, 69, 89-109.

[16] C. Lemaréchal (1974), "An algorithm for minimizing convex functions," In J.L. Rosenfeld (ed.), *Proceedings IFIP '74 Congress* 17, North-Holland, Amsterdam, 552-556.

[17] C. Lemaréchal (1975), "An extension of Davidon methods to nondifferentiable problems," In M.L. Balinski and P. Wolfe (eds.), *Mathematical Programming Study* 3, North-Holland, Amsterdam, 95-109.

[18] C. Lemaréchal, A. Nemirovskij and Y. Nesterov (1995), "New variants of bundle methods," *Mathematical Programming*, 69, 111-147.

[19] C. Lemaréchal, F. Oustry and C. Sagastizábal (1997), "The $\mathcal{U}$-Lagrangian of a convex function," *Transaction of AMS, accepted for publication.*

[20] C. Lemaréchal and C. Sagastizábal (1997), "Practical aspects of the Moreau-Yosida regularization: theoretical preliminaries," *SIAM Journal on Optimization*, 7(2).

[21] C. Lemaréchal and C. Sagastizábal (1997), "Variable metric bundle methods: from conceptual to implementable form," *Mathematical Programming*, 76, 393-410.

[22] M.M. Mäkelä (1993), "Issues of implementing a Fortran subroutine package NSOLIB for nonsmooth optimization," *Technical report 5/93*, University of Jyväskylä - Department of Mathematics - Laboratory of Scientific Computing.

[23] M.M. Mäkelä and P. Neittaanmäki. (1992), *Nonsmooth optimization*, World Scientific, Singapore, New Jersey, London, Hong Kong.

[24] R. Mifflin (1982), "A modification and an extension of Lemaréchal's algorithm for nonsmooth minimization," In D.C. Sorensen and R.J.B. Wets (eds.), *Mathematical Programming Study* 17, North-Holland, Amsterdam, 77-90.

[25] R. Mifflin (1996), "A quasi-second-order proximal bundle algorithm," *Mathematical Programming*, 73, 51-72.

[26] R. Mifflin and C. Sagastizábal (1998), "$\mathcal{VU}$-Decomposition derivatives for convex max-functions," *Submitted.*

[27] R. Mifflin, D. Sun and L. Qi (1998), "Quasi-Newton bundle-type methods for nondifferentiable convex optimization," *SIAM Journal on Optimization*, 8, 583-603.

[28] L. Qi (1994), "Second-order analysis of the Moreau-Yosida regularization of a convex function," *Technical report AMR 94/20*, University of New South Wales.

[29] L. Qi and X. Chen (1997), "A preconditioning proximal Newton method for nondifferentiable convex optimization," *Mathematical Programming*, 76, 411-429.

[30] R.T. Rockafellar (1976), "Monotone operators and the proximal point algorithm," *SIAM Journal on Control and Optimization*, 14, 877-898.

[31] H. Schramm and J. Zowe (1992), "A version of bundle idea for minimizing a nonsmooth function: conceptual idea, convergence analysis, numerical results," *SIAM Journal on Optimization*, 2(1), 121-152.

[32] N.Z. Shor. (1985), *Minimization methods for nondifferentiable functions*, Springer-Verlag, Berlin, Heidelberg, New York, Tokyo.

[33] P. Wolfe (1975), "A method of conjugate subgradients for minimizing nondifferentiable functions," In M.L. Balinski and P. Wolfe (eds.), *Mathematical Programming Study* 3, North-Holland, Amsterdam, 145-173.

Iterative methods for ill-conditioned linear systems from optimization

Nicholas I. M. Gould (n.gould@rl.ac.uk)
Computational Science and Engineering Department,
Rutherford Appleton Laboratory, Chilton, Oxfordshire, OX11 0QX, England, EU

Abstract

Preconditioned conjugate-gradient methods are proposed for solving the ill-conditioned linear systems which arise in penalty and barrier methods for nonlinear minimization. The preconditioners are chosen so as to isolate the dominant cause of ill conditioning. The methods are stablized using a restricted form of iterative refinement. Numerical results illustrate the approaches considered.

Keywords: penalty functions, ill-conditioning, preconditioning, conjugate gradients.

1 Introduction

Let A and H be, respectively, full-rank m by n ($m \leq n$) and symmetric n by n real matrices. Suppose furthermore that any nonzero coefficients in this data are *modest*, that is the data is $O(1)$.[1] We consider the iterative solution of the linear system

$$(H + A^T D^{-1} A)x_* = b \tag{1.1}$$

where b is modest and D is a positive definite diagonal matrix with small entries. Systems of the form (1.1) occur throughout optimization.

[1] We abuse notation here, since by $O(a)$, we mean a quantity which is at most a small constant > 1 times a.

G. Di Pillo and F. Giannessi (eds.), Nonlinear Optimization and Related Topics, 123–141.

Example 1.1. Given the equality constrained optimization problem

$$\underset{x \in \mathbb{R}^n}{\text{minimize}} \quad f(x) \quad \text{subject to} \quad c_i(x) = 0 \quad (i = 1, \ldots, m) \tag{1.2}$$

and a positive penalty parameter μ, the quadratic penalty function is the composite function

$$\phi(x, \mu) = f(x) + \frac{1}{2\mu} \sum_{i=1}^{m} c_i(x)^2. \tag{1.3}$$

The quadratic penalty method traces a local minimizer $x(\mu)$ of ϕ for a sequence of μ converging to zero. Under mild conditions, the limit of the sequence of $x(\mu)$ gives a local solution of (1.2) [11]. Each sequential minimization is ideally performed using Newton's method. Writing $A^T(x) \stackrel{\text{def}}{=} \nabla_x c(x)$, the Newton equations are

$$\left(H(x, y^q(x, \mu)) + \frac{1}{\mu} A^T(x) A(x) \right) \Delta x = -g(x, y^q(x, \mu)), \tag{1.4}$$

where $g(x, y)$ and $H(x, y)$ are respectively the gradient $\nabla_x \ell(x, y)$ and Hessian matrix $\nabla_{xx} \ell(x, y)$ of the Lagrangian function $\ell(x, y) = f(x) - y^T c(x)$, and the Lagrange multiplier estimates $y^q(x, \mu) = -c(x)/\mu$. As it can be shown that $y_q(x(\mu), \mu)$ converge to Lagrange multipliers at the solution of (1.2), (1.4) is precisely of the form (1.1) with $D = \mu I$. □

Example 1.2. Given the inequality constrained optimization problem

$$\underset{x \in \mathbb{R}^n}{\text{minimize}} \quad f(x) \quad \text{subject to} \quad c_i(x) \geq 0 \quad (i = 1, \ldots, m) \tag{1.5}$$

and a positive barrier parameter μ, the (logarithmic) barrier function is the composite function

$$\psi(x, \mu) = f(x) + \frac{1}{2\mu} \sum_{i=1}^{m} \log c_i(x). \tag{1.6}$$

The barrier method traces a local minimizer $x(\mu)$ of ψ for a sequence of μ converging to zero. As for example 1.1., under mild conditions, the limit of the sequence of $x(\mu)$ gives a local solution of (1.5) [11], and each sequential minimization may be performed by Newton's method. The Newton equations are now

$$\left(H(x, y^b(x, \mu)) + A^T(x) D^{-1}(x, \mu) A(x) \right) \Delta x = -g(x, y^b(x, \mu)), \tag{1.7}$$

where the Lagrange multiplier estimates $y_i^b(x, \mu) = \mu / c_i(x)$ converge to Lagrange multipliers at the solution of (1.5) as x approaches $x(\mu)$, and the diagonal matrix $D(x, \mu)$ has diagonal entries $c_i(x)/y_i^b(x, \mu)$. If the solution to (1.5) is strictly complementary, some entries of $D(x, \mu)$ converge to zero while the remainder approach infinity. Combining the terms for which $d_{ii}(x, \mu)$ approaches infinity (and thus for which $d_{ii}^{-1}(x, \mu)$ converges to zero) and the Hessian $H(x, y^b(x, \mu))$ in a composite H, (1.7) is again of the form (1.1) with D now being made up from those terms for which $d_{ii}(x, \mu)$ converges to zero. □

At a first glance, the presence of the rank m term $A^TD^{-1}A$ would suggest that accurate solutions to (1.1) are hard to obtain when D is small. Indeed, the eigenvalue spectrum splits into two segments, m large eigenvalues corresponding to the $A^TD^{-1}A$ term and the remaining $n-m$ modest eigenvalues from the portion of H lying in the null-space of A [22]. Fortunately, a number of authors ([4], [16]) have noted that accurate solutions are possible despite this ill-conditioning. A simple way of seeing this is to introduce "auxiliary" variables

$$y_* = D^{-1}Ax_* \tag{1.8}$$

into (1.1) to obtain the augmented system

$$\begin{pmatrix} H & A^T \\ A & -D \end{pmatrix} \begin{pmatrix} x_* \\ y_* \end{pmatrix} = \begin{pmatrix} b \\ 0 \end{pmatrix}. \tag{1.9}$$

The conditioning of this problem then depends on the matrix

$$K(D) = \begin{pmatrix} H & A^T \\ A & -D \end{pmatrix}, \tag{1.10}$$

and, for small D is related to that of the perturbation

$$K \equiv K(0) = \begin{pmatrix} H & A^T \\ A & 0 \end{pmatrix}. \tag{1.11}$$

It is the matrix (1.11) which often reflects the "real" conditioning of the underlying problem (see [24] for a discussion of this for optimization applications). Significantly, M. Wright [25] and S. Wright [27] have recently shown that it is actually possible to obtain accurate solutions *directly* from (1.1) in many cases so long as a backward stable factorization of $H + A^TD^{-1}A$ is used—such an approach is commonplace, at least in linear programming circles, since then H vanishes and the computationally convenient Cholesky factors of the positive-definite matrix $A^TD^{-1}A$ obtained (see, for example, [1], or [26] for details).

The system (1.9) then suggests an attractive way of accurately determining x_*. Notice that (1.10) is symmetric and, for sufficiently small D, indefinite. Thus any of the stable symmetric, indefinite factorizations (see, [6], [5], or [12], in the dense case, and [10], or [9], in the sparse case) may be applied.

However, there are two difficulties with such an approach. The first is simply that, just because (1.10) is well conditioned, and just because there are stable methods for solving (1.9), it does not follow that we can compute all the components of the solution with high relative accuracy. Indeed, we would normally only expect to compute the large components of the solution to high accuracy, and to do so might require that we perform iterative refinement (see, for example, [20, Chapter 11]). However, (1.9) suggests that y_* is modest, while, if we consider (1.8), it seems likely[2] that x_* is actually $O(\|D\|)$ rather than simply $O(1)$—we shall say that data of $O(\|D\|)$ is *small*. Thus, while we can compute y_* to high relative accuracy using the method we have suggested, the same is not likely to be true for x_*. In order to avoid this defect, suppose that y_{est} is a highly accurate approximation of y_*, and let $y_{\text{cor}} = y_* - y_{\text{est}}$. Then (1.9) may simply be rearranged to give

$$\begin{pmatrix} H & A^T \\ A & -D \end{pmatrix} \begin{pmatrix} x_* \\ y_{\text{cor}} \end{pmatrix} = \begin{pmatrix} b - A^T y_{\text{est}} \\ D y_{\text{est}} \end{pmatrix}. \tag{1.12}$$

But now, so long as y_{est} is a good approximation of y_*, y_{cor} must be small, and x_* will not be swamped by y_{cor} if a stable method is used to solve (1.12). This suggests we should compute y_* from (1.9), and then recover x_* from (1.12). This is reminiscent of iterative refinement, but since only part of the residual is used, we prefer to call it *iterative semi-refinement*. Note, that even this is not without difficulty since there may be significant cancellation when forming $b - A^T y_{\text{est}}$. A similar method was proposed in the case $H = I$ and $D = 0$ for least squares problems by Businger and Golub [7] and analysed by Golub and Wilkinson [15]. Their conclusions were that such a method only performed well when the required x_* is significantly smaller than y_*, which is fortunately the case of interest here.

The second disadvantage only becomes apparent when m and n are large. For then, it sometimes happens that the factors of $K(D)$ are considerably denser than $K(D)$ itself. Indeed, this fill-in may ultimately cause the factors to exhaust the available computer memory and thus for the factorization to fail. For example, consider the variable-dimension problem `CVXQP3.SIF` from the CUTE test set [3]. Discarding the simple bounds, and factorizing the resulting matrix (1.10) with $D = 10^{-8} I$ using the Harwell Subroutine Library [19] sparse symmetric-indefinite factorization code `MA27` for increasing values of n, we find the following:

n	m	nnz K(D)	nonzeros in factors
100	75	783	7365
1000	750	7981	6200099
10000	7500	79981	> 15721736 (failed)

Table 1.1: Applying `MA27` to the matrix $K(D)$ obtained from `CVXQP3.SIF`. Here "nnz $K(D)$" is the number of nonzeros in the upper triangular part of $K(D)$, while "nonzeros in factors" is the number of real words of storage required to hold the factors.

Thus, we see that the factorization of even moderate-sized problems may be out of the question, and we are forced to consider alternatives.

[2]This is, of course, not rigorous since Ax_* may actually be small because x_* lies close to the null-space of A with x_* being modest.

In this paper, we are concerned with alternative, iterative methods for solving (1.1). Two possibilities immediately present themselves. One could apply an iterative method directly to (1.1), or alternatively, one could use an iterative method to solve (1.9). Both alternatives have disadvantages. It is likely that the bad conditioning of (1.1) will adversely affect an iterative method unless a sophisticated preconditioner is applied. On the other hand, the indefinite nature of (1.10) restricts the choice of iterative methods that might be applied to (1.9); MINRES [23] is probably the method of choice in this case, but the conflict between the form of permissible preconditioners (symmetric and positive definite) and the form of (1.10) (symmetric and indefinite) is unfortunate (see, however, [13], for some possibilities). We will not explore this possibility further in this paper and shall concentrate on iterative methods for (1.1).

In Section 2, we shall consider conjugate-gradient methods for the problem. We start by reviewing the traditional preconditioned conjugate-gradient method, and discuss the form of preconditioners we shall allow. We then indicate that quantities generated by such preconditioners are modest despite the potential for large values. We also explain why some form of iterative refinement is needed when applying the preconditioner to ensure that the iterates behave properly. Finally, we propose variants on the basic method which have different matrix-vector product requirements, and indicate how economies may be made in the iterative refinement without jeopardising the iteration. In Section 3, we perform numerical comparisons between these various methods, using a variety of simple preconditioners. We draw our conclusions and suggest further developments in Section 4.

2 Preconditioned conjugate gradients

2.1 The traditional conjugate-gradient method

We shall suppose in what follows that $H + A^T D^{-1} A$ is positive definite. In this case, the method of (preconditioned) conjugate gradients may be described as follows. We suppose that the *preconditioner* W is a "simple"[3] positive definite matrix which approximates $H + A^T D^{-1} A$. and define the gradient

$$g(x) = (H + A^T D^{-1} A)x - b. \tag{2.13}$$

Then we may solve (1.1) by applying Algorithm 2.1.

This iteration is the traditional conjugate-gradient method applied to the system

$$W^{-1}(H + A^T D^{-1} A)x_* = W^{-1} b, \tag{2.21}$$

[3]Simple here means that $Wx = b$ is easy to solve.

Algorithm 2.1: The preconditioned conjugate-gradient method for (1.1)

Given $x = 0$, $g\ (\equiv g(0)) = -b$, $r = W^{-1}g$, $p = -r$ and $\sigma = r^Tg$, perform the following iteration until convergence:

Form $(H + A^TD^{-1}A)p$

$$\alpha = \sigma / p^T(H + A^TD^{-1}A)p \tag{2.14}$$

$$x_{\text{new}} = x + \alpha p \tag{2.15}$$

$$g_{\text{new}} = g + \alpha(H + A^TD^{-1}A)p \quad (\equiv g(x_{\text{new}})\ ) \tag{2.16}$$

$$r_{\text{new}} = W^{-1}g_{\text{new}} \tag{2.17}$$

$$\sigma_{\text{new}} = g_{\text{new}}^T r_{\text{new}} \tag{2.18}$$

$$\beta = \sigma_{\text{new}} / \sigma \tag{2.19}$$

$$p_{\text{new}} = -r_{\text{new}} + \beta p \tag{2.20}$$

see, for instance, [14, Section 4.8.5.1]. The iteration is normally stopped when the "preconditioned" residual $\sigma = g^Tr \equiv g^TW^{-1}g$ is smaller than some predefined tolerance (possibly relative to the initial "preconditioned" residual). In order to mimic the coefficient matrix $H + A^TD^{-1}A$, we choose preconditioners of the form $M + A^TD^{-1}A$. Luenberger [21, Chapter 12], suggested that, at the very least, the preconditioner should reflect the terms which lead to the ill-conditioning, and proposed a preconditioner of the form $I + A^TD^{-1}A$. The vector r_{new} required in (2.17) must be found as the solution to the system

$$(M + A^TD^{-1}A)r_{\text{new}} = g_{\text{new}}. \tag{2.22}$$

It may appear at first sight that (2.22) will be hard to solve accurately since the matrix $M + A^TD^{-1}A$ has both modest and large eigenvalues. However, as we saw in the introduction, this effect is illusory since, by introducing

$$s_{\text{new}} = D^{-1}Ar_{\text{new}}, \tag{2.23}$$

equation (2.22) may be rewritten as

$$\begin{pmatrix} M & A^T \\ A & -D \end{pmatrix} \begin{pmatrix} r_{\text{new}} \\ s_{\text{new}} \end{pmatrix} = \begin{pmatrix} g_{\text{new}} \\ 0 \end{pmatrix}. \tag{2.24}$$

Provided that M is a good approximation to H and D is small, the coefficient matrix for (2.24) approximates (1.11), which we have suggested reflects the natural conditioning of the problem. Hence both r_{new} and s_{new} are at most modest (they may be smaller) so long as g_{new} is.

2.2 Applying a special conjugate-gradient method to (1.1)

We now show that all computed quantities remain (at most) modest. To see this, suppose that x, g, p, s and

$$q = D^{-1}Ap \tag{2.25}$$

are modest. This is certainly the case at the start of the iteration, since $x = 0$, and $g = -b$, while the latter is modest by the assumptions we made on the problem data at the start of Section 1. Furthermore, $p = -r$, where r and s may be found from

$$\begin{pmatrix} M & A^T \\ A & -D \end{pmatrix} \begin{pmatrix} r \\ s \end{pmatrix} = -\begin{pmatrix} g \\ 0 \end{pmatrix}, \tag{2.26}$$

which has a modest right-hand side, and thus both p and s are modest. Finally, $q = -s$ which is modest because s is. Turning to the iteration itself, (2.15) shows that x_{new} is modest since x and p are,[4] since (2.16) may be replaced by the equivalent

$$g_{\text{new}} = g + \alpha(Hp + A^T q). \tag{2.27}$$

It also follows, as we indicated at the end of the previous paragraph, that r_{new} and s_{new} are modest as g_{new} is. Finally (2.20) and (2.23) show that

$$q_{\text{new}} = -D^{-1}Ar_{\text{new}} + \beta q = -s_{\text{new}} + \beta q, \tag{2.28}$$

and hence q_{new} is also modest. It is worth noting that (2.25) suggests that p is actually small rather than modest, while (2.23) suggests the same is true for s_{new}.

Any worry that the D^{-1} term in the matrix-vector product $(H + A^T D^{-1} A)p$, which occurs in (2.14) and (2.16), might swamp the remaining contribution is unfounded since the product can be rewritten as the modest $Hp + A^T q$. The recurrence (2.28) is also significant algorithmically since (2.16) may be replaced by the less expensive (2.27).

In summary, we propose the following special algorithm, Algorithm 2.2, for solving (1.1).

Notice that there is no longer any need for matrix-vector products involving A, but just with A^T. However, also note that Algorithm 2.2 requires an extra vector of storage above that of its naive predecessor Algorithm 2.1.

[4]This is not strictly rigorous since we are making an implicit assumption here that α and, later, β are modest. In numerical tests, this always seems to be the case.

Algorithm 2.2: A special conjugate-gradient method for problem (1.1)

Given $x = 0$, set $g = -b$, solve

$$\begin{pmatrix} M & A^T \\ A & -D \end{pmatrix} \begin{pmatrix} r \\ s \end{pmatrix} = \begin{pmatrix} g \\ 0 \end{pmatrix} \tag{2.29}$$

and set $p = -r$, $q = -s$ and $\sigma = r^T g$. Perform the following iteration until convergence:

$$\text{Form } Hp + A^T q$$

$$\alpha = \sigma / p^T (Hp + A^T q) \tag{2.30}$$

$$x_{\text{new}} = x + \alpha p \tag{2.31}$$

$$g_{\text{new}} = g + \alpha (Hp + A^T q) \tag{2.32}$$

$$\text{Solve } \begin{pmatrix} M & A^T \\ A & -D \end{pmatrix} \begin{pmatrix} r_{\text{new}} \\ s_{\text{new}} \end{pmatrix} = \begin{pmatrix} g_{\text{new}} \\ 0 \end{pmatrix} \tag{2.33}$$

$$\sigma_{\text{new}} = g_{\text{new}}^T r_{\text{new}} \tag{2.34}$$

$$\beta = \sigma_{\text{new}} / \sigma \tag{2.35}$$

$$\begin{pmatrix} p_{\text{new}} \\ q_{\text{new}} \end{pmatrix} = -\begin{pmatrix} r_{\text{new}} \\ s_{\text{new}} \end{pmatrix} + \beta \begin{pmatrix} p \\ q \end{pmatrix} \tag{2.36}$$

2.3 An alternative conjugate-gradient method for (1.1)

There is one remaining concern over Algorithm 2.2, namely that the potentially small solution components r_{new} in (2.33) may not be computed very accurately, at least relative to the remaining modest components s_{new}. We have seen this problem in Section 1, and the cure in this case is essentially the same. The easiest solution is simply to apply iterative refinement to (2.29) and, at every iteration, to (2.33). A slightly less expensive idea is to note that, as s_{new} is modest, (2.23) indicates that r_{new} is likely of $O(\|D\|)$, and thus it may be preferable to apply iterative semi-refinement rather than iterative refinement to each system. However, it turns out that there is a significantly less expensive alternative.

Suppose that we transform (2.33) to

$$\begin{pmatrix} M & A^T \\ A & -D \end{pmatrix} \begin{pmatrix} r_{\text{new}} \\ u_{\text{new}} \end{pmatrix} = \begin{pmatrix} g_{\text{new}} - A^T z_{\text{new}} \\ D z_{\text{new}} \end{pmatrix}. \tag{2.37}$$

where

$$s_{\text{new}} = z_{\text{new}} + u_{\text{new}}$$

and z_{new} is a good approximation to s_{new}. Suppose further that the previous right-hand side

$$\begin{pmatrix} v \\ w \end{pmatrix} \stackrel{\mathrm{def}}{=} \begin{pmatrix} g - A^T z \\ Dz \end{pmatrix} \tag{2.38}$$

was small. Our aim is to choose z_{new} so that the right-hand side of (2.37) is also small. Consider the first component of this vector. We have that

$$\begin{aligned} v_{\mathrm{new}} \equiv &\ g_{\mathrm{new}} - A^T z_{\mathrm{new}} \\ = &\ g + \alpha(Hp + A^T q) - A^T z_{\mathrm{new}} \\ = &\ g - A^T z + \alpha Hp - A^T(z_{\mathrm{new}} - z - \alpha q) \\ = &\ v + \alpha Hp - A^T(z_{\mathrm{new}} - z - \alpha q) \end{aligned} \tag{2.39}$$

using (2.32) and (2.38). Now suppose we pick

$$z_{\mathrm{new}} = z + \alpha q. \tag{2.40}$$

Then, combining (2.39) and (2.40), we deduce that

$$v_{\mathrm{new}} = v + \alpha Hp. \tag{2.41}$$

But, as we have already noted, (2.25) suggests that p is small, and thus v_{new} is small if v is. Turning to the second component of the right-hand side of (2.37), (2.38) and (2.40) show that

$$w_{\mathrm{new}} \equiv Dz_{\mathrm{new}} = D(z + \alpha q) = w + \alpha Dq, \tag{2.42}$$

which is small, if w is, since q is modest. Furthermore, (2.23) and (2.38) give

$$r^T g = r^T w + r^T A^T z = r^T w + s^T Dz = r^T w + s^T v. \tag{2.43}$$

Since (2.14) may be written as the less expensive

$$\alpha = g^T r / (p^T Hp + q^T Dq), \tag{2.44}$$

and as the matrix-vector product Dq is needed to form the denominator of (2.44), the update (2.42) may be performed efficiently. Combining the above recurrences, we arrive at the following alternative to Algorithm 2.2, Algorithm 2.3.

Notice that there is now no longer any need for matrix-vector products involving A or A^T, nor do we maintain $g(x)$. However, also note that Algorithm 2.3 requires two more vectors of storage than Algorithm 2.2 and three more than Algorithm 2.1.

Algorithm 2.3 needs a suitable starting vector z. Probably the simplest choice is to proceed as we did in Section 1 by first solving

$$\begin{pmatrix} M & A^T \\ A & -D \end{pmatrix} \begin{pmatrix} r \\ z \end{pmatrix} = \begin{pmatrix} -b \\ 0 \end{pmatrix}$$

to obtain z (discarding the inaccurate r). In this case, the initial v and w are both small.

Algorithm 2.3: An alternative conjugate-gradient method for (1.1)

Given $x = 0$ and an appropriate z for which $v = -b - A^T z$ and $w = Dz$ are small, solve

$$\begin{pmatrix} M & A^T \\ A & -D \end{pmatrix} \begin{pmatrix} r \\ u \end{pmatrix} = \begin{pmatrix} v \\ w \end{pmatrix} \tag{2.45}$$

and set $s = z + u$, $p = -r$, $q = -s$ and $\sigma = r^T v + s^T w$. Perform the following iteration until convergence:

$$\text{Form } Hp \text{ and } Dq$$

$$\alpha = \sigma / (p^T Hp + q^T Dq) \tag{2.46}$$

$$\begin{pmatrix} x_{\text{new}} \\ z_{\text{new}} \\ v_{\text{new}} \\ w_{\text{new}} \end{pmatrix} = \begin{pmatrix} x \\ z \\ v \\ w \end{pmatrix} + \alpha \begin{pmatrix} p \\ q \\ Hp \\ Dq \end{pmatrix} \tag{2.47}$$

$$\text{Solve } \begin{pmatrix} M & A^T \\ A & -D \end{pmatrix} \begin{pmatrix} r_{\text{new}} \\ u_{\text{new}} \end{pmatrix} = \begin{pmatrix} v_{\text{new}} \\ w_{\text{new}} \end{pmatrix} \tag{2.48}$$

$$s_{\text{new}} = z_{\text{new}} + u_{\text{new}} \tag{2.49}$$

$$\sigma_{\text{new}} = r_{\text{new}}^T v_{\text{new}} + s_{\text{new}}^T w_{\text{new}} \tag{2.50}$$

$$\beta = \sigma_{\text{new}} / \sigma \tag{2.51}$$

$$\begin{pmatrix} p_{\text{new}} \\ q_{\text{new}} \end{pmatrix} = -\begin{pmatrix} r_{\text{new}} \\ s_{\text{new}} \end{pmatrix} + \beta \begin{pmatrix} p \\ q \end{pmatrix} \tag{2.52}$$

2.4 Cost of the algorithm

In Table 2.2, we compare the cost per iteration of the three algorithms given in this section, neglecting the (possibly not inconsiderable) cost of applying the preconditioner. Which variant proves to be the most cost effective depends on the density of the matrices involved.

Algorithm	vectors required	matrix-vector products	inner-products	axpys
2.1	$4(n)$, $1(m)$	H, A, A^T, D	$2(n)$	$4(n)$
2.2	$4(n)$, $2(m)$	H, A^T	$2(n)$	$4n$, $1(m)$
2.3	$4(n)$, $4(m)$	H, D	$2(n)$, $2(m)$	$5(n)$, $2\ (m)$

Table 2.2: Costs per iteration of Algorithms 2.1–2.3. The parentheses indicate the lengths of vectors involved.

2.5 Stablizing the conjugate-gradient method for (1.1)

In the development of the previous algorithm, there was a tacit assumption that if we balance the right-hand side of (2.48), then the solution to (2.48) will itself be balanced. Of course, this may not necessarily be the case, and we may still be forced to apply some form of iterative refinement if we wish to generate accurate solutions. Since we prefer iterative semi-refinement, we propose the following algorithm.

Algorithm 2.4: iterative semi-refinement: in (v, w, z), out (r, u, v, w, z)

1. Solve

$$\begin{pmatrix} M & A^T \\ A & -D \end{pmatrix} \begin{pmatrix} r \\ u \end{pmatrix} = \begin{pmatrix} v \\ w \end{pmatrix}$$

2. If necessary, update

$$\begin{pmatrix} v \\ w \\ z \end{pmatrix} \leftarrow \begin{pmatrix} v - A^T u \\ w + Du \\ z + u \end{pmatrix}$$

and resolve

$$\begin{pmatrix} M & A^T \\ A & -D \end{pmatrix} \begin{pmatrix} r \\ u \end{pmatrix} = \begin{pmatrix} v \\ w \end{pmatrix}$$

It is easy to see that we may replace (2.45) and (2.48) by calls to Algorithm 2.4. Indeed, the only relationship we need to check is that (2.50) is valid if Step 2 of Algorithm 2.4 is applied. If we (temporarily) denote the quantities computed in this step with a bar, we trivially have that $\bar{r} = r$ and $\bar{s} = s$, while

$$\bar{\sigma} = \bar{r}^T \bar{v} + \bar{s}^T \bar{w} = r^T (v - A^T u) + s^T (w + Du) = \sigma - u^T (Ar - Ds) = \sigma$$

follows from (2.26). In summary, our preferred algorithm may be stated as Algorithm 2.5.

The remaining issue is to decide when Step 2 of Algorithm 2.4 is required. We have experimented with a number of possibilities, but have seen little, if any, improvement over the simplest expedient of executing Step 2 whenever $\|r\|$ is significantly smaller than $\|u\|$ in Step 1. We have settled on executing Step 2 whenever

$$\|r\| \leq \|D\|^{0.5} \|u\|$$

Algorithm 2.5: A stablized conjugate-gradient method for (1.1)

Given $x = 0$, let $v = -b$, $w = 0$ and $z = 0$, apply Algorithm 2.4 to obtain (r, u, v, w, z), and set $s = z + u$, $p = -r$, $q = -s$ and $\sigma = r^T v + s^T w$. Perform the following iteration until convergence:

Form Hp and Dq

$$\alpha = \sigma/(p^T Hp + q^T Dq) \tag{2.53}$$

$$\begin{pmatrix} x_{\text{new}} \\ z_{\text{new}} \\ v_{\text{new}} \\ w_{\text{new}} \end{pmatrix} = \begin{pmatrix} x \\ z \\ v \\ w \end{pmatrix} + \alpha \begin{pmatrix} p \\ q \\ Hp \\ Dq \end{pmatrix} \tag{2.54}$$

Apply Algorithm 2.4 to obtain $(r_{\text{new}}, u_{\text{new}}, v_{\text{new}}, w_{\text{new}}, z_{\text{new}})$

$$s_{\text{new}} = z_{\text{new}} + u_{\text{new}} \tag{2.55}$$

$$\sigma_{\text{new}} = r_{\text{new}}^T v_{\text{new}} + s_{\text{new}}^T w_{\text{new}} \tag{2.56}$$

$$\beta = \sigma_{\text{new}}/\sigma \tag{2.57}$$

$$\begin{pmatrix} p_{\text{new}} \\ q_{\text{new}} \end{pmatrix} = -\begin{pmatrix} r_{\text{new}} \\ s_{\text{new}} \end{pmatrix} + \beta \begin{pmatrix} p \\ q \end{pmatrix} \tag{2.58}$$

as an acceptable heuristic.

We remark that Algorithm 2.4 could equally well be applied to Algorithm 2.2. However, Algorithm 2.2, unlike Algorithm 2.3, makes no effort to balance the right-hand sides of the preconditioning systems (2.29) and (2.33), and thus Step 2 of Algorithm 2.4 will likely be necessary at every iteration.

3 Numerical Experiments

In this section we compare Algorithms 2.1–2.3 using a variety of preconditioners on a number of standard optimization test problems. The problems we use arise as the larger convex quadratic programs—that is problems of the form

$$\underset{x \in \mathbb{R}^n}{\text{minimize}} \ \tfrac{1}{2} x^T Hx + c^T x \ \text{ subject to } \ Ax = b \ \text{ and } \ l \leq x \leq u$$

for given vectors b, c, l and u—within the CUTE test set [3]. For those examples which lie in a class of similar problems, a representative is chosen. From these examples, we discard the data b and c and add a specified amount (in our examples, 0.1) to the i-th diagonal of H if either of l_i or u_i is finite—the intention here is to try to mimic the type of Hessian matrix which might have arisen from a mixed penalty-barrier function for the quadratic program, and the entries on the diagonal correspond to barrier terms for the simple bounds $l \leq x \leq u$. We now construct the vector b in (1.1) by specifying x_* and setting $y_* = D^{-1}Ax_*$ and forming $b = Hx_* + A^Ty_*$. Since we are presuming that b should be modest, this requires that the same is true of y_*, and so suggests that we should choose $x_* = O(\|D\|)$. In our experiments, we pick $D = \mu I$ for some small μ and $x_* = \mu e$, where e is the vector of ones. We report on those experiments for which $\mu = 10^{-8}$. Similar results were obtained with random small D.

All experiments were performed in double precision on an IBM RISC System/6000 3BT workstation with 64 Megabytes of RAM, using the xlf90 compiler and optimization level -O3. All matrix factorizations are performed using the forthcoming Harwell Subroutine Library sparse symmetric-indefinite factorization module HSL_MA27. The iteration is stopped as soon as the test

$$\sigma \leq \max(\epsilon_r \cdot \sigma_0, \epsilon_a)$$

is satisfied. Here σ_0 is the initial value of σ, while ϵ_r and ϵ_a are small constants—in our experiments, $\epsilon_r = 10^{-12}$, while ϵ_a is set to the machine precision ($\approx 10^{-16}$). A run is considered unsuccessful whenever more than $2(n - m + 1)$ iterations have occurred, or more than 1800 CPU seconds are needed, or more than one million words of real storage is required for the factors of the preconditioner.

We first illustrate the need for preconditioning. In Table 3.3, we compare the solution obtained using the basic Algorithm 2.1 without preconditioning, with those from the same algorithm using Luenberger's preconditioner $M = I$, as well as Algorithm 2.2 with the same preconditioner. A single step of iterative refinement is applied, since there are frequent failures when the preconditioning systems without such a precaution.

We make the following observations. Firstly, as one might expect, even a naive preconditioner $M = I$ provides a dramatic reduction in the number of iterations performed. Of course, a reduction in the iteration count does not necessarily result in a reduction in the total CPU time required (see, for example, AUG3DCQP), since the cost of forming and applying the preconditioner is often non-negligible. When it is possible to form the factors of the preconditioner in Algorithm 2.1, this algorithm is often, but not always, faster than Algorithm 2.2. However, the number of failures due to the density of the factors of $W = I + A^TD^{-1}A$ limits the use of such a method. In this sense, Algorithm 2.2 is to be preferred, since (at least for these examples) the algorithm always succeeded. Perhaps the best strategy would be to use Algorithm 2.1 so long as a factorization of $M + A^TD^{-1}A$ is possible, keeping Algorithm 2.2 in reserve for the remaining problems. However, using Algorithm 2.1 with more sophisticated preconditioners is even more likely to fail through lack of storage space.

Next, we consider using slightly more general, but nonetheless rather unsophisticated, preconditioners. The first is simply to use $M = H$, which is equivalent to

			Algorithm 2.1						Algorithm 2.2		
			$W = I$			$M = I$			$M = I$		
Problem	n	m	err	iter	cpu	err	iter	cpu	err	iter	cpu
AUG2DCQP	20200	10000	-16	484	40	-16	3	5	-16	3	3
AUG2DQP	20200	10000	-12	1422	118	-15	13	7	-15	13	6
AUG3DCQP	27543	8000	-15	144	16		mem	16	-15	3	18
AUG3DQP	27543	8000	-11	353	38		mem	16	-15	11	22
BLOCKQP1	20006	10001	-8	2	0		mem	0	-8	1	2
BLOWEYA	20002	10002	-8	8112	771		mem	0	-8	10	3
CVXQP1	15000	7500		maxit	1202	-16	2167	260	-16	2173	523
GOULDQP2	19999	9999	-10	295	23	-14	39	5	-14	39	10
KSIP	10021	10001	-8	14	2	-10	32	8	-9	31	11
MOSARQP1	30000	10000	-10	479	55	-14	64	23	-14	64	36
SOSQP1	20000	10001	-21	1	0		mem	0	-12	2	1
STCQP2	8193	4095	-11	779	46	-15	157	11	-15	157	18
UBH1	18009	12000	-8	10082	740	-8	3589	402	-8	3178	770
YAO	20002	10000	-10	93	7	-14	14	2	-14	14	4

Table 3.3: Primitive preconditioners. Key: "n, m" = number of columns and rows of A, "err" = $log_{10}\|x - x_*\|$, "iter" = number of iterations, "cpu" = CPU time required in seconds, "maxit" indicates that more than $2(n - m + 1)$ iterations were required, "mem" indicates that the factors required more than one million words of storage.

solving the problem using a direct method. At the other extreme, the second preconditioner used is simply to pick M as the diagonal of H. A third choice is barely more sophisticated, in which M is a band matrix whose entries within the band correspond to those of H, except that the i-th and j-th diagonal entries for each term h_{ij} outside the band are augmented by $\sqrt{|h_{ij}|}$ [2]: we consider both diagonal and tridiagonal bands enhanced in this way. The effect of these preconditioners on Algorithm 2.2 is illustrated in Table 3.4; the final columns from the previous table are repeated for reference purposes.

Now we see the effect, as well as the cost, of more sophisticated preconditioning. In general, the more M is allowed to reflect the true structure of H, the better (in general) is the convergence behaviour, and conversely, the more chance that there will be insufficient room to hold the factors. In particular, the choice $M = H$ is most effective when there is sufficient space, while the diagonal and, especially, the enhanced diagonal preconditioners curb the wilder behaviour of the identity preconditioners with little loss in reliability. Ideally, a form of incomplete factorization, in which entries/fill-ins from H, but not A or D, might be removed, would seem to be a useful possibility, and we are currently investigating such ideas.

Finally, in Table 3.5, we examine whether the stablized Algorithm 2.5 is effective.

	$M = I$			$M = H$			$M =$ diagonal of H			$M =$ enhanced diagonal of H			$M =$ enhanced tridiagonal of H		
Problem	err	iter	cpu	err	iter	cpu	err	iter	cpu	err	iter	cpu	err	iter	cpu
AUG2DCQP	-16	3	3	-16	1	2	-16	1	2	-16	1	2	-16	1	2
AUG2DQP	-15	13	6	-15	1	2	-15	1	2	-15	1	2	-15	1	2
AUG3DCQP	-15	3	18	-15	1	17	-15	1	17	-15	1	17	-15	1	17
AUG3DQP	-15	11	22	-15	1	17	-15	1	17	-15	1	17	-15	1	16
BLOCKQP1	-8	1	2		cond	2	-8	3	2	-8	1	2		cond	2
BLOWEYA	-8	10	3		mem	110		mem	90	-9	35	10		mem	115
CVXQP1	-16	2173	523		mem	15		mem	9		mem	9		mem	9
GOULDQP2	-14	39	10	-15	2	2	-15	19	6	-14	19	6	-15	2	2
KSIP	-9	31	11	-11	2	4	-11	2	4	-11	2	4	-11	2	4
MOSARQP1	-14	64	36	-15	2	6	-14	9	9	-15	12	11	-15	12	11
SOSQP1	-12	2	1	-13	2	2	-13	2	2	-12	1	1	-13	2	1
STCQP2	-15	157	18		mem	6	-15	35	5	-14	34	6	-14	33	6
UBH1	-8	3178	770	-13	2	2	-13	2	2	-13	2	2	-13	2	2
YAO	-14	14	4	-16	1	1	-16	1	1	-16	1	2	-16	1	1

Table 3.4: Using Algorithm 2.2 with more sophisticated preconditioners. Key: "err" $= log_{10}\|x - x_*\|$, "iter" = number of iterations, "cpu" = CPU time required in seconds, "cond" indicates that the matrix was too ill-conditioned for iterative refinement to converge, "mem" indicates that the factors required more than one million words of storage.

We observe that the method is no less effective or reliable than its predecessors, but offers a significant CPU time saving in some cases. For example, compare CVXQP1,

	$M = I$			$M = H$			$M =$ diagonal of H			$M =$ enhanced diagonal of H			$M =$ enhanced tridiagonal of H		
Problem	err	iter	cpu	err	iter	cpu	err	iter	cpu	err	iter	cpu	err	iter	cpu
AUG2DCQP	-17	3 (3)	3	-17	1 (2)	2	-17	1 (2)	2	-17	1 (2)	2	-17	1 (2)	2
AUG2DQP	-15	13 (2)	4	-16	1 (2)	2	-16	1 (2)	2	-16	1 (2)	2	-16	1 (2)	2
AUG3DCQP	-15	3 (3)	18	-15	1 (2)	17	-15	1 (2)	17	-15	1 (2)	17	-15	1 (2)	17
AUG3DQP	-15	11 (2)	20	-16	1 (2)	17	-16	1 (2)	17	-16	1 (2)	17	-16	1 (2)	17
BLOCKQP1	-8	2 (1)	2		c (0)	2	-7	2 (2)	2	-8	1 (2)	2		c (0)	2
BLOWEYA	-8	10 (11)	4		m (0)	109		m (0)	90	-10	17 (11)	5		m (0)	115
CVXQP1	-13	2456 (16)	364		m (0)	15		m (0)	9		m (0)	9		m (0)	9
GOULDQP2	-14	39 (2)	7	-15	1 (2)	2	-14	19 (2)	4	-14	19 (2)	4	-15	1 (2)	2
KSIP	-8	28 (2)	7	-8	1 (1)	4	-8	1 (1)	4	-8	1 (1)	4	-8	1 (1)	4
MOSARQP1	-14	64 (2)	23	-15	1 (1)	6	-14	9 (1)	8	-15	11 (1)	8	-15	11 (1)	8
SOSQP1	-20	1 (2)	1	-20	1 (2)	1	-20	1 (2)	1	-20	1 (2)	1	-20	1 (2)	1
STCQP2	-16	160 (3)	12		m (0)	6	-15	35 (4)	3	-14	34 (5)	3	-14	33 (4)	3
UBH1	-8	4536 (4)	674	-11	1 (2)	2	-11	1 (2)	2	-11	1 (2)	2	-11	1 (2)	2
YAO	-14	14 (2)	3	-16	1 (1)	1	-16	1 (1)	1	-16	1 (1)	1	-16	1 (1)	1

Table 3.5: Using Algorithm 2.5 with more sophisticated preconditioners. Key: "err" $= log_{10}\|x - x_*\|$, "iter" = number of iterations, with the number of iterative semi-refinements in brackets, "cpu" = CPU time required in seconds, "c" indicates that the matrix was too ill-conditioned for iterative refinement to converge, "m" indicates that the factors required more than one million words of storage.

MOSARQP1, STCQP2 and UBH1 when $M = I$, and BLOWEYA, GOULDQP2, MOSARQP1 and STCQP2 when M is the enhanced diagonal of H, in Tables 3.4 and 3.5. Notice, in particular, how few iterations require iterative semi-refinement when $M = I$. For the other cases, the total number of iterations is rarely high enough for these savings to become apparent in the CPU times, but for those examples where a significant reduction in CPU time occurs, there is a corresponding drop in the number of iterative semi-refinements required (see, for instance, STCQP2 when M is the enhanced diagonal of H).

4 Comments and perspectives

In this paper, we have shown that the iterative solution of highly ill-conditioned structured linear systems from optimization is possible even when the application of a direct method to the original system, or to its related augmented system, fails through lack of memory. We believe that the preconditioner must reflect the dominant form of ill-conditioning, while efforts to reflect the remaining structure of the coefficient matrix are also worthwhile. When applying such preconditioners, care must be taken to ensure that the solution is computed accurately. This may be achieved by some form of iterative refinement, and our preferred Algorithm 2.5 aims to lessen the need for refinement at every iteration.

We have not yet examined the best choice for the matrix M in the preconditioner, although it is clear that some compromise between the goal of controlling the non-dominant conditioning, and the cost of forming and manipulating the resulting preconditioner must be reached. Our current belief is that some form of incomplete factorization in which entries or fill-ins from H, but not from A or D, may be dropped is worth investigating in the future.

Of course, we have made a tacit assumption that $H + A^TD^{-1}A$ is positive definite throughout this paper. In optimization applications, systems of the form (1.1) most readily appear as necessary optimality conditions for the problem

$$\underset{x \in \mathbb{R}^n}{\text{minimize}} \ \tfrac{1}{2} x^T(H + A^TD^{-1}A)x - b^Tx.$$

We may remove the restriction that $H + A^TD^{-1}A$ be positive definite in these cases so long as we introduce some other means of "convexifying" the problem; the most popular technique is to impose a "trust-region" constraint of the form $\|x\| \leq \Delta$, for some $\Delta > 0$. Such ill-conditioned trust-region problems have been considered by Coleman and Hempel [8] when the dimension is small, but the challenge is to construct effective iterative methods for solving larger problems. Our belief is that this is possible by combining the methods proposed here with the Generalized Lanczos Trust-Region method recently proposed by Gould, Lucidi, Roma and Toint [18].

Similar methods for quadratic programming (where $D = 0$) are currently being considered [17].

Acknowledgement

The author is grateful to Iain Duff for his comments and suggestions on an earlier draft of this paper. Useful conversations with Jorge Nocedal and Steve Wright at the 1998 Erice workshop on Nonlinear Optimization and Applications are also appreciated.

References

[1] E. D. Andersen, J. Gondzio, C. Mészáros, and X. Xu. Implementation of interior point methods for large scale linear programming. In T. Terlaky, editor, *Interior Point Methods in Mathematical Programming*, pages 189–252, Dordrecht, The Netherlands, 1996. Kluwer Academic Publishers.

[2] M. A. Aziz and A. Jennings. A robust incomplete Choleski conjugate gradient algorithm. *International Journal on Numerical Methods in Engineering*, 20:949–966, 1984.

[3] I. Bongartz, A. R. Conn, N. I. M. Gould, and Ph. L. Toint. CUTE: Constrained and unconstrained testing environment. *ACM Transactions on Mathematical Software*, 21(1):123–160, 1995.

[4] C. G. Broyden and N. F. Attia. A smooth sequential penalty function method for nonlinear programming. In A. V. Balakrishnan and M. Thomas, editors, *11th IFIP Conference on System Modelling and Optimization*, number 59 in Lecture Notes in Control and Information Sciences, pages 237–245, Heidelberg, Berlin, New York, 1984. Springer Verlag.

[5] J. R. Bunch and L. C. Kaufman. Some stable methods for calculating inertia and solving symmetric linear equations. *Mathematics of Computation*, 31:163–179, 1977.

[6] J. R. Bunch and B. N. Parlett. Direct methods for solving symmetric indefinite systems of linear equations. *SIAM Journal on Numerical Analysis*, 8(4):639–655, 1971.

[7] P. Businger and G. H. Golub. Linear least squares solutions by Housholder transformations. *Numerische Mathematik*, 7:269–276, 1965.

[8] T. F. Coleman and C. Hempel. Computing a trust region step for a penalty function. *SIAM Journal on Scientific Computing*, 11(1):180–201, 1990.

[9] I. S. Duff and J. K. Reid. The multifrontal solution of indefinite sparse symmetric linear equations. *Transactions of the ACM on Mathematical Software*, 9(3):302–325, 1983.

[10] I. S. Duff, J. K. Reid, N. Munksgaard, and H. B. Neilsen. Direct solution of sets of linear equations whose matrix is sparse, symmetric and indefinite. *Journal of the Institute of Mathematics and its Applications*, 23:235–250, 1979.

[11] A. V. Fiacco and G. P. McCormick. *Nonlinear Programming: Sequential Unconstrained Minimization Techniques.* J. Wiley and Sons, Chichester, England, 1968. Reprinted as *Classics in Applied Mathematics 4*, SIAM, Philadelphia, USA, 1990.

[12] R. Fletcher. Factorizing symmetric indefinite matrices. *Linear Algebra and its Applications*, 14:257–272, 1976.

[13] P. E. Gill, W. Murray, D. B. Ponceléon, and M. A. Saunders. Preconditioners for indefinite systems arising in optimization. *SIAM Journal on Matrix Analysis and Applications*, 13(1):292–311, 1992.

[14] P. E. Gill, W. Murray, and M. H. Wright. *Practical Optimization.* Academic Press, London, 1981.

[15] G. H. Golub and J. H. Wilkinson. Note on iterative refinement of least squares solutions. *Numerische Mathematik*, 9:139–148, 1966.

[16] N. I. M. Gould. On the accurate determination of search directions for simple differentiable penalty functions. *IMA Journal of Numerical Analysis*, 6:357–372, 1986.

[17] N. I. M. Gould, M. E. Hribar, and J. Nocedal. On the solution of equality constrained quadratic problems arising in optimization. Technical Report RAL-TR-98-069, Rutherford Appleton Laboratory, Chilton, Oxfordshire, England, 1998.

[18] N. I. M. Gould, S. Lucidi, M. Roma, and Ph. L. Toint. Solving the trust-region subproblem using the Lanczos method. *SIAM Journal on Optimization*, 9(2):504–525, 1999.

[19] Harwell Subroutine Library. *A catalogue of subroutines (release 12).* AEA Technology, Harwell, Oxfordshire, England, 1995.

[20] N. J. Higham. *Accuracy and Stability of Numerical Algorithms.* SIAM, Philadelphia, USA, 1996.

[21] D. G. Luenberger. *Linear and Nonlinear Programming.* Addison-Wesley Publishing Company, Reading, Massachusetts, USA, second edition, 1984.

[22] W. Murray. Analytical expressions for eigenvalues and eigenvectors of the Hessian matrices of barrier and penalty functions. *Journal of Optimization Theory and Applications*, 7:189–196, 1971.

[23] C. C. Paige and M. A. Saunders. Solution of sparse indefinite systems of linear equations. *SIAM Journal on Numerical Analysis*, 12(4):617–629, 1975.

[24] S. M. Robinson. Generalized equations and their solutions. 2. applications to non-linear programming. *Mathematical Programming Studies*, 19:200–221, 1982.

[25] M. H. Wright. Ill-conditioning and computational error in interior methods for nonlinear programming. *SIAM Journal on Optimization*, 9(1):84–111, 1999.

[26] S. J. Wright. *Primal-Dual Interior-Point Methods.* SIAM, Philadelphia, USA, 1997.

[27] S. J. Wright. Effects of finite-precision arithmetic on interior-point methods for nonlinear programming. Technical Report MCS-P705-0198, Argonne National Laboratory, Argonne, Illinois, USA, 1998.

An Algorithm for Solving Nonlinear Programs with Noisy Inequality Constraints[1]

Michael Hintermüller (michael.hintermueller@kfunigraz.ac.at)
Institute of Mathematics, Karl-Franzens University of Graz, Graz, Austria

Abstract

An algorithm for solving nonlinear programming problems with noisy inequality constraints is developed. The efficiency of the new algorithm relies on the usage of surrogate functions for the unknown exact constraints. These surrogates are based on the bundle idea well-known from nonsmooth optimization and exhibit some very desirable differentiability properties. The overall method is a quasi-Newton type procedure which reduces to a powerful (locally) superlinearly convergent algorithm in the noiseless case provided the bundle is controlled appropriately. It is proved that first order Kuhn-Tucker conditions are satisfied asymptotically. Also a qualification result for the solution obtained by the algorithm is given. Finally, some encouraging numerical tests are reported.

Keywords: Algorithm, bundle, noisy function, quasi-Newton, sequential quadratic programming.

1 Introduction

In this paper, we consider the nonlinear programming problem

$$\begin{aligned} f(x) &= \min! \\ c_i(x) &\leq 0 \qquad i \in I, \end{aligned} \tag{1}$$

where $f : \mathbb{R}^n \to \mathbb{R}$, $c_i : \mathbb{R}^n \to \mathbb{R}$ for all $i \in I$ are twice continuously differentiable functions, and $I \subset \mathbb{N}$ is a finite set. Actually, at a point $x \in \mathbb{R}^n$ we have knowledge of $f(x)$, approximate constraint values $\tilde{c}_i^x$, and error estimates $\tilde{\varepsilon}_i^x$ for all $i \in I$, which ensure $|\tilde{c}_i^x - c_i(x)| \leq \tilde{\varepsilon}_i^x$ for all x. Moreover, approximate differential information $\tilde{a}_i^x$, $i \in I$, is at our disposal. Concerning the *noisy data* we invoke the following assumptions:

[1]Work supported by the Austrian Academy of Sciences.

G. Di Pillo and F. Giannessi (eds.), Nonlinear Optimization and Related Topics, 143–167.

Assumption 1.1 *For each $i \in I$ we have $\tilde{c}_i : \mathbb{R}^n \times \mathbb{R}^n \to \mathbb{R}$ twice continuously differentiable with respect of the first argument and $\tilde{\varepsilon}_i : \mathbb{R}^n \to \mathbb{R}$ such that*

$$\tilde{c}_i(x, y) \mid_{y=x} = \tilde{c}_i^x, \quad \nabla_x \tilde{c}_i(x, y) \mid_{y=x} = \tilde{a}_i^x, \quad \tilde{\varepsilon}_i(x) = \tilde{\varepsilon}_i^x.$$

Assumption 1.2 *The error-function $\tilde{\varepsilon}_i(\cdot)$ is uniformly bounded, i.e there exists $\varepsilon > 0$ such that*

$$|\tilde{\varepsilon}_i(y)| \leq \varepsilon \quad \text{for all } y \in \mathbb{R}^n \text{ and } i \in I.$$

Assumption 1.3 *For all $i \in I$ there exists a neighbourhood $U_i \subset \mathbb{R}^n$ such that*

$$|\tilde{c}_i(x, y) - c_i(x)| \leq \tilde{\varepsilon}_i(y) \quad \text{for all } y \in \mathbb{R}^n \text{ and } x \in y + U_i.$$

Assumption 1.4 *For all $i \in I$ there exists a positive constant K_i' such that*

$$\|\nabla^2_{xx} \tilde{c}_i(\cdot, y)\| \leq K_i' \quad \text{for all } y \in \mathbb{R}^n.$$

Remark 1.1 *Without loss of generality one may assume that K_i' also denotes the bound of the second order derivative of the exact constraint function $c_i(\cdot)$.*

Our assumption on $c_i(\cdot)$ being corrupted by some noise is realistic in many practical applications. Values of process output characteristics gained by measurement are usually noise-corrupted. In problems of optimal control one may regard the functions $c_i(x)$, $i \in I$, as obtained by solving (partial) differential equations such as initial value problems or boundary value problems. As usual one will apply a numerical algorithm to solve the underlying differential equation. In this situation, it is well known that errors caused by discretization, truncation, computer calculation and so on have to be accepted. Standard solvers may provide upper estimates of these errors. Hence, for an arbitrary $x \in \mathbb{R}^n$ we obtain the value of the corresponding approximative solution $\tilde{c}_i^x$ of the differential equation, some gradient information (more general differential information) $\tilde{a}_i^x$ and finally an error-estimate $\tilde{\varepsilon}_i^x$.

If one applies a standard algorithm of nonlinear, constrained and differentiable optimization, see for example [9], [10], [12], no convergence will be achieved. The problems are caused by line search. Usually the original problem (1) is approximated by a sequence of quadratic problems (SQP) to obtain descent directions. The noisy data $\tilde{c}_i^x$ and $\tilde{a}_i^x$ may represent such poor values that the SQP-solution — if it exists — yields no useful direction and therefore the line search breaks down. This behaviour may occur even far away from the optimal solution.

The *simplex method for function minimization* of Nelder and Mead [8] was one of the first algorithms being able to cope with noisy function values. The method exhibits a few disadvantages. First, it has only a slow rate of convergence due to the lack of gradient information. False convergence may occur, and the method is restricted to small sized problems. The methods of Winfield [17] and Glad and Goldstein [4] need a large amount of memory during the calculation. Moreover, the computational effort may vary up to the sixth power of the problem dimension. Elster and Neumaier [2],[3] propose *a grid algorithm for bound constrained optimization of noisy functions* which is well suited for small sized problems. The *pattern search methods* of Torczon [15],[16], which may be extended to noisy functions as well, are again restricted to problems of low dimension.

Unlike the methods encountered above, the algorithm developed in this paper can handle large problems. The method is based on suitably chosen surrogate functions for the exact and unknown constraints. The computation of the surrogates is based on the *bundle concept* of nonsmooth optimization. The new algorithm is a quasi-Newton like procedure and can be reduced to a (locally) superlinearly convergent method in the noiseless case. To achieve this, one has to control the bundle appropriately. The paper is organized as follows: In Section 2 the surrogate functions for the unknown constraints are introduced and analysed. Section 3 is devoted to an efficient way of computing the surrogate functions by using the dual problem. Moreover, a suitable reset strategy is developed. The algorithm is displayed in Section 4, and the convergence analysis including qualification results is given in Section 5. Finally, Section 6 contains numerical aspects for practical implementation and a report on preliminary tests.

2 Bundle based approximations from above

Since we only have noisy data of the constraints of (1) at our disposal an appropriate model for the feasible set has to be found. A Taylor expansion and Assumption 1.1 - 1.4 yield for all $x \in y + U_i$

$$\tilde{c}_i^y + \tilde{\varepsilon}_i^y + \tilde{a}_i^{yT}(x - y) + \frac{K_i'}{2}\|x - y\|^2 \geq c_i(x). \tag{2}$$

This local behaviour can be extended to a global one by finding a suitable constant $K_i \geq K_i'$. It is essential to bound the distance between the noisy differential information and the exact gradient since this distance basically influences the choice of K_i.

Let r_i denote the radius of a ball contained in U_i given by Assumption 1.3, and let $a_i(\cdot) = \nabla c_i(\cdot)$ in the sequel.

Lemma 2.1 *Suppose that Assumption 1.1 - 1.4 hold. Then*

$$\|\tilde{a}_i^y - a_i(y)\| \leq \max\left\{\sqrt{2\,\tilde{\varepsilon}_i^y K_i'},\ \frac{2\tilde{\varepsilon}_i^y}{r_i}\right\} + \min\left\{r_i K_i',\ \sqrt{2\,\tilde{\varepsilon}_i^y K_i'}\right\}$$

for all $y \in \mathbb{R}^n$ *and* $i \in I$.

Proof: Let $y \in I\!R^n$ and $i \in I$ be arbitrarily fixed. By Assumptions 1.1 - 1.4 we have

$$\tilde{c}_i^y + \tilde{\varepsilon}_i^y + \tilde{a}_i^{y^T}(x-y) + \frac{K_i'}{2}\|x-y\|^2 \geq \tilde{c}_i(x,y) + \tilde{\varepsilon}_i^y \geq c_i(x) \tag{3}$$

for all $x \in y + U_i$. A Taylor expansion and application of the bound to the second order derivative of $c_i(\cdot)$ allow to estimate $c_i(x)$ in (3) resulting in

$$c_i(x) - a_i(y)^T(x-y) + \frac{K_i'}{2}\|x-y\|^2 \geq c_i(y) \geq \tilde{c}_i^y - \tilde{\varepsilon}_i^y. \tag{4}$$

Hence, (3) and (4) yield

$$\left(a_i(y) - \tilde{a}_i^y\right)^T(x-y) \leq 2\,\tilde{\varepsilon}_i^y + K_i'\|x-y\|^2 \tag{5}$$

for all $x \in I\!R^n$ satisfying $\|x-y\| \leq r_i$. Fixing x on a line segment by putting $x - y := \alpha\, r_i(a_i(y) - \tilde{a}_i^y)\|a_i(y) - \tilde{a}_i^y\|^{-1}$ we obtain under consideration of (5)

$$\alpha\, r_i\|a_i(y) - \tilde{a}_i^y\| \leq 2\,\tilde{\varepsilon}_i^y + K_i'\alpha^2 r_i^2$$

for all $\alpha \in [0,\ 1]$ and hence

$$\|a_i(y) - \tilde{a}_i^y\| \leq \inf_{\alpha \in (0,\ 1]}\left\{2\,\tilde{\varepsilon}_i^y\alpha^{-1}r_i^{-1} + K_i'\alpha\, r_i\right\}. \tag{6}$$

Some elementary calculations show that the infimum on the right hand side of (6) is attained for $\alpha = \min\{r_i^{-1}\sqrt{2\,\tilde{\varepsilon}_i^y K_i'^{-1}},\ 1\}$ proving the assertion of Lemma 2.1.□

We are now able to ensure the existence of a global constant satisfying relation (2) for all $x \in I\!R^n$.

Lemma 2.2 *Suppose that Assumption 1.1 - 1.4 hold. Then for each $i \in I$ there exists some positive constant K_i such that*

$$\tilde{c}_i^y + \tilde{\varepsilon}_i^y + \tilde{a}_i^{y\,T}(x-y) + \frac{K_i}{2}\|x-y\|^2 \geq c_i(x) \quad \textit{for all } x,\, y \in I\!R^n. \tag{7}$$

Proof: By Lemma 2.1, we can find for each $i \in I$ some constant K_i satisfying $K_i \geq K_i' + 2\,r_i^{-1}\|a_i(y) - \tilde{a}_i^y\|$ for all $y \in I\!R^n$. Using (4) we obtain that (7) is valid for $\|x-y\| \leq r_i$. If, on the contrary, $\|x-y\| > r_i$, then we have $K_i \geq K_i' + 2\,\|x - y\|^{-1}\|a_i(y) - \tilde{a}_i^y\|$ and hence

$$\left(a_i(y) - \tilde{a}_i^y\right)^T(x-y) \leq \|a_i(y) - \tilde{a}_i^y\|\,\|x-y\| \leq \left(\frac{K_i}{2} - \frac{K_i'}{2}\right)\|x-y\|^2.$$

Therefore we have

$$\tilde{c}_i^y + \tilde{\varepsilon}_i^y + \tilde{a}_i^{y\,T}(x-y) + \frac{K_i}{2}\|x-y\|^2 \geq \tilde{c}_i^y + \tilde{\varepsilon}_i^y + a_i(y)^T(x-y) + \frac{K_i'}{2}\|x-y\|^2$$

and since

$$\tilde{c}_i^y + \tilde{\varepsilon}_i^y \geq c_i(y) \geq c_i(x) - a_i(y)^T(x-y) - \frac{K_i'}{2}\|x-y\|^2,$$

the bound (7) follows for $\|x-y\| > r_i$, too. □

As we can see, the function $\bar{p}_i : I\!R^n \times I\!R^n \to I\!R$,

$$\bar{p}_i(x;y) := \tilde{c}_i^y + \tilde{\varepsilon}_i^y + \tilde{a}_i^{y^T}(x-y) + \frac{K_i}{2}\|x-y\|^2$$

turns out to be an *approximation from above* to $c_i(x)$ for some $y \in I\!R^n$ and for all $i \in I$. More generally, we are able to deduce the following property: Let $Y \subset I\!R^n$ be a finite set. Elements of Y are denoted by y^j. For arbitrary $\bar{x} \in I\!R^n$ we have

$$c_i(\bar{x}) + a_i(\bar{x})^T(x-\bar{x}) - \frac{K_i}{2}\|x-\bar{x}\|^2 \leq c_i(x) \leq \bar{p}_i(x;y^j) \tag{8}$$

for all $x \in I\!R^n$, $y^j \in Y$ and $i \in I$. For convenience we use $\tilde{c}_i^j$, $\tilde{a}_i^j$ and $\tilde{\varepsilon}_i^j$ instead of $\tilde{c}_i^{y^j}$, $\tilde{a}_i^{y^j}$ and $\tilde{\varepsilon}_i^{y^j}$ in the sequel.

Following the analogous concept of nonsmooth optimization (see for example [7], [18]), Y is related to *the bundle* in order to represent a set of mutually different points of $I\!R^n$ which are used to build up approximations to the noisy constraint functions. Considering (8) the following interpretation is useful: The point y^j represents the information $(\tilde{c}_i^j, \tilde{a}_i^j, \tilde{\varepsilon}_i^j)$. Therefore, it will be convenient to call the index set J *the bundle* (instead of Y) where each $j \in J$ stands for $(y^j, \tilde{c}_i^j, \tilde{a}_i^j, \tilde{\varepsilon}_i^j)$.

With regard to (8) we have

$$c_i(\bar{x}) + a_i(\bar{x})^T(x-\bar{x}) - \frac{K_i}{2}\|x-\bar{x}\|^2 - \bar{p}_i(x;y^j) \leq 0 \tag{9}$$

for all $x \in I\!R^n$ and $j \in J$. Maximization of the left hand side in (9) yields the optimal value

$$\begin{aligned} \bar{b}_i^j(c_i(\bar{x}), a_i(\bar{x}), \bar{x}) := \; & c_i(\bar{x}) - \tilde{c}_i^j - \tilde{\varepsilon}_i^j + \frac{1}{4K_i}\|a_i(\bar{x}) - \tilde{a}_i^j\|^2 + \\ & + \frac{1}{2}(a_i(\bar{x}) + \tilde{a}_i^j)^T(y^j - \bar{x}) - \frac{K_i}{4}\|\bar{x} - y^j\|^2. \end{aligned} \tag{10}$$

Moreover, we have $\bar{b}_i^j(c_i(\bar{x}), a_i(\bar{x}), \bar{x}) \leq 0$ for all $j \in J$.

Next consider the minimization problem

$$\left\{ \begin{array}{rcl} -c_i^J & = & \min! \\ (c_i^J, a_i^J) & \in & \bar{\Omega}_i^J(\bar{x}) := \left\{ (c_i^J, a_i^J) \mid \bar{b}_i^j(c_i^J, a_i^J, \bar{x}) \leq 0 \text{ for all } j \in J \right\}. \end{array} \right. \tag{11}$$

Note that we have replaced the exact but unknown data $(c_i(\bar{x}), a_i(\bar{x}))$ by (c_i^J, a_i^J) to obtain the feasible set $\bar{\Omega}_i^J(\bar{x}) \subset I\!R^{n+1}$ for (11). Observe that problem (11) is convex.

Let $(\bar{c}_i^J(\tilde{x}), \bar{a}_i^J(\tilde{x}))$ denote the optimal solution of (11) with $\bar{\Lambda}_i^J(\bar{c}_i^J(\tilde{x}), \bar{a}_i^J(\tilde{x}), \tilde{x})$ being the corresponding set of optimal multipliers. For convenience, we sometimes shall use $\bar{\Lambda}_i^J(\tilde{x})$ instead. By $\bar{\mathcal{L}}_i(c_i^J, a_i^J, \tilde{x}, \lambda_i) := -c_i^J + \sum_{j\in J} \lambda_i^j \bar{b}_i^j(c_i^J, a_i^J, \tilde{x})$ we denote the Lagrangian function of problem (11). The first order Kuhn-Tucker conditions of (11) are

$$\begin{aligned} \sum_{j\in J} \bar{\lambda}_i^j &= 1 \\ \bar{a}_i^J(\tilde{x}) &= K_i\tilde{x} - \sum_{j\in J} \bar{\lambda}_i^j (K_i y^j - \tilde{a}_i^j) \\ \bar{\lambda}_i^j \bar{b}_i^j(\bar{c}_i^J(\tilde{x}), \bar{a}_i^J(\tilde{x}), \tilde{x}) &= 0 \quad \text{for all } j \in J, \\ \bar{b}_i^j(\bar{c}_i^J(\tilde{x}), \bar{a}_i^J(\tilde{x}), \tilde{x}) &\leq 0 \quad \text{for all } j \in J, \\ \bar{\lambda}_i^j &\geq 0 \quad \text{for all } j \in J, \end{aligned} \tag{12}$$

where $\bar{\lambda}_i \in \bar{\Lambda}_i^J(\bar{c}_i^J(\tilde{x}), \bar{a}_i^J(\tilde{x}), \tilde{x})$. The following theorem ensures the existence of the optimal solution of (11).

Theorem 2.1 *(a) The unique optimal solution $(\bar{c}_i^J(\tilde{x}), \bar{a}_i^J(\tilde{x}))$ of problem (11) exists.*

(b) For the optimal $\bar{c}_i^J(\tilde{x})$ we have: $\bar{c}_i^J(\tilde{x}) \geq c_i(\tilde{x})$ for all $i \in I$.

Proof: (a) For convenience we define

$$\bar{e}_i^j(a_i^J, \tilde{x}) := \frac{1}{4K_i}\|a_i^J - \tilde{a}_i^j\|^2 + \frac{1}{2}(a_i^J + \tilde{a}_i^j)^T(y^j - \tilde{x}) - \frac{K_i}{4}\|\tilde{x} - y^j\|^2 - \tilde{c}_i^j - \tilde{\varepsilon}_i^j.$$

Problem (11) is equivalent to the following problem

$$\bar{\phi}_i(a_i^J, \tilde{x}) = \min! \quad \text{for all} \quad a_i^J \in \mathbb{R}^n,$$

where $\bar{\phi}_i(a_i^J, \tilde{x}) := \max\{\bar{e}_i^j(a_i^J, \tilde{x}) \mid j \in J\}$. Observe that $\bar{\phi}_i(a_i^J, \tilde{x})$ is strictly convex (with respect of a_i^J) and

$$\bar{\phi}_i(a_i^J, \tilde{x}) \geq \frac{1}{4K_i}\|a_i^J - \tilde{a}_i^j\|^2 - \frac{1}{2}\|a_i^J\|\,\|y^j - \tilde{x}\| + \bar{u}_i^j(\tilde{x})$$

for any $j \in J$ and $a_i^J \in \mathbb{R}^n$, where $\bar{u}_i^j(\tilde{x}) := \frac{1}{2}\tilde{a}_i^{jT}(y^j - \tilde{x}) - \frac{K_i}{4}\|\tilde{x} - y^j\|^2 - \tilde{c}_i^j - \tilde{\varepsilon}_i^j$. Thus, $\bar{\phi}_i(a_i^J, \tilde{x}) \to +\infty$ for $\|a_i^J\| \to +\infty$. Therefore $\bar{a}_i^J(\tilde{x})$, the unique minimum point of $\bar{\phi}_i$, exists.

(b) By (9) we find that $(c_i(\tilde{x}), a_i(\tilde{x})) \in \mathbb{R}^{n+1}$ is feasible for (11). Hence, $-\bar{c}_i^J(\tilde{x}) \leq -c_i(\tilde{x})$ proving the assertion.□

Remark 2.1 *For all $j \in J$ we obtain the bounds $c_i(y^j) \leq \bar{c}_i^J(y^j) \leq \tilde{c}_i^j + \tilde{\varepsilon}_i^j$.*

Due to the derivation of (11) we will regard $(\bar{c}_i^J(\tilde{x}), \bar{a}_i^J(\tilde{x}))$ as approximations to $c_i(\tilde{x})$ and its gradient vector $a_i(\tilde{x})$ for all $i \in I$. Moreover, observe that $(\bar{c}_i^J(\tilde{x}), \bar{a}_i^J(\tilde{x}))$ depends on $\bar{\lambda}_i$.

For later use we define the optimal value function of (11) at $\tilde{x}$ by

$$-\bar{\gamma}_i^J(\tilde{x}) := \inf\{-c_i^J \mid \bar{b}_i^j(c_i^J, a_i^J, \tilde{x}) \leq 0 \text{ for all } j \in J\}$$

for all $i \in I$.

Since $a_i(\tilde{x})$ is the gradient of $c_i(x)$ at $x = \tilde{x}$ for all $i \in I$, it is a subject of interest to investigate differential relations between $\bar{c}_i^J(\tilde{x})$ and $\bar{a}_i^J(\tilde{x})$. The following lemma proves that problem (11) satisfies several regularity conditions. This enables us to apply well-known results from sensitivity analysis in order to obtain differentiability relations.

Lemma 2.3 *Problem (11) satisfies the following regularity conditions:*

(a) Mangasarian-Fromovitz-condition. For all $i \in I$ there exist $(\hat{c}_i, \hat{a}_i^T)^T \in \mathbb{R}^{n+1}$ such that $(\hat{c}_i, \hat{a}_i^T)\nabla_\omega \bar{b}_i^j(\bar{c}_i^J(\tilde{x}), \bar{a}_i^J(\tilde{x}), \tilde{x}) < 0$ for all $j \in J$, where $\omega = (c_i^J, a_i^{J^T})$.

(b) Strong second order sufficient conditions. For all $(c_i, a_i^T)^T \in C_i$ with $(c_i, a_i^T) \neq 0$ we have $\max_{\bar{\lambda}_i \in \bar{\Lambda}_i^J(\tilde{x})}(c_i, a_i^T)\, \nabla^2_{\omega\omega}\bar{\mathcal{L}}(\bar{c}_i^J, \bar{a}_i^J, \tilde{x}, \bar{\lambda}_i)\,(c_i, a_i^T)^T > 0$, where C_i is the cone of critical directions.

(c) Uniqueness of the optimal set. The set of optimal solutions of problem (11) at $\tilde{x}$ is a singleton.

(d) Inf-compactness. There exists a number κ and a compact set $\mathcal{K} \subset \mathbb{R}^{n+1}$ such that $\kappa > -\bar{c}_i^J(\tilde{x})$ and $\{(c_i^J, a_i^{J^T})^T \in \mathbb{R}^{n+1} \mid \bar{b}_i^j(c_i^J, a_i^J, x) \leq 0 \wedge -c_i^J \leq \kappa\} \subset \mathcal{K}$ for all x in a sufficiently small neighbourhood of $\tilde{x}$.

Proof:

(a) Cf. (10) we find that

$$\nabla_\omega \bar{b}_i^j(\bar{c}_i^J(\tilde{x}), \bar{a}_i^J(\tilde{x}), \tilde{x}) = \begin{pmatrix} 1 \\ \frac{1}{2K_i}(\bar{a}_i^J(\tilde{x}) - \tilde{a}_i^j) - \frac{1}{2}\tilde{x} + \frac{1}{2}y^j \end{pmatrix}$$

for all $j \in J$. If we choose $\hat{a}_i := 0$, for instance, then for $\hat{c}_i < 0$ the Mangasarian-Fromovitz-condition is satisfied for all $j \in J$.

(b) We have that for all $(c_i, a_i^T)^T \in C_i$ with $(c_i, a_i^T) \neq 0$ the condition of (b) is equivalent to

$$\max_{\lambda_i \in \bar{\Lambda}_i^J(\tilde{x})} (c_i, a_i^T) \begin{pmatrix} 0 & 0 \\ 0 & \frac{1}{2K_i} I_n \end{pmatrix} \begin{pmatrix} c_i \\ a_i \end{pmatrix} = \frac{1}{2K_i}\|a_i\|^2 > 0, \tag{13}$$

Since the right hand side is independent of $\lambda_i \in \bar{\Lambda}_i^J(\tilde{x})$, we only have to check

a_i. A closer look at C_i reveals:

$$C_i = \{(c_i, a_i^T)^T \in I\!R^{n+1} \mid (c_i, a_i^T)\begin{pmatrix} 1 \\ \frac{1}{2K_i}(\bar{a}_i^J - \tilde{a}_i^j) - \frac{1}{2}\tilde{x} + \frac{1}{2}y^j \end{pmatrix} \leq 0 \wedge$$
$$\wedge\ (c_i, a_i^T)\begin{pmatrix} -1 \\ 0 \end{pmatrix} \leq 0 \text{ for all } j \in J\}$$

Elements of C_i therefore satisfy

$$0 \geq -c_i \geq \frac{1}{2} a_i^T \left(\frac{1}{K_i}(\bar{a}_i^J - \tilde{a}_i^j) - \tilde{x} + y^j \right) \tag{14}$$

for all $j \in J$. Problems in (13) will only arise if $a_i = 0$. But then, by (14), we will obtain $c_i = 0$ which contradicts $(c_i, a_i^T) \neq 0$.

(c) Let $\mathcal{M}(\tilde{x})$ be the set of optimal solutions of problem (11) at $\tilde{x}$. Then the uniqueness follows from the fact that (b) proves the optimal solution $(\bar{c}_i^J(\tilde{x}), \bar{a}_i^J(\tilde{x}))$ to be an isolated, hence strict, local minimizer. Taking this fact into consideration in combination with the convexity of (11) yields the uniqueness.

(d) Theorem 2.1 ensures that the unique solution of (11) at $\tilde{x}$ exists. Hence, choose $\kappa > -\bar{c}_i^J(\tilde{x})$. Let $\bar{B}(\tilde{x}; 1)$ denote the closed ball with radius 1 and center $\tilde{x}$. For $\kappa \geq -c_i^J$ we obtain from (10) that

$$\frac{1}{4K_i}\|a_i^J - \tilde{a}_i^j\|^2 + \frac{1}{2}(a_i^J + \tilde{a}_i^j)^T(y^j - x) - \frac{K_i}{4}\|x - y^j\|^2 - \tilde{c}_i^j - \tilde{\varepsilon}_i^j \leq \kappa \tag{15}$$

for $x \in \bar{B}(\tilde{x}; 1)$. Thus the set of $a_i^J \in I\!R^n$ satisfying (15) is bounded. This yields the boundedness of c_i^J by $\bar{b}_i^j(c_i^J, a_i^J, x) \leq 0$ for $x \in \bar{B}(\tilde{x}; 1)$ and for all $j \in J$. From the boundedness of a_i^J, c_i^J and the compactness of $\bar{B}(\tilde{x}; 1)$ we deduce that there exists a compact $\mathcal{K} \subset I\!R^{n+1}$ such that

$$\{(c_i^J, a_i^{JT})^T \in I\!R^{n+1} \mid \bar{b}_i^j(c_i^J, a_i^J, x) \leq 0 \wedge -c_i^J \leq \kappa\} \subset \mathcal{K}$$

for all x in a sufficiently small neighbourhood of $\tilde{x}$.□

The next result is concerned with first order directional differentiability.

Theorem 2.2 *For all $i \in I$ we have that*

(a) $\bar{\gamma}_i^J(\tilde{x}) = \bar{c}_i^J(\tilde{x})$ is well-defined,

(b) $\bar{\gamma}_i^{J\prime}(\tilde{x}; v) = v^T \bar{a}_i^J(\tilde{x})$ for all $v \in I\!R^n$, where $\bar{\gamma}_i^{J\prime}(\tilde{x}; v) = \lim_{t\downarrow 0} \frac{\bar{\gamma}_i^J(\tilde{x}+tv) - \bar{\gamma}_i^J(\tilde{x})}{t}$.

Proof:

(a) The assertion follows by definition and Lemma 2.3.

(b) We will apply [14, Theorem 4.1] to prove the second part. First observe that $(\bar{c}_i^J(\tilde{x}), \bar{a}_i^J(\tilde{x})) \in \mathcal{M}(\tilde{x})$. By Lemma 2.3 the assumptions postulated in [14, Theorem 4.1] are satisfied. Therefore, $\bar{\gamma}_i^J(x)$ is directionally differentiable at $\tilde{x}$ and takes the form

$$\bar{\gamma}_i^{J\prime}(\tilde{x}; v) = -\max\{v^T \nabla_{\tilde{x}} \bar{\mathcal{L}}_i(\bar{c}_i^J, \bar{a}_i^J, \tilde{x}, \bar{\lambda}_i) \mid \bar{\lambda}_i \in \bar{\Lambda}_i^J(\bar{c}_i^J(\tilde{x}), \bar{a}_i^J(\tilde{x}), \tilde{x})\} \quad (16)$$

for all $v \in I\!R^n$. Calculation of the right hand side of (16) yields by using (12)

$$\begin{aligned}
\bar{\gamma}_i^{J\prime}(\tilde{x}; v) &= -v^T\Bigg(\sum_{j\in J} \bar{\lambda}_i^j \Bigg[\frac{1}{2}\Bigg(K_i\tilde{x} - \sum_{l\in J}\bar{\lambda}_i^l(K_i y^l - \bar{a}_i^l) - \bar{a}_i^j\Bigg) + \frac{K_i}{2}(y^j - \tilde{x}) - \\
&\quad - \frac{1}{2}\Bigg(K_i\tilde{x} - \sum_{l\in J}\bar{\lambda}_i^l(K_i y^l - \bar{a}_i^l) + \bar{a}_i^j\Bigg) - \frac{K_i}{2}(\tilde{x} - y^j)\Bigg]\Bigg) \\
&= -v^T\Bigg(\sum_{j\in J} \bar{\lambda}_i^j \Bigg[\frac{1}{2}\Bigg(K_i\tilde{x} - \sum_{l\in J}\bar{\lambda}_i^l(K_i y^l - \bar{a}_i^l)\Bigg) - \frac{\bar{a}_i^j}{2} - \\
&\quad - \frac{1}{2}\Bigg(K_i\tilde{x} - \sum_{l\in J}\bar{\lambda}_i^l(K_i y^l - \bar{a}_i^l)\Bigg) - \frac{\bar{a}_i^j}{2} - K_i\tilde{x} + K_i y^j\Bigg]\Bigg) \\
&= -v^T\Bigg(-K_i\tilde{x} + \sum_{j\in J}\bar{\lambda}_i^j(K_i y^j - \bar{a}_i^j)\Bigg) = v^T \bar{a}_i^J(\tilde{x}) \quad \text{for all } v \in I\!R^n. \square
\end{aligned}$$

Considering the continuity property of $\mathcal{M}(\tilde{x})$, the set of optimal solutions of (11), at $\tilde{x}$ we obtain

Theorem 2.3 *For all $i \in I$ the optimal value function $\bar{\gamma}_i^J(x)$ is continuously differentiable at $\tilde{x}$ and $\nabla_x \bar{\gamma}_i^J(x)_{|x=\tilde{x}} = \bar{a}_i^J(\tilde{x})$.*

Proof: Theorem 2.2(b) yields that $\bar{a}_i^J(\tilde{x})$ is the Gateaux-derivative of $\bar{\gamma}_i^J$ at $\tilde{x}$. Then [14, Theorem 2.1] proves $\mathcal{M}(x)$ to be upper semicontinuous at $\tilde{x}$. Since $\mathcal{M}(x)$ is single-valued for all x in a neighbourhood of $\tilde{x}$, $\mathcal{M}(x)$ is continuous at $\tilde{x}$. Hence, $\bar{\gamma}_i^J$ is continuously (Gateaux) differentiable at $\tilde{x}$.$\square$

In view of Assumption 1.4 and SQP we need some second order differential properties, too. In addition, we will take advantage of this information when modelling a reset-strategy. The second order directional derivatives of the optimal value functions $\bar{\gamma}_i^J(x)$ at $\tilde{x}$ are defined by $\bar{\gamma}_i^{J\prime\prime}(\tilde{x}; v) = \lim_{t\downarrow 0} \frac{\bar{\gamma}_i^J(\tilde{x}+tv) - \bar{\gamma}_i^J(\tilde{x}) - t\bar{\gamma}_i^{J\prime}(\tilde{x};v)}{t^2}$.

Theorem 2.4 *The second order directional derivative $\bar{\gamma}_i^{J\prime\prime}(\tilde{x}; v)$ exists for all $i \in I$ and satisfies*

$$-\frac{K_i}{2}\|v\|^2 \le \bar{\gamma}_i^{J\prime\prime}(\tilde{x}; v) \le \frac{K_i}{2}\|v\|^2 \quad \textit{for all } v \in I\!R^n.$$

Proof: At first we take a look at

$$\begin{aligned}\xi_{\lambda_i}(u,v) &:= \frac{1}{2}u^T\nabla^2_{\omega\omega}\bar{\mathcal{L}}_i(\bar{c}_i^J(\bar{x}),\bar{a}_i^J(\bar{x}),\bar{x},\lambda_i)u + u^T\nabla^2_{\omega\bar{x}}\bar{\mathcal{L}}_i(\bar{c}_i^J(\bar{x}),\bar{a}_i^J(\bar{x}),\bar{x},\lambda_i)v + \\ &\quad + \frac{1}{2}v^T\nabla^2_{\bar{x}\bar{x}}\bar{\mathcal{L}}_i(\bar{c}_i^J(\bar{x}),\bar{a}_i^J(\bar{x}),\bar{x},\lambda_i)v\end{aligned}$$

In terms of our situation ξ_{λ_i} turns out to be

$$\begin{aligned}\xi_{\lambda_i}(u,v) &= \frac{1}{4K_i}u^T\begin{pmatrix}0 & 0\\ 0 & I_n\end{pmatrix}u - \frac{1}{2}u^T\begin{pmatrix}0\\ I_n\end{pmatrix}v - \frac{K_i}{4}v^TI_nv \\ &= \frac{1}{4K_i}\|u_2\|^2 - \frac{1}{2}u_2^Tv - \frac{K_i}{4}\|v\|^2,\end{aligned}$$

where $u := (u_1, u_2^T)^T$, $u_1 \in I\!R$, $u_2 \in I\!R^n$. Obviously, $\xi_{\lambda_i}(u,v)$ does not depend on λ_i. Lemma 2.3 allows the application of [14, Theorem 4.2] which reduces to

$$\bar{\gamma}_i^{J''}(\bar{x};v) = \min\{\xi_{\lambda_i}(u,v) \mid u \in \bar{\Sigma}(v)\},$$

where

$$\bar{\Sigma}(v) := \left\{u \in I\!R^{n+1} \mid u^T\nabla_\omega\bar{b}_i^j(\bar{c}_i^J(\bar{x}),\bar{a}_i^J(\bar{x}),\bar{x}) + v^T\nabla_{\bar{x}}\bar{b}_i^j(\bar{c}_i^J(\bar{x}),\bar{a}_i^J(\bar{x}),\bar{x}) \le 0,\ j \in J\right\}$$

with

$$\begin{aligned}\nabla_\omega\bar{b}_i^j(\bar{c}_i^J(\bar{x}),\bar{a}_i^J(\bar{x}),\bar{x}) &= \left(1, \frac{1}{2K_i}(\bar{a}_i^J(\bar{x}) - \bar{a}_i^j)^T - \frac{1}{2}(\bar{x} - y^j)^T\right)^T \\ \nabla_{\bar{x}}\bar{b}_i^j(\bar{c}_i^J(\bar{x}),\bar{a}_i^J(\bar{x}),\bar{x}) &= -\frac{1}{2}(\bar{a}_i^J(\bar{x}) + \bar{a}_i^j) - \frac{K_i}{2}(\bar{x} - y^j).\end{aligned}$$

Next we fix a feasible point $\hat{u} \in \bar{\Sigma}(v)$ in order to obtain an upper bound on $\xi_{\lambda_i}(u,v)$. We put $\hat{u}_2 := -K_iv \in I\!R^n$ and fix $\hat{u}_1 := v^T\bar{a}_i^J(\bar{x})$. Then $\hat{u} = (\hat{u}_1, \hat{u}_2^T)^T \in \bar{\Sigma}(v)$ due to

$$\hat{u}_1 - K_iv^T\left(\frac{1}{2K_i}(\bar{a}_i^J(\bar{x}) - \bar{a}_i^j) - \frac{1}{2}(\bar{x} - y^j)\right) + v^T\left(-\frac{1}{2}(\bar{a}_i^J(\bar{x}) + \bar{a}_i^j) - \frac{K_i}{2}(\bar{x} - y^j)\right) \le 0.$$

Hence $\xi_{\lambda_i}(\hat{u},v) = \frac{K_i}{2}\|v\|^2$ implying $\bar{\gamma}_i^{J''}(\bar{x};v) \le \frac{K_i}{2}\|v\|^2$.
On the other hand, let $\bar{u} = (\bar{u}_1, \bar{u}_2)$ be the unconstrained minimizer of $\xi_{\lambda_i}(u,v)$. Naturally, one obtains $\bar{u}_2 = K_iv$ which results in $\bar{\gamma}_i^{J''}(\bar{x};v) \ge -\frac{K_i}{2}\|v\|^2$.
Since $v \in I\!R^n$ is arbitrary in both cases above, the assertion is proved. □

3 Dual problem and reset strategy

In general one has to solve $|I|$-many problems of the type (11) in order to compute a proximal model of (1). However, we can not be content as a closer look at the constraint functions $\bar{b}_i^j(c_i^J, a_i^J, x)$ reveals a quadratic structure with respect to a_i^J. Simplification is achieved by transition to the corresponding dual problem that results in a linearly constrained programming problem with a quadratic objective function. There exist several algorithms for efficiently solving the resulting problems; see [5], [6].

The Lagrangian dual problem of (11) is

$$\bar{\mathcal{L}}_i(\bar{c}_i^J(\bar{x}, \lambda_i), \bar{a}_i^J(\bar{x}, \lambda_i), \bar{x}, \lambda_i) = \max_{\lambda_i \in \Lambda(J)} ! \,, \tag{17}$$

where $\Lambda(J) = \{\lambda_i^j \geq 0 \mid j \in J, \ \lambda_i{}^T e = 1\}$, with $e = (1, \ldots, 1)^T \in I\!R^{|J|}$. Moreover we use $(\bar{c}_i^J(\bar{x}), \bar{a}_i^J(\bar{x})) =: (\bar{c}_i^J(\bar{x}, \bar{\lambda}_i), \bar{a}_i^J(\bar{x}, \bar{\lambda}_i))$. The following theorem proves a tight relation between the primal problem (11) and a problem equivalent to the Lagrangian dual problem (17) (we only dismiss a constant term in the objective function).

Theorem 3.1 *Let $\bar{\lambda}_i \in I\!R^{|J|}$ be optimal solution of*

$$\begin{aligned} \lambda_i^T \bar{v}_i^J(\bar{x}) + \frac{1}{2}\lambda_i^T \bar{Z}_i^{JT} \bar{Z}_i^J \lambda_i &= \min! \\ \lambda_i^T e &= 1 \\ \lambda_i^j &\geq 0 \quad \textit{for all } j \in J, \end{aligned} \tag{18}$$

where $\bar{v}_i^J(\bar{x}) := \left(\bar{v}_i^1, \ldots, \bar{v}_i^{|J|}\right)(\bar{x})$ with

$$\bar{v}_i^j(\bar{x}) := \bar{c}_i^j + \bar{\varepsilon}_i^j + \bar{x}^T \bar{a}_i^j + \frac{K_i}{4}\|y^j\|^2 - \frac{1}{4K_i}\|\bar{a}_i^j\|^2 - K_i \bar{x}^T y^j - \frac{1}{2}\bar{a}_i^{jT} y^j$$

for all $j \in J$, and $\bar{Z}_i^J := \frac{1}{\sqrt{2K_i}}\left(K_i y^1 - \bar{a}_i^1, \ldots, K_i y^{|J|} - \bar{a}_i^{|J|}\right)$. Then $\bar{\lambda}_i$ satisfies

$$\sum_{j\in J} \bar{\lambda}_i^j = 1, \quad \bar{\lambda}_i^j \geq 0 \quad \textit{for all } j \in J, \tag{19}$$

$$\bar{a}_i^J(\bar{x}) = K_i \bar{x} + \sum_{j \in J} \bar{\lambda}_i^j (\bar{a}_i^j - K_i y^j) \tag{20}$$

$$\bar{b}_i^j(\bar{c}_i^J(\bar{x}), \bar{a}_i^J(\bar{x}), \bar{x}) \leq 0 \quad \textit{for all } j \in J, \tag{21}$$

$$\sum_{j \in J} \bar{\lambda}_i^j \bar{b}_i^j(\bar{c}_i^J(\bar{x}), \bar{a}_i^J(\bar{x}), \bar{x}) = 0 \quad \textit{for all } j \in J. \tag{22}$$

Proof: First observe that problem (18) is convex with linear constraints. Hence, by [1] the Kuhn-Tucker conditions are both necessary and sufficient. Let

$$L(\lambda_i, \pi, \pi_0) := \frac{1}{2}\lambda_i^T \bar{Z}_i^{JT} \bar{Z}_i^J \lambda_i + \lambda_i^T \bar{v}_i^J(\bar{x}) - \pi^T \lambda_i + \pi_0(\lambda_i^T e - 1)$$

denote the Lagrangian function of (18). An optimal solution $\bar{\lambda}_i \in I\!R^{|J|}$ of (18) with corresponding optimal multipliers $\bar{\pi} \in I\!R^{|J|}$ and $\bar{\pi}_0 \in I\!R$ satisfies the first order Kuhn-Tucker conditions

$$\begin{aligned}
\bar{Z}_i^{J^T} \bar{Z}_i^J \bar{\lambda}_i + \bar{v}_i^J(\tilde{x}) - \bar{\pi} + \bar{\pi}_0 e &= 0 && (23)\\
\bar{\lambda}_i^j &\geq 0 \quad \text{for all } j \in J, && (24)\\
-\bar{\pi}^j \bar{\lambda}_i^j &= 0 \quad \text{for all } j \in J, && (25)\\
\bar{\pi}^j &\geq 0 \quad \text{for all } j \in J, && (26)\\
\bar{\lambda}_i^T e &= 1. && (27)
\end{aligned}$$

Multiplying condition (23) with $\bar{\lambda}_i^T$ yields (after some simple calculations)

$$\bar{\pi}_0 = -\bar{c}_i^J(\tilde{x}) + \frac{1}{2}\bar{a}_i^J(\tilde{x})^T \tilde{x} + \frac{K_i}{4}\|\tilde{x}\|^2 - \frac{1}{4K_i}\|\bar{a}_i^J(\tilde{x})\|^2.$$

By inserting $\bar{\pi}_0$ in (23) we obtain (again after some simple manipulations) for $j \in J$

$$\begin{aligned}
\bar{\pi}^j &= \bar{\pi}_0 + (\bar{Z}_i^{J^T} \bar{Z}_i^J \bar{\lambda}_i)^j + \bar{v}_i^j(\tilde{x})\\
&= -\bar{c}_i^J(\tilde{x}) - \bar{e}_i^j(\bar{a}_i^J(\tilde{x}), \tilde{x}) = -\bar{b}_i^j(\bar{c}_i^J(\tilde{x}), \bar{a}_i^J(\tilde{x}), \tilde{x}).
\end{aligned}$$

This proves (20) – (22). Condition (19) directly follows from (24) and (27).□

As already mentioned, the results of Theorem 2.2, Theorem 2.3 and of Theorem 2.4 are basic for designing an efficient reset strategy. By Theorem 2.2, Theorem 2.3 and (11) $\bar{c}_i^J(\tilde{x})$ and $\bar{a}_i^J(\tilde{x})$ may be regarded as *aggregate approximations* that assemble and smooth all of the noisy information obtained so far. Let

$$\bar{p}_i^a(x; \tilde{x}) := \bar{c}_i^J(\tilde{x}) + \bar{a}_i^J(\tilde{x})^T(x - \tilde{x}) + \frac{K_i}{2}\|x - \tilde{x}\|^2$$

denote the corresponding aggregate quadratic approximation. The following theorem investigates the relationship between $\bar{p}_i^a(x; \tilde{x})$, $\bar{c}_i^J(x)$ and $c_i(x)$, respectively.

Theorem 3.2 *Suppose that Assumption 1.1 – 1.4 hold. Then*

$$c_i(x) \leq \bar{c}_i^J(x) \leq \bar{p}_i^a(x; \tilde{x}) \quad \textit{for all } x \in I\!R^n \textit{ and } i \in I.$$

Proof: First observe that for $j \in J$ and for all $x \in I\!R^n$

$$\bar{v}_i^j(x) = \bar{v}_i^j(\tilde{x}) + \bar{v}_i^j(x) - \bar{v}_i^j(\tilde{x}) = \bar{v}_i^j(\tilde{x}) - (K_i y^j - \tilde{a}_i^j)^T(x - \tilde{x}). \qquad (28)$$

Let $\bar{\lambda}_i(\tilde{x})$ denote the optimal multiplier vector at $\tilde{x}$ being optimal solution of (18) according to Theorem 3.1. In order to recover the Lagrangian dual problem of (11) at x we have to add the constant $\frac{1}{2}K_i\|x\|^2$ to the objective of (18) at x. Then then Lagrangian dual problem becomes

$$\begin{aligned}
\frac{K_i}{2}\|x\|^2 + \lambda_i^T \bar{v}_i^J(x) + \frac{1}{2}\lambda_i^T \bar{Z}_i^{J^T} \bar{Z}_i^J \lambda_i &= \text{min!} \\
\lambda_i^T e &= 1 \\
\lambda_i^j &\geq 0 \quad \text{for all } j \in J.
\end{aligned} \tag{29}$$

Hence, $\bar{\lambda}_i(\bar{x})$ is feasible for (29). Let $\bar{w}^x : \mathbb{R}^{|J|} \to \mathbb{R}$,

$$\bar{w}^x(\lambda) := \frac{K_i}{2}\|x\|^2 + \lambda^T \bar{v}_i^J(x) + \frac{1}{2}\lambda^T \bar{Z}_i^{J^T} \bar{Z}_i^J \lambda$$

denote the dual objective value of (29) at λ. Then, by (28), (12) and with $\hat{Z}_i^J := \sqrt{2K_i}\bar{Z}_i^J$

$$\begin{aligned}
\bar{w}^x(\bar{\lambda}_i(\bar{x})) &= \frac{1}{2}\bar{\lambda}_i(\bar{x})^T \bar{Z}_i^{J^T} \bar{Z}_i^J \bar{\lambda}_i(\bar{x}) + \bar{\lambda}_i(\bar{x})^T \bar{v}_i^J(\bar{x}) - \bar{\lambda}_i(\bar{x})^T \hat{Z}_i^{J^T}(x - \bar{x}) + \frac{K_i}{2}\|x\|^2 \\
&= \bar{w}^{\bar{x}}(\bar{\lambda}_i(\bar{x})) + \bar{a}_i^J(\bar{x})^T(x - \bar{x}) + \frac{K_i}{2}\|x - \bar{x}\|^2.
\end{aligned} \tag{30}$$

Here, $\bar{w}^{\bar{x}}(\bar{\lambda}_i(\bar{x}))$ is the optimal value of (29) at $\bar{x}$. Lemma 2.3 states that the convex problem (11) satisfies the Mangasarian-Fromovitz constraint qualification. Therefore, by a result given in [1] the optimal dual objective value and the optimal primal objective value coincide. Hence, $\bar{c}_i^J(\bar{x}) = \bar{w}^{\bar{x}}(\bar{\lambda}_i(\bar{x}))$. Thus we can continue (30) and obtain

$$\bar{w}^x(\bar{\lambda}_i(\bar{x})) = \bar{c}_i^J(\bar{x}) + \bar{a}_i^J(\bar{x})^T(x - \bar{x}) + \frac{K_i}{2}\|x - \bar{x}\|^2. \tag{31}$$

The relation $\bar{c}_i^J(x) \leq \bar{p}_i^a(x;\bar{x})$ then follows from $\bar{w}^x(\bar{\lambda}_i(\bar{x})) \geq \bar{w}^x(\bar{\lambda}_i(x)) = \bar{c}_i^J(x)$, where $\bar{\lambda}_i(x)$ denotes the optimal solution of (29).

The second inequality, i.e $c_i(x) \leq \bar{c}_i^J(x)$, $x \in \mathbb{R}^n$, follows by Theorem 2.1(b) applied at $x \in \mathbb{R}^n$.□

Theorem 3.2 ensures that $\bar{p}_i^a(x;\bar{x})$ again is an approximation from above. We will refer to it as *aggregate approximation from above.* Moreover, $\bar{p}_i^a(\bar{x};\bar{x}) \leq \bar{p}_i(y^j;\bar{x})$ for all $j \in J$. Therefore, if $|J|$ exceeds a predetermined number $J_{\max}$ an additional index $\bar{j}$ is introduced with $\bar{c}_i^{\bar{j}} := \bar{c}_i^J(\bar{x}), \bar{a}_i^{\bar{j}} := \bar{a}_i^J(\bar{x}), y^{\bar{j}} := \bar{x}$ and $\bar{\varepsilon}_i^{\bar{j}} := 0$. One may apply the following *reset strategy.* Choose a subset $J^s \subset J$ (a reasonable choice consists in a selection of active constraints with large $\bar{\lambda}_i^j$). Then put $J := J^s \cup \{\bar{j}\}$. Note one may also choose $J^s = \emptyset$. In order to obtain a well-defined reset-strategy there are two important conditions which have to be satisfied. First, in view of (10) the exact $c_i(x')$ and $a_i(x')$ at $x' \in \mathbb{R}^n$ have to be feasible for (11) at x'. Second, if we replace some or all $j \in J$ by $\bar{j}$ then the optimal solution of (11) must not change.

In order to see the first assertion above, consider Theorem 3.2. By a Taylor expansion one obtains for all $x \in \mathbb{R}^n$

$$c_i(x') + a_i(x')^T(x - x') - \frac{K_i}{2}\|x - x'\|^2 - \bar{p}_i^a(x;\tilde{x}) \le 0.$$

Maximization with respect of x yields $\bar{b}_i^{\bar{j}}(c_i(x'), a_i(x'), x') \le 0$.

For the second statement consider the following

Lemma 3.1 *Subproblem (11) is equivalent to the following reduced subproblem*

$$\begin{aligned} -c_i^J &= \min! \\ (c_i^J, a_i^J) &\in \bar{\Omega}_i^{J^a}(\tilde{x}) \end{aligned} \tag{32}$$

where $J^a := J^s \cup \{\bar{j}\}$ with J^s being any (possibly empty) subset of J.

Proof: By convexity of (32) the Kuhn Tucker conditions are both sufficient and necessary. Hence, if we choose $\tilde{\lambda}_i^{\bar{j}} := 1$ and $\tilde{\lambda}_i^j := 0$ for $j \in J^s$, then

$$\begin{aligned}
&\tilde{\lambda}_i^j \ge 0 \quad \text{for all } j \in J^a \text{ and } \sum_{j \in J^a} \tilde{\lambda}_i^j = \tilde{\lambda}_i^{\bar{j}} = 1 \\
&\bar{a}_i^{J^a}(\tilde{x}) = K_i\tilde{x} - \sum_{j \in J^s} \tilde{\lambda}_i^j(K_i y^j - \bar{a}_i^j) - \tilde{\lambda}_i^{\bar{j}}(K_i\tilde{x} - \bar{a}_i^J(\tilde{x})) = \bar{a}_i^J(\tilde{x}) \\
&\bar{c}_i^{J^a}(\tilde{x}) = -\sum_{j \in J^s} \tilde{\lambda}_i^j \bar{e}_i^j(a_i^{J^a}(\tilde{x});\tilde{x}) - \tilde{\lambda}_i^{\bar{j}}\left(\frac{1}{4K_i}\|\bar{a}_i^{J^a}(\tilde{x}) - \bar{a}_i^J(\tilde{x})\|^2 - \bar{c}_i^J(\tilde{x})\right) = \bar{c}_i^J(\tilde{x}) \\
&\tilde{\lambda}_i^j \bar{b}_i^j(\bar{c}_i^{J^a}(\tilde{x}), \bar{a}_i^{J^a}(\tilde{x}), \tilde{x}) = 0 \quad \text{for all } j \in J^s \text{ since } \tilde{\lambda}_i^j = 0 \text{ for } j \in J^s, \\
&\tilde{\lambda}_i^{\bar{j}} \bar{b}_i^{\bar{j}}(\bar{c}_i^{J^a}(\tilde{x}), \bar{a}_i^{J^a}(\tilde{x}), \tilde{x}) = 0 \quad \text{since } \bar{b}_i^{\bar{j}}(\bar{c}_i^{J^a}(\tilde{x}), \bar{a}_i^{J^a}(\tilde{x}), \tilde{x}) = 0.
\end{aligned}$$

Thus $(\bar{c}_i^J(\tilde{x}), \bar{a}_i^J(\tilde{x}))$ solves (32). Hence, Lemma 2.1(a) yields that (11) and (32) have the same unique optimal solution.□

4 The algorithm

For approximating the original nonlinear problem (1) we use the well-known SQP-approach. From now on we will replace $\tilde{x}$ by x^k, the k-th iterate. For building up the linear approximations to the (unknown) constraints of (1), we will use the solutions $\bar{c}_i^k$, $\bar{a}_i^k$ of the $|I|$ problems (18)[2]. We shall remember that one problem of the kind of (18) has to be solved in order to obtain the corresponding approximation of the i-th constraint in (1). The required quadratic problem at the k-th iteration becomes

[2]Notation: Let J^k be the bundle during the k-th iteration and x^k the actual iterate. Then we set $\bar{c}_i^k := \bar{c}_i^{J^k}(x^k)$ and $\bar{a}_i^k := \bar{a}_i^{J^k}(x^k)$.

$$
\begin{aligned}
f^k + \nabla f^{k^T} d + \tfrac{1}{2} d^T B^k d + \tfrac{1}{2}\rho^k \delta^2 &= \min! \\
(1-\delta)\bar{c}_i^k + \bar{a}_i^{kT} d &\le 0 \qquad i \in I_k^+ \\
\bar{c}_i^k + \bar{a}_i^{kT} d &\le 0 \qquad i \in I_k^0 \\
0 \le \delta &\le 1,
\end{aligned} \tag{33}
$$

where $d := x - x^k$, $I_k^+ := \{i \in I \mid \bar{c}_i^k \ge 0\}$ and $I_k^0 := I \setminus I_k^+$. The matrix $B^k \in I\!R^{n \times n}$ is positive definite. The usage of δ with pertinent penalty-parameter ρ shall avoid inconsistencies. Let (d^k, δ^k) denote the optimal solution of (33), and let $\hat{\lambda}^k \in I\!R^{|I|}$, $\pi_1^k \in I\!R$, $\pi_2^k \in I\!R$ denote the corresponding optimal multipliers. The parameter ρ^k is increased whenever δ^k is larger than some given threshold value $\bar{\delta} < 1$. After an increase of ρ^k problem (33) is solved again. It can be shown that this loop is finite provided there exists a direction $\bar{d} \in I\!R^n$ such that $\bar{a}_i^{k^T} \bar{d} < 0$ for $\bar{c}_i^k \ge 0$.

After having obtained a search direction d^k from (33), the algorithm performs a line search to calculate a suitable step-size α^k (> 0) for which a sufficient descent is achieved. Let $J^k(\alpha) := J^k \cup \{k(\alpha)\}$ denote the trial bundle in the line search of the k-th iteration. The index $k(\alpha)$ corresponds to $x_\alpha^k := x^k + \alpha d^k$ and its noisy information and error bound. Moreover, we suppose $J^k(\alpha) = J^k$ if $\alpha = 0$. Since we have to consider both the descent in f and the violation of constraints we will make use of a penalty function. One option is the *exact penalty function*

$$
\varphi^k(\alpha) := f(x^k + \alpha d^k) + \sum_{i \in I} \mu_i^k \max\{\bar{c}_i^{J^k(\alpha)}(x^k + \alpha d^k), 0\},
$$

where $\mu^k \in I\!R^{|I|}$, $\mu^k > 0$, is the vector of penalty parameters. As we can see, an evaluation of $\varphi^k(\cdot) : I\!R \to I\!R$ in α requires to solve problems of the kind as (18) with $\tilde{x} := x^k + \alpha d^k$ and $J := J^k(\alpha)$ as the actual bundle. This choice of J indicates that we always consider the noisy information of the actual point where we solve (18) to be represented by the bundle.

In a first order sense we make use of the following approximation of φ^k:

$$
\bar{\varphi}^k(\alpha) := f(x^k) + \alpha \nabla f(x^k)^T d + \sum_{i \in I} \mu_i^k \max\{\bar{c}_i^k + \alpha\, \bar{a}_i^{k^T} d^k, 0\}.
$$

This enables us to define the expected descent by $\eta^k := \bar{\varphi}^k(1) - \bar{\varphi}^k(0)$ which can be proved to be negative provided that $\mu_i^k > \hat{\lambda}_i^k$, for all $i \in I$, and $d^k \ne 0$. The step length α^k is then chosen such that

$$
\varphi^k(\alpha^k) - \varphi^k(0) \le \nu \alpha^k \eta^k < 0, \quad \text{with} \quad 0 < \nu \le \frac{1}{2}. \tag{34}
$$

Now we are able to outline the overall algorithm:

Algorithm 4.1 *Given* $0 < \nu \le \frac{1}{2}$, $0 < \bar{\delta} < 1$, $\bar{\rho} > 1$, $J_{\max} \ge 2$ *with* $J_{\max} \in I\!N$:

1. *Start:* *Put* $k := 1$. *Choose initially* $x^1 \in \mathbb{R}^n$, $B^1 \in \mathbb{R}^{n \times n}$ *positive definite,* $\rho^1 \in \mathbb{R}$, $\mu^1 > 0 (\in \mathbb{R}^{|I|})$, $y^1 := x^1$, $J^1 := \{1\}$. *Then put* $\bar{c}_i^1 = \tilde{c}_i^1$ *and* $\bar{a}_i^1 = \tilde{a}_1^1$ *for all* $i \in I$.

2. *QP-Phase: Obtain* (d^k, δ^k) *as optimal solution of (33). If* $\delta^k > \bar{\delta}$, *put* $\rho^k := \bar{\rho}\rho^k$ *and start with 2. again.*

3. *Stopping criterion: If some stopping rule is satisfied, then stop; otherwise go to step 4.*

4. *Penalty update: Adjust* μ_i^k *such that* $\mu_i^k > \hat{\lambda}_i^k$ *for all* $i \in I$.

5. *Line search: Set* $\alpha_0 := 1$. *Generate a sequence* (α_l) *by* $\alpha_l = \omega\alpha_{l-1}$, $0 < \omega < 1$. *Compute* $x_l^k := x^k + \alpha_l d^k$ *and* $(\bar{c}_i^{J^k(\alpha_l)}(x_l^k), \bar{a}_i^{J^k(\alpha_l)}(x_l^k))$. *Let* ℓ *be the first index for which* α_ℓ *satisfies (34). Then set* $\alpha^k := \alpha_\ell$.

6. *Step: Put* $x^{k+1} := x^k + \alpha^k d^k$, $y^{k+1} := x^{k+1}$, *and* $J := J^k \cup \{k+1\}$.

7. *Matrix update: Calculate* $\bar{c}_i^J(x^k)$ *and* $\bar{a}_i^J(x^k)$ *by solving (18) for all* $i \in I$. *Update* B^k *in order to obtain* B^{k+1} *(positive definite).*

8. *Reset: If* $|J| = J_{\max}$, *choose* $J^s \subset J$ *with* $|J^s| \leq J_{\max} - 2$ *and put* $J^{k+1} := J^s \cup \{\tilde{j}\}$ *and relate the aggregate information* $\bar{c}_i^{k+1}$, $\bar{a}_i^{k+1}$, $\bar{\varepsilon}_i^{x^{k+1}} := 0$ *and* $y^{\tilde{j}} := x^{k+1}$ *to* $\tilde{j}$; *otherwise set* $J^{k+1} = J$. *Put* $k := k+1$ *and go to step 2.*

The following lemma proves that the line search step of Algorithm 4.1 is a finite procedure at each iteration level.

Lemma 4.1 *Suppose the following assumptions hold:*

(i) $\mu_i^k > \hat{\lambda}_i^k + \kappa_i$ *for all* $i \in I$ *with* $\kappa_i > 0$,

(ii) $0 \leq \delta^k \leq \bar{\delta} < 1$,

(iii) $f \in C^2$ *with* $\xi := \sup\{\|\nabla^2 f(x+h)\| \mid \|h\| \leq \|d^k\|\}$.

Then the line search test

$$\varphi^k(\alpha) - \varphi^k(0) \leq \nu\alpha\eta^k \quad \text{with } 0 < \nu \leq \frac{1}{2}, \tag{35}$$

holds for all $0 < \alpha \leq \bar{\alpha}^k$ *with*

$$\bar{\alpha}^k := \min\left\{\frac{2(\nu-1)\eta^k}{\|d^k\|^2(\xi + \sum_{i \in I}\mu_i^k K_i)}, 1\right\}.$$

Proof: We first observe that

$$\varphi^k(\alpha) \le \varphi^k_{J^k}(\alpha) := f(x^k + \alpha d^k) + \sum_{i\in I} \mu_i^k \max\{\bar{c}_i^{J^k}(x^k + \alpha d^k), 0\}$$

which is due to $J^k \subset J^k(\alpha)$. Moreover, we know by Theorem 3.2 that

$$\bar{c}_i^{J^k}(x^k + \alpha d^k) \le \bar{c}_i^k + \alpha\, \bar{a}_i^{kT} d^k + \frac{K_i}{2}\alpha^2 \|d^k\|^2.$$

Therefore, we find for $\alpha \in [0,1]$ that

$$\begin{aligned}
\varphi^k_{J^k}(\alpha) &\le f^k + \alpha \nabla f^{kT} d^k + \frac{1}{2}\alpha^2 \xi \|d^k\|^2 + \sum_{i\in I} \mu_i^k \max\{\bar{c}_i^k + \alpha\, \bar{a}_i^{kT} d^k + \frac{K_i}{2}\alpha^2\|d^k\|^2, 0\} \\
&\le \bar{\varphi}^k(\alpha) + \frac{1}{2}\alpha^2 \|d^k\|^2 (\xi + \sum_{i\in I} \mu_i^k K_i) \\
&\le \alpha\left(\bar{\varphi}^k(1) - \bar{\varphi}^k(0)\right) + \bar{\varphi}^k(0) + \frac{1}{2}\alpha^2 \|d^k\|^2 (\xi + \sum_{i\in I} \mu_i^k K_i)
\end{aligned}$$

due to the convexity of $\bar{\varphi}^k$. A straightforward calculation reveals that (35) holds for all $\alpha \in (0, \bar{\alpha}^k)$ with

$$\bar{\alpha}^k := \min\left\{1, \frac{2(\nu - 1)\eta^k}{\|d^k\|^2(\xi + \sum_{i\in I}\mu_i^k K_i)}\right\}. \square$$

5 Convergence analysis

Based on Lemma 4.1, which proves the inner loop (line search) of Algorithm 4.1 to be a finite procedure, convergence of the outer loop (step 2. - step 8.) will be proved by the following theorem. We will show that the first order Kuhn-Tucker conditions of the proximal model of (1) are satisfied asymptotically.

Theorem 5.1 *Let $f \in C^2$. Suppose that the following assumptions hold:*

(i) $(x^k)_{k\in\mathbb{N}}$, $(d^k)_{k\in\mathbb{N}}$, $(B^k)_{k\in\mathbb{N}}$ and $(\mu^k)_{k\in\mathbb{N}}$ are bounded,

(ii) $\mu_i^k \ge \hat{\lambda}_i^k + \kappa_i$ for all $k \in \mathbb{N}$ and for all $i \in I$, where $\kappa_i > 0$,

(iii) $\mu_i^k = \mu_i^\infty \in \mathbb{R}$ for all sufficiently large k and for all $i \in I$,

(iv) $\delta^k \le \bar{\delta} < 1$ for all k.

Then the following conditions hold:

$$\lim_{k\to\infty} \nabla f^k + \sum_{i\in I} \hat\lambda_i^k \bar a_i^k = 0 \tag{36}$$

$$\lim_{k\to\infty} \max\{\bar c_i^k, 0\} = 0 \quad \textit{for all } i \in I, \tag{37}$$

$$\lim_{k\to\infty} \hat\lambda_i^k \bar c_i^k = 0 \quad \textit{for all } i \in I, \tag{38}$$

$$\hat\lambda_i^k \geq 0 \quad \textit{for all } i \in I \textit{ and } k \in \mathbb{N}. \tag{39}$$

Proof: First observe that (39) is immediate since $\hat\lambda^k \in \mathbb{R}^{|I|}$ denotes the optimal multiplier vector of (33) (and thus is nonnegative).

Next, we will show that $\lim_k \eta^k = 0$. For this matter assume that $\liminf \eta^k \leq 2\bar\eta < 0$. Consider iterations k where the expected descent η^k remains under $\bar\eta < 0$. Let D be the bound on (d^k). If we choose

$$\bar\alpha := \min\left\{\frac{2(\nu-1)\bar\eta}{D^2(\xi + \sum_{i\in I}\mu_i^\infty K_i)},\, 1\right\} > 0$$

and consider the way how α^k is determined, we find that $\alpha^k \in (0, \bar\alpha]$ is bounded away from 0 for all k. More precisely, we have $\alpha^k \geq \omega\bar\alpha$. Lemma 4.1 concludes that

$$\varphi^k(\alpha^k) - \varphi^k(0) \leq \nu\alpha^k\eta^k \leq \nu\omega\bar\alpha\bar\eta < 0. \tag{40}$$

Consequently, the sequence $(\varphi^k(\alpha^k) - \varphi^k(0))$ is bounded away from 0 for $\eta^k \leq \bar\eta < 0$. On the other hand, by assumption (iii) we have $\varphi^k(\alpha^k) = \varphi^{k+1}(0)$ for k large enough. Hence the sequence $(\varphi^k(0))$ is monotonically decreasing for sufficiently large k and bounded below. Consequently we find $\lim_{k\to\infty} \varphi^k(\alpha^k) - \varphi^k(0) = 0$ in contradiction to (40). Hence

$$\lim_{k\to\infty} \eta^k = 0. \tag{41}$$

The first order Kuhn-Tucker conditions of (33) and assumptions (ii) and (iv) yield

$$\begin{aligned} \eta^k &\leq -d^{k^T} B^k d^k - \sum_{i\in I} (\mu_i^k - \hat\lambda_i^k)\, \beta_i^k \max\{\bar c_i^k, 0\} \\ &\leq (\bar\delta - 1) \sum_{i\in I} \kappa_i \max\{\bar c_i^k, 0\} \leq 0 \end{aligned} \tag{42}$$

with $\beta_i^k = 1 - \delta^k$ for $i \in I_k^+$ and $\beta_i^k = 1$ otherwise. Hence, (41) and assumption (iv) imply

$$\lim_{k\to\infty} \max\{\bar c_i^k, 0\} = 0 \quad \text{for all } i \in I. \tag{43}$$

Moreover, by (41) and (43) we obtain from (42) that $\lim_k d^{k^T} B^k d^k = 0$. Since (B^k) is bounded, we obtain

$$\lim_{k\to\infty} B^{k^{1/2}} d^k = 0. \tag{44}$$

Define $\mathcal{L}(x^k, \hat{\lambda}^k) := f(x^k) + \sum_{i\in I} \hat{\lambda}_i^k \bar{c}_i^k$. For proving assertion (36) we consider the gradient vector $\nabla_x \mathcal{L}(x^k, \hat{\lambda}^k)$ and the fact that d^k is optimal solution of (33) with optimal multipliers $\hat{\lambda}_i^k$, $i \in I$. Thus

$$\|\nabla f^k + \sum_{i\in I} \hat{\lambda}_i^k \bar{a}_i^k\| = \|B^k d^k\| \le \|B^{k^{1/2}}\| \, \|B^{k^{1/2}} d^k\|.$$

Applying (44) yields (36).
For complementary slackness (38) consider

$$\begin{aligned} \bar{\varphi}^k(0) - f^k - \sum_{i\in I} \hat{\lambda}_i^k \bar{c}_i^k &= \sum_{i\in I} \left(\mu_i^k \max\{\bar{c}_i^k, 0\} - \hat{\lambda}_i^k \bar{c}_i^k\right) \\ &\ge \mu_i^k \max\{\bar{c}_i^k, 0\} - \hat{\lambda}_i^k \bar{c}_i^k \ge -\hat{\lambda}_i^k \bar{c}_i^k, \end{aligned} \tag{45}$$

for any $i \in I$, where we used assumption (ii). Next observe that

$$\begin{aligned} \max\{\bar{c}_i^k + \bar{a}_i^{k^T} d^k, 0\} &\le \delta^k \max\{\bar{c}_i^k, 0\} \qquad \text{for } i \in I_k^+, \\ \max\{\bar{c}_i^k + \bar{a}_i^{k^T} d^k, 0\} &= 0 \qquad \text{for } i \in I_k^0, \end{aligned} \tag{46}$$

since (d^k, δ^k) is optimal solution of (33) with pertinent optimal multipliers $(\hat{\lambda}^k, \pi_1^k, \pi_2^k)$. We further obtain with $F^k = f^k + \nabla f^{k^T} d^k + \frac{1}{2} d^{k^T} B^k d^k + \frac{1}{2}\rho^k \delta^{k^2}$ that

$$\begin{aligned} \bar{\varphi}^k(0) &\ge f^k + \sum_{i\in I} \hat{\lambda}_i^k \bar{c}_i^k \ge F^k + \sum_{i\in I} \hat{\lambda}_i^k (\beta_i^k \bar{c}_i^k + \bar{a}_i^{k^T} d^k) - \pi_1^k \delta^k + \pi_2^k (\delta^k - 1) \\ &\ge f^k + \nabla f^{k^T} d^k + \sum_{i\in I_k^0} \mu_i^k \max\{\bar{c}_i^k + \bar{a}_i^{k^T} d^k, 0\} + \\ &\quad + \sum_{i\in I_k^+} \mu_i^k \left[(\max\{\bar{c}_i^k + \bar{a}_i^{k^T} d^k, 0\} - \delta^k \max\{\bar{c}_i^k, 0\})\right] \\ &\ge \bar{\varphi}^k(1) - \sum_{i\in I_k^+} \mu_i^k \delta^k \max\{\bar{c}_i^k, 0\} \end{aligned}$$

where we additionally used (46) and the positive definiteness of B^k. ¿From the boundedness of (μ^k), (41) and (43) we conclude

$$\lim_{k\to\infty} \left(\bar{\varphi}^k(0) - f^k - \sum_{i\in I} \hat{\lambda}_i^k \bar{c}_i^k\right) = 0.$$

If we consider this together with (43),(45) and (39), then (38) is obtained.□

Remark 5.1 *Assumption (i) of Theorem 5.1 includes the boundedness of* $(x^k)_{k\in\mathbb{N}}$. *Therefore, we may weaken Assumption 1.4 in the sense that we only need knowledge of* K_i', $i \in I$, *on a sufficiently large bounded set.*

By Theorem 5.1 we found that our algorithm generates sequences $(x^k)_{k\in \mathbb{N}}$ and $(\hat{\lambda}^k)_{k\in \mathbb{N}}$ satisfying asymptotically the first order Kuhn-Tucker conditions of the proximal model. Moreover, since $\bar{c}_i^k \geq c_i(x^k)$ for all k and for all $i \in I$ we obtain asymptotic feasibility for (1). Therefore, we want to measure the quality of the proximals $\bar{c}_i^k$ and $\bar{a}_i^k$ with respect to the exact values $c_i(x^k)$, $a_i(x^k)$ of (1).

Theorem 5.2 *Suppose that Assumption 1.1 – 1.4 hold. Then for all $k \in \mathbb{N}$ and for all $i \in I$*

(a) $0 \leq \bar{c}_i(x^k) - c_i(x^k) \leq 2\,\bar{\varepsilon}_i^k$,

(b) $\|a_i(x^k) - \bar{a}_i^k\| \leq 2\sqrt{2\,K_i\,\bar{\varepsilon}_i^k}$.

Proof: Theorem 2.1(b) yields $\bar{c}_i^k \geq c_i(x^k)$. Since we always have $(x^k, \bar{c}_i^k, \bar{a}_i^k, \bar{\varepsilon}_i^k)$ to be represented by the bundle when calculating $\bar{c}_i^k$, we obtain $\bar{c}_i^{x^k} + \bar{\varepsilon}_i^{x^k} - \bar{c}_i^k \geq 0$. Expansion by $\pm c_i(x^k)$ yields

$$0 \leq \left(\bar{c}_i^{x^k} - c_i(x^k)\right) + \left(c_i(x^k) - \bar{c}_i^k\right) + \bar{\varepsilon}^{x^k} \leq c_i(x^k) - \bar{c}_i(x^k) + 2\,\bar{\varepsilon}_i^k$$

for all k. This proves (a).

By Theorem 3.2 we know that for all $x \in \mathbb{R}^n$

$$\bar{c}_i^k + \bar{a}_i^{k^T}(x - x^k) + \frac{K_i}{2}\|x - x^k\|^2 \geq c_i(x^k) + a_i(x^k)^T(x - x^k) - \frac{K_i}{2}\|x - x^k\|^2 \quad (47)$$

for all $k \in \mathbb{N}$ and $i \in I$. Remodelling (47) yields

$$\left(\bar{a}_i^k - a_i(x^k)\right)^T x + K_i\|x - x^k\|^2 \geq c_i(x^k) - \bar{c}_i^k + (\bar{a}_i^k - a_i(x^k))^T x^k. \quad (48)$$

For $\bar{x}_i = x^k + \frac{1}{2}K_i^{-1}(a_i(x^k) - \bar{a}_i^k)$ the minimum in the left hand side of (48) is attained. If we introduce $\bar{x}_i$, then we obtain

$$\begin{aligned} 0 &\leq \bar{c}_i^k - c_i(x^k) - \left(a_i(x^k) - \bar{a}_i^k\right)^T \left(\frac{a_i(x^k) - \bar{a}_i^k}{2\,K_i}\right) + \frac{1}{4\,K_i}\|a_i(x^k) - \bar{a}_i^k\|^2 \\ &\leq 2\,\bar{\varepsilon}_i^k - \frac{1}{4\,K_i}\|a_i(x^k) - \bar{a}_i^k\|^2 \end{aligned}$$

which proves (b).□

Remark 5.2 *Whenever we detect $\bar{\varepsilon}^{x_l^k} = 0$ for all $i \in I$ in the line search step of Algorithm 4.1, then we may choose $J^k(\alpha_l) = \{k(\alpha_l)\}$ yielding $\bar{c}_i^{k(\alpha_l)} = c_i(x_l^k)$ and $\bar{a}_i^{k(\alpha_l)} = a_i(x_l^k)$. If (34) is satisfied, then we may reset $J^k := \emptyset$ in step 6. Hence, if $\varepsilon = 0$ (see Assumption 1.2) then the algorithm is similar to a variable metric method for constrained optimization which can be analysed like in [9]. It is known that these methods exhibit a (locally) superlinear rate of convergence.*

6 Numerical aspects and tests

Before we shall give our preliminary numerical results, a few numerical aspects will be addressed.

In many practical situations the constants K_i, $i \in I$, are unknown a priori. Therefore the algorithm shall incorporate a mechanism that adjusts K_i appropriately. In our implementation this is done by using

$$\underline{p}_i(x^k; y^j) := \tilde{c}_i^j - \tilde{\varepsilon}_i^j + \tilde{a}_i^{j^T}(x^k - y^j) - \frac{K_i}{2}\|x^k - y^j\|^2$$

with $j \in J^k$. We have $\underline{p}_i(x^k; y^j) \leq c_i(x^k) \leq \bar{p}_i(x^k; y^j)$ for all $j \in J^k$ and for all k. Now, if K_i is unknown we compute a sequence $(K_i^k)_{k\in\mathbb{N}}$ with start-up values $K_i^1 \geq 0$. Whenever we detect that $\underline{p}_i(x^k; y^j) > \bar{p}_i(x^k; y^j)$ for some $j \in J^k$ (note that K_i is replaced by K_i^k), then K_i^k is enlarged sufficiently. In the case that the bundle already contains aggregate information a *total reset* is performed, i.e we dismiss the aggregate information and – if there is reason to do so – other bundle points. By Assumption 1.4 it is ensured that K_i^k, $i \in I$, will be updated only finitely many times.

A second aspect is concerned with the updating of B^k. The well-known BFGS-formula is suitable. As usual it has to be modified to guarantee positive definite matrices and, here in addition, bounded condition number in a single updating step. So we use (remember $\nabla_x \mathcal{L}(x^k, \lambda) = \nabla f(x^k) + \sum_{i\in I} \lambda_i \bar{a}_i^k$)

$$w' := \nabla_x \mathcal{L}(x^{k+1}, \hat{\lambda}^{k+1}) - \nabla_x \mathcal{L}(x^k, \hat{\lambda}^{k+1}), \quad s := x^{k+1} - x^k$$

$$\Theta := \begin{cases} 1 & \text{if } s^T w' \geq 0.1 s^T B^k s \ \wedge \ s^T w' \leq 10 s^T B^k s \\ \frac{9 s^T B^k s}{s^T B^k s - s^T w'} & \text{if } s^T w' > 10 s^T B^k s \\ \frac{0.9 s^T B^k s}{s^T w' - s^T B^k s} & \text{if } s^T w' < 0.1 s^T B^k s \end{cases}$$

$$w := \Theta w' + (1 - \Theta) B^k s$$

The updated matrix is calculated by means of

$$B^{k+1} := B^k + \frac{w w^T}{w^T s} - \frac{B^k s s^T B^k}{s^T B^k s}.$$

We shall now report on computational testing of the algorithm with a double precision FORTRAN-code on a DECstation 5000/25 with relative accuracy $\varepsilon_M \approx 1.1 \times 10^{-16}$.

We use the following stopping rule (step 3. of Algorithm 4.1):

$$d^{k^T} B^k d^k \leq \texttt{tol}^2, \quad \sum_{j\in J^k} \max\{\tilde{c}_i^k, 0\} \leq \texttt{tol},$$

$$\|\nabla_x \mathcal{L}(x^k, \hat{\lambda}^k)\|^2 \leq \texttt{tol}, \quad \sum_{j\in J^k} \hat{\lambda}^k \mid \tilde{c}_i^k \mid \leq \texttt{tol},$$

where tol > 0 denotes a user-defined stopping tolerance.

The adjustment of penalty parameters (step 4. of Algorithm 4.1) is realized by putting

$$\mu_i^k := \begin{cases} \xi_1 \hat{\lambda}_i^k & \text{if } \mu_i^{k-1} < \xi_2 \hat{\lambda}_i^k,\ i \in I \\ \mu_i^{k-1} & \text{otherwise,} \end{cases}$$

where $1 < \xi_2 < \xi_1$.

The parameters had values $\nu = 0.1$, $\bar{\delta} > 0.9$, $\bar{\rho} = 10$, $\xi_2 = 1.1$, $\xi_1 = 10$ and the stopping tolerance tol $= 10^{-7}$.

All of the test problems come from [11] and [13]. We use the same numbers as in the references, where I stands for [11] and II for [13]. We shall mention that I66 does not satisfy Assumption 1.4 since the constraints involve exp-terms. But, by the boundedness of the iteration sequence, Remark 5.1 applies.

We generated perturbations by $\tilde{c}_i(x) := c_i(x) + \varepsilon_i^x \sin(\sum_{l=1}^n q_{i,l}^x x_l + q_{i,0}^x \pi)$, where $c_i(x)$ denotes the exact constraint, $\varepsilon_i^x \in \varepsilon \cdot \mathrm{U}[0,1]$ and $q_{i,l}^x \in q \cdot \mathrm{U}[-1,1]$ for all $l = 0, \ldots, n$. By $\mathrm{U}[a,b]$ we denote uniformly distributed reals in $[a,b]$.

In the following tables we denote by f^* the optimal objective value of the noisy problem, by x^* the corresponding optimal point; $f^\dagger$ and $x^\dagger$ represent the corresponding optimal values of the noiseless problem. Moreover $rc = \max\{\tilde{c}_i^* - c_i(x^*) \mid i \in I\}$, and $K_+^* := \max\{K_i^* | i \in I\}$.

Table 1: Test results for $\varepsilon := 10^{-3}$and $q := 1$.

Test	n	$\|I\|$	$J_{\max}$	#it	$f^* - f^\dagger$	$\\|x^* - x^\dagger\\|$	rc	K_+^*
$I19$	2	2	5	13	0.4300104	0.4219E-3	0.2903E-3	20.00
$I43$	4	3	10	21	3.6766E-4	5.4318E-3	1.1781E-4	25.78
$I66$	3	2	10	14	5.9522E-4	8.6694E-3	6.8515E-4	1.21
$II386$	15	11	15	70	0.4986E-3	0.9180E-3	0.9201E-4	928.12

Table 1 shows the results for moderate ε and q. In all cases, sufficiently large K_i-values are reached. Therefore, the residuals $\tilde{c}_i^* - c_i(x^*)$, $i \in I$, are all nonnegative. Hence, the optimal objective value of the proximal problem is greater (or at best equal) to the exact optimal objective value. Moreover, one can observe that the maximal residuals are all smaller than $\varepsilon = 10^{-3}$. This behaviour is justified by our qualification theorem.

Table 2 is concerned with varying ε values. We set $q := 1$ in all cases. The corresponding $J_{\max}$-values can be found in Table 1. In all cases, sufficiently large K_i-values are reached.

In the following Table 3 we give the results for the worst possible case concerning the gradient perturbation. A further enlargement of q causes inconsistencies that can not be smoothed out by adjusting $J_{\max}$.

Table 2: Test results for variable ε choices with $q := 1$.

Test	$\varepsilon = 10^{-2}$			$\varepsilon = 10^{-1}$			$\varepsilon = 1$		
	#it	$f^* - f^\dagger$	rc	#it	$f^* - f^\dagger$	rc	#it	$f^* - f^\dagger$	rc
$I19$	15	4.53521	0.20E-2	31	20.1925	0.14E-1	48	125.334	0.9E-1
$I43$	16	0.23E-2	0.17E-2	15	0.31E-1	0.14E-1	16	0.41339	0.11334
$I66$	6	0.11E-1	0.14E-1	7	0.90E-1	0.10458	76	0.69435	0.28141
$II386$	72	0.60E-2	0.70E-3	89	0.10949	0.67E-2	232	15.5265	0.13333

One can observe that K_i expectedly enlarges proportionally to $\varepsilon\, q^2$. Usually K_i is adjusted during the starting iterations. However, in some cases (and dependent on $J_{\max}$) a fine tuning later on takes place.

Table 3: Test results for $\varepsilon := 10^{-3}$ and largest possible q.

Test	q	$J_{\max}$	#it	$f^* - f^\dagger$	$\|x^* - x^\dagger\|$	rc	K_+^*
$I19$	1E4	10	42	0.1659926	0.164E-3	0.101E-3	2.73E5
$I43$	1E5	20	106	0.3833185	0.303618	0.293E-4	1.45E7
$I66$	1E4	20	253	0.5932E-2	0.305328	0.538E-3	2.60E5
$II386$	1E4	15	217	3.1849092	0.245839	0.783E-4	2.35E6

Finally, in Table 4 we investigate the dependence on $J_{\max}$. We consider the largest possible noise corruption in the gradient. Hence, the corresponding q-values can be found in Table 3. Table 4 shows no obvious tendency concerning $J_{\max}$. Of course, depending on the noise level and on the curvature of the exact constraint a minimum number of bundle elements is of need to avoid inconsistency. One may conclude that there exists a typical number $J_{\max}$ for each problem. Moreover, depending on how often a total reset has to be performed in order to find an appropriate K_i the number of iterations varies. This is a reason why, in some cases, a larger $J_{\max}$-value also enlarges the number of iterations.

Table 4: Test results for $\varepsilon = 10^{-3}$ and variable $J_{\max}$.

Test	q	$J_{\max}$								
		20	30	40	50	60	70	80	90	100
$I19$	1E4	107	163	116	116	111	121	116	167	116
$I43$	1E5	106	122	113	122	122	122	122	122	122
$I66$	1E4	253	185	167	220	153	151	167	151	151
$II386$	1E4	206	204	221	215	162	173	188	173	173

References

[1] M. S. Bazaraa, C. M. Shetty (1976), *Foundations of Optimization*, Lecture Notes in Economics and Mathematical Systems, Springer, NY.

[2] C. Elster, A. Neumaier (1995), "A Grid Algorithm for Bound Constrained Optimization of Noisy Functions", IMA J. Num. An., 15, pp. 585-608.

[3] C. Elster, A. Neumaier (1997), "A Trust Region Method for the Optimization of Noisy Functions", Computing, 58, pp. 31-46.

[4] T. Glad, A. Goldstein (1977), "Optimization of Functions Whose Values are Subjected to Small Errors", BIT, 17, pp. 160-169.

[5] K. C. Kiwiel (1986), "A Method for Solving Certain Quadratic Programming Problems Arising in Nonsmooth Optimization", IMA J. of Num. An., 6, pp. 137-152.

[6] K. C. Kiwiel (1989), "A Dual Method for Certain Positive Semidefinite Quadratic Programming Problems", SIAM J. Sci. Stat. Comput., 10, pp. 175-186.

[7] K. C. Kiwiel (1990), "Proximity Control in Bundle Methods for Convex Nondifferentiable Minimization", Math. Prog., 46, pp. 105-122.

[8] J. A. Nelder, R. Mead (1965), "A Simplex Method for Function Minimization", Computer J., 7, pp. 308-313.

[9] M. J. D. Powell (1982), "Variable Metric Methods for Constrained Optimization", Math. Prog., The State of the Art, Springer, pp. 288-311.

[10] B. Rustem (1993), "Equality and Inequality Constrained Optimization Algorithms with Convergent Stepsizes", JOTA, 76, pp. 429-450.

[11] K. Schittkowski, W. Hock (1981), *Test Examples for Nonlinear Programming Codes*, Springer, NY.

[12] K. Schittkowski (1983), "On the Convergence of a Sequential Quadratic Programming Method with an Augmented Lagrangian Line Search Function", Math. Operationsforsch. und Statist., Ser. Optimization, 14, pp. 197-216.

[13] K. Schittkowski (1987), *More Test Examples for Nonlinear Programming Codes*, Springer, NY.

[14] A. Shapiro (1988), "Sensitivity Analysis of Nonlinear Programs and Differentiability Properties of Metric Projections", SIAM J. Cont. and Opt., 26, pp. 628-645.

[15] V. Torczon (1991), "On the Convergence of the Multidirectional Search Algorithm", SIAM J. Optim., 1, pp. 123-145.

[16] V. Torczon (1997), "On the Convergence of Pattern Search Algorithms", SIAM J. Optim., 7, pp. 1-25.

[17] D. Winfield (1973), "Function Minimization by Interpolation in a Data Table", J. Inst. Maths Applics, 12, pp. 339-347.

[18] J. Zowe (1985), "Nondifferentiable Optimization", Computational Mathematical Programming, Ed. K. Schittkowski, Series F: Computer and System Sciences, 15, Springer, NY, pp. 323-356.

Generic Existence, Uniqueness and Stability in Optimization Problems[1]

Alexander Ioffe (ioffe@math.technion.ac.il)
Department of Mathematics, Technion, Haifa 32000, Israel

Roberto Lucchetti (rel@komodo.ing.unico.it)
Politecnico di Milano, Piazzale Gerbetto 6, 22100 Como, Italia

Abstract

Some results concerning generic existence, uniqueness and stability for classes of optimization problems are considered. Specifically, we show that problems of convex and quasi-convex programming with inequality constraints have these properties on the class of data consisting of continuous convex (resp. quasi-convex and bounded below on bounded sets) functions with the topology of uniform convergence on bounded sets. On the other hand, it is shown that the simplest problem of calculus of variations is not generically uniquely solvable on the class of problems with autonomous integrands.

Keywords: well-posed optimization problem, generic variational principle, mathematical programming, calculus of variations.

1 Introduction

The absolute majority of existence theorems for solutions of optimization problems are based on the Weierstrass principle saying that a lower semi-continuous function attains its minimum on every compact set. As a rule, several assumptions are needed to make the principle work in specific problems. To an even greater extent this

[1]This research was supported in part by Israel-USA BS Fund under grant 94237. The second author is grateful to the Department of Mathematics of Haifa for its warm hospitality during the preparation of the paper.

G. Di Pillo and F. Giannessi (eds.), Nonlinear Optimization and Related Topics, 169–182.

is valid for principles which guarantee uniqueness and stability of solutions under perturbation. The combination of these three properties is often referred to as *well-posedness* of the problem and we shall follow this tradition in the paper.

However, it may happen that, when we pass from the study of the behavior of an individual problem to the study of the "mass" behavior of a big cluster of similar problems, a positive answer to the question of whether existence, uniqueness and stability are "typical" for problems of the cluster can be obtained under assumptions which are not usually sufficient to conclude that an arbitrary individual problem may possess one or all of these properties.

The word "typical" can be interpreted in two different ways: either as "generic" or as "with probability one". In certain important cases (e.g. when we have a family of problems parametrized by a finite dimensional parameter) the probability approach may be possible and preferable (e.g. [7]). In other cases, especially when really big clusters of problems are concerned (e.g. the collection of all minimization problems for lower semi-continuous extended-real-valued functions) no natural probability distribution can be associated with the cluster while good metric structures and appropriate genericity concepts can be found. In either case, though, not many results have been known (e.g. [3, 4, 6, 8] and references therein).

However, it turns out that, at least in the context of the genericity approach, a very general principle can be established. This is the *generic variational principle* recently proved in [5] by Ioffe and Zaslavskii. This principle is a generalization of the well known variational principle of Deville-Godefroy-Zizler [2] and can be applied to a broader variety of optimization problems, from finite dimensional mathematical programming to the calculus of variations and optimal control.

In the next section we remind the generic variational principle; the rest of the paper is dedicated to some of its applications.

2 Generic Variational Principle

We shall consider two complete metric spaces (X, ρ) and $(\mathcal{A}, d)$, the first being called the *domain space* and the second the *data space*. We shall further assume that to every $a \in \mathcal{A}$ a lower semicontinuous function f_a on X is associated with values in $\overline{I\!R} = [-\infty, \infty]$ and none of these functions is identically $+\infty$. The following is the basic hypothesis about the functions which we adopt in the principle.

(**H**) *There is a dense subset $\mathcal{B} \subset \mathcal{A}$, such that for any $a \in \mathcal{B}$, any $\varepsilon > 0$, and any $\gamma > 0$ there exist a nonempty open set $U \subset \mathcal{A}$, $x \in X$, $\alpha \in I\!R$ and $\eta > 0$ such that U belongs to the ball of radius ε around a and for any $b \in U$*

(i) $\inf f_b > -\infty$;

(ii) *if $z \in X$ is such that $f_b(z) \leq \inf f_b + \eta$, then $\rho(z, x) \leq \gamma$ and $|f_b(z) - \alpha| \leq \gamma$.*

We shall say, following tradition (see e.g. [4]), that the problem of minimizing f_a

is *well posed* if it has a unique solution $\bar{x}$ and for any sequence $\{a_n\} \subset \mathcal{A}$ converging to a, $\inf f_{a_n} > -\infty$ starting with a certain n and any sequence $\{z_n\} \subset X$ such that $f_{a_n}(z_n) - \inf f_{a_n} \to 0$ converges to $\bar{x}$ and, moreover, $\inf f_{a_n} \to f_a(\bar{x}) = \inf f_a$.

Theorem 1. *Assume* (**H**)*. Then $\mathcal{A}$ contains a dense G_δ subset $\mathcal{A}'$ such that for any $a \in \mathcal{A}'$ the problem of minimizing f_a is well-posed.*

3 Application to convex and quasi-convex minimization

It was proved in [1] that the problem of minimizing an extended-real-valued l.s.c. convex function on a Banach space is generically well-posed with respect to the topology of so-called bounded epi-convergence. In [5] it was shown that the latter follows from the generic variational principle. Here we shall demonstrate that the same conclusion can be drawn from the generic variational principle for other classes of convex minimization problems and also for corresponding classes of quasi-convex minimization.

So let X be a Banach space and let $C(X)$ denote the collection of continuous functions on X with the topology of uniform convergence on bounded subsets of X. This is a topology metrizable by the metric:

$$d(f,g) = \sum_{n=1}^{\infty} 2^{-n} \frac{\|f-g\|_n}{1+\|f-g\|_n},$$

where $\|f-g\|_n = \sup\{|f(x)-g(x)| : \|x\| \le n\}$.

Standard arguments show that $(C(X), d)$ is a complete metric space.

We shall denote further by $CC(X)$ the subspace of $C(X)$ consisting of convex functions and by $CQC(X)$ the subspace consisting of quasi-convex functions bounded below on every bounded set. Both are closed subspaces of $C(X)$. Indeed, a pointwise limit of a sequence of convex (resp. quasi-convex) functions is a convex (resp. quasi-convex) function and, on the other hand, the limit of a converging sequence of elements of $C(X)$ bounded below on some bounded set is also bounded below on the set. As a result, they are complete metric spaces with respect to the above introduced metric.

The first problem we consider is the following:

(P1) $$\text{minimize } f_0(x) \text{ s.t. } x \in X,$$

where $f_0 \in CC(X)$, or $f_0 \in CQC(X)$. Here is the result:

Theorem 2. *The problem* **(P1)** *is generically well-posed either on $CC(X)$ or on $CQC(X)$.*

Proof. We have to verify that the hypothesis (**H**) is valid. To this end we set $\mathcal{A} = CC(X)$ (in the convex case), $\mathcal{A} = CQC(X)$ (in the quasi-convex case) and in both cases we define $\mathcal{B} = \{f \in \mathcal{A} : \inf f > -\infty\}$. Then $\mathcal{B}$ is dense in $\mathcal{A}$. For, given $f \in \mathcal{A}$ and $\varepsilon > 0$, we choose n so big that $2^{-(n-1)} < \varepsilon$, set $\alpha = \inf\{f(x) : \|x\| \leq n\}$ and take $g(x) = \max\{\alpha, f(x)\}$. Then $g \in \mathcal{B}$ and $d(f,g) < \varepsilon$.

Now, take $f \in \mathcal{B}$, $0 < \varepsilon < 1$ and $0 < \gamma < 1$. Choose a natural n and a positive $\delta < \gamma$ such that $d(f,g) < \varepsilon/2$ if $|f(x) - g(x)| \leq \delta$ for $\|x\| \leq n$ and then an $\bar{x} \in X$ such that $f(\bar{x}) < \inf f + \delta/2$. Set

$$\bar{f}(x) = \max\{f(x), f(\bar{x}) + \delta_1\|x - \bar{x}\|\},$$

where $\delta_1 = (2\|\bar{x}\| + n)^{-1}(\delta/2)$. Then $|f(x) - \bar{f}(x)| \leq \delta$ if $\|x\| \leq \|\bar{x}\| + n$ and therefore $d(f, \bar{f}) < \varepsilon/2$.

Set finally

$$U = \{g \in \mathcal{A} : \quad |g(x) - \bar{f}(x)| < (\delta_1\gamma)/3, \text{ if } \|x\| \leq \|\bar{x}\| + n + \gamma\}.$$

Let $g \in U$. Then $d(g, f) < \varepsilon$. We have further that for any x with $\|x - \bar{x}\| = \gamma$,

$$g(x) > \bar{f}(x) - (\delta_1\gamma)/3 \geq \bar{f}(\bar{x}) + 2(\delta_1\gamma)/3 \geq g(\bar{x}) + (\delta_1\gamma)/3.$$

It follows (due to convexity or quasi-convexity alike) that $\inf g = \inf\{g(x) : \|x - \bar{x}\| < \gamma\}$ which proves the property (i) of (**H**) and, on the other hand, that $g(x) - g(\bar{x}) > (\delta_1\gamma)/3$ if $\|x - \bar{x}\| > \gamma$, so that it remains to take $\eta = (\delta_1\gamma)/3$ to see that the property (ii) also holds with $\alpha = f(\bar{x})$. The theorem has been proved.

We turn next to constrained minimization and consider the problem

(P2) $\qquad$ minimize $f_0(x)$ s.t. $f_i(x) \leq 0,\ i = 1, ..., k$.

The data of the problem are $(k+1)$-tuples $a = (f_0, f_1, ..., f_k)$ of either convex continuous functions on X (which, as above, is a Banach space) or quasi-convex functions on X which are bounded below on every bounded subset of X. Thus the data space could be identified either as $(CC(X))^{k+1}$ or as $(CQC(X))^{k+1}$.

But this choice of the data space can hardly be considered good as the admissible set $\mathcal{F}(a) = \{x : f_1(x) \leq 0, ..., f_k(x) \leq 0\}$ can be empty. Moreover, it is quite obvious that nonemptiness of $\mathcal{F}$ cannot be a generic, even densely valid, property with respect to any reasonable topology: consider for instance a problem where one function f_i is coercive in the sense that $f_i(x) \to \infty$ as $\|x\| \to \infty$ and also $f_i(x) \geq 1$, $\forall x$. Then close problems still have an empty constraint set. On the other hand the set $\{a : \mathcal{F}(a) \neq \emptyset\}$ is not closed either in $(CC(X))^{k+1}$ or in $(CQC(X))^{k+1}$, even when endowed with the very fine topology of uniform convergence. Therefore, in either case it is natural to take as the data space $\mathcal{A}$ the closure of $\{a : \mathcal{F}(a) \neq \emptyset\}$.

As to the distance in the data space, we make a natural choice:

$$\mathbf{d}(a,b) = \sum_{i=0}^{k} d(f_i, g_i), \quad (a = (f_0, ..., f_k),\ b = (g_0, ..., g_k)),$$

where $d(f,g)$ is the distance in $C(X)$.

To avoid confusion we shall denote, whenever necessary, the data spaces for convex and quasiconvex problems by $\mathcal{A}_c$ and $\mathcal{A}_{qc}$ respectively.

The functions f_a are defined in the standard way:

$$f_a(x) = \begin{cases} f_0(x), & \text{if } f_i(x) \leq 0,\ i = 1, ..., k; \\ \infty, & \text{otherwise.} \end{cases}$$

Theorem 3. *The problem* **(P2)** *is generically well-posed on* $\mathcal{A}_c$.

Proof. We shall write $\mathcal{A}$ instead of $\mathcal{A}_c$ in the proof. Set

$$\mathcal{B} = \{a = (f_0, ..., f_k) \in \mathcal{A} : \lim_{\|x\| \to \infty} f_0(x) = \infty;\ \mathcal{F}(a) \neq \emptyset\}.$$

Then $\mathcal{B}$ is dense in $\mathcal{A}$. Indeed, given an $\varepsilon > 0$ and an $a = (f_0, ..., f_k) \in \mathcal{A}$, we first take an $a' = (f_0', ..., f_k') \in \mathcal{A}$ such that $\mathcal{F}(a') \neq \emptyset$ and $\mathbf{d}(a, a') < \varepsilon/2$, then choose n so big that $2^{-(n-1)} < \varepsilon/2$ and set $\bar{a} = (\bar{f}_0, \bar{f}_1, ..., \bar{f}_k)$, where

$$\bar{f}_0 = \max\{f_0'(x), m + (\|x\| - n)^+)^2\};\ \bar{f}_i(x) = f_i'(x) i = 1, ..., n,$$

and $m = \inf_{\|x\| \leq n} f_0(x)$. Then $\mathbf{d}(a, \bar{a}) < \varepsilon$.

Now let $a = (f_0, ..., f_k) \in \mathcal{B}$. This means that $\mathcal{F}(a) \neq \emptyset$ and $\inf f_a > -\infty$. Take n so big that $f_0(x) \geq \inf f_a + 1$ if $\|x\| \geq n - 1$. Call k a Lipschitz constant for f_0 on the ball of radius n. Suppose now we are given $\varepsilon > 0$ and $\gamma > 0$. We fix a number $0 < \tau < 1$ with the properties that

$$\tau \leq \frac{\varepsilon}{2(k+1)}, \tag{1}$$

$$\tau + k\tau\frac{\tau^2}{2} < \gamma \tag{2}$$

and

$$d(f, g) < \frac{\varepsilon}{2(k+1)} \tag{3}$$

if $|f(x) - g(x)| \leq \frac{\tau^2}{4}$. Next we choose an $\bar{x} \in \mathcal{F}(a)$ such that $f_0(\bar{x}) < \inf f_a + \tau^2/4$. Then $\|\bar{x}\| < n - 1$ and so $\|\bar{x}\| + \tau \leq n$. Set

$$\bar{f}_0(x) = f_0(x) + \tau\|x - \bar{x}\|,\ \bar{f}_i(x) = f_i(x) + \tau(\|x - \bar{x}\| - \tau/2).$$

Then for $\bar{a} = (\bar{f}_0, ..., \bar{f}_k)$ we have $\mathbf{d}(a, \bar{a}) < \varepsilon/2$ (because of (1)). Set further

$$U = \{b = (g_0, ..., g_k) :\ |g_i(x) - \bar{f}_i(x)| \leq \tau^2/4, \text{ if } \|x\| \leq n\}.$$

Then, by (3), $\mathbf{d}(b, \bar{a}) < \varepsilon/2$ for all $b \in U$ and therefore $\mathbf{d}(b, a) < \varepsilon$. Moreover, we see that $g_i(\bar{x}) \leq \bar{f}_i(\bar{x}) + \tau^2/4 = f_i(\bar{x}) - \tau^2/4 < 0$, that is $\bar{x} \in \mathcal{F}(b)$ for any $b \in U$.

Take a $z \in \mathcal{F}(b)$. If $\|z\| \leq n$ and $\|z - \bar{x}\| \geq \tau$ then we have, for $i = 1, ..., k$ $0 \geq g_i(z) \geq \bar{f}_i(z) - \tau^2/4 \geq f_i(z) + \tau^2/4$ which means that $z \in \mathcal{F}(a)$. Moreover,

$$\begin{aligned} g_0(z) &\geq \bar{f}_0(z) - \tau^2/4 \geq f_0(z) + 3\tau^2/4 \geq \inf f_a + 3\tau^2/4 \\ &> f_0(\bar{x}) + \tau^2/2 = \bar{f}_0(\bar{x}) + \tau^2/2 \geq g_0(\bar{x}) + \tau^2/4 \geq \inf f_b + \tau^2/4. \end{aligned} \tag{4}$$

Thus

$$z \in \mathcal{F}(b),\ \|z\| \leq n,\ g_0(z) \leq \inf f_b + \tau^2/4 \ \Rightarrow\ \|z - \bar{x}\| < \tau. \tag{5}$$

On the other hand, suppose $\|z\| > n$. Remember that $\|\bar{x}\| \leq n - 1$, and call w the element in the segment $[z, \bar{x}]$ such that $\|w\| = n$. Then $\|w - \bar{x}\| > \tau$, $w \in \mathcal{F}(b)$ (as $\mathcal{F}(b)$ is convex) and therefore, as follows from (4) and (5),

$$g_0(w) > \inf f_b + \tau^2/4 \quad \&\ g_0(w) \geq g_0(\bar{x}).$$

The latter, due to convexity, implies that $g_0(z) \geq g_0(w)$ and we conclude that the implication

$$g_0(z) \leq \inf f_b + \tau^2/4 \ \Rightarrow\ \|z - \bar{x}\| < \tau$$

is valid for all $z \in \mathcal{F}(b)$.

Finally, if $\|z - \bar{x}\| < \tau$, then $\|\bar{x}\| \leq n$ and $\|z\| \leq n$ and using k-Lipschitzianity of f_0 on the n-ball,

$$g_0(z) \geq \bar{f}_0(z) - \tau^2/4 \geq f_0(z) - \tau^2/2 \geq f_0 - k\tau - \tau^2/2 \tag{6}$$

while if $g_0(z) \leq \inf f_b + \tau^2/4$, then

$$g_0(z) \leq \inf f_b + \tau^2/4 \leq g_0(\bar{x}) + \tau^2/4 \leq \bar{f}_0(\bar{x}) + \tau^2/2 \leq \inf f_a + 3\tau^2/4. \tag{7}$$

The proof is completed, if we choose $\eta = \frac{\tau^2}{4}$. Indeed, in this case we have $\|z - \bar{x}\| < \tau < \gamma$ and, from (2), (6) and (7), $\|g_0(z) - \bar{f}(\bar{x})\| < \gamma$.

Theorem 4. *Suppose that X is a reflexive Banach space. Then the problem* **(P2)** *is generically well-posed on $\mathcal{A}_{qc}$.*

In the proof of Theorem 3, given an objective function f_0, we made a perturbation by adding to it a convex function; here this idea does not apply since a sum of a quasi-convex function and a convex function need not be quasi-convex. Thus we have to use another construction.

Let P and Q be convex closed sets with $P \subset \text{int Q}$. Set

$$\mu_{P,Q}(x) = \inf\{\lambda \geq 0 :\ x \in \lambda Q + (1 - \lambda)P\}.$$

if $P = \{u\}$ is a singleton, this is the Minkowski function of Q with respect to u.

Proposition 1. *$\mu_{P,Q}$ is a convex continuous function such that $\mu_{P,Q}(x) = 0$ if $x \in P$, $\mu_{P,Q}(x) < 1$ if $x \in$ int Q and $\mu_{P,Q} > 1$ if $x \notin Q$.*

Proof. Standard.

Proposition 2. *Let f be a quasi-convex continuous function on X. Let ξ and η be a couple of real numbers such that $\inf f < \xi < \eta \leq \sup f$. Set for any λ, $Q_\lambda(f) = \{x : f(x) \leq \lambda\}$. Then*

$$f_{\xi,\eta}(x) = \begin{cases} f(x), & \text{if } f(x) > \eta; \\ \xi + (\eta - \xi)\mu_{Q_\xi,Q_\eta}(x), & \text{if } \xi \leq f(x) \leq \eta; \\ f(x), & \text{if } f(x) < \xi. \end{cases}$$

is a quasi-convex continuous function.

*More generally, let **C** be a finite sequence of pairs (c_k, C_k), $k = 1, ..., m$, where $\inf f < ... < c_k < c_{k+1} < ... < c_m \leq \sup f$, $C_1 = Q_{c_1}$, $C_m = Q_{c_m}$ and $C_k \subset \text{int } C_{k+1}$. Then the function*

$$f^{\mathbf{C}}(x) = \begin{cases} f(x), & \text{if } f(x) \leq c_1; \\ c_k + (c_{k+1} - c_k)\mu_{C_k,C_{k+1}}(x), & \text{if } x \in C_{k+1} \backslash C_k,\ k = 1, ..., m; \\ f(x), & \text{if } f(x) \geq c_m \end{cases}$$

is quasi-convex continuous and $|f(x) - f^{\mathbf{C}}(x)| \leq c_m - c_1$ for all x.

The proof of the proposition is also straightforward. Observe that the Propositions 1 and 2 are valid in every Banach space.

Proof of Theorem 4. We can now return to the proof of the theorem. Again, let us agree to write $\mathcal{A}$ instead of $\mathcal{A}_{qc}$ in the proof. Define $\mathcal{B}$ as the collection of $a = (f_0, ..., f_k) \in \mathcal{A}$ such that f_0 is coercive (that is $\lim_{\|x\| \to \infty} f_0(x) = \infty$, as in the Proof of Theorem 3), $\inf_x \max\{f_1(x), ..., f_k(x)\} < 0$, and for any $i = 1, ..., k$ the set $Q_0(f_i)$ is the closure of $\{x : f_i(x) < 0\}$, that is to say, the latter is the interior of $Q_0(f_i)$. Again we verify easily that $\mathcal{B}$ is dense in $\mathcal{A}$. The first requirement can be satisfied by a slight change of f_0 exactly as in beginning of the proof of Theorem 3. To get the second requirement satisfied, we can choose a small positive σ and set $\bar{f}_i(x) = f_i'(x) - \sigma$ for all $i = 1, ..., k$ (where, as in the proof of Theorem, $a' = (f_0', ..., f_k')$ is an element in $\mathcal{A}$ sufficiently close to a and such that $\mathcal{F}(a') \neq \emptyset$), finally, to fulfil the third requirement it is enough to replace every f_i' $(i = 1, ..., k)$ by $\bar{f}_i = (f_i')_{\xi,\eta}$, with $\eta = 0$ and ξ being a negative number sufficiently close to zero, as explained in Proposition 2. With these three requirements satisfied for a given a, it clearly holds that

$$\inf f_a := \inf\{f_0(x) : f_i(x) \leq 0\} = \inf\{f_0(x) : f_i(x) < 0\}.$$

So let $a = (f_0, ...f_k) \in \mathcal{B}$. As f_0 is coercive, $\inf f_a$ is finite and sublevel sets of f_a are bounded, hence weakly compact. It follows that $\inf f_a$ is attained at a certain point $\bar{z} \in \mathcal{F}(a)$.

Now suppose we are given $\varepsilon > 0$ and $\gamma > 0$. There is no loss of generality in assuming that $\gamma \leq \varepsilon/2$. Choose

- an $n > 0$ such that $f_0(x) > \inf f_a + 1$ if $\|x\| \geq n$ and $2^{-n} < \varepsilon/2(k+1)$;

- a $\delta > 0$ such that $\delta < \gamma$, $8\delta/\gamma < \varepsilon/4$ and $d(f,g) < \varepsilon/4(k+1)$ if $|f(x) - g(x)| < \delta$ for $\|x\| \leq n$; it will be further specified in the course of the proof.

We shall separately consider two possible situations.

a) (degenerate case). Suppose $\inf f_a = \inf f_0$. Take an $\bar{x} \in \text{int } \mathcal{F}(\text{a})$ with $f_0(\bar{x}) < \inf f_a + \frac{4\delta}{\gamma}$ and define

$$\bar{f}_0 = \max\{f(x), f(\bar{x}) + \frac{4\delta}{\gamma}\|x - \bar{x}\|\}.$$

Without loss of generality, we can also suppose that $f_i(\bar{x}) + 2\delta < 0$ for $i = 1, ..., k$, as $\bar{x} \in \text{int } \mathcal{F}(\text{a})$ and $a \in \mathcal{B}$, and also $3\delta + \frac{8\delta}{\gamma} < \gamma$. Set finally $\bar{f}_i(x) = f_i(x) + \delta$ $(i = 1, ..., k)$ and

$$U = \{b = (g_0, ..., g_k) \in \mathcal{A} : \; |\bar{f}_i(x) - g_i(x)| < \delta, i = 0, ..., k, \text{ if } \|x\| \leq n\}.$$

Then $\mathbf{d}(a, b) < \varepsilon$ and $\bar{x} \in \mathcal{F}(b)$ for any $b \in U$. We have too, if $\|z - \bar{x}\| = \gamma$:

$$g_0(z) \geq \bar{f}(z) - \delta \geq f_0(\bar{x}) + 3\delta \geq g_0(\bar{x}) + 2\delta.$$

Thus the condition $g_0(z) \leq \inf f_b + \eta$, where $\eta = 2\delta$, ensures $\|z - \bar{x}\| < \gamma$ (as in the proof of Theorem 2). To conclude, observe that

$$|\inf f_b - \inf f_a| \leq \delta + \frac{4\delta}{\gamma}$$

so that

$$|g_0(z) - \bar{f}(\bar{x})| < \delta + \frac{8\delta}{\gamma} + 2\delta < \gamma,$$

and this ends the proof in the degenerate case.

b) (general case). Suppose $\inf f_a > \inf f_0$. Set $\lambda = \inf f_a$ and for any ξ, denote by Q_ξ the ξ-sublevel set of f_0: $Q_\xi = \{x : \; f_0(x) \leq \xi\}$. By the assumptions the interiors of Q_λ and $\mathcal{F}(a)$ are nonempty and neither of the sets meets the interior of the other. Therefore there is a hyperplane H separating Q_λ and $\mathcal{F}(a)$. To be more specific: there are $z^* \in X^*$, $\|z^*\| = 1$ and $\alpha \in I\!R$ such that $H = \{x : \; \langle z^*, x\rangle = \alpha\}$ and $Q_\lambda \subset H^- = \{x : \; \langle z^*, x\rangle \leq \alpha\}$, $\mathcal{F}(a) \subset H^+ = \{x : \; \langle z^*, x\rangle \geq \alpha\}$.

Assume that the chosen δ satisfies $\inf f_0 < \lambda - 3\delta$. For any $\xi < \lambda$ the set Q_ξ belongs to the interior of H^- and, moreover, the distance from this set to H is positive as the space is assumed reflexive. Let $\rho > 0$ be the distance from $Q_{\lambda-\delta}$. We then have that

$$\langle z^*, x\rangle \leq \alpha - \rho$$

$\forall x \in Q_{\lambda-\delta}$. Finally, for any $x \in \mathcal{F}(a)$ set

$$S_x = \text{cl conv}(\{x\} \bigcup Q_{\lambda-\delta}),$$

and let D stands for the diameter of $Q_{\lambda+1}$.

Then elementary geometric arguments give that

$$\text{diam } (S_x \bigcap \mathcal{F}(a)) \leq \delta,$$

if the distance from x to H is not greater than $\delta\rho/2D$.

Now we choose an $\bar{x} \in \text{int } \mathcal{F}(a)$ such that $f_0(\bar{x}) < \lambda + \delta/2$ and the distance from $\bar{x}$ to H is not greater than $\beta = \delta\rho/4D$ and set

$$c_1 = \lambda - 3\delta, \quad c_2 = f_0(\bar{x}), \qquad c_3 = c_2 + 3\delta/2, \qquad c_4 = \lambda + 2\delta;$$
$$C_1 = Q_{c_1}, \quad C_2 = \text{cl conv}(\{\bar{x}\} \bigcup Q_{\lambda-2\delta}), \quad C_3 = \text{cl conv}(B(\bar{x},\beta) \bigcup Q_{\lambda-\delta}), \quad C_4 = Q_{c_4}.$$

Here $B(\bar{x},\beta)$ is the ball of radius β around $\bar{x}$. We further set $\mathbf{C} = (c_i, C_i,\ i = 1,2,3,4)$ and define $\bar{f}_0$ as $f_0^{\mathbf{C}}$. Then $|\bar{f}_0(x) - f(x)| \leq 5\delta$ for all x by Proposition 2 and $f_0(\bar{x}) = \bar{f}_0(\bar{x})$.

We have also: $C_3 = \bigcup_{x \in B(\bar{x},\beta)} S_x$. Therefore the diameter of $C_3 \bigcap \mathcal{F}(a)$ is not greater than δ. On the other hand, $\bar{x}$ belongs to this set, so that $C_3 \bigcap \mathcal{F}(a) \subset B(\bar{x},\delta)$.

We next define $\bar{f}_i$ for $i = 1,...,k$ as in the first (degenerate) case: as $\bar{x} \in \text{int } \mathcal{F}(\text{a})$ there is a $\sigma > 0$ with $f_i(\bar{x}) + 2\sigma \leq 0$ for all $i = 1,...k$, so we set $\bar{f}_i(x) = f_i(x) + \sigma$. As above, we can assume without loss of generality that $\sigma < \delta/4$ and define U as the collection of $b = (g_0,...,g_k) \in \mathcal{A}$ such that $|\bar{f}_i(x) - g_i(x)| < \sigma$ for all x with $\|x\| \leq n$ and all $i = 0,...,k$.

We see as in the first case that $\bar{x} \in \mathcal{F}(b)$, $\mathcal{F}(b) \subset \mathcal{F}(a)$ and

$$\inf f_b \leq g_0(\bar{x}) \leq \bar{f}_0(\bar{x}) + \delta/2 = f_0(\bar{x}) + \delta/2 \leq \lambda + \delta/2$$

for every $b \in U$. On the other hand if $x \in \mathcal{F}(b)$ and $\|x - \bar{x}\| > \delta$, then $x \in \mathcal{F}(a)$ and therefore $x \notin C_3$ which means that $\bar{f}_0(x) > f_0(\bar{x}) + 3\delta/2$. This implies that

$$g_0(x) > \bar{f}_0(x) - \delta/4 \geq \bar{f}_0(\bar{x}) + 5\delta/4 \geq \lambda + 5\delta/4 \geq \inf f_b + \delta/4.$$

The case $\|x\| > n$ is considered as in the proof of Theorem 2. Thus to conclude the proof we have to choose δ small enough to make sure that $5\delta < \varepsilon/2(k+1)$ (in which case also $\sigma < \varepsilon/2(k+1)$) and $\eta < \delta/4$. Thus $\mathbf{d}(a,b) < \varepsilon$ for every $b \in U$ and $|g_0(x) - \bar{f}(\bar{x})| < \gamma$ if $g_0(x) \leq \inf f_b + \eta$.

4 An example in calculus of variations

In this section we are concerned with generic well–posedness of the problem:

$$\textbf{(P3)} \qquad \text{minimize } I_L(x) \stackrel{\text{def}}{=} \int_0^1 L(t, x(t), x'(t))dt, \quad x(0) = x_o,\ x(1) = x_1,$$

where the *Lagrangean* L fulfills some regularity conditions and I_L is defined on the arcs belonging to the Sobolev space $W^{1,2}(0,1)$ and fulfilling the boundary conditions. To prove a general existence result for the calculus of variations problem **(P3)**, usually one needs assumptions as convexity and superlinearity of the Lagrangean L, with respect to the third variable. In the article [5] it is shown that both these assumptions are *not* needed if existence (and also well-posedness) is required only generically, for a suitable topology making the space of Lagrangeans a complete metric space. However the proof given there cannot be adapted to cover the case of autonomous problems, i.e. when one considers only Lagrangeans that do not depend on the time variable t. In this section we show that actually the genericity result does not hold in the class of autonomous problems. To this end, let us introduce the topology used in the [5]: Given a complete metric space X, consider the space $LSC^+(X)$ of the lower semicontinuous non-negative extended–real valued functions defined on X. For every $p \geq 1$ consider the following uniform structure with basis formed by the sets:

$$U_p(\varepsilon) = \{(f,g): \; |f(x) - g(x)| \leq \varepsilon(\|x\|^p + 1), \; \forall\, x \in X\},$$

with the convention that $\infty - \infty = 0$. The space $LSC^+(X)$ so topologized will be denoted by $LSC_p^+(X)$, and it can be shown that $LSC_p^+(X)$ is a complete metrizable space. We shall consider here the following space for the Lagrangean functions:

$$\mathcal{L} \stackrel{\text{def}}{=} \{L \in LSC_2^+(\mathbb{R} \times \mathbb{R}) : L \geq 0 \quad \text{and } L(\cdot,\cdot) \text{ is lower semicontinuous}\},$$

endowed with the complete metric described above (but with x replaced by (x,u) and $\|x\|^p$ by $|x|^p + |u|^p$) for $p = 2$. Here is the main result:

Theorem 5 *There exist a Lagrangean $L_0 \in \mathcal{L}$, a fixed pair of boundary conditions (x_o, x_1) and a neighbourhood $\mathcal{N}$ of L_0 in $\mathcal{L}$ such that for every $L \in \mathcal{N}$ the following alternative holds :*

- *either* **(P3)** *does not have a solution;*
- *or* **(P3)** *has more than one solution.*

As a result, the problems around I_{L_0} are not well–posed, and so the situation is completely different from the non-autonomous case.

The Lagrangean L_0 we are referring to in the theorem and the (trivial) boundary conditions are the famous Bolza example of non-existence when there is lack of convexity in the last variable:

$$L_0(x,u) = x^2 + (u^2 - 1)^2.$$

It is actually easy to show that zero is the infimum of the problem

$$\text{minimize} \quad I_{L_0}(x) = \int_0^1 x^2(t) + (x'^2(t) - 1)^2 dt, \quad \text{s.t. } x(0) = x(1) = 0,$$

(e.g. take an $x(\cdot)$ with $|x'(t)| = 1$, $|x(t)| \leq \varepsilon$ for all t), but clearly no function x can achieve this value.

Finally, for a given $L \in \mathcal{L}$, let us denote by L^* its Fenchel conjugate *with respect to the second variable*. Observe that functions L around L_0 have bounded sublevel sets. Let $v(L)$ denote the value function, that is the minimal value of the functional $J_L(\cdot)$ in the corresponding problem (we do not vary boundary conditions!) To prove the theorem, we need the following lemmas.

Lemma 1 *The value function $v : \mathcal{L} \to I\!R$ is continuous in $\mathcal{L}$ at L_0.*

Proof. Fix an $\varepsilon > 0$ and consider a function $\bar{x}$ such that $I_{L_0}(\bar{x}) < \frac{\varepsilon}{2}$. Now, choose $\delta > 0$ so small that $\int_0^1(\bar{x}^2(t) + \bar{x}'^2(t) + 1)dt \leq \varepsilon/2\delta$. Then for any $x(\cdot)$ such that $(L, L_0) \in U_2(\delta)$ we have $J_L(\bar{x}) \leq \varepsilon$. It remains to remember that elements of $\mathcal{L}$ are assumed nonnegative.

Lemma 2 *$\forall\varepsilon > 0$, $\forall K > 0$, there exists a neighbourhood $\mathcal{N}$ of L_0 such that $\forall L \in \mathcal{N}$:*

$$|L^{**}(x,u) - L_0^{**}(x,u)| < \varepsilon$$

$\forall(x,u)$ such that $\|x\| \leq K$ and $\|u\| \leq K$.

Proof. We have

$$L_0^{**}(x,u) = \begin{cases} x^2, & \text{if } |u| \leq 1; \\ L_0(x,u), & \text{if } |u| \geq 1, \end{cases}$$

For any $L \in \mathcal{L}$ set

$$\tilde{L}(x,u) = \begin{cases} \min\{L(x,u), \alpha L(x,1) + (1-\alpha)L(x,-1)\}, & \text{if } u = 2\alpha - 1 \ (0 \leq \alpha \leq 1); \\ L(x,u), & \text{if } |u| \geq 1, \end{cases}$$

Then $\tilde{L} \in \mathcal{L}$ and $L^{**} \leq \tilde{L} \leq L$ and therefore $(\tilde{L})^{**} = L^{**}$. In particular for $L = L_0$ we have $\tilde{L}_0 = L_0^{**}$.

Set $r = \max_{|u|\leq 1} |L(x,u) - L_0(x,u)|$. Then for any $u \in [-1,1]$:

$$\tilde{L}(x,u) \leq \min\{L_0(x,u) + r, \alpha L(x,1) + (1-\alpha)L(x,-1) + r\} = L_0^{**}(x,u) + r$$

(where as above $u = 2\alpha - 1$) and, likewise, $\tilde{L}(x,u) \geq L_0^{**}(x,u) - r$, that is

$$\max_{|u|\leq 1} |\tilde{L}(x,u) - L_0^{**}(x,u)| \leq \max_{|u|\leq 1} |L(x,u) - L_0(x,u)|.$$

Now, suppose $|L(x,u) - L_0(x,u)| \leq \delta(x^2 + u^2 + 1)$ for all x, u. As

$$|\tilde{L}(x,u) - L_0^{**}(x,u)| \begin{cases} = & |L(x,u) - L_0(x,u)|, & \text{if } |u| \geq 1; \\ \leq & \max_{|u|\leq 1} |\tilde{L}(x,u) - L_0(x,u)|, & \text{if } |u| \leq 1; \end{cases},$$

from which we conclude that

$$|\tilde{L}(x,u) - L_0^{**}(x,u)| \leq \delta(x^2 + u^2 + 2).$$

¿From the inequality

$$L_0^{**}(x,u) - \delta(x^2 + u^2 + 2) \leq \tilde{L}(x,u)$$

we get that

$$\tilde{L}(x,u) \geq (1-\delta)L_0^{**}(x,u) - 4\delta$$

(as $u^4 \geq u^2$ for $|u| \geq 1$ and $u^2 \leq 1$ for $|u| \leq 1$). As the function in the right-hand side of the last inequality is convex continuous in u, it follows that

$$L^{**}(x,u) \geq (1-\delta)L_0^{**}(x,u) - 4\delta.$$

Moreover

$$L_0^{**}(x,u) \geq \tilde{L}(x,u) - \delta(x^2+u^2+2) \geq L^{**}(x,u) - \delta(x^2+u^2+2).$$

Now, given a $K > 0$, choose $\delta < \varepsilon(3K^2 + (K^2-1)^2 + 4)^{-1}$. The desired result follows if we define $\mathcal{N}$ as the collection of $L \in \mathcal{L}$ for which $|L(x,u) - L_0(x,u)| < \delta(x^2+u^2+1)$ for all x, u.

Now, for a given continuous function $x : [0,1] \to I\!R$ and for $c \in I\!R$, define:

$$E_c^+ \stackrel{\text{def}}{=} \{t \in [0,1] : x(t) > c\},$$

$$E_c^- \stackrel{\text{def}}{=} \{t \in [0,1] : x(t) < c\}.$$

Call m, M the minimum and the maximum points of $x(\cdot)$, respectively.

Lemma 3. *Let $x \in C[0,1]$ be such that $x(0) = x(1) = 0$ and suppose x changes sign on $[0,1]$. Moreover, let us assume that, for all c, at least one between E_c^+ and E_c^- is a (possibly degenerate) interval. Then x is monotone between m and M, and monotone in the opposite way in the other intervals.*

Proof. Suppose, w.l.o.g. $0 < m < M < 1$, and by the sake of contradiction, that there are $m < p < q < M$ such that $x(p) > x(q)$. Taking any $c \in (x(q), x(p))$ leads to a contradiction. The behaviour of x in the other intervals can be checked in the same way.

Proof of Theorem 5. Suppose the problem **(P3)** has solution x, for some $L \in \mathcal{L}$. We shall prove that, if L is sufficiently close to L_0, this solution cannot be unique. At first, let us prove that there is a neighbourhood $\mathcal{N}$ of L_0, such that, if $L \in \mathcal{N}$, then $|x'(t)| \geq \frac{1}{2}$ almost everywhere. ¿From a well-known fact in calculus of variations, since x is a solution of **(P3)**, then $L^{**}(x(t), x'(t)) = L(x(t), x'(t))$ a.e.. As $L_0(x(t), u) > \frac{1}{2}$ for $|u| < \frac{1}{2}$, then $L(x(t), u) > \frac{1}{2}$, for L sufficiently close to L_0 and $|u| < \frac{1}{2}$. Since $L_0^{**}(x(t), u) = 0$ for $|u| \leq \frac{1}{2}$, by Lemma 2, $L^{**}(x(t), u) < \frac{1}{2}$, for $|u| \leq \frac{1}{2}$. But then $L(x(t), u) \neq L^{**}(x(t), u)$, if $|u| \leq \frac{1}{2}$. As a result $|x'(t)| > \frac{1}{2}$ a.e. and the claim is proved. We prove now that there exists $c \in I\!R$ such that both E_c^+ and E_c^- have at least two different connected components.

First of all, observe that, if $|L_0(x,u) - L(x,u)| \leq \sigma(x^2 + u^2 + 1)$, then

$$I_L(x) \geq I_{L_0}(x) - \sigma(\|x\|^2_{W^{1,2}} + 1).$$

As $I_L(x) \to \infty$ if $\|x\|_{W^{1,2}} \to \infty$ and $v(L)$ is upper bounded around L_0, we see that there is a neighbourhood $\mathcal{N}$ di L_0 such that $\forall L \in \mathcal{N}$, $\forall x$ minimizing I_L:

$$I_L(x) \geq I_{L_0}(x) - \varepsilon \quad \& \quad v(L) < \varepsilon,$$

for a sufficiently small $\varepsilon > 0$. If $x \geq 0$ ($x \leq 0$) and if there is no c as in the claim, it is easy to see that, being $|x'(t)| \geq \frac{1}{2}$, then

$$v(L) = I_L(x) \geq I_{L_0}(x) - \varepsilon \geq \int_0^1 x^2 - \varepsilon \geq \frac{1}{48} - \varepsilon,$$

a contradiction with the fact that $v(L) < \varepsilon$. We show now that we get the same contradiction also in the case if x changes sign. In this case we use Lemma 4 supposing, w.l.o.g., that $m < M$. Let $(0,a)$ be the set E_0^-. As $|x'(t)| > \frac{1}{2}$ a.e. we get:

$$\varepsilon \geq v(L) \geq \int_0^a x^2 \geq \frac{1}{12}a^3,$$

from which we get $a \leq \sqrt[3]{12\varepsilon}$. Applying the same argument to $E_0^+ = (a,1)$ shows that also $1 - a \leq \sqrt[3]{12\varepsilon}$. This is a contradiction, if ε is sufficiently small. As a result, there must be c such that both E_c^- and E_c^+ have at least two distinct connected components. It follows, always supposing $m < M$ and $c > 0$, that there are $0 < a < b < d < 1$ such that

$$(0,a) \in E_c^-, (a,b) \in E_c^+, (b,d) \in E_c^-,$$

and $x(a) = x(b) = x(d) = c$. Define now

$$z(t) = \begin{cases} x(t) & 0 \leq t \leq a \\ x(t+b-a) & a \leq t \leq d-b+a \\ x(t+b-d) & d-b+a \leq t \leq d \\ x(t) & t \geq d \end{cases}$$

It is easy to check that z is a solution, different from x, of the problem **(P3)**, with Lagrangean L. This finishes the proof.

To conclude, we make the following observation: Theorem 5 deals with a Lagrangean $L : \mathbb{R} \times \mathbb{R} \to \mathbb{R}$. It is clear that the argument used in its proof utterly fails for problems with Lagrangean functions $L : \mathbb{R}^n \times \mathbb{R}^n \to \mathbb{R}$, $n > 1$. So, we formulate the conjecture:

Conjecture If $n > 1$, the problem **(P3)** is generically well-posed on every component of the space $W^{1,p}$ on which it is proper.

References

[1] G. Beer and R. Lucchetti, The epi-distance topology: continuity and stability results with application to convex optimization problems, Math. Operation Res., **17** (1992), 715-726.

[2] R. Deville, G. Godefroy and V. Zizler, *Smoothness and Renormings in Banach Spaces*, Longman 1993.

[3] R. Deville and J. Revalski, Porosity of ill-posed problems, Proc. Amer. Math. Soc., to appear.

[4] A. Dontchev and T. Zolezzi, *Well-Posed Optimization Problems*, Lecture Notes Math., vol. **1354**, Springer Verlag 1993.

[5] A. Ioffe and A. Zaslavskii, Variational principles and well-posedness in optimization and calculus of variations, SIAM J. Control Optimization, to appear.

[6] P. Kenderov and J. Revalski, Generic well-posedness of optimization problems and the Banach-Mazur game, in *Recent Developments in Well-Posed Optimization Problems*, R. Lucchetti and J. Revalski, eds., MIA, vol **331**, Kluwer 1995, pp. 117-136.

[7] R.T. Rockafellar and J. Spingarn, The generic nature of optimality conditions in nonlinear programming, Math. Oper. Res., **4** (1979), 425-430.

[8] T. Zolezzi, Extended well-posedness of optimization problems, J. Optimization Theory Appl., **91** (1996), 257-268.

On a class of bilevel programs

Martine Labbé (mlabbe@ulb.ac.be)
SMG, Institut de Statistique et Recherche Opérationnelle,
Université Libre de Bruxelles
Boulevard du Triomphe
B1050 Bruxelles, Belgique

Patrice Marcotte (marcotte@iro.umontreal.ca)
DIRO, Université de Montréal
CP 6128, succursale Centre-Ville
Montréal (QC) H3C 3J7, Canada

Gilles Savard (gilles@crt.umontreal.ca)
Département de Mathématiques et Génie Industriel
École Polytechnique
CP 6079, succursale Centre-Ville
Montréal (QC) H3C 3A7, Canada

Abstract

The optimal setting of taxes or subsidies on goods and services can be naturally modelled as a bilinear bilevel program. We analyze this class of hierarchical problems both from the theoretical and algorithmical points of view, devoting special attention to the problem of setting profit-maximizing tolls on a transportation network.

Keywords: Bilevel programming, nonlinear programming, transportation, networks, pricing.

G. Di Pillo and F. Giannessi (eds.), Nonlinear Optimization and Related Topics, 183–206.

1 Introduction

Bilevel programming is the adequate framework for modelling those optimization situations where a subset of decision variables is not controlled by the principal optimizer (the 'leader'), but rather by a second agent (the 'follower') who optimizes his own objective function with respect to this subset of variables. If one denotes by x (respectively y) the decision vector of the leader (respectively the follower), a bilevel program can be expressed mathematically in the form:

$$\begin{aligned} \min_{x,y} \quad & f(x,y) \\ \text{subject to} \quad & (x,y) \in X \\ & y \in \mathcal{S}(x), \end{aligned} \tag{1}$$

where $\mathcal{S}(x)$ denotes the set of optimal solutions of the mathematical program parameterized in the leader's vector x:

$$\mathcal{S}(x) = \begin{aligned} \arg\min_{y} \quad & g(x,y) \\ \text{subject to} \quad & (x,y) \in Y. \end{aligned} \tag{2}$$

Whenever the objective function g of the lower level program is differentiable and convex in y, and the set Y is convex, a feasible vector y is in the set $\mathcal{S}(x)$ if and only if $(x,y) \in Y$ and the subvector y satisfies the first order condition:

$$\langle \nabla_y g(x,y), y - y' \rangle \leq 0$$

for all y' such that $(x,y') \in Y$. Letting $Y(x) = \{y : (x,y) \in Y\}$, one can reduce, under the abovementioned conditions, the bilevel program (1) to the standard mathematical program:

$$\begin{aligned} \min_{x,y} \quad & f(x,y) \\ \text{subject to} \quad & (x,y) \in X \\ & y \in Y(x) \\ & \langle \nabla_y g(x,y), y - y' \rangle \leq 0 \qquad \forall y' \in Y(x). \end{aligned} \tag{3}$$

The above program is subsumed by the more general form:

$$\begin{aligned} \min_{x,y} \quad & f(x,y) \\ \text{subject to} \quad & (x,y) \in X \\ & y \in Y(x) \\ & \langle G(x,y), y - y' \rangle \leq 0 \qquad \forall y' \in Y(x), \end{aligned} \tag{4}$$

where the vector function G need not be a gradient mapping with respect to the variable y. This generalization allows to model situations where the lower level vector

y must achieve some equilibrium state described by a variational inequality parameterized in the leader variable x. For instance, the vector x might represent design variables for an urban road network, and y the traffic equilibrium corresponding to that x. Note that this more general form can be reduced to the standard form, since a vector y is solution of the lower level variational inequality if and only if it globally minimizes, with respect to the argument y, the function $\text{gap}(x, y)$ defined as (see Fukushima [4]):

$$\text{gap}(x, y) = \max_{y' \in Y(x)} \langle G(x, y), y - y' \rangle - \frac{1}{2}\|y - y'\|^2.$$

These and similar models have been studied by several authors. The interested reader is referred to the recent books by Shimizu, Ishizuka and Bard [10], and that of Luo, Pang and Ralph [9], the latter one addressing mathematical programs with equilibrium (or variational) constraints. Much attention has been devoted to the case where the functions f and g are linear, and the sets X, $Y(x)$ are polyhedra. Even in this simple situation, the problem that consists in determining whether a solution is locally optimal is strongly NP-hard (see Hansen, Jaumard and Savard [6]) and exact algorithms have to rely on enumeration procedures, such as branch-and-bound. In the nonlinear case, (weak) optimality conditions have been derived and descent algorithms for nonsmooth formulations have been proposed (see Kocvara and Outrata [7] for instance). These however may fail to yield even local optimum for the bilevel program, without strong assumptions.

An interesting class of bilevel problems is obtained by letting y denote the follower's vector corresponding to goods or activities; y_1 represents the subvector of goods (or activities) subject to taxation, y_2 the subvector of goods not subject to taxes, and x_j the taxation level for the good or activity j. The leader, typically a government, wishes to maximize taxation revenues, while assuming that customers (individuals, industries, businesses) achieve an equilibrium state in accordance with taxes set at the upper level. If the equilibrium state can be represented as the solution set of a linear program, we obtain a bilinear bilevel program, which we record in the following 'vertical' format:

$$\begin{array}{rl} \max\limits_{x} & xy_1 \\ \min\limits_{y_1, y_2} & (c_1 + x)y_1 + c_2 y_2 \\ \text{subject to} & A_1 y_1 + A_2 y_2 = b \\ & y_1, y_2 \geq 0. \end{array} \tag{5}$$

Note that, for given lower level reaction vector (y_1, y_2) the above program is a particular case of an *inverse optimization* problem where one is required to select a tax vector x such that (i) (y_1, y_2) is optimal with respect to this tax vector and (ii) the profit xy_1 is maximal. As we will see in Section 3, this problem is endowed with a mathematical structure very similar to that of the lower level problem.

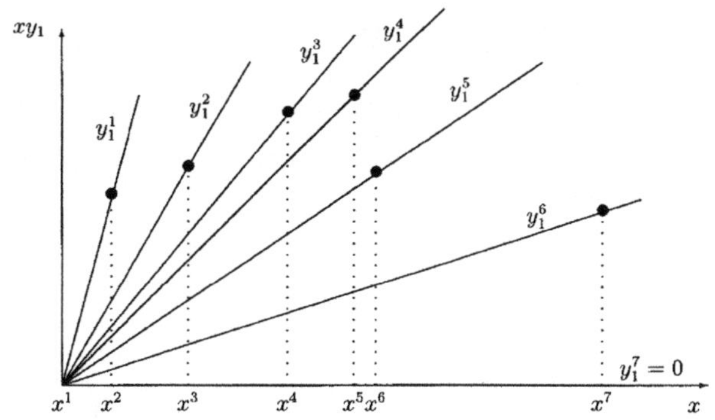

Figure 1: The profit function.

¿From the leader's perspective, the objective function xy_1 is discontinuous and piecewise linear, as illustrated by the two-dimensional example of Figure 1 where the slope y_1^i is the first component of the vector (y_1^i, y_2^i), which itself is the optimal reaction of the follower to a tax value lying in the interval $[x^i, x^{i+1}]$.

This paper is about the investigation of the above program. The second section is devoted to a general study of bilevel taxation programs. The third section discusses algorithmic approaches to taxation programs, focusing on efficient primal-dual heuristic procedures. The fourth section specializes the results to toll setting problems, i.e., problems involving a single or a multicommodity network flow structure at the lower level; such problems were initially considered by Labbé, Marcotte and Savard [8].

2 A bilinear taxation model and its properties

Let us consider the mathematical program (5). For simplicity, we assume that the polyhedron $\{(y_1, y_2) : A_1y_1 + A_2y_2 = b, y_1, y_2 \geq 0\}$ is nonempty, bounded and that the set $\{y_2 : A_2y_2 = b, y_2 \geq 0\}$ is nonempty. The latter two conditions ensures that, for any value of the tax vector x, the lower level LP

$$\begin{aligned} \min_{y_1,y_2} \quad & (c_1 + x)y_1 + c_2y_2 \\ \text{subject to} \quad & A_1y_1 + A_2y_2 = b \\ & y_1, y_2 \geq 0 \end{aligned} \tag{6}$$

is bounded. It also constitutes a necessary and sufficient condition for the objective function of the taxation problem (5) to be finite.

Since the lower level problem is, for a fixed tax vector x, a linear program, one may replace the lower level problem of (5) by its primal-dual optimality conditions. This operation yields the mathematical program with linear and complementarity constraints given by:

$$
\begin{aligned}
\max_{x,y_1,y_2,\lambda} \quad & xy_1 \\
\text{subject to} \quad & A_1y_1 + A_2y_2 = b \\
& y_1, y_2 \geq 0 \\
& \lambda A_1 \leq c_1 + x \\
& \lambda A_2 \leq c_2 \\
& (c_1 + x - \lambda A_1)y_1 = 0 \\
& (c_2 - \lambda A_2)y_2 = 0.
\end{aligned} \tag{7}
$$

¿From the complementarity slackness and primal feasibility conditions one obtains the equivalent form:

$$
\begin{aligned}
\max_{x,y_1,y_2,\lambda} \quad xy_1 &= (\lambda A_1 - c_1)y_1 \\
&= \lambda(b - A_2y_2) - c_1y_1 \\
&= \lambda b - (c_1y_1 + c_2y_2) \\
\text{subject to} \quad & A_1y_1 + A_2y_2 = b \\
& y_1, y_2 \geq 0 \\
& \lambda A_1 \leq c_1 + x \\
& \lambda A_2 \leq c_2 \\
& (c_1 + x - \lambda A_1)y_1 = 0 \\
& (c_2 - \lambda A_2)y_2 = 0.
\end{aligned} \tag{8}
$$

Since the leader wishes to maximize profit, it is in his interest to set taxes as high as possible. Consequently, one should have:

$$x_i = (\lambda A_1 - c_1)_i \tag{9}$$

whenever the variable y_{1i} is positive. For the remaining indices, one is free to select any tax value that satisfies:

$$x_i \geq (\lambda A_1 - c_1)_i.$$

One might for instance set those taxes to $+\infty$ or, simply, set:

$$x = \lambda A_1 - c_1 \tag{10}$$

to obtain a simple, closed form expression for the tax vector. This allows to get rid of the feasibility and complementarity terms involving x in formulation (8). Once these constraints are dropped, one is left with the single-level mathematical program:

$$\boxed{\begin{array}{lllll} \max\limits_{y_1,y_2,\lambda} & \lambda b & - & (c_1y_1 + c_2y_2) \\ \text{subject to} & \lambda A_2 \le c_2 & & A_1y_1 + A_2y_2 = b \\ & & & y_1, y_2 \ge 0 \\ & & \boxed{(c_2 - \lambda A_2)y_2 = 0} & \end{array}} \tag{11}$$

If one relaxes the complementarity constraint, the above program decomposes into the two linear programs:

$$\begin{array}{ll} \max\limits_{\lambda} & \lambda b \\ \text{subject to} & \lambda A_2 \le c_2 \end{array} \tag{12}$$

and

$$\begin{array}{ll} \min\limits_{y_1,y_2} & c_1y_1 + c_2y_2 \\ \text{subject to} & A_1y_1 + A_2y_2 = b \\ & y_1, y_2 \ge 0. \end{array} \tag{13}$$

The LP (13) corresponds to the solution of the lower level problem (6) with taxes set at zero. The dual of the LP (12) is simply the same linear program (6) where the choice of activities is restricted to untaxed ones or, equivalently, where taxes are set to arbitrary high levels.

Returning to the single-level formulation (11), one can penalize the complementarity constraint into the objective function, yielding a bilinear problem separable in the dual vector λ on the one hand, and in the primal vector (y_1, y_2) on the other hand:

$$\begin{array}{ll} \max\limits_{y_1,y_2,\lambda} & \lambda b - (c_1y_1 + c_2y_2) - M(c_2 - \lambda A_2)y_2 \\ \text{subject to} & \lambda A_2 \le c_2 \\ & A_1y_1 + A_2y_2 = b \\ & y_1, y_2 \ge 0. \end{array} \tag{14}$$

An optimal solution of the mathematical program (13), together with an optimal dual vector λ, yields a feasible solution to (14) with zero objective function value and is obtained by setting the tax vector x to zero. Indeed, for such a solution, the first two terms of the objective function correspond respectively to the dual and primal objective function values of the LP (13), while the third term is the complementarity constraint associated with the same program. Furthermore, since strong duality holds for the lower level problem, there exists an extremal solution to the corresponding dual problem. Let $\{\lambda^l\}_{l\in L}$ denotes the set of extremal solutions of the dual polyhedron, and $\{y_1^j, y_2^j\}_{j\in J}$ the set of extremal solutions of the primal polyhedron. Whenever

$$M \geq M^* = \max_{(l,j)\in L\times J} \frac{\lambda^l b - (c_1 y_1^j + c_2 y_2^j)}{(\lambda^l A_2 - c_2) y_2^j},$$

the objective of the penalized problem becomes negative, since an optimal solution occurs at an extreme point. From the above considerations, we conclude that an optimal solution to the penalized problem satisfies the complementarity constraint whenever M exceeds M^*.

The penalized bilinear program (14) can be transformed into a *linear* bilevel program. Indeed, for fixed primal variables y_1 and y_2, one can replace the objective function λb of the resulting LP by its dual objective, yielding the linear bilevel program:

$$\begin{aligned} \max_{y_1,y_2} \quad & -c_1 y_1 - (M+1) c_2 y_2 + c_2 y_2' \\ \text{subject to} \quad & A_1 y_1 + A_2 y_2 = b \\ & y_1, y_2 \geq 0 \\ \\ \min_{y_2'} \quad & c_2 y_2' \\ \text{subject to} \quad & A_2 y_2' = b + M A_2 y_2 \\ & y_2' \geq 0. \end{aligned} \tag{15}$$

Next we replace $A_2 y_2$ by $b - A_1 y_1$, perform the change of variables $y_2' = (M+1) y_2''$ and set $\epsilon = 1/(M+1)$ to end up with the bilevel program:

$$\begin{aligned} \max_{y_1,y_2} \quad & -\epsilon c_1 y_1 + c_2 y_2'' - c_2 y_2 \\ \text{subject to} \quad & A_1 y_1 + A_2 y_2 = b \\ & y_1, y_2 \geq 0 \\ \\ \min_{y_2''} \quad & c_2 y_2'' \\ \text{subject to} \quad & A_1 y_1 + A_2 y_2'' = b + \epsilon A_1 y_1 \\ & y_2'' \geq 0. \end{aligned} \tag{16}$$

There is an interesting economic interpretation of the linear bilevel program (16). First note that, without loss of generality, one may assume that the cost vector c_1 is zero. It suffices to observe that formulation (5) can be rewritten as:

$$\begin{aligned} \max_{x} \quad & x y_1 \\ \min_{y_1,y_1',y_2} \quad & x y_1 + c_1 y_1' + c_2 y_2 \\ & A_1 y_1 + A_2 y_2 = b \\ & y_1' - y_1 = 0 \\ & y_1, y_1', y_2 \geq 0, \end{aligned} \tag{17}$$

which is a program with the desired structure since the vector of taxable variables y_1 has a zero coefficient in the lower level objective function. Under the assumption that $c_1 = 0$ the leader, in (16), positions himself by selecting a (y_1, y_2)-vector such that any marginal deviation from this proposed solution by the follower results in a large deterioration for the follower. This gap between the leader's proposal and the follower's second best alternative provides room for taxation. It is obviously in the leader's interest to maintain this gap as wide as possible. This is exactly what program (16) achieves. Indeed the lower level strives to satisfy a marginal increase in demand $\epsilon A_1 y_1$ at the least possible cost, while setting the vector y_1 at the level prescribed by the leader.

If ϵ is set to zero, a valued of $c_2 y_2''$ that is less or equal to that of $c_2 y_2$ can clearly be achieved by the follower, since the value $c_2 y_2$ corresponds to the feasible lower level program $y_2'' = y_2$. This means that, at best, the leader's objective is zero. The leader may achieve this optimal value zero by setting y_2 to y_2'', where y_2'' is an optimal response to any feasible vector y_1 selected by the leader. If ϵ is positive, the follower optimally adjusts himself to the increased demand $\epsilon A y_1$ by adopting a recourse involving the sole vector y_2''. In the linear bilevel formulation (16), the leader wants to maximize the added cost of this recourse to the follower.

If the lower level equality in (16) is equivalently expressed as

$$A_1[(1-\epsilon)y_1] + A_2 y_2'' = b,$$

the aim of the lower level is simply to substitute, at minimum cost, y_2''-variables to compensate for a marginal decrease ϵ in the supply of the upper level resource y_1, while meeting the demand vector b. This alternative constitutes the follower's marginal best recourse. Since the penalty scheme is exact, there exists a value ϵ^* such that the above marginal analysis is exact, i.e., there exists an optimal recourse that corresponds to an extremal solution of the polyhedron $\{(y_1, y_2) : A_1 y_1 + A_2 y_2 = b, y_1, y_2 \geq 0\}$ and this extremal solution remains optimal when ϵ stays within some range $(0, \epsilon^*]$. As stated earlier, the optimal tax vector x can easily be recovered from the dual vector associated with the constraint of the lower level program through the equation $x = \lambda A_1 - c_1 = \lambda A_1$. (Recall that c_1 has been set to zero.) This economic interpretation will be illustrated on a network example in Section 4.

For fixed y_1, the optimal solution of the mathematical program (16) is easily obtained by solving two linear programs parameterized in y_1. It follows that the objective of (16) is continuous, as the difference of two convex, continuous functions of the vector y_1. This is to be contrasted with the original formulation (5) where the profit function is a piecewise linear but discontinuous function of the decision variable x, and where the discontinuities occur when perturbations in the taxation vector x yield a change of the optimal lower level basic solution.

We close this section with considerations on the computational complexity of the taxation problem. In its original form, we have not been able to prove that (5) is NP-hard, although we strongly suspect it is. However we could prove that a variant of (5)

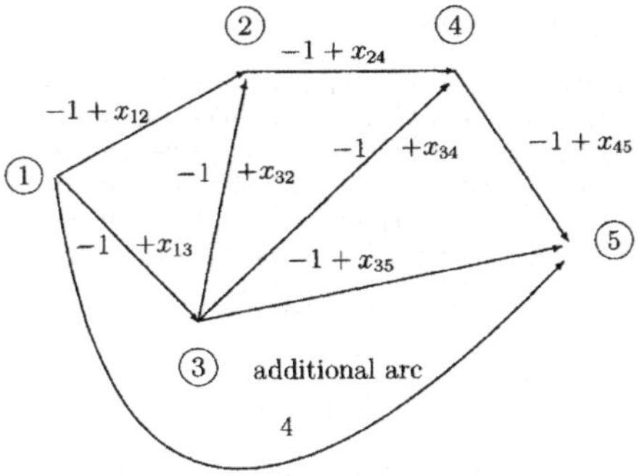

Plain arcs: arcs of the original network

Figure 2: The reduction yielding a Hamiltonian path from the solution of a toll problem

where the tax vector x is bounded from below by some vector l is indeed NP-hard. The proof relies on a reduction from the 'Hamiltonian Path' problem in a directed graph to the taxation problem with lower bound constraints.

Let us consider a directed graph $G = (N, A)$ with two distinguished nodes s and t. If there exists a Hamiltonian path from s to t in G, then this path can be recovered from an optimal solution to the following taxation problem: The leader seeks to maximize revenues from taxes set on the arcs of G, while the follower wishes to send one unit of flow from s to t, i.e., looks for a shortest $s - t$ path in G. The costs on all arcs of G are set to -1, and the lower bound on each arc of G to 2. An additional arc $s - t$ is appended to the graph. This additional arc is not subject to tax and bears a cost of $|N| - 1$. Let $\mathcal{P}$ denote the shortest 'after tax' path. If the number of arcs in $\mathcal{P}$ is k, the total profit (tax) T on $\mathcal{P}$ must satisfy the inequality: $-k + T \leq |N| - 1$, i.e., $T \leq |N| - 1 + k$. The profit T is maximized by selecting k as large as possible, while maintaining $\mathcal{P}$ loopless. Hence the solution of the taxation problem consists in setting the tax to 2 on the arcs of a Hamiltonian path (if such path exists) and to an arbitrarily large value ($|N|$ is suitably large) on the other arcs of the graph. (See Figure 2 where $s = 1$ and $t = 5$.)

3 Solution algorithms

The main difficulty in solving bilevel programs stems from the complementarity constraint arising when formulating the lower level first-order optimality conditions. Linear or bilinear bilevel programs, as ordinary LPs, admit extreme point solutions and can be viewed either as continuous or discrete optimization programs. For instance, one can derive from the formulation (11) a mixed integer formulation (MIP) by replacing the complementarity constraint $(c_2 - \lambda A_2)y_2 = 0$ by the linear constraints

$$\begin{array}{rcll} M\alpha & \geq & c_2 - \lambda A_2 & \\ N\beta & \geq & y_2 & \\ \alpha_i + \beta_i & \leq & 1 & \forall i \\ \alpha_i, \beta_i & \in & \{0,1\} & \forall i, \end{array} \tag{18}$$

where $M \geq \max_i \max_{l \in L} (c_2 - \lambda^l A_2)_i$ and $N \geq \max_i \max_{j \in J} y^j_{2i}$. This yields the mixed integer program:

$$\begin{array}{ll} \max\limits_{\lambda, y_1, y_2} & \lambda b - (c_1 y_1 + c_2 y_2) \\ \text{subject to} & A_1 y_1 + A_2 y_2 = b \\ & y_1, y_2 \geq 0 \\ & \lambda A_2 \leq c_2 \\ & M\alpha \geq c_2 - \lambda A_2 \\ & N\beta \geq y_2 \\ & \alpha_i + \beta_i \leq 1 \qquad \forall i \\ & \alpha_i, \beta_i \in \{0,1\} \qquad \forall i. \end{array} \tag{19}$$

So far we have only considered problems where the tax vector x is unconstrained. For the sake of realism, upper bound constraints can be introduced. This leads to the primal-dual formulation:

$$\begin{array}{ll} \max\limits_{\lambda, y_1, y_2} & \lambda b - (c_1 y_1 + c_2 y_2) \\ \text{subject to} & A_1 y_1 + A_2 y_2 = b \\ & y_1, y_2 \geq 0 \\ & \lambda A_2 \leq c_2 \\ & \lambda A_1 - c_1 \leq u \\ & M\alpha \geq c_2 - \lambda A_2 \\ & N\beta \geq y_2 \\ & \alpha_i + \beta_i \leq 1 \qquad \forall i \\ & \alpha_i, \beta_i \in \{0,1\} \qquad \forall i. \end{array} \tag{20}$$

Branch-and-bound procedures can then be constructed to solve the above MIP. The number of integer variables can be significantly reduced by noting that the inequality

$\alpha_i + \beta_i \leq 1$ can be replaced by the equality $\alpha_i + \beta_i = 1$. This allows to set $\beta_i = 1 - \alpha_i$ and to remove the now redundant constraint $\alpha_i + \beta_i \leq 1$. Actually, whenever the index set P of positive variables y_{2j} $(j \in P)$ in an optimal solution is given, an optimal λ-vector can be obtained by solving the linear program:

$$\begin{aligned} \max_{\lambda} \quad & \lambda b \\ \text{subject to} \quad & (\lambda A_2 - c_2)_j \leq 0 \quad j \notin P \\ & (\lambda A_2 - c_2)_j = 0 \quad j \in P \\ & \lambda A_1 \leq c_1 + u, \end{aligned} \tag{21}$$

whose dual

$$\begin{aligned} \min_{y_1, y_2} \quad & (c_1 + u) y_1 + c_2 y_2 \\ \text{subject to} \quad & A_1 y_1 + A_2 y_2 = b \\ & y_1 \geq 0 \\ & y_{2j} \quad \text{unrestricted for } j \in P \\ & y_{2j} \quad \geq 0 \text{ for } j \notin P \end{aligned} \tag{22}$$

has a structure similar to that of the original lower level LP. The case where upper bounds are absent from the problem formulation corresponds to setting $u = +\infty$, or simply to removing the vector y_1 from the formulation.

In practice, it is frequently the case that the size of the taxed vector y_1 is much smaller than the size of y_2. Then the knowledge of the vector y_1 (not only the index set of its positive components, but also their values) allows to recover the values of y_2 by solving the LP:

$$\begin{aligned} \min_{y_2} \quad & c_2 y_2 \\ \text{subject to} \quad & A_2 y_2 = b - A_1 y_1 \\ & y_2 \geq 0. \end{aligned} \tag{23}$$

The optimal tax level $x = \lambda A_1 - c_1$ can again be recovered by solving (21) or (22). If the LP (23) is infeasible or unbounded, then it is impossible to have the given vector y_1 in an optimal solution. Note that branching on the component y_{1j} will be facilitated if it is binary-valued. This is the case if the lower level is a shortest path problem for instance. This situation will be investigated in more detail in Section 4.

The introduction of general constraints on the tax vector x is more problematic. Consider for instance the lower bound constraint $x \geq l$. One cannot simply insert

$$\lambda A_1 - c_1 \geq l \tag{24}$$

within the formulation (20), since $x \geq l$ and $x \geq \lambda A_1 - c_1$ (see (8)) do not imply that (24) is satisfied. In other words, setting $x = \lambda A_1 - c_1$ and imposing the constraint (24) might unduly restrict the feasible set and lead to suboptimal or even infeasible problems. Indeed, in presence of a lower bound constraint, one cannot get rid of the complementarity constraint $(c_1 + x - \lambda A_1)y_1 = 0$ (see (7)) by setting x to $\lambda A_1 - c_1$. However one may introduce a nonnegative 'slack' vector s_1, set $x = \lambda A_1 - c_1 + s_1$ and add the constraints $\lambda A_1 - c_1 + s_1 \geq l$, $s_1 \geq 0$ and $s_1 y_1 = 0$ to (11). This would lead to additional binary variables associated with the complementarity constraint $s_1 y_1 = 0$ in the MIP formulation (20). In the presence of lower bounds, it is yet possible to derive an equivalent of problem (22). Indeed, let us append the bound constraints $l \leq x \leq u$ to the formulation (8) and let Q (respectively P) denote the index set of positive components of the vector y_1 (respectively y_2). Given these index sets, one may rewrite the dual part of (8) as

$$\begin{array}{lll} \max_{x,\lambda} & \lambda b & \\ \text{subject to} & \lambda A_1 \leq c_1 + u & \\ & (\lambda A_1)_j \geq c_{1j} + l_j & j \in Q \\ & \lambda A_2 \leq c_2 & \\ & (\lambda A_2)_j = c_{2j} & j \in P \\ & x_j = (\lambda A_1 - c_1)_j & j \in Q. \end{array}$$

After removal of the last constraint, which is no longer required, one obtains a linear program whose dual is

$$\begin{array}{ll} \min\limits_{y_1, y'_1, y_2} & (c_1 + u)y_1 - \sum\limits_{j \in Q}(c_{1j} + l_j)y'_{1j} + c_2 y_2 \\ \text{subject to} & A_1 y_1 - A_1 y'_1 + A_2 y_2 = b \\ & y_1, y'_1 \geq 0 \\ & y'_{1j} = 0 \qquad j \notin Q \\ & y_{2j} \qquad \text{unrestricted for } j \in P \\ & y_{2j} \geq 0 \qquad j \notin P. \end{array} \tag{25}$$

The structure of the original problem has been preserved in (25). Let λ^* denote the optimal solution of (25) and set $x^*_j = (\lambda^* A_1 - c_1)_j$ if $j \in Q$ and $x^*_j = l_j$ if $j \notin Q$. If the lower level optimal solution corresponding to x^* is distinct from (y_1, y_2), then one concludes that no value of the taxation vector x can make the vector (y_1, y_2) optimal for the lower level.

Usually, exact methods can only address small instances of combinatorial or mixed-integer programs. Consequently, the remaining of this section is devoted to heuristic procedures that are expected to produce near-optimal solutions. These are based on the exact penalty, primal-dual formulation (14), and inspired from a heuristic procedure that performed extremely favourably for finding good solutions to linear bilevel programs (see Gendreau, Marcotte and Savard [5]). First we recall this procedure. Consider the linear bilevel program:

$$\begin{array}{rl} \min\limits_{y_1} & d_1 y_1 + d_2 y_2 \\ \min\limits_{y_2} & c_2 y_2 \\ \text{subject to} & A_1 y_1 + A_2 y_2 = b \\ & y_1, y_2 \geq 0. \end{array} \tag{26}$$

After writing the optimality conditions of the lower level program and penalizing the complementarity constraints, we obtain the single-level bilinear program:

$$\begin{array}{rl} \min\limits_{y_1, y_2, \lambda} & d_1 y_1 + d_2 y_2 + M(c_2 - \lambda A_2) y_2 \\ \text{subject to} & A_1 y_1 + A_2 y_2 = b \\ & y_1, y_2 \geq 0 \\ & \lambda A_2 \leq c_2. \end{array} \tag{27}$$

The general idea of the procedure is to iteratively solve (27) for increasing values of the penalty parameter M. Starting from the 'ideal' leader solution corresponding to $M = 0$, one aims towards the induced region, i.e., the union of polyhedral faces where the lower level duality gap $(c_2 - \lambda A_2) y_2$ is zero. For a given value of M, one can perform Gauss-Seidel iterates, alternatively fixing the values of the vectors λ and (y_1, y_2) (or y_2 and (λ, y_1)) until no progress is achieved. All subproblems involved preserve the structure, either in primal or dual form, of the original lower level program. In [5], a slightly different approach has been adopted: for a fixed vector y_1, the dual vector λ is set to the dual vector corresponding to the optimal solution of the lower level LP. This results in the following algorithm

HEURISTIC LINBIL

- Fix y_1. Solve lower level. Get dual solution λ.
- Fix λ. Solve (27) for y_1 and y_2. If $(c_2 - \lambda A_2) y_2 = 0$, STOP.
- Increase M and repeat.

Similar approaches can be adapted to the penalty formulation (14) of the taxation problem. One can perform Gauss-Seidel iterations for increasing values of the penalty parameter M or mimic the above algorithm:

HEURISTIC BILINBIL

- Fix the tax vector x. Solve lower level for y_1 and y_2.
- Fix y_1 and y_2. Solve penalized problem (14) for λ. If $(c_2 - \lambda A_2) y_2 = 0$, STOP.
- Set $x = \lambda A_1 - c_1$, increase M and repeat.

Note that, at the second step of the above procedure, one has to solve the linear program:

$$\begin{aligned} \max_{\lambda} \quad & \lambda(b + MA_2y_2) \\ \text{subject to} \quad & \lambda A_2 \leq c_2 \end{aligned} \tag{28}$$

whose dual program:

$$\begin{aligned} \min_{y_2'} \quad & c_2y_2' \\ \text{subject to} \quad & A_2y_2' = b + MA_2y_2 \\ & y_2' \geq 0 \end{aligned} \tag{29}$$

has a familiar structure.

The algorithm has been enhanced in the following manner: Whenever y_2 has been obtained at the end of the first step, one can derive the corresponding 'optimal' tax vector $x = \lambda A_1 - c_1$ by setting the vector λ to the dual optimal vector associated with the primal problem (22), and recording the best solution thus encountered. This procedure actually solves instances of inverse optimization problems (see Burton and Toint [3]) where one is required to modify the cost vector c_1 by adding to it a tax x in such a way that a prespecified vector be lower-level optimal and that the associated profit be as high as possible. Note that this strategy cannot be implemented meaningfully in the case of linear bilevel programs.

4 A toll setting problem

In this section we consider a network application where the upper level is concerned with maximizing profits obtained from tolls set on a subset of the arcs of a road network. Assuming that time and money costs are expressed in the same units, the users of the road network minimize their individual travel cost. Neglecting congestion and assuming that demand is fixed, users of the network are simply assigned to shortest paths linking their respective departure and arrival nodes.

4.1 Formulation, properties and counterproperties of the Toll Setting Problem

Let $G = (\bar{N}, \bar{A})$ be a graph with node set $\bar{N}$ and arc set $\bar{A}$. We endow G with a cost vector $c = \{c_a\}_{a \in \bar{A}}$, lower and upper bound vectors $l = \{l_a\}_{a \in \bar{A}_1}$ and $u = \{u_a\}_{a \in \bar{A}_1}$, a demand vector $d = \{d^k\}_{k \in K}$, where K denotes the set of origin-destination couples. Let us partition $\bar{A} = \bar{A}_1 \cup \bar{A}_2$, where $\bar{A}_1$ denotes the set of toll arcs and $\bar{A}_2$ the set of 'free' arcs. With each node i of the network we associate two sets of arcs: the forward star

$$i^+ = \{a = (i, j) \in \bar{A}\},$$

and the backward star

$$i^- = \{a = (j, i) \in \bar{A}\}.$$

This allows to formulate the toll setting problem as the bilinear bilevel program:

$$\begin{array}{lll}
\max\limits_{x} & \sum\limits_{a \in \bar{A}_1} x_a y_a & \\
\min\limits_{y} & \sum\limits_{a \in \bar{A}_1} (c_a + x_a) y_a + \sum\limits_{a \in \bar{A}_2} c_a y_a & \\
\text{subject to} & \sum\limits_{a \in i^-} y_a^k - \sum\limits_{a \in i^+} y_a^k = \begin{cases} -d^k & i = \text{origin of } k \\ d^k & i = \text{destination of } k \\ 0 & \text{else} \end{cases} & i \in \bar{N}, k \in K \\
& y_a = \sum\limits_{k \in K} y_a^k & \\
& y_a^k \geq 0 & a \in \bar{A}, k \in K \\
& l_a \leq x_a \leq u_a & a \in \bar{A}_1,
\end{array} \tag{30}$$

where tolls are imposed on the total flow of the arcs in the set $\bar{A}_1$, while the origin-destination ('commodity') flows are 'toll-free'. Letting $y_1^k = (y_a^k)_{a \in \bar{A}_1}$, $y_2^k = (y_a^k)_{a \in \bar{A}_2}$ and

$$b_i^k = \begin{cases} -d^k & i = \text{ origin of } k \\ d^k & i = \text{ destination of } k \\ 0 & \text{else,} \end{cases}$$

one obtains the matrix form of the above formulation:

$$\begin{array}{lll}
\max\limits_{x} & x y_1 & \\
\min\limits_{y_1, y_2} & (c_1 + x) y_1 + c_2 y_2 & \\
\text{subject to} & A_1 y_1^k + A_2 y_2^k = b^k & k \in K \\
& y_1^k, y_2^k \geq 0 & k \in K \\
& l \leq x \leq u & \\
& y_1 = \sum\limits_{k \in K} y_1^k & \\
& y_2 = \sum\limits_{k \in K} y_2^k &
\end{array} \tag{31}$$

where A_1 (respectively A_2) denotes the node-arc incidence matrix associated with the set of arcs $\bar{A}_1$ (respectively $\bar{A}_2$). This leads to the 'nonstandard' format where taxed activities y_1^k are subject to a common tax x:

$$\begin{array}{lll}
 & \max_{x} & \sum_{k\in K} x^k y_1^k \\
\forall k \in K: & \min_{y_1,y_2} & (c_1 + x^k)y_1^k + c_2 y_2^k \\
 & \text{subject to} & A_1 y_1^k + A_2 y_2^k = b^k \\
 & & y_1^k, y_2^k \geq 0 \\
 & & l \leq x^k \leq u \\
 & & x^k = x^1.
\end{array} \tag{32}$$

Although almost identical, these formulations suggest slightly different strategies for solving multicommodity bilinear bilevel programs.

Taking advantage of the fact that, for a given commodity flow, either $y_a^k = 0$ or $y_a^k = d^k$ it is also possible to formulate the above toll problem as a mixed integer program involving $|K| \times |\bar{A}_1|$ binary variables. Let us perform the change of variables $y^k = d^k y'^k$ and replace each commodity's lower level linear program by its optimality conditions, as usual. This yields the following mixed integer programming formulation:

$$\begin{array}{lll}
\max_{x,x^k,y'_1,y'_2,\lambda} & \sum_{k\in K} d^k (ex^k) & \\
\text{subject to} & A_1 y_1'^k + A_2 y_2'^k = b^k / d^k & k \in K \\
 & y_1'^k, y_2'^k \geq 0 & k \in K \\
 & \lambda^k A_1 \leq c_1 + x & k \in K \\
 & \lambda^k A_2 \leq c_2 & k \in K \\
 & c_1 y_1'^k + x^k e + c_2 y_2'^k = (\lambda^k b^k)/d^k & k \in K \\
 & -M y_1'^k \leq x^k \leq M y_1'^k & k \in K \\
 & -M(1 - y_1'^k) \leq x^k - x \leq M(1 - y_1'^k) & k \in K \\
 & y_1'^k \in \{0,1\} & k \in K \\
 & l \leq x \leq u, &
\end{array} \tag{33}$$

where e denotes the vector whose coefficients are all equal to one. In the above formulation, the bilinear term $d^k(x^k y'^k)$ has been replaced by the linear term $d^k(ex^k)$. Similarly, the bilinear term in the lower level complementarity condition has been 'linearized'. This is justified by the fact that whenever an arc a lies on a shortest path with respect to the commodity k, i.e., $y_a'^k = 1$, the associated revenue is equal to $d^k x_a^k$. Note that in this case the commodity toll x^k is forced by the seventh constraint to match the toll x common to all commodities whose shortest paths go through arc a; if the shortest path does *not* go through arc a, then the sixth constraint forces the commodity tax x_a^k to take the value zero. The toll x_a^k may then differ from x_a; however this is still consistent with the fact the corresponding revenue is then zero for that very commodity.

We now provide some intuition into the network toll problem by means of examples and counterexamples, all on single-commodity instances. First consider the network of Figure 3 where the dotted arcs correspond to toll arcs and the arc costs are shown next to the corresponding arcs, and demand is set to one between origin node 1 and destination node 5. The upper bound obtained by solving the 0-toll and ∞-toll linear programs (12) and (13) is equal to 22 (shortest path cost with no toll arcs) $-$ 6 (shortest path cost with zero tolls) $=$ 16. It is not difficult to convince oneself that, at the optimum of the toll problem, the path 1–2–3–4–5 will be a lower-level shortest path. If the tax vector is constrained to be nonnegative, then an optimal solution can be determined by solving problem (25) where the component y_1 associated with the upper bound constraint has been removed. In network terms, this corresponds to solving for a shortest path from node 1 to node 5 on the network of Figure 4, where forward toll arcs have been removed and backward arcs on the lower level optimal path have been incorporated into the network, together with reverse costs. The shortest path in the modified network is 1–2–4–3–5, and its dual vector (node potentials), as shown on Figure 4, is given by $\lambda = (0, 2, 9, 11, 21)$. The optimal toll vector x is then given by:

$$\begin{aligned} x = \lambda A_1 - c_1 &= (\lambda_3 - \lambda_2 - c_{23}, \lambda_5 - \lambda_4 - c_{45}) \\ &= (9 - 2 - 2, 21 - 11) \\ &= (5, 10). \end{aligned}$$

The total profit of 15 is one less than the upper bound 16.

It is instructive to interpret the solution of this example within the framework of the linear bilevel program (16). In order to conform with the requirement that the initial cost of the toll arcs be zero, we first replace the toll arc (2,3) by an untolled arc $(2, 2')$ with cost 2 and a toll arc $(2', 3)$ with zero cost (see Figure 5). Next we reduce the flow on toll arcs by ϵ. Since flow conservation is no longer satisfied, we must reroute this flow along untolled arcs, at lowest cost. This is achieved by solving the transshipment problem illustrated in Figure 5, where a supply of ϵ, indicated by $+\epsilon$ is issued at nodes $2'$ and 4, and a demand of ϵ, indicated by $-\epsilon$, is issued at nodes 3 and 5. An optimal solution to this problem is obtained by sending a flow of ϵ along paths $2'$–2–4–3 and 4–3–5, respectively. The added cost of this rerouting is equal to ϵ times $(-2 + 9 - 2)$ (cost of first path) plus $(-2 + 12)$ (cost of second path), i.e., 15ϵ, and the optimal profit 15 is obtained by dividing by ϵ. The dual variables, or node potentials, corresponding to this solution are

$$\lambda_1 = 0 \quad \lambda_2 = 2 \quad \lambda_{2'} = 4 \quad \lambda_3 = 9 \quad \lambda_4 = 11 \quad \lambda_5 = 21,$$

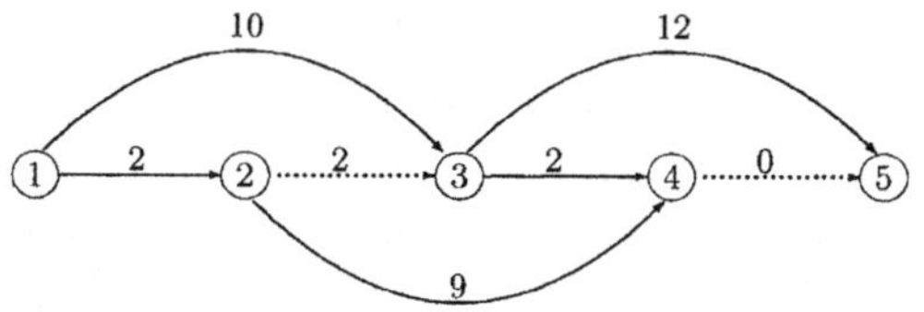

Figure 3: Network I

where the dual variable of node 1 has been arbitrarily set to zero. The optimal tax levels are, again, given by the expression

$$x = \lambda A_1 - c_1 = \lambda A_1,$$

since $c = 0$, i.e.,

$$x_{23} = x_{2'3} = \lambda_3 - \lambda_{2'} = 9 - 4 = 5 \quad x_{45} = \lambda_5 - \lambda_4 = 21 - 11 = 10.$$

This matches the total profit $15\epsilon/\epsilon$ obtained previously.

Of course, the above solution of the transshipment problem is only valid insofar as flows remain nonnegative. In the above example, this requirement is satisfied if and only if ϵ is less than or equal to $1/2$, which is consistent with our marginal analysis. Therefore, for this problem, the exact penalty factor M in (15) can be set to $(1/\epsilon) - 1 = 1$. We conjecture that this upper bound holds for all toll-setting problems involving a single origin-destination pair.

While it is sensible to compute optimal tolls with respect to the shortest zero-toll path, this need not always lead to an optimal solution. Indeed, let us incorporate into the network of Figure 3 a toll arc of cost $6\frac{1}{2}$ from node 1 to node 5. The optimal solution would then consist in setting the toll x_{15} to $15\frac{1}{2}$ and to suitably high values on the other toll arcs, thus raising the profit by $1/2$.

In the example of Figure 3, thc solution is not modified if lower bounds on the toll vector are set to $-\infty$, i.e., removed. This need not be always be the case, as can be observed on the problem instance illustrated in Figure 6, where one unit of flow is required between origin node 1 and destination node 4. With unconstrained tolls, an optimal solution corresponds to a shortest path in the modified network of Figure 7. The dual vector corresponding to the tree solution in Figure 7 being $\lambda = (0, 4, 2, 6)$, an optimal toll vector is given by

$$x = (4 - 0, 2 - 4, 6 - 2) = (4, -2, 4).$$

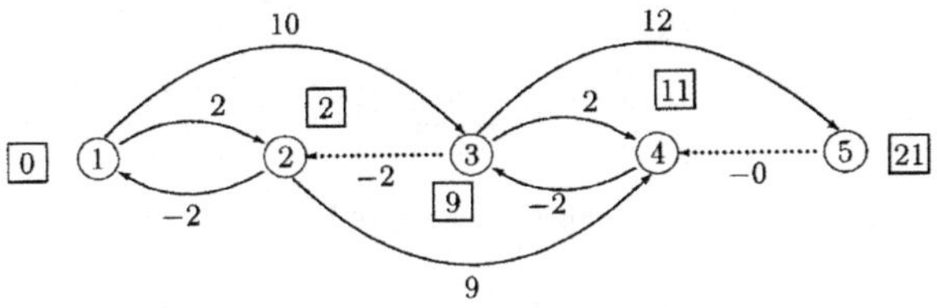

Figure 4: Modified network I

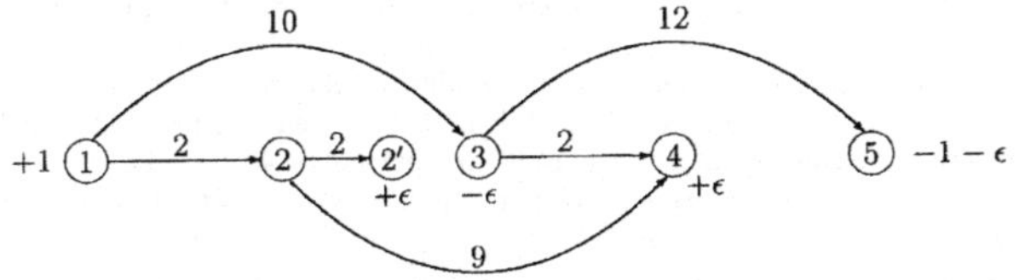

Figure 5: Economic interpretation

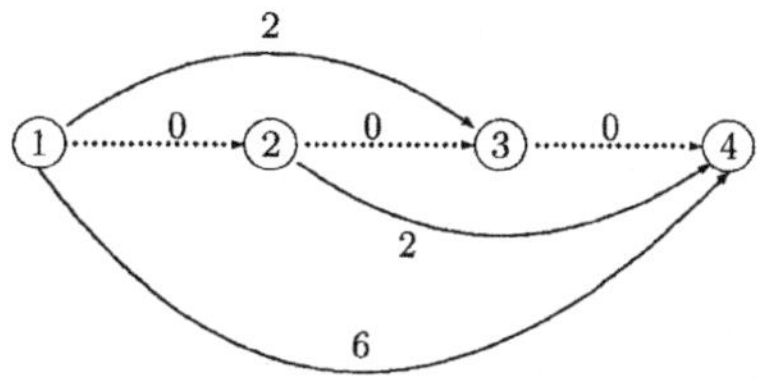

Figure 6: Network II

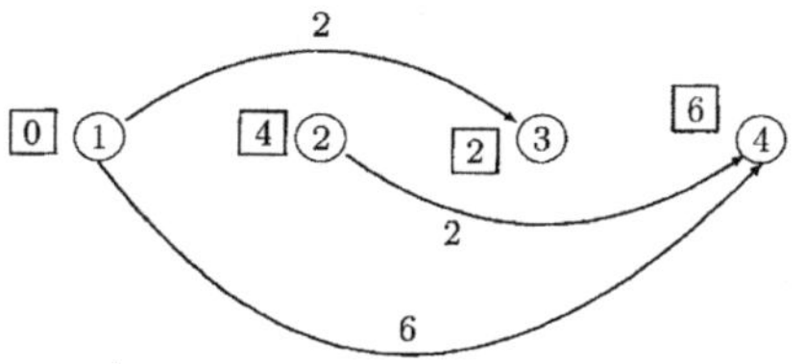

Figure 7: Modified network II: x unconstrained

This solution is not unique. Indeed, after normalizing λ_1 to zero, any vector λ such that $\lambda_2 \geq 4$, $\lambda_3 \leq 2$ and $\lambda_4 = 6$ is dual optimal. Note that any optimal solution involves a *negative* toll $\lambda_3 - \lambda_2$ on the middle arc (2,3). In the presence of nonnegativity constraints, the shortest path on the modified network (see Figure 8) is 1–3–2–4, leading to the dual vector $\lambda = (0, 2, 2, 4)$ and to the optimal tax vector $(2, 0, 2)$. This solution is not unique, since it is possible to optimally toll the paths 1–2–4 and 1–3–4. The total profit, in the presence of nonnegativity constraints, has decreased from 6 (which is the upper bound) to 4.

Note that, even in the abscence of nonnegativity constraints on tolls, it might be impossible to derive toll values such that a given subset of arcs belongs to the optimal solution of the lower level. Indeed it might be that making a given path the shortest lower level path requires to set some tolls to negative values, thus creating a negative cycle, clearly a nonoptimal solution for the leader. In this situation, no tolling policy is compatible with the selected path being a solution to the lower level problem.

Finally, as mentioned previously (see discussion leading to problem (23)) one can recover the lower level solution if one knows the *values* of the coefficients of the tax vector y_1. Since, in the case of a shortest path problem, this amounts to knowing the

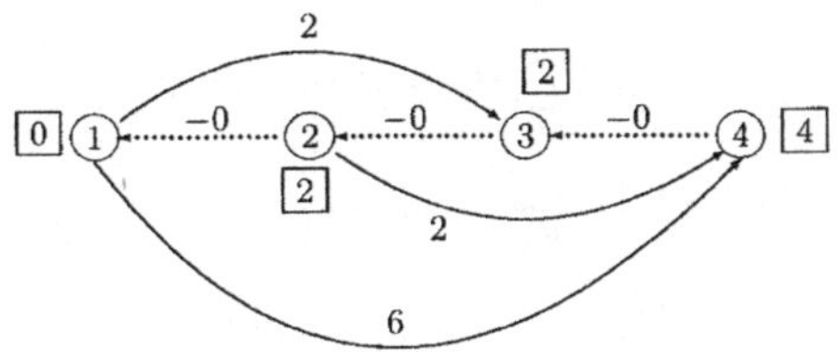

Figure 8: Modified network II: x nonnegative

indices of the positive coefficients, this information suffices to recover the lower level path to be tolled by solving the transportation problem (22).

4.2 Algorithms

For simplicity we restrict our attention to problems that do not contain lower bounds on the tax vector x. (See discussion about lower bounds in Section 3.)

Exact algorithms for solving the toll problem can be constructed around formulations (20) or (33). In practice we expect the second to involve a smaller number of integer variables. If heuristic BILINBIL is implemented, multicommodity flow problems of the form (29) are to be solved at each iteration. Since these problems do not involve upper bounds, they reduce to shortest path problems and are easy to solve. An efficient implementation of the enhanced form suggested just after Algorithm BILINBIL, however, involves solving linear programs of the form (31) for fixed values of the total flow vector y_1. These problems are multicommodity flow problems with fixed (actually equal) lower and upper bounds on some of the arcs, i.e., difficult problems. While these constraints could be penalized into the objective, a different approach, based on the 'nonstandard' formulation (32), has been adopted and is now described. Let us penalize the last constraint of (32), which enforces the equality of commodity toll vectors x^k:

$$\begin{array}{lll} & \max\limits_{x} & \sum\limits_{k \in K}(x^k y_1^k - M_1 \|x^k - x^1\|^2) \\ \forall k \in K: & \min\limits_{y_1, y_2} & (c_1 + x^k) y_1^k + c_2 y_2^k \\ & \text{subject to} & A_1 y_1^k + A_2 y_2^k = b^k \\ & & y_1^k, y_2^k \geq 0 \\ & & x^k \leq u. \end{array} \tag{34}$$

¿From the lower level LP optimality conditions, one may derive the equivalent single-level formulations:

$$\max_{x} \sum_{k \in K} (x^k y_1^k - M_1 \|x^k - x^1\|^2)$$

$$\begin{aligned}
\text{subject to} \quad & A_1 y_1^k + A_2 y_2^k = b^k && k \in K \\
& y_1^k, y_2^k \geq 0 && k \in K \\
& \lambda^k A_1 \leq c_1 + x^k && k \in K \\
& \lambda^k A_2 \leq c_2 && k \in K \\
& (c_1 + x^k - \lambda^k A_1) y_1^k = 0 && k \in K \\
& (c_2 - \lambda^k A_2) y_2^k = 0 && k \in K \\
& x^k \leq u && k \in K
\end{aligned} \tag{35}$$

and, after setting $x^k = \lambda^k A_1 - c_1$ and penalizing the remaining complementarity term:

$$\max_{x} \sum_{k \in K} [\lambda^k b^k - (c_1 y_1^k + c_2 y_2^k) - M_1 \|(\lambda^k - \lambda^1) A_1\|^2 - M_2 (c_2 - \lambda^k A_2) y_2^k]$$

$$\begin{aligned}
\text{subject to} \quad & A_1 y_1^k + A_2 y_2^k = b^k && k \in K \\
& y_1^k, y_2^k \geq 0 && k \in K \\
& (c_2 - \lambda^k A_2) y_2^k = 0 && k \in K \\
& \lambda^k A_1 - c_1 \leq u && k \in K.
\end{aligned} \tag{36}$$

This suggests, for solving this multicommodity problem, the following nonlinear adaptation of algorithm BILINBIL:

HEURISTIC MULTIKOM

Repeat, for increasing values of the penalty parameter M_1, the following procedure:

- Fix the vectors x^k. Solve lower level for y_1^k and y_2^k.
- Fix y_1^k and y_2^k. Solve penalized problem (36) for λ^k. If $\sum_{k \in K} (c_2 - \lambda^k A_2) y_2^k = 0$, STOP.
- Set $x^k = \lambda^k A_1 - c_1$, increase M_2 and repeat.

At the second step of the above procedure, the convex quadratic program is solved using Frank and Wolfe's linearization algorithm. This allows to retain the network structure of the linear subproblems. Remember that the aim of the procedure is to induce basis changes, as in metaheuristics such as Simulated Annealing and Tabu search. Subsequently, a profit-maximizing toll policies that is compatible with these lower-level basis can be derived, exactly as in the single-commodity case.

We conclude this section by mentioning that these heuristics have been tested on both single and multicommodity network flow problems, where the single-commodity lower-level problems were of the transportation type. In all cases, solutions either optimal or within one percent of optimality could be obtained in a small fraction of the computing time required to solve these problems to optimality, using the mixed-integer code CPLEX. The interested reader is referred to the Ph.D. thesis of Luce Brotcorne [2] for a detailed presentation of the numerical results.

5 Conclusion

In this paper, we have considered a particular class of bilevel programs that provides a suitable framework for the analysis of tarification problems. Although of a nature quite similar to that of the much studied linear bilevel program, this bilinear form lends itself to efficient algorithms. A weak point of this approach is the assumption that, given equally desirable solutions, the follower selects the one that yields the highest profit for the leader. The remedy to this situation is to have a more disaggregated description of the lower level model, either through a more detailed LP or a nonlinear model, thus smoothing out the 'bang-bang' behaviour of the optimal solution of a crude LP. Both avenues will be considered in the future and we expect that the main algorithmic ideas developed in this paper may be adapted to this generalized framework.

6 Acknowledgments

This research was partially supported by NSERC (Canada), FCAR (Québec) and a RIB grant from the Brussels capital region.

References

[1] Anandalingan, G. and White, D. J., "A solution method for the linear Stackelberg problem using penalty functions", *IEEE Transactions on Automatic Control* **35** (1990) 1170–1173.

[2] Brotcorne, L., Approches opérationnelles et stratégiques des problèmes de trafic routier, Ph.D. thesis, Université Libre de Bruxelles, February 1998.

[3] Burton, D. and Toint, Ph. L., "On an instance of the inverse shortest path problem", *Mathematical Programming* **32** (1992) 45–61.

[4] Fukushima, M., "Equivalent differentiable optimization problems and descent methods for asymmetric variational inequality problems", *Mathematical Programming* **53** (1992) 99–110.

[5] Gendreau, M., Marcotte, P. and Savard, G., "A Hybrid tabu-ascent algorithm for the linear bilevel programming problem", *Journal of Global Optimization* **8** (1996) 217–232.

[6] Hansen, P., Jaumard, B. and Savard, G., "New branch-and-bound rules for linear bilevel programming", *JOTA* **22** (1992) 1194–1217.

[7] Kočvara, M. and Outrata, J.V., "A nonsmooth approach to optimization problems with equilibrium constraints", in *Complementarity and variational problems. State of the art,* M.C. Ferris and J.S. Pang eds., SIAM, Philadelphia (1997).

[8] Labbé, M., Marcotte, P. and Savard, G., "A bilevel model of taxation and its application to optimal highway pricing", *Management Science* **44** (1998) 1595–1607 .

[9] Luo, Z.Q., Pang, J.S. and Ralph, D., *Mathematical Programs with Equilibrium Constraints,* Cambridge University Press, Cambridge (1996).

[10] Shimizu, K., Ishizuka, Y, and Bard, J.F., *Nondifferentiable and Two-Level Mathematical Programming,* Kluwer, Boston (1997).

Separation methods for Vector Variational Inequalities. Saddle point and gap function

Giandomenico Mastroeni (e-mail: mastroen@dm.unipi.it)
Department of Mathematics, University of Pisa, Via Buonarroti 2, 56127 Pisa

Abstract

The image space approach is applied to the study of vector variational inequalities. Exploiting separation arguments in the image space, Lagrangian type optimality conditions and gap functions for vector variational inequalities are derived.

Keywords: Vector Variational Inequalities, Image Space, Separation.

1 Introduction

The theory of variational inequalities finds applications in many fields of optimization: from the classical optimality conditions for constrained extremum problems to the equilibrium conditions for network flow, economic and mechanical engineering equilibrium problems [8, 10]. In recent years, variational inequalities, that were first introduced in a scalar form, have been generalized to the vector case [7].

In this paper, by means of the image space analysis, we extend the theory of the gap functions [5, 16] to vector variational inequalities (in short, VVI) defined by the following problem:

$$\text{find } x \in K \quad s.t. \quad F(x)(y-x) \not\leq_{C\setminus\{0\}} 0, \quad \forall y \in K,$$

where $F : X \longrightarrow R^{p\times n}$, $K \subseteq X \subseteq R^n$, C is a convex cone in R^p; in the definition of a VVI we have used the notation: $x \not\geq_C y$ iff $x - y \notin C$. When $p = 1$ and $C = R_+$, the VVI collapses to the classic variational inequality (VI).
Given the vector optimization problem:

$$\min_{C\setminus\{0\}} h(x) \quad s.t. \quad x \in K, \qquad (P)$$

G. Di Pillo and F. Giannessi (eds.), Nonlinear Optimization and Related Topics, 207–217.

where $h : X \longrightarrow R^p$, in the hypotheses that h is a (componentwise) convex differentiable function on the convex set K, it is known [6] that, if we put $F := \nabla h$, then the VVI is a sufficient optimality condition for (P).

The image space analysis can be applied everytime the problem, we want to deal with, is expressed under the form of the impossibility of a suitable system

$$f(x,y) \in C \setminus \{0\}, \quad g(y) \in D, \quad y \in X, \qquad S(x)$$

where $f : X \times X \longrightarrow R^p$, $X \subseteq R^n$, C is a convex cone in R^p, $g : X \longrightarrow R^m$, D is a closed convex cone in R^m. The space R^{p+n} in which the function (f, g) runs, is called the image space associated to $S(x)$. The impossibility of $S(x)$ is stated by means of separation arguments in the image space, proving that two suitable subsets of the image space lie in disjoint level sets of a separating functional.

We recall that a gap function $p : K \longrightarrow R$ is a non-negative function that fulfils the condition $p(x) = 0$ if and only if x is a solution of VI on K. This definition, which originally has been given for scalar variational inequalities, can be extended to the vector case, so that solving a VVI is equivalent to minimize p on the feasible set K. In Section 2 we will analyse the general features of the image space approach for generalized systems. In Section 3 we will consider the general applications to the VVI while, in Section 4, following the approach introduced in [8], we will show how the separations tecniques in the image space, allow to define a gap function for a VVI.

We recall the main notations and definitions that will be used in the sequel. Let $M \subseteq R^p$. $intM, clM$, will denote the interior and the closure of M, respectively.
Let $y \in R^p$, $y := (y_1, \dots, y_p)$; $y_{(1-)} := (y_2, \dots, y_p)$,
$y_{(i-)} := (y_1, \dots, y_{i-1}, \; y_{i+1}, \dots, y_p)$, $i = 2, \dots, p-1$, $y_{(p-)} := (y_1, \dots, y_{p-1})$.
$\langle \cdot, \cdot \rangle$ is the scalar product in R^p, $y \geq 0$ iff $y_i \geq 0$, i=1,...,p. $R^p_+ := \{x \in R^p : x \geq 0\}$.
Let $D \subseteq R^m$ be a convex cone, the positive polar of D is the set $D^* := \{x^* \in R^m : \langle x^*, x \rangle \geq 0, \; \forall x \in D\}$. A closed convex cone D is said pointed if $D \cap (-D) = \{0\}$.
Let $g : R^n \longrightarrow R^m$. g is said D-function on the convex set $K \in R^n$ iff:

$$g(\lambda x_1 + (1-\lambda)x_2) - \lambda g(x_1) - (1-\lambda)g(x_2) \in D, \quad \forall x_1, x_2 \in K, \; \forall \lambda \in (0,1).$$

$x^* \in K$ is said a vector minimum point (in short v.m.p.) for (P) iff the following system is impossible:

$$h(x^*) - h(y) \in C \setminus \{0\}, \quad y \in K.$$

2 A separation scheme for generalized systems

In this section we will present the image space analysis for the generalized system $S(x)$, giving particular attention to the linear separation arguments and the regularity conditions that will allow to state the impossibility of $S(x)$. Suppose that the feasible set is defined by $K := \{y \in X : \; g(y) \in D\}$, and consider the following problem:

find $x^* \in K$, s.t. $S(x^*)$ is impossible.

It is easy to see that vector optimization problems and vector variational inequalities can be formulated as the impossibility of the system $S(x)$ choosing a suitable function $f(x,y)$.

The following result is an immediate consequence of the definition of an optimal solution of a vector optimization problem and the statement of a VVI.

Proposition 2.1 *1. Let $f(x,y) := h(x) - h(y)$, then x^* is a v.m.p. for (P) iff $S(x^*)$ is impossible.*

2. Let $f(x,y) := F(x)(x-y)$, then x^* is a solution of VVI iff the system $S(x^*)$ is impossible.

Define the following subsets of the space R^{p+m}, that we will call "the image space" associated to the system $S(x)$:

$$\mathcal{K}(x) := \{(u,v) \in R^{p+m} : \ u = f(x,y), \quad v = g(y), \quad y \in X\},$$

$$\mathcal{H} := \{(u,v) \in R^{p+m} : \ u \in C \setminus \{0\}, \quad v \in D\}.$$

The impossibility of $S(x)$ can be formulated in terms of the sets $\mathcal{K}(x)$ and $\mathcal{H}$.

Proposition 2.2 *$S(x)$ is impossible iff*

$$\mathcal{K}(x) \cap \mathcal{H} = \emptyset. \tag{1}$$

Let $\mathcal{E}(x) := \mathcal{K}(x) - cl\mathcal{H}$;

Proposition 2.3 [4] *If the cone $\mathcal{H}$ fulfils the condition $\mathcal{H} = \mathcal{H} + cl\mathcal{H}$, then (1) is equivalent to the condition*

$$\mathcal{E}(x) \cap \mathcal{H} = \emptyset. \tag{2}$$

Remark 2.1 In [3] it has been proved that if C is an open or closed convex cone, then $\mathcal{H} = \mathcal{H} + cl\mathcal{H}$, provided that D is a closed convex cone.
Moreover, it is known ([9], Lemma 3.1) that $\mathcal{E}(x)$ is a convex set when g is a D–function and $f(x,y)$ is a (clC)–function with respect to y.

Condition (2) can be proved showing that $\mathcal{E}(x)$ and $\mathcal{H}$ lie in two disjoint level sets of a suitable functional; when the functional can be chosen to be linear we say that $\mathcal{E}(x)$ and $\mathcal{H}$ admit a linear separation.

Definition 2.1 *The sets $\mathcal{E}(x)$ and $\mathcal{H}$ admit a linear separation iff $\exists(\mu^*, \lambda^*) \in C^* \times D^*$, $(\mu^*, \lambda^*) \neq 0$, such that*

$$\langle \mu^*, f(x,y) \rangle + \langle \lambda^*, g(y) \rangle \leq 0, \quad \forall y \in X. \tag{3}$$

The existence of a separating hyperplane doesn't guarantee that $\mathcal{E}(x) \cap \mathcal{H} = \emptyset$. In order to ensure the disjunction of the two sets, some restrictions on the choice of the multipliers (μ^*, λ^*) must be imposed.

Proposition 2.4 *Let clC be a pointed cone and assume that the sets $\mathcal{E}(x)$ and $\mathcal{H}$ admit a linear separation.*
i) If $\mu^* \in intC^*$ then $\mathcal{E}(x) \cap \mathcal{H} = \emptyset$.
ii) Suppose that C is an open cone. If $\mu^* \neq 0$ then $\mathcal{E}(x) \cap \mathcal{H} = \emptyset$.

Proof. We recall (see e.g. [2]) that clC is pointed iff $\text{int}C^* \neq \emptyset$ and that

$$intC^* = \{x^* \in C^* : \langle x, x^* \rangle > 0, \quad \forall x \in clC, x \neq 0\}.$$

i) Ab absurdo, suppose that $\mathcal{E}(x) \cap \mathcal{H} \neq \emptyset$. This implies that $\mathcal{K}(x) \cap \mathcal{H} \neq \emptyset$ and, therefore, $\exists z \in K$ such that $f(x,z) \in C \setminus \{0\}$, then, taking into account that $\mu^* \in \text{int}C^*$, we have $0 < \langle \mu^*, f(x,z) \rangle \leq \langle \mu^*, f(x,z) \rangle + \langle \lambda^*, g(z) \rangle \leq 0$, and we achieve the absurdity.
ii) Ab absurdo, suppose that $\mathcal{E}(x) \cap \mathcal{H} \neq \emptyset$. Following the proof of part i) $\exists z \in K$ such that $f(x,z) \in C = intC$, then, taking into account that $\mu^* \neq 0$, we have $0 < \langle \mu^*, f(x,z) \rangle \leq \langle \mu^*, f(x,z) \rangle + \langle \lambda^*, g(z) \rangle \leq 0$, and we achieve the absurdity. □

Remark 2.2 In particular, if we define $f(x,y) := h(x) - h(y), C = R^p_+$ (resp. C open cone), and $\mathcal{E}(x)$ and $\mathcal{H}$ admit a linear separation with $\mu > 0$ (resp. $\mu \neq 0$), then x is a v.m.p. for (P).

The following Theorem gives sufficient conditions that guarantee that the hypotheses of the Proposition 2.4 are fulfilled.

Theorem 2.1 *Suppose that the sets $\mathcal{E}(x)$ and $\mathcal{H}$ admit a linear separation.*

1. *Let $C := R^p_+$. Assume that, for every $i = 1, \ldots, p$, the following system is possible:*

$$f_{i^-}(x,y) > 0, \quad g(y) \in intD, \quad y \in X. \qquad S_i(x)$$

then in (3) we can suppose that $\mu > 0$.

2. *If there exists $\bar{y} \in X$ such that $g(\bar{y}) \in intD$, then in (3) we can suppose that $\mu \neq 0$.*

Proof. *1.* Ab absurdo, suppose that, $\exists i \in \{1, \ldots, p\}$ such that $\mu^*_i = 0$; then $(\mu^*_{i^-}, \lambda^*) \neq 0$ and, since $S_i(x)$ is possible, $\exists \bar{y} \in X$ such that

$$0 < \langle \mu^*_{i^-}, f_{i^-}(x, \bar{y}) \rangle + \langle \lambda^*, g(\bar{y}) \rangle \leq -\langle \mu^*_i, f_i(x, \bar{y}) \rangle = 0,$$

which is absurd.
2. Ab absurdo, suppose that $\mu^* = 0$ in (3); then, $\lambda^* \neq 0$ and, since $g(\bar{y}) \in intD$, it is

$$0 < \langle \lambda^*, g(y) \rangle \leq 0,$$

which is absurd. □

Remark 2.3 The condition given in the statement *1*, which has been also considered in [12], in a slightly different form, is a generalization of the Slater condition for scalar optimization problems [13, 14] taken as the assumption of the statement *2*.

3 Linear separation and saddle point conditions

As observed in [3, 4], linear separation is closely related to the Lagrangian-type optimality conditions. Following the line considered in [3, 4] we will characterize the linear separation in terms of a saddle point condition of the Lagrangian function associated to the system $S(x^*)$, defined by $L : C^* \times D^* \times X \longrightarrow R$,

$$L(x^*; \mu, \lambda, y) := -[\langle \mu, f(x^*, y)\rangle + \langle \lambda, g(y)\rangle].$$

Proposition 3.1 *Suppose that* $f(x^*, x^*) = 0$ *and* $g(x^*) \in D$. *Then* $\mathcal{E}(x^*)$ *and* $\mathcal{H}$ *admit a linear separation iff* $\exists(\mu^*, \lambda^*) \in C^* \times D^*$, $(\mu^*, \lambda^*) \neq 0$, *such that the point* (μ^*, λ^*, x^*) *is a saddle point for* $L(x^*; \mu, \lambda, y)$ *on* $(C^* \times D^*) \times X$.

Proof. Suppose that $\mathcal{E}(x^*)$ and $\mathcal{H}$ admit a linear separation. ¿From (3) we obtain that $\langle \lambda^*, g(x^*)\rangle \leq 0$, which implies that $\langle \lambda^*, g(x^*)\rangle = 0$, since $g(x^*) \in D$ and $\lambda^* \in D^*$. Therefore

$$0 = L(x^*; \mu^*, \lambda^*, x^*) \ \leq L(x^*; \mu^*, \lambda^*, y), \quad \forall y \in X.$$

It remains to show that $L(x^*; \mu, \lambda, x^*) \leq 0, \quad \forall(\mu, \lambda) \in (C^* \times D^*)$. We have that $L(x^*; \mu, \lambda, x^*) = -\langle \lambda, g(x^*)\rangle$ which is negative, $\forall \lambda \in D^*$, and the necessity part of the statement is proved.

Sufficiency. Suppose that (μ^*, λ^*, x^*) is a saddle point for $L(x^*; \mu, \lambda, y)$ on $(C^* \times D^*) \times X$, that is

$-\langle \lambda, g(x^*)\rangle \leq \ -\langle \lambda^*, g(x^*)\rangle \leq -[\langle \mu^*, f(x^*, y)\rangle + \langle \lambda^*, g(y)\rangle], \quad \forall(\mu, \lambda, y) \in (C^* \times D^*) \times X$.

Computing the first inequality for $\lambda = 0$, we obtain $\langle \lambda^*, g(x^*)\rangle \leq 0$ and, therefore, $\langle \lambda^*, g(x^*)\rangle = 0$. The first inequality coincides with (3) and the proposition is proved. □

Remark 3.1 We observe that the saddle value, $L(x^*; \mu^*, \lambda^*, x^*)$, is equal to zero. This property will be useful in the analysis of the gap function associated to a vector variational inequality defined in Section 5.

Proposition 3.2 *Let* X *be an open convex set in* R^n. *Assume that*

1. $f(x^*, y)$ *is a differentiable C-function, with respect to* y, *such that* $f(x^*, x^*) = 0$;

2. g *is a differentiable D-function;*

Then (μ^*, λ^*, x^*) *is a saddle point for* $L(x^*; \mu, \lambda, y)$ *on* $(C^* \times D^*) \times X$ *iff it is a solution of the following system (S)*

$$\begin{cases} \nabla_y L(x^*; \mu, \lambda, y) = 0 \\ \langle \lambda, g(y) \rangle = 0 \\ g(y) \in D, \mu \in C^*, \quad \lambda \in D^*, \quad y \in X. \end{cases}$$

Proof. Suppose that (μ^*, λ^*, x^*) is a saddle point for $L(x^*; \mu, \lambda, y)$ on $(C^* \times D^*) \times X$, that is

$-\langle \lambda, g(x^*) \rangle \leq -\langle \lambda^*, g(x^*) \rangle \leq -[\langle \mu^*, f(x^*, y) \rangle + \langle \lambda^*, g(y) \rangle], \quad \forall (\mu, \lambda, y) \in (C^* \times D^*) \times X.$

First of all we prove that $g(x^*) \in D$. Ab absurdo suppose that $g(x^*) \notin D = (D^*)^*$; then $\exists \bar{\lambda} \in D^*$ such that $\langle \bar{\lambda}, g(x^*) \rangle < 0$. Since D^* is a cone, then $\alpha \bar{\lambda} \in D^*$, $\forall \alpha \geq 0$ and $-\alpha \langle \bar{\lambda}, g(x^*) \rangle \longrightarrow +\infty$, $\alpha \longrightarrow +\infty$; this contradicts the first inequality in the saddle point condition.

Computing the first inequality for $\lambda = 0$, we obtain $\langle \lambda^*, g(x^*) \rangle \leq 0$ and, therefore,

$$\langle \lambda^*, g(x^*) \rangle = 0. \tag{4}$$

The second inequality implies that x^* is a global minimum point of $L(x^*; \mu^*, \lambda^*, y)$, since $f(x^*, x^*) = 0$. Then

$$\nabla_y L(x^*; \mu^*, \lambda^*, x^*) = 0. \tag{5}$$

(5), together with (4) and the relation $(\mu^*, \lambda^*) \in (C^* \times D^*)$, allows to achieve the necessity part of the statement.

Sufficiency. Suppose that (μ^*, λ^*, x^*) is a solution of (S). Since $L(x^*; \mu^*, \lambda^*, y)$ is a convex function in the variable y, then $\nabla_y L(x^*; \mu^*, \lambda^*, x^*) = 0$ implies that

$$L(x^*; \mu^*, \lambda^*, x^*) \leq L(x^*; \mu^*, \lambda^*, y), \quad \forall y \in X.$$

Taking into account the complementarity relation $\langle \lambda^*, g(x^*) \rangle = 0$ and the condition $\lambda \in D^*$, we obtain

$$-\langle \lambda, g(x^*) \rangle \leq -\langle \lambda^*, g(x^*) \rangle, \quad \forall (\mu, \lambda) \in (C^* \times D^*),$$

and the statement is proved. □

4 Separation methods for Vector Variational Inequalities

The results stated in the previous sections can be applied in the analysis of a VVI, allowing to obtain Kuhn–Tucker type optimality conditions. Consider the vector variational inequality:

$$\text{find } x \in K \quad s.t. \quad F(x)(y - x) \not\leq_{C \setminus \{0\}} 0, \quad \forall y \in K := \{y \in X : \ g(y) \in D\},$$

where $F : X \longrightarrow R^{p \times n}$, $g : X \longrightarrow R^m$, D is a closed convex cone in R^m, g is a D–function and cl$C$ is a convex pointed cone in R^p.

Following the scheme introduced in Section 2, we define the following subsets of the space R^{p+m}, that we will call the image space associated to VVI:

$$\mathcal{K}(x) := \{(u,v) \in R^{p+m} : \; u = F(x)(x-y), \quad v = g(y), \quad y \in X\},$$

$$\mathcal{H} := \{(u,v) \in R^{p+m} : \; u \in C \setminus \{0\}, \quad v \in D\}.$$

The next result is analogous to Propositions 2.2 and 2.3.

Proposition 4.1 *i) $x^* \in K$ is a solution of VVI iff*

$$\mathcal{K}(x^*) \cap \mathcal{H} = \emptyset. \tag{6}$$

ii) If $\mathcal{H}$ is a convex cone that fulfils the condition $\mathcal{H} := \mathcal{H} + cl\mathcal{H}$, then (6) is equivalent to the condition

$$\mathcal{E}(x^*) \cap \mathcal{H} = \emptyset, \tag{7}$$

where $\mathcal{E}(x^*) := \mathcal{K}(x^*) - cl\mathcal{H}$.

As observed in the Remark 2.1, if g is a D–function, then $\mathcal{E}(x^*)$ is a convex set [9]. Therefore, using separation tecniques in the image space, it is possible to obtain necessary and (or) sufficient Lagrangian–type optimality conditions for VVI.
Let $f : X \times X \longrightarrow R^p$, $\quad f(x,y) = F(x)(x-y)$.

Theorem 4.1 *Assume that $C := R^p_+$, X is an open convex set in R^n, and*

1. *g is a differentiable D–function;*
2. *for every $i = 1, \dots, p$, the following system is possible*

$$f_{i^-}(x^*,y) > 0, \quad g(y) \in intD, \quad y \in X. \qquad S_i(x^*)$$

Then $x^* \in K$ is a solution of VVI iff $\exists(\mu,\lambda) \in (C^* \times D^*)$, $(\mu,\lambda) \neq 0$, such that (μ, λ, x^*) is a solution of the following system (S)

$$\begin{cases} \mu F(x) - \lambda \nabla g(x) = 0 \\ \langle \lambda, g(x) \rangle = 0 \\ g(x) \in D, \; \mu \in C^*, \; \lambda \in D^*, \; x \in X. \end{cases}$$

Proof. We observe that $C^* = R^p_+$. Suppose that x^* is a solution of VVI. Then (7) holds. Since $f(x,y)$ is a linear function in the variable y and g is a D-function, the set $\mathcal{E}(x^*)$ is convex (see Remark 2.1). Therefore $\mathcal{E}(x^*)$ and $\mathcal{H}$ admit a linear separation. By Proposition 3.1, we have that $\exists(\mu^*,\lambda^*) \in C^* \times D^*$ such that the point (μ^*,λ^*,x^*) is a saddle point for the Lagrangian function $L(x^*;\mu,\lambda,y) = -[\langle \mu, f(x^*,y)\rangle + \langle \lambda, g(y)\rangle]$.

By Proposition 3.2, we obtain that (μ^*, λ^*, x^*) is a solution of the system (S).
Sufficiency. Let (μ^*, λ^*, x^*) be a solution of the system (S); by Proposition 3.2 we get that (μ^*, λ^*, x^*) is a saddle point for $L(x^*; \mu, \lambda, y)$, and, therefore, $\mathcal{E}(x^*)$ and $\mathcal{H}$ admit a linear separation. Taking into account Theorem 2.1, condition *2* implies that $\mu^* > 0$. Proposition 2.4 ensures that $\mathcal{E}(x^*) \cap \mathcal{H} = \emptyset$ and, therefore, x^* is a solution of VVI. □

In the last part of the section we will consider the, so called, weak case, in which C is an open convex cone. We will see that much less restrictive conditions are required in order to obtain an analogous result to Theorem 4.1 for the weak case. In particular, we will show that the classical Slater condition will be a sufficient regularity assumption on the constraint function g.

Theorem 4.2 *Assume that C is an open convex cone, X is an open convex set in R^n and that*
(a) g is a differentiable D-function;
(b) there exists $y \in X$ such that $g(y) \in intD$.
Then x^* is a solution of VVI iff $\exists(\mu, \lambda) \in C^* \times D^*, (\mu, \lambda) \neq 0$, such that (μ, λ, x^*) is a solution of the system

$$\begin{cases} \mu F(x) - \lambda \nabla g(x) = 0 \\ \langle \lambda, g(x) \rangle = 0 \\ g(x) \in D,\ \mu \in C^*,\ \lambda \in D^*, x \in X. \end{cases}$$

Proof. Necessity. The proof is analogous to the one of Theorem 4.1.
Sufficiency. The proof is analogous to the one of Theorem 4.1, replacing the hypothesis *2* with the hypothesis *(b)*, and $\mu^* > 0$ with $\mu^* \neq 0$. □

5 Gap functions for Vector Variational Inequalities

Given the variational inequality:

$$\text{find } x^* \in K \text{ s.t. } \langle F(x^*), y - x^* \rangle \geq 0, \quad \forall x \in K, \qquad (VI)$$

where $F : K \longrightarrow R^n$, $K \subseteq R^n$, a gap function $p : K \longrightarrow R$ is a non-negative function that fulfils the condition $p(x) = 0$ if and only if x is a solution of VI. Therefore, solving a VI is equivalent to the minimization of the gap function on the feasible set K. A first example of gap function was given by Auslender [1] who considered the function $p(x) := \sup_{y \in K} \langle F(x), x - y \rangle$. Similarly to the scalar case, a gap function can be defined for a vector variational inequality.

Definition 5.1 *A function $p : K \longrightarrow R$ is a gap function for VVI iff*
i) $p(x) \geq 0, \quad \forall x \in K$;
ii) $p(x) = 0$ if and only if x is a solution of VVI.

Consider the following function $\psi : K \longrightarrow R$:

$$\psi(x) := \min_{(\mu,\lambda)\in S} \sup_{y\in X} [\langle \mu, f(x,y)\rangle \; + \; \langle \lambda, g(y)\rangle],$$

where $S := \{(\mu, \lambda) \in (C^* \times D^*) : \|(\mu, \lambda)\| = 1\}$.
Let $F : X \longrightarrow R^{p\times n}$, $f : X \times X \longrightarrow R^p, f(x,y) = F(x)(x - y)$, and $\Omega := \{x \in K : \psi(x) = 0\}$. The saddle point condition, that characterizes the separation in the image space (see the Proposition 3.1), allows to prove that $\psi(x)$ is a gap function for VVI.

Theorem 5.1 *Let g be a D-function on the convex set $X \subseteq R^n$.*

1. *Assume that $C := R^p_+$ and that, for every $i = 1, \ldots, p$ and $\forall x^* \in \Omega$, the following system is possible*

$$f_{i^-}(x^*, y) > 0, \quad g(y) \in intD, \quad y \in X; \qquad\qquad S_i(x^*)$$

   then $\psi(x)$ is a gap function for VVI.

2. *Assume that C is an open convex cone and that*

$$\exists \bar{y} \in X \text{ such that } g(\bar{y}) \in intD; \tag{8}$$

   then $\psi(x)$ is a gap function for VVI.

Proof. *1.* It is easy to prove that $\psi(x) \geq 0, \ \forall x \in K$; in fact, if $(\mu, \lambda) \in (C^* \times D^*)$, then

$$\langle \mu, f(x,x)\rangle + \langle \lambda, g(x)\rangle = \langle \lambda, g(x)\rangle \geq 0.$$

Suppose that x^* is a solution of VVI. Since $f(x,y)$ is a linear function in the variable y and g is a D–function, the set $\mathcal{E}(x^*)$ is convex (see Remark 2.1). Therefore $\mathcal{E}(x^*)$ and $\mathcal{H}$ admit a linear separation. Without loss of generality we can suppose that the coefficients of the separating hyperplane $(\mu^*, \lambda^*) \in S$. From Proposition 3.1, we have that (μ^*, λ^*, x^*) is a saddle point for $L(x^*; \mu, \lambda, y) := -[\langle \mu, f(x^*, y)\rangle \; + \; \langle \lambda, g(y)\rangle]$ on $(C^* \times D^*) \times X$ and the saddle value $L(x^*; \mu^*, \lambda^*, x^*) = 0$ (see Remark 3.1). Recalling that the saddle point condition can be characterized by suitable minimax problems [15], we have

$$\min_{(\mu,\lambda)\in C^*\times D^*} \sup_{y\in X} [\langle \mu, f(x,y)\rangle \; + \; \langle \lambda, g(y)\rangle] = L(x^*; \mu^*, \lambda^*, x^*) = 0. \tag{9}$$

Since $(\mu^*, \lambda^*) \in S$, taking into account (9), we obtain that $\psi(x^*) = 0$.

Vice–versa, suppose that $\psi(x^*) = 0$. Then $\exists(\mu^*, \lambda^*) \in S$, such that

$$\langle \mu^*, f(x^*, y)\rangle + \langle \lambda^*, g(y)\rangle \leq 0, \quad \forall y \in X.$$

For Theorem 2.1, the possibility of the system $S_i(x^*)$ for $i = 1, \ldots, p$, implies that $\mu^* > 0$. Applying Proposition 2.4 , we obtain that x^* is a solution of VVI.
2. The proof is analogous to the one of *1* using (8) instead of the fact that $S_i(x^*)$ is possible for $i = 1, \ldots, p$ and replacing the condition $\mu^* > 0$ with $\mu^* \neq 0$. □

Remark 5.1 We observe that $h_x(\mu, \lambda) := \sup_{y \in X}[\langle \mu, f(x, y)\rangle + \langle \lambda, g(y)\rangle]$, being the supremum of a collection of linear functions, is a convex function, so that $\psi(x) = \min_{(\mu,\lambda)\in S} h_x(\mu, \lambda)$ is the optimal value of a parametric problem on a compact set, with a convex objective function.

The gap function ψ that we have analysed in this section, in general, is not differentiable. Following the line adopted in [5, 16], adding a suitable regularizing term $H(x, y) : X \times X \longrightarrow R$ to the function $\langle \mu, f(x, y)\rangle + \langle \lambda, g(y)\rangle$, it is possible to obtain a differentiable gap function for VVI. To this aim, scalarization methods (see e.g. [11]) for VVI can be a further useful tool to carry out the analysis.

References

[1] Auslender A. (1976) Optimization. Methodes Numeriques, Masson, Paris;

[2] Berman A. (1973) Cones, matrices and mathematical programming, Lecture Notes in Economics and Mathematical Systems, Springer-Verlag, Berlin, Germany;

[3] Castellani M., Mastroeni G., and Pappalardo M. (1996) On regularity for generalized systems and applications, in *"Nonlinear Optimization and Applications"* , G. Di Pillo, F.Giannessi (eds.), Plenum Press, New York, *pp. 13-26*;

[4] Dien P.H., Mastroeni G., Pappalardo M., and Quang P.H. (1994) Regularity Conditions for Constrained Extremum Problems via Image Space: the Linear Case, in *Generalized Convexity*, Lecture Notes in Economics and Mathematical Systems, S. Komlosi, T. Rapcsak and S. Schaible (eds.), Springer Verlag, Berlin, Germany, *pp.145-152*;

[5] Fukushima M. (1992) Equivalent differentiable optimization problems and descent methods for asymmetric variational inequality problems,*Mathematical Programming*, **Vol.53**, *pp.99-110*;

[6] Giannessi F. (1998) On Minty Variational Principle, in *"New Trends in Mathematical Programming"*, F.Giannessi, S.Komlosi, T.Rapcsak (eds.), Kluwer;

[7] Giannessi F. (1980) Theorems of the Alternative, Quadratic Programs and Complementarity Problems, in *"Variational Inequalities and Complementarity Problems"*,R.W. Cottle, F. Giannessi and J.L. Lions (eds.), Wiley, New York, *pp.151-186*;

[8] Giannessi F. (1995) Separation of sets and Gap Functions for Quasi-Variational Inequalities, in *"Variational Inequalities and Network Equilibrium Problems"*, F.Giannessi and A.Maugeri (eds.),Plenum Publishing Co, *pp.101-121*;

[9] Giannessi F. (1984) Theorems of the Alternative and Optimality Conditions, *Journal of Optimization Theory and Applications*, **Vol.42**, *pp. 331-365*;

[10] Harker P.T.,Pang J.S. (1990) Finite–Dimensional Variational Inequalities and Nonlinear Complementarity Problem: a Survey of Theory, Algorithms and Applications, *Mathematical Programming*, **Vol.48**, *pp.161-220*;

[11] Lee G.M., Kim D.S., Lee B.S. and Yen N.D. (1998) Vector Variational Inequality as a tool for studying Vector Optimization Problems, *Nonlinear Analysis*, **Vol.34**, *pp. 745-765*;

[12] Maeda T. (1994) Constraint Qualifications in Multiobjective Optimization Problems: Differentiable Case, *Journal of Optimization Theory and Applications*, **Vol.80**, *pp. 483-500*;

[13] Mangasarian O.L. (1969) Nonlinear Programming, New York, Academic Press;

[14] Mangasarian O.L. and Fromovitz S. (1967) The Fritz–John Necessary Optimality Condition in the Presence of Equality and Inequality Constraints, *Journal of Mathematical Analysis and Applications*, **Vol. 7**, *pp. 37-47*;

[15] Rockafellar R.T. (1970) Convex Analysis, Princeton University Press, Princeton;

[16] Zhu D.L.,Marcotte P. (1994) An extended descent framework for variational inequalities, *Journal of Optimization Theory and Applications*, **Vol.80**, *pp.349-366*.

Functions with Primal-Dual Gradient Structure and $\mathcal{U}$-Hessians

Robert Mifflin (mifflin@math.wsu.edu)
Mathematics, Washington State University, Pullman, WA 99164-3113[1]

Claudia Sagastizábal (sagastiz@ele.puc-rio.br)
DEE/PUC-Rio. Rua Marquês de São Vicente, 225. 22453-000 Rio de Janeiro RJ, Brazil [2]

Abstract

This paper introduces examples of convex functions in a general class of functions having what we call primal-dual gradient structure. The class contains finitely defined maximum value functions and maximum eigenvalue functions as well as other infinitely determined max-functions. For such a function there is a space decomposition that allows us to identify a subspace $\mathcal{U}$ on which the function appears smooth. Moreover, the special structure combined with sufficiency conditions implies the existence of smooth trajectories on which the function has a certain second order expansion. This results in an explicit expression for the Hessian of a related $\mathcal{U}$-Lagrangian.

Keywords: Convex optimization, second-order derivatives, $\mathcal{U}$-Lagrangian.

1 Introduction

1.1 Motivation

The issue of defining a second order expansion for a convex lower semicontinuous function f has captured the attention of many researchers for several years. Concerning first order approximations, it is well known that the notion of gradient is

[1]Research supported by the National Science Foundation under Grant No. DMS-9703952 and by FAPERJ (Brazil) under Grant No. E-26/171.393/97.
[2]Research supported by FAPERJ (Brazil) under Grant No.E26/150.205/98.

G. Di Pillo and F. Giannessi (eds.), Nonlinear Optimization and Related Topics, 219–233.

replaced by a set of subgradients collecting all of the first order information about f. When going to second order, the answer is not so clear. A wealth of possible choices can be found in the literature: B-derivatives ([Rob87]); proto-derivatives ([Roc89a], [Pol90]); epiderivatives ([Roc88]); pseudo-derivatives ([CC90], [Com91]); second-order subderivatives ([Iof91]); graphical derivatives and sub-Hessians ([Pen94]); $\mathcal{U}$-Hessians ([LOS97]).

Most of the second-order objects listed above have given rise to elegant and powerful theories, including generalized second order optimality conditions ([BTZ82], [HU79], [HU82], [Roc89b]) and epigraphical convergence ([RW98]). Unfortunately, few of these notions have led to rapidly convergent minimization algorithms that are implementable.

The $\mathcal{U}$-Hessian introduced in [LOS97] is a rare exception. The general idea is to decompose $\mathbb{R}^n$ into two orthogonal subspaces $\mathcal{V}$ and $\mathcal{U}$ in such a way that (locally) all of f's nonsmoothness is concentrated in $\mathcal{V}$. As a result, f has some kind of Hessian on $\mathcal{U}$ (a "$\mathcal{U}$-Hessian") and can be expanded up to second order along certain trajectories ("smooth trajectories"), at least whenever f is **structured enough**. In particular, $\mathcal{VU}$-theory yields a superlinearly convergent bundle-type method ([Ous97], [Ous98]) when applied to minimization of a max-eigenvalue function (mef), provided enough eigenvalues and eigenvectors at each point are known. For the case of more general convex functions of one variable there is a rapidly convergent $\mathcal{VU}$-algorithm [LM82] that requires only one subderivative at each iterate.

In this paper we give a more precise meaning to the wording "structured enough" by introducing examples of functions having *primal-dual gradient* (pdg) structure as given in Definition 1 below from [MS98]. The class of pdg functions is quite large and includes mef 's as well as finitely defined ([MS99], [BTB93]) and other infinitely defined maximum value functions. In [MS98] we show in detail how pdg structure provides us with a basic index set whose associated vectors span a subspace of $\mathcal{V}$ and generate an implicit function therein from which a smooth trajectory tangent to $\mathcal{U}$ can be defined. Under some sufficiency conditions, playing a role similar to that of constraint qualification conditions in nonlinear programming, f has a second order expansion along such a trajectory.

The paper is organized as follows. We recall the main elements of the $\mathcal{VU}$-decomposition theory in § 3. Since $\mathcal{VU}$-theory is rather involved, throughout this study we employ two prototype functions, $\mathcal{E}_1$ and $\mathcal{E}_2$. These functions, introduced in § 2, represent the pdg class well and allow us to perform calculations and illustrate our results without entering into the deep technicalities of the proofs in [MS99] and [MS98]. Dual feasible basic index sets are studied in § 4 while smooth trajectories and $\mathcal{V}$-optimality conditions are considered in Section 5. Finally, in § 6, we give an explicit expression for the $\mathcal{U}$-Hessian.

1.2 Primal-dual gradient structure

Consider a convex function f having the max-function form

$$f(x) := \max\{F(x,t) : t \in \mathcal{T}\} \text{ for } x \in \mathbb{R}^n \tag{1}$$

where $\mathcal{T}$ is a closed subset of $\mathbb{R}^{n_\mathcal{T}}$ and $F(x,t)$ and its first and second order partial derivatives with respect to components of x are continuous on $\mathbb{R}^n \times \mathcal{T}$. Since for a convex function f as in (1) $\partial f(x)$ equals the set of Clarke generalized gradients at x [Cla83], if $\mathcal{T}$ is compact, the subdifferential is the following convex hull:

$$\partial f(x) = \text{conv}\{\nabla_x F(x,t_a) \text{ for all } t_a \in \mathcal{T} \text{ such that } f(x) = F(x,t_a)\}\,. \tag{2}$$

When $\mathcal{T}$ is not a finite set, the maximizing t_a's corresponding to x in (2) may have a convex hull containing an infinite number of extreme points (for instance, **mef** 's as in Section 3 of [MS98] with $\mathcal{T} = \{t \in \mathbb{R}^n : t^T t = 1\}$). This can lead to $\partial f(x)$ having a continuum of extreme points. To apply $\mathcal{VU}$-theory, it is crucial to properly describe all of the subgradients. As a result, multipliers forming the convex combinations in (2) need to satisfy certain conditions. This is the purpose of the functions f_i and φ_ℓ and associated multipliers α_i introduced next. In particular, the presence of the functions φ_ℓ and corresponding multipliers allows the subdifferential to have a continuum of extreme points.

Definition 1 We say that a convex function $f : \mathbb{R}^n \to \mathbb{R}$ has *primal-dual gradient structure* about $\bar{x} \in \mathbb{R}^n$ if the following conditions hold:

There exists *a ball* about $\bar{x}$, $B(\bar{x})$, $m_1 + 1 + m_2$ *primal functions*

$$\{f_i(x)\}_{i=0}^{m_1} \quad \text{and} \quad \{\varphi_\ell(x)\}_{\ell=1}^{m_2}$$

that are C^2 on $B(\bar{x})$ and a *dual multiplier set* $\Delta \subset \mathbb{R}^{m_1+1+m_2}$ such that

(i) $\bar{x} \in \mathcal{P} := \{x \in B(\bar{x}) : \varphi_\ell(x) = 0 \text{ for } \ell = 1,\ldots,m_2\}$ and $f_i(\bar{x}) = f(\bar{x})$ for $i = 0,1,\ldots,m_1$;

(ii) for each $x \in \mathcal{P}$

$$f(x) = \max\{f_i(x) : i = 0,1,\ldots,m_1\}\,;$$

(iii) Δ is a closed convex set such that

(a) if $\alpha := (\alpha_0,\cdots,\alpha_{m_1},\alpha_{m_1+1},\cdots,\alpha_{m_1+m_2}) \in \Delta$ then $(\alpha_0,\cdots,\alpha_{m_1})$ is an element of the unit simplex Δ_1 in $\mathbb{R}^{m_1+1}$ given by

$$\Delta_1 := \{(\alpha_0,\alpha_1,\cdots,\alpha_{m_1}) : \sum_{i=0}^{m_1} \alpha_i = 1\,, \alpha_i \geq 0\,, i = 0,1,\ldots,m_1\}\,;$$

(b) for each $i = 0,1,\cdots,m_1$ $\mathbf{1}_{i+1} \in \Delta$, where $\mathbf{1}_j$ is the jth unit vector in $\mathbb{R}^{m_1+1+m_2}$

and

(c) for each $\ell = 1, 2, \cdots, m_2$ there exists $\alpha^\ell \in \Delta$ such that $\alpha^\ell_{m_1+\ell} \neq 0$ and $\alpha^\ell_{m_1+i} = 0$ for $i \in \{1, 2, \cdots, m_2\} \backslash \{\ell\}$;

(iv) for each $x \in \mathcal{P}$, $g \in \partial f(x)$ if and only if

$$g = \sum_{i=0}^{m_1} \alpha_i \nabla f_i(x) + \sum_{i=m_1+1}^{m_1+m_2} \alpha_i \nabla \varphi_{i-m_1}(x), \tag{3}$$

where the multipliers $\alpha_0, \alpha_1, \ldots, \alpha_{m_1+m_2}$ satisfy

Complementary slackness: $\alpha_i = 0$ if $f_i(x) < f(x)$ and $i \leq m_1$

and

Dual feasibility: $\alpha = (\alpha_0, \alpha_1, \ldots, \alpha_{m_1+m_2}) \in \Delta$.

□

For a finite max-function, $m_2 = 0$, there are no φ_ℓ functions and $\Delta = \Delta_1$. When f is a mef, we have shown in [MS98] that $m_2 \neq 0$ and Δ corresponds to the set of positive semidefinite matrices having unit traces. In the next section we give our two prototype examples and show that they have pdg structure.

2 Example functions

In [MS99] $\mathcal{VU}$-theory is developed fully for some bivariate example functions that are maximums of two or three C^2-functions. Here we introduce and analyze two infinitely defined max-functions.

For the example parameter $p = 1, 2$ consider the following particular instance of (1):

$$\begin{array}{rccl} \mathcal{E}_p : & \mathbb{R}^2 & \to & \mathbb{R} \\ & (x_1, x_2) & \mapsto & \max\{t_1^2 x_1^2 + 2(t_1 t_2)^p x_2 : t_1^2 + t_2^2 = 1\}. \end{array}$$

When $p = 1$, the resulting function is the maximum eigenvalue of the matrix $\begin{bmatrix} x_1^2 & x_2 \\ x_2 & 0 \end{bmatrix}$:

$$\mathcal{E}_1(x_1, x_2) = \frac{1}{2} x_1^2 + \frac{1}{2}\sqrt{x_1^4 + 4x_2^2}.$$

When $p = 2$, after some algebra we obtain

$$\mathcal{E}_2(x_1, x_2) = \begin{cases} x_1^2 & \text{if } x_2 \leq \frac{1}{2} x_1^2 \\ \frac{1}{2} x_2 (1 + \dfrac{x_1^2}{2x_2})^2 & \text{otherwise} \end{cases}$$

(the change of variables $(sin\theta, cos\theta) \leftrightarrow (t_1, t_2)$ or $(\tau, 1-\tau) \leftrightarrow (t_1^2, t_2^2)$ with $0 \leq \tau \leq 1$ can be helpful in the calculations). Note that the second function is neither a `mef` nor a maximum of a finite number of C^2-functions.

Both functions are convex and are minimized at the kink $\bar{x} = (\bar{x}_1, \bar{x}_2) = (0,0)^T$. Since for convex functions the set of Clarke generalized gradients [Cla83] equals the subdifferential, to compute $\partial\mathcal{E}_p(0,0)$ we calculate the respective gradients where they exist and then take limits and convex hulls:

$$\nabla\mathcal{E}_1(x_1,x_2) = \left(x_1 + \frac{x_1^3}{\sqrt{x_1^4+4x_2^2}}, \frac{2x_2}{\sqrt{x_1^4+4x_2^2}}\right) \quad \rightarrow \quad \begin{cases} \left(0, \dfrac{x_2}{|x_2|}\right) & \text{if } x_1 \to 0\,, x_2 \neq 0 \\ (2x_1, 0) & \text{if } x_2 \to 0\,, x_1 \neq 0 \end{cases}$$

so that

$$\partial\mathcal{E}_1(x_1,0) = \begin{cases} \{(2x_1,0)\} & \text{if } x_1 \neq 0 \\ \operatorname{conv}\{(0,0)\,,(0,\pm 1)\} & \text{if } x_1 = 0\,. \end{cases} \tag{4}$$

Similarly,

$$\nabla\mathcal{E}_2(x_1,x_2) = \begin{cases} (2x_1, 0) & \text{if } x_2 < \frac{1}{2}x_1^2 \\ (1+\dfrac{x_1^2}{2x_2})\left(x_1, \dfrac{1}{2}(1-\dfrac{x_1^2}{2x_2})\right) & \text{if } x_2 > \frac{1}{2}x_1^2 \end{cases}$$

and

$$\partial\mathcal{E}_2(x_1,0) = \begin{cases} \{(2x_1,0)\} & \text{if } x_1 \neq 0 \\ \operatorname{conv}\left\{(0,0)\,,\left(0,\dfrac{1}{2}\right)\right\} & \text{if } x_1 = 0\,. \end{cases} \tag{5}$$

Hence, neither of these functions has a Hessian at $\bar{x}$. However, they do have $\mathcal{U}$-Hessians, as will be shown in Section 6 of this paper.

To show that our protoype functions have pdg structure about $\bar{x} = (0,0)^T$, it is sufficient to let $B(\bar{x}) = \mathbb{R}^2$, $m_1 = m_2 = 1$ and to consider the following three primal functions and their gradients:

$$\begin{aligned} f_0(x_1,x_2) &:= x_1^2\,, & f_1(x_1,x_2) &:= 0\,, & \varphi_1(x_1,x_2) &:= \frac{1}{p}x_2\,; \\ \nabla f_0(x_1,x_2) &= (2x_1,0)\,, & \nabla f_1(x_1,x_2) &= (0,0)\,, & \nabla\varphi_1(x_1,x_2) &= \frac{1}{p}(0,1)\,. \end{aligned} \tag{6}$$

Because $\mathcal{P} = \{(x_1,x_2) \in \mathbb{R}^2 : x_2 = 0\}$ and $\mathcal{E}_p(x_1,0) = x_1^2 \geq 0$, items (i)-(ii) in Definition 1 hold trivially. As for the dual multiplier set $\Delta \in \mathbb{R}^3$, it can be expressed as follows:

$$(\alpha_0,\alpha_1,\alpha_2) \in \Delta \iff \text{the matrix} \begin{bmatrix} \alpha_0 & \frac{1}{2}\alpha_2 \\ \frac{1}{2}\alpha_2 & \alpha_1 \end{bmatrix} \begin{array}{l} \text{is positive semidefinite} \\ \text{with trace equal to 1} \\ \text{and with } \alpha_2 \geq 0 \text{ if } p = 2. \end{array} \tag{7}$$

To see that definition (7) satisfies the requirements of Definition 1(iii) note that

(a) a positive semidefinite matrix with unit trace has nonnegative diagonal elements that sum to one,

(b) the first two unit vectors in $\mathbb{R}^3$ correspond to the unit trace positive semidefinite matrices $\begin{bmatrix} 1 & 0 \\ 0 & 0 \end{bmatrix}$ and $\begin{bmatrix} 0 & 0 \\ 0 & 1 \end{bmatrix}$

and

(c) the positive vector $(\frac{1}{2}, \frac{1}{2}, 1)$ corresponds to the unit trace positive semidefinite matrix $\begin{bmatrix} \frac{1}{2} & \frac{1}{2} \\ \frac{1}{2} & \frac{1}{2} \end{bmatrix}$.

The key to establishing (iv) in Definition 1 with the above definition of Δ is to note that the unit trace matrix $\begin{bmatrix} \alpha_0 & \frac{1}{2}\alpha_2 \\ \frac{1}{2}\alpha_2 & 1-\alpha_0 \end{bmatrix}$ is positive semidefinite if and only if

$$\alpha_2^2 \le 1 \quad \text{and} \quad 0 \le \frac{1}{2} - \frac{1}{2}\sqrt{1-\alpha_2^2} \le \alpha_0 \le \frac{1}{2} + \frac{1}{2}\sqrt{1-\alpha_2^2} \le 1\,. \tag{8}$$

For $(x_1, x_2) \in \mathcal{P}$, $x_2 = 0$ and expression (3) combined with (6) gives

$$g = (2\alpha_0 x_1, \frac{1}{p}\alpha_2)\,.$$

Suppose $x_1 \neq 0$. Then, since $f_0(x_1, 0) = x_1^2 > 0 = f_1(x_1, 0)$, complementary slackness requires $\alpha_0 = 1 - \alpha_1 = 1$ and then the positive semidefiniteness requirement (8) gives $\alpha_2 = 0$. Thus,

$$g = (2x_1, 0) \quad \text{if}\, x_1 \neq 0$$

and, from (4) and (5), this is the correct gradient when $x_1 \neq 0$. We note in passing that this gradient is the reason that the zero function f_1 is needed in the structure of $\mathcal{E}_p$ given by (6).

Alternatively, suppose $x_1 = 0$. Then

$$g = (0, \frac{1}{p}\alpha_2) = \begin{cases} (0, \alpha_2) & \text{if}\, p = 1 \\ (0, \frac{1}{2}\alpha_2)\ \text{with}\, \alpha_2 \ge 0 & \text{if}\, p = 2\,. \end{cases}$$

Since in this case complementary slackness places no zero restriction on α_0 or α_1, (8) allows $-1 \le \alpha_2 \le 1$ and we obtain the set of subgradients given by (4) and (5), respectively, when $x_1 = 0$.

Thus, condition (iv) holds and both $\mathcal{E}_1$ and $\mathcal{E}_2$ have pdg structure about $\bar{x} = (0,0)^T$.

3 $\mathcal{VU}$-space decomposition theory

$\mathcal{VU}$-theory from [LOS97] proposes to overcome f's nondifferentiability at a given $\bar{x} \in \mathbb{R}^n$ (not necessarily a minimizer of f) by making a space decomposition as follows:

Given an arbitrary subgradient $g \in \partial f(\bar{x})$, consider the orthogonal subspaces

$$\mathcal{V} := \operatorname{lin}(\partial f(\bar{x}) - g) \qquad \text{and} \qquad \mathcal{U} := \mathcal{V}^{\perp}, \tag{9}$$

so that $\mathbb{R}^n = \mathcal{U} \oplus \mathcal{V}$. Here $\operatorname{lin} Y$ denotes the linear hull of a set Y. The subspace $\mathcal{V}$ approximates locally a subset on which f's nonsmoothness is concentrated (the graph of f is "$\mathcal{V}$-shaped"). On its orthogonal complement the graph of f appears to be smooth or "$\mathcal{U}$-shaped".

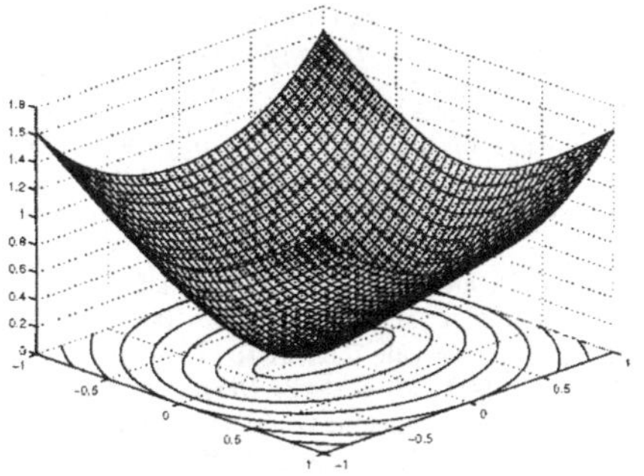

Figure 1: Graph of $\mathcal{E}_1$

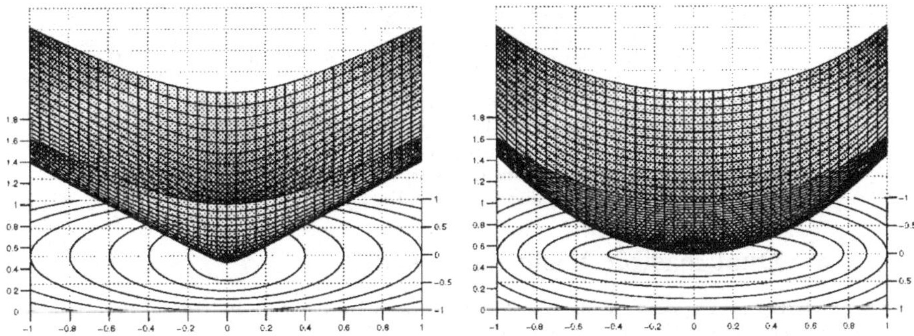

Figure 2: $\mathcal{V}$ and $\mathcal{U}$ views, respectively, of the graph of $\mathcal{E}_1$

For our example functions at $\bar{x} = (0,0)^T$, recalling from (4) and (5) that

$$\partial \mathcal{E}_p(0,0) = \begin{cases} \{0\} \times [-1,1] & \text{if}\, p = 1 \\ \{0\} \times [0, \frac{1}{2}] & \text{if}\, p = 2\,, \end{cases} \tag{10}$$

we see that the relevant subspaces are

$$\mathcal{V} = 0 \times \mathbb{R} \quad \text{and} \quad \mathcal{U} = \mathbb{R} \times 0\,. \tag{11}$$

Figures 1 and 2 illustrate this decomposition for $\mathcal{E}_1$.

The apparent smoothness of f on the $\mathcal{U}$-subspace suggests that it may have some kind of Hessian. Actually, it is the *$\mathcal{U}$-Lagrangian* that contains the relevant second order information about f near $\bar{x}$:

$$L_{\mathcal{U}}(u) = \min_{v \in \mathcal{V}} \{f(\bar{x} + u \oplus v) - \langle \bar{g}_{\mathcal{V}}, v \rangle_{\mathcal{V}}\} \quad \text{for } u \in \mathcal{U}\,. \tag{12}$$

Note that $L_{\mathcal{U}}(\cdot)$ depends on both $\bar{x} \in \mathbb{R}^n$ and $\bar{g}_{\mathcal{V}}$, the $\mathcal{V}$-component of a given subgradient $\bar{g}^T \in \partial f(\bar{x})$.

The $\mathcal{U}$-Lagrangian is a convex function that is differentiable at $u = 0$ with $\nabla L_{\mathcal{U}}(0) = \bar{g}_{\mathcal{U}}$, the $\mathcal{U}$-component of $\bar{g}$. Since, from (9), the $\mathcal{U}$-component of any subgradient $g \in \partial f(\bar{x})$ is $\bar{g}_{\mathcal{U}}$, we call $\nabla L_{\mathcal{U}}(0)$ the *$\mathcal{U}$-gradient* of f. When $L_{\mathcal{U}}$ has a Hessian at 0, we say that $\nabla^2 L_{\mathcal{U}}(0)$ is the *$\mathcal{U}$-Hessian* of f at $\bar{x}$ relative to $\bar{g}_{\mathcal{V}}$ and we denote it by $\mathrm{H}_{\mathcal{U}} f(\bar{x})$. So the dependence on $\bar{g}_{\mathcal{V}}$ must be kept in mind. For our example functions we will choose $\bar{g}^T = (0,0) \in \partial \mathcal{E}_p(0,0)$ so that $\bar{g}_{\mathcal{V}} = 0$.

Smoothness of $L_{\mathcal{U}}$ depends on f through the *W*inning set of minimizers

$$W(u) := \operatorname{Argmin}_{v \in \mathcal{V}} \{f(\bar{x} + u \oplus v) - \langle \bar{g}_{\mathcal{V}}, v \rangle_{\mathcal{V}}\}\,. \tag{13}$$

Indeed, whenever $\nabla^2 L_{\mathcal{U}}(0)$ exists, those $\bar{x} + u \oplus w$ with $w \in W(u)$ yield the following second order expansion of f:

$$\begin{aligned} f(\bar{x} + u \oplus w) &= f(\bar{x}) + \langle \nabla L_{\mathcal{U}}(0), u \rangle_{\mathcal{U}} + \langle \bar{g}_{\mathcal{V}}, w \rangle_{\mathcal{V}} + \frac{1}{2} \left\langle u, \nabla^2 L_{\mathcal{U}}(0) u \right\rangle_{\mathcal{U}} + o(|u|^2) \\ &= f(\bar{x}) + \langle \bar{g}, u \oplus w \rangle + \frac{1}{2} \langle u, \mathrm{H}_{\mathcal{U}} f(\bar{x}) u \rangle_{\mathcal{U}} + o(|u \oplus w|^2)\,. \end{aligned} \tag{14}$$

Accordingly, we call such $\bar{x} + u \oplus w \in \bar{x} + u \oplus W(u)$ *smooth trajectories.* In the sequel we give abbreviated results from [MS98] which show how to determine smooth trajectories and a $\mathcal{U}$-Hessian for a pdg structured function satisfying certain so-called $\mathcal{V}$-optimality conditions. These objects depend on the primal functions f_i and φ_l and on the dual multiplier set Δ introduced in Definition 1.

4 Dual feasible basic index sets

When f has pdg structure about $\bar{x}$ it is possible to characterize $\mathcal{V}$ in terms of the gradients of the primal functions in Definition 1 as follows:

$$\mathcal{V} = \operatorname{lin}\left(\{\nabla f_i(\bar{x}) - \nabla f_0(\bar{x})\}_{i=0}^{m_1} \cup \{\nabla \varphi_\ell(\bar{x})\}_{\ell=1}^{m_2}\right)\,. \tag{15}$$

This is Lemma 4.1 in [MS98]. For our example functions $\mathcal{E}_p$ with gradients given by (6) at $\bar{x} = (0,0)^T$ (15) becomes

$$\mathcal{V} = \operatorname{lin}\left(\left\{(0,0)\,, (0,0)\,, \frac{1}{p}(0,1)\right\}\right) = \operatorname{lin}((0,1))\,,$$

which agrees with (11).

A linearly independent set of generators from among the vectors $\nabla f_i(\bar{x}) - \nabla f_0(\bar{x})$ and $\nabla \varphi_\ell(\bar{x})$ is crucial for constructing a smooth trajectory tangent to $\mathcal{U}$. The following definition identifies the corresponding indices.

Definition 2 An index set K of the form $K = K_f \cup K_\varphi \subseteq \{0, 1, \ldots, m_1\} \cup \{m_1 + 1, \ldots, m_1 + m_2\}$ with $0 \in K_f$ is called a *basic index set* if

(i) the $n+1$ dimensional vectors $\{(\nabla f_i(\bar{x}), 1)\}_{i \in K_f} \cup \{(\nabla \varphi_{i-m_1}(\bar{x}), 0)\}_{i \in K_\varphi}$ are linearly independent.

K is called a *dual feasible basic index set* relative to $\bar{g}^T \in \partial f(\bar{x})$ if, in addition,

(ii) the linear system with variables α_i

$$\begin{array}{lcl} \sum_{i \in K_f} \alpha_i \nabla f_i(\bar{x})^T + \sum_{i \in K_\varphi} \alpha_i \nabla \varphi_{i-m_1}(\bar{x})^T & = & \bar{g} \\ \sum_{i \in K_f} \alpha_i & = & 1 \end{array}$$

has a (unique) solution $\{\alpha_i = \bar{\alpha}_i\}_{i \in K}$ such that together with $\bar{\alpha}_i := 0$ for all $i \notin K$

$$(\bar{\alpha}_0, \bar{\alpha}_1, \ldots, \bar{\alpha}_{m_1+m_2}) \in \Delta .$$

□

The definition assumes that, if necessary, the f_i's are re-indexed so that the nonempty set K_f contains $i = 0$.

We adopt the following convention: given a set of column vectors $\{v_1, \ldots, v_\ell\}$, we denote the corresponding matrix by $[\{v_1, \ldots, v_\ell\}]$.

Thus, if K is a basic index set, the linear independence condition in Definition 2 (i) implies that the $n \times (|K_f| - 1 + |K_\varphi|)$ matrix

$$\bar{V} := [\{\nabla f_i(\bar{x})^T - \nabla f_0(\bar{x})^T\}_{0 \neq i \in K_f} \cup \{\nabla \varphi_{i-m_1}(\bar{x})^T\}_{i \in K_\varphi}] \tag{16}$$

has full column rank. Moreover, $\bar{V}$ is a basis matrix for the subspace of $\mathcal{V}$ defined by

$$\mathcal{V}_K := \text{lin}\left(\{\nabla f_i(\bar{x}) - \nabla f_0(\bar{x})\}_{i \in K_f} \cup \{\nabla \varphi_{i-m_1}(\bar{x})\}_{i \in K_\varphi}\right),$$

having dimension $|K_f| - 1 + |K_\varphi|$. Note that if K is a singleton, the matrix $\bar{V}$ is vacuous and $\mathcal{V}_K$ is only the zero vector.

We let the augmented matrix $[\bar{V}|\bar{Z}]$ be a basis matrix for $\mathcal{V}$, where $\bar{Z}$ is vacuous if $\mathcal{V}_K \subseteq \mathcal{V}$ is the whole subspace. Hence, both $\bar{V}$ and $\bar{Z}$ depend on the particular basic index set K under consideration. Furthermore, letting $\bar{U}$ be a basis matrix for $\mathcal{U}$, $[\bar{U}|\bar{V}|\bar{Z}]$ is a basis matrix for $\mathbb{R}^n$.

For each of our prototype functions at $\bar{x} = (0,0)^T$ there are two basic index sets of interest, one having K_φ empty:

$$K_0 := \{0\} \quad \text{and} \quad K_{0,2} := \{0, 2\} .$$

Both of these sets are dual feasible relative to $\bar{g} = (0,0)^T \in \partial\mathcal{E}_p(0,0)$, because in all cases Definition 2 (ii) yields $\bar{\alpha} := (\bar{\alpha}_0, \bar{\alpha}_1, \bar{\alpha}_2) = (1,0,0)$ which is an element of the dual multiplier set defined in (7). However, note that the corresponding matrices $\bar{V}$ from (16) are vacuous and $\begin{bmatrix} 0 \\ \frac{1}{p} \end{bmatrix}$, respectively, so that

$$\mathcal{V}_{K_0} = \{(0,0)^T\} \quad \subsetneq \quad \mathcal{V}_{K_{0,2}} = \mathcal{V}. \tag{17}$$

5 Finding smooth trajectories

We already mentioned that, to give formulas for the $\mathcal{U}$-Lagrangian (12) and its derivatives, we need to identify smooth trajectories tangent to $\mathcal{U}$. We achieve this in two steps. First, based on the primal functions associated with our pdg structured f, we define a system of nonlinear equations, depending on a basic index set, whose solution generates a candidate for an element of $W(u)$. Then, we give conditions sufficient for producing a successful candidate.

5.1 Candidate trajectories

Rather than describing the whole set $W(u)$ from (13), we characterize candidates for being elements of this Winning set via an implicit function that provides the desired smoothness and tangency properties.

We do this, relative to a particular basic index set K, by changing notation and replacing $u \oplus v$ of Section 3 by $\bar{U}u + \bar{V}v + \bar{Z}z$, where $(u,v,z) \in \mathbb{R}^{\dim\mathcal{U}} \times \mathbb{R}^{\dim\mathcal{V}_K} \times \mathbb{R}^{\dim\mathcal{V}\setminus\mathcal{V}_K}$ and $[\bar{U}|\bar{V}|\bar{Z}]$ is a basis matrix for $\mathbb{R}^n$ as defined in Section 4. As in [MS98] we only consider $z = 0$, so for u near $0 \in \mathbb{R}^{\dim\mathcal{U}}$ we are interested in trajectories $x(u) \in B(\bar{x})$ of the form

$$x(u) := \bar{x} + \bar{U}u + \bar{V}v_K(u)\,, \tag{18}$$

where the function $v_K : \mathbb{R}^{\dim\mathcal{U}} \mapsto \mathbb{R}^{\dim\mathcal{V}_K}$ is defined in the next theorem which combines Theorems 5.1 and 5.2 from [MS98].

Theorem 3 *Let f have* pdg *structure about $\bar{x}$ and suppose $K = K_f \cup K_\varphi$ is a nonsingleton basic index set as described in Definition 2(i). For u sufficiently near $0 \in \mathbb{R}^{\dim\mathcal{U}}$*

(i) the nonlinear system with variables (u,v)

$$\begin{cases} f_i(\bar{x}+\bar{U}u+\bar{V}v) \quad -f_0(\bar{x}+\bar{U}u+\bar{V}v) = 0\,, & 0 \neq i \in K_f \\ \varphi_{i-m_1}(\bar{x}+\bar{U}u+\bar{V}v) = 0\,, & i \in K_\varphi \end{cases}$$

has a unique solution $v = v_K(u)$ with $\bar{V}v_K(u)$ in the subspace $\mathcal{V}_K \subseteq \mathcal{V}$. Also, $v_K(u)$ and the corresponding trajectory $x(u)$ have continuous Jacobians $Jv_K(u)$ and $Jx(u)$, respectively. In particular, $v_K(0) = 0$, $x(0) = \bar{x}$, $Jv_K(0) = 0$ and $Jx(0) = \bar{U}$, so $v_K(u) = o(|u|)$ and $x(u)$ is tangent to $\mathcal{U}$ at $\bar{x}$.

(ii) If, in addition, $K = K_f \cup K_\varphi$ is a dual feasible basic index set relative to $\bar{g}^T \in \partial f(\bar{x})$, with corresponding multipliers $\{\bar{\alpha}_i\}$ from Definition 2(ii), then the linear system with variables α_i

$$\begin{cases} \bar{V}^T \left(\sum_{i\in K_f} \alpha_i \nabla f_i(x(u))^T + \sum_{i\in K_\varphi} \alpha_i \nabla \varphi_{i-m_1}(x(u))^T - \bar{g}\right) & = 0 \\ \sum_{i\in K_f} \alpha_i & = 1 \\ \alpha_i & = 0 \quad \text{for } i \notin K \end{cases}$$

has a unique solution $\alpha = \alpha(u)$ satisfying $\alpha_i(0) = \bar{\alpha}_i$ for $i = 0, 1, \cdots, m_1 + m_2$. □

Explicit expressions for the Jacobians $Jv_K(u)$ and $Jx(u)$ and the multipliers $\alpha_i(u)$ are given in [MS98]. When K is a singleton, $v_K(u)$ is vacuous and $x(u)$ is simply defined to be $\bar{x} + \bar{U}u$.

A result of Theorem 3 is that f will have a second order expansion as in (14) written with $w = \bar{V}v_K(u)$, if we show that $\bar{V}v_K(u) \in W(u)$ and that a $\mathcal{U}$-Hessian exists. Our concluding Theorem 5, depending on Definition 4 below, establishes conditions on a basic index set K for such an expansion along $x(u)$ to hold.

Consider the structure functions for $\mathcal{E}_p$ given in (6) and take $\bar{U} = \begin{bmatrix} 1 \\ 0 \end{bmatrix}$ as the basis matrix for $\mathcal{U}$ corresponding to $\bar{x} = (0,0)^T$. Then solving the two systems in Theorem 3 for both dual feasible index sets K_0 and $K_{0,2}$ relative to $\bar{g} = (0,0)^T \in \partial\mathcal{E}_p(0,0)$ gives

$$v_K(u) = \begin{cases} \text{vacuous} & \text{if } K = K_0 \\ 0 & \text{if } K = K_{0,2} \end{cases}, \quad x(u) = (u, 0)^T \quad \text{and } \alpha(u) = (1, 0, 0) \tag{19}$$

for all $u \in \mathbb{R}$.

5.2 $\mathcal{V}$-optimality conditions

Following [MS98], to express the $\mathcal{U}$-Hessian of f in terms of the structure Hessians $\nabla^2 f_i$ and $\nabla^2 \varphi_\ell$ on a smooth trajectory $x(u)$, we need some conditions in addition to pdg structure. They play a role similar to that of constraint qualification conditions in nonlinear programming.

We formulate these qualifications in the form of conditions to be satisfied by a dual feasible basic index set.

Definition 4 Consider a dual feasible basic index set $K = K_f \cup K_\varphi$ relative to $\bar{g}^T \in \partial f(\bar{x})$. Let $x(u)$ and $\alpha(u)$ be the associated trajectory and multiplier functions from Theorem 3 (i) and (ii), respectively. Consider the following conditions holding for all u sufficiently near $0 \in \mathbb{R}^{\dim \mathcal{U}}$:

$$\text{(i)} \begin{cases} f_i(x(u)) \le f_0(x(u)) & \text{for all } i \in \{1, \ldots, m_1\} \backslash K_f \\ \varphi_{i-m_1}(x(u)) = 0 & \text{for all } i \in \{m_1+1, \ldots, m_1+m_2\} \backslash K_\varphi . \end{cases}$$

(ii) $(\alpha_0(u), \alpha_1(u), \ldots, \alpha_{m_1+m_2}(u)) \in \Delta$ where Δ is the dual multiplier set from Definition 1.

(iii) $g(u) - \bar{g}$ is orthogonal to $\mathcal{V} \backslash \mathcal{V}_K$ where

$$g(u) := \sum_{i \in K_f} \alpha_i(u) \nabla f_i(x(u))^T + \sum_{i \in K_\varphi} \alpha_i(u) \nabla \varphi_{i-m_1}(x(u))^T . \qquad (20)$$

Note that (i) and Theorem 3(i) imply that for $u \in \mathbb{R}^{\dim \mathcal{U}}$ small enough

$$\begin{cases} f_i(x(u)) \leq f_0(x(u)) & \text{for all } i \in \{0, \ldots, m_1\} \\ \varphi_\ell(x(u)) = 0 & \text{for all } \ell \in \{1, \ldots, m_2\} \end{cases}$$

and, hence, from Definition 1, $x(u) \in \mathcal{P}$ and $f(x(u)) = f_0(x(u))$. Also, (iii) and Theorem 3 (ii) imply that $g(u) - \bar{g}$ is orthogonal to $\mathcal{V}$ which clearly holds if $\mathcal{V}_K = \mathcal{V}$. Accordingly, we say that K satisfies **transversality** if $\mathcal{V}_K = \mathcal{V}$, and relative to $x(u)$, K satisfies

primal feasibility if (i) holds,

dual feasibility if (ii) holds and

weak transversality if (iii) holds.

□

For our prototype functions $\mathcal{E}_p$ and the two dual feasible basic index sets K_0 and $K_{0,2}$ having the same associated trajectory $x(u) = (u, 0)^T$ from (19), condition (i) holds, since it becomes

$$f_1(u, 0) = 0 \leq u^2 = f_0(u, 0) \quad \text{and, if } K = K_{0,2}, \quad \varphi_1(u, 0) = \frac{1}{p} 0 = 0 .$$

Dual feasibility is satisfied by both of these K-sets, because the common corresponding multiplier vector $\alpha(u) = (1, 0, 0)$ is in the Δ-set defined in (7). Finally, to check transversality, note that, from (20), $g(u) = (2u, 0)^T$. Thus, $g(u) - \bar{g} = (2u, 0)^T - (0, 0)^T \in \mathcal{U} = \mathcal{V}^T$ and we conclude from (17) that K_0 satisfies weak transversality, while $K_{0,2}$ satisfies the stronger transversality condition.

Our function $\mathcal{E}_1$ is an example of a maximum eigenvalue function that does not satisfy the even stronger transversality assumption of [Ous98], [Ous97] and [OW95], because the full index set $\{0, 1, 2\}$ is not a basic index set at $\bar{x} = (0, 0)^T$.

For $\mathcal{E}_2$, the subgradient $\bar{g}^T = (0, 0)$ is not in the relative interior of $\partial \mathcal{E}_2(0, 0)$ given by (10), so $\mathcal{E}_2$ is an example where the strong assumption of $\bar{g}^T$ being in the relative interior of $\partial f(\bar{x})$ is not required.

6 Expressions for the $\mathcal{U}$-Lagrangian and its derivatives

The following expressions for the $\mathcal{U}$-Lagrangian (12) and its derivatives are established in the proof of Theorem 6.3 in [MS98]. They depend on the result that the candidate $\bar{V}v_K(u)$ from Theorem 3 is an element of the Winning set $W(u)$ from (13) when K satisfies all three $\mathcal{V}$-optimality conditions in Definition 4.

Theorem 5 *Let f have* pdg *structure about $\bar{x}$. Suppose $\bar{K} = \bar{K}_f \cup \bar{K}_\varphi$ is a dual feasible basic index set relative to $\bar{g}^T \in \partial f(\bar{x})$ satisfying primal and dual feasibility and weak transversality relative to $x(u) = \bar{x} + \bar{U}u + \bar{V}v_{\bar{K}}(u)$ where $v_{\bar{K}}(u)$ is defined in Theorem 3 if $K = \bar{K}$ is not a singleton, and $\bar{V}v_{\bar{K}}(u) := 0$ otherwise.*

Then, for u sufficiently near $0 \in \mathbb{R}^{\dim \mathcal{U}}$:

(i) The vector $\bar{V}v_{\bar{K}}(u)$ is an element of $W(u)$. Equivalently, $x(u) \in \bar{x}+\bar{U}u+W(u)$.

(ii) (a) If $\bar{K}_\varphi$ is nonempty, $0 = \varphi_{i-m_1}(x(u))$ for each $i \in \bar{K}_\varphi$.

(b) The $\mathcal{U}$-Lagrangian is given by

$$\begin{aligned} L_{\mathcal{U}}(u) &= f(x(u)) - \bar{g}^T\bar{V}v_{\bar{K}}(u) \\ &= f_i(x(u)) - \bar{g}^T\bar{V}v_{\bar{K}}(u) \end{aligned}$$

for each $i \in \bar{K}_f$, where

$$\bar{g}^T = \sum_{i\in\bar{K}_f} \bar{\alpha}_i \nabla f_i(\bar{x}) + \sum_{i\in\bar{K}_\varphi} \bar{\alpha}_i \nabla \varphi_{i-m_1}(\bar{x}),$$

$\bar{\alpha}_i \geq 0$ for each $i \in \bar{K}_f$ and $\sum_{i\in\bar{K}_f} \bar{\alpha}_i = 1$.

In particular, $L_{\mathcal{U}}(0) = f(\bar{x})$.

(iii) The gradient of $L_{\mathcal{U}}$ is given by $\nabla L_{\mathcal{U}}(u) = g(u)^T\bar{U}$, where $g(u)$ is given in (20) with $K = \bar{K}$ and $\bar{U}$ is a basis matrix for $\mathcal{U}$.

In particular, $\nabla L_{\mathcal{U}}(0) = \bar{g}^T\bar{U} = g^T\bar{U}$ for all $g^T \in \partial f(\bar{x})$.

(iv) The Hessian of $L_{\mathcal{U}}$ is given by

$$\nabla^2 L_{\mathcal{U}}(u) = Jx(u)^T \left(\sum_{i\in\bar{K}_f} \alpha_i(u)\nabla^2 f_i(x(u)) + \sum_{i\in\bar{K}_\varphi} \alpha_i(u)\nabla^2\varphi_{i-m_1}(x(u)) \right) Jx(u).$$

In particular,

$$\nabla^2 L_{\mathcal{U}}(0) = \bar{U}^T \left(\sum_{i\in\bar{K}_f} \bar{\alpha}_i\nabla^2 f_i(\bar{x}) + \sum_{i\in\bar{K}_\varphi} \bar{\alpha}_i\nabla^2\varphi_{i-m_1}(\bar{x}) \right) \bar{U}.$$

□

Taking into account (6) and (19) together with $g(u) = (2u, 0)^T$ and $\bar{U} = \begin{bmatrix} 1 \\ 0 \end{bmatrix}$, Theorem 5 applied to $\mathcal{E}_p$ with $\bar{g} = (0,0)^T \in \partial\mathcal{E}_p(0,0)$ and with $\bar{K}$ equal to K_0 or $K_{0,2}$ gives the respective quadratic, linear and constant expressions for the $\mathcal{U}$-Lagrangian and its derivatives:

$$L_{\mathcal{U}}(u) = \mathcal{E}_p(x(u)) = u^2\,, \quad \nabla L_{\mathcal{U}}(u) = 2u\,, \quad \nabla^2 L_{\mathcal{U}}(u) = 2\,.$$

References

[BTB93] A. Ben-Tal and M. Bendsøe. A new method for optimal truss topology design. *SIAM Journal on Optimization*, 3:322–358, 1993.

[BTZ82] A. Ben-Tal and J. Zowe. Necessary and sufficient optimality conditions for a class of nonsmooth minimization problems. *Mathematical Programming*, 24(1):70–91, 1982.

[CC90] R. Cominetti and R. Correa. A generalized second-order derivative in nonsmooth optimization. *SIAM Journal on Control and Optimization*, 28(4):789–809, 1990.

[Cla83] F.H. Clarke. *Optimization and Nonsmooth Analysis.* Wiley, 1983. Reprinted by Siam, 1990.

[Com91] R. Cominetti. On pseudo-differentiability. *Trans. Amer. Math. Soc.*, 2324:843–865, 1991.

[HU79] J.-B. Hiriart-Urruty. Refinements of necessary optimality conditions in nondifferentiable programming I. *Applied Mathematics and Optimization*, 5:63–82, 1979.

[HU82] J.-B. Hiriart-Urruty. Refinements of necessary optimality conditions in nondifferentiable programming II. *Mathematical Programming Study*, 19:120–139, 1982.

[Iof91] A.D. Ioffe. Variational analysis of a composite function: a formula for the lower second order epi-derivative. *Journal of Mathematical Analysis and Applications*, pages 379–405, 1991.

[LM82] C. Lemaréchal and R. Mifflin. Global and superlinear convergence of an algorithm for one-dimensional minimization of convex functions. *Mathematical Programming*, 24:241–256, 1982.

[LOS97] C. Lemaréchal, F. Oustry, and C. Sagastizábal. The $\mathcal{U}$-Lagrangian of a convex function. *Transactions of the AMS*, 1997. Accepted for publication.

[MS98] R. Mifflin and C. Sagastizábal. On $\mathcal{VU}$-theory for functions with primal-dual gradient structure. Submitted, 1999.

[MS99] R. Mifflin and C. Sagastizábal. $\mathcal{VU}$-decomposition derivatives for convex max-functions. In R. Tichatschke and M. Théra, editors, *Ill-posed problems and variational inequalities*, Lecture Notes. Springer, 1999. Accepted for publication.

[Ous97] F. Oustry. A second-order bundle method to minimize the maximum eigenvalue function. Submitted, 1997.

[Ous98] F. Oustry. The $\mathcal{U}$-Lagrangian of the maximum eigenvalue function. *SIAM Journal on Optimization*. To appear.

[OW95] M. L. Overton and R.S. Womersley. Second derivatives for optimizing eigenvalues of symmetric matrices. *Journal of Mathematical Analysis and Applications*, 3:667–718, July 1995.

[Pen94] J.-P. Penot. Subhessians, superhessians and conjugation. *Nonlinear Analysis: Theory, Methods and Applications*, 23(6):689–702, 1994.

[Pol90] R.A. Poliquin. Proto-differentiation of subgradient set-valued mappings. *Canadian Journal of Mathematics*, 42(3):520–532, 1990.

[Rob87] S.M. Robinson. Local structure of feasible sets in nonlinear programming, part III: stability and sensitivity. *Mathematical Programming Study*, 30:45–66, 1987.

[Roc88] R.T. Rockafellar. First and second-order epi-differentiability in nonlinear programming. *Trans. Amer. Math. Soc.*, 307:75–107, 1988.

[Roc89a] R.T. Rockafellar. Proto-differentiability of set-valued mappings and its applications in optimization. *Annales de l'Institut Henri Poincaré, Analyse non linéaire*, 6:449–482, 1989.

[Roc89b] R.T. Rockafellar. Second-order optimality conditions in nonlinear programming obtained by way of epi-derivatives. *Mathematics of Operations Research*, 14:462–484, 1989.

[RW98] R.T. Rockafellar and R.J.-B. Wets. *Variational Analysis*. Number 317 in Grund. der math. Wiss. Springer-Verlag, 1998.

Quadratic and Multidimensional Assignment Problems

Panos M. Pardalos[1] (pardalos@ophelia.ise.ufl.edu)
Department of Industrial and Systems Engineering, University of Florida, Gainesville, FL 32611 USA

Leonidas S. Pitsoulis[1] (leonidas@dragon.princeton.edu)
Department of Operations Research and Financial Engineering, Princeton University, Princeton, NJ 08544 USA

Abstract

In this paper we discuss various aspects regarding nonlinear assignment problems. In particular, we focus on Quadratic Assignment Problems and data association problems which are formulated as Multidimensional Assignment Problems. Computational results with two applications, the turbine balancing problem and multitarget multisensor tracking problem, are presented.

Keywords: Combinatorial optimization, assignment problems, heuristics, turbine balancing problem, data association problem.

1 Introduction

Nonlinear Assignment Problems (NAP's) are NP-hard combinatorial optimization problems that are natural extensions of the classical Linear Assignment Problem (LAP). NAPs have numerous applications in different fields, ranging from the assignment of facilities to locations in location theory, to the tracking of elementary particles in high energy physics. We begin this section with the most basic of all assignment problems, the linear assignment problem. Consider that we are given n facilities and locations, and a matrix $\mathbb{R}^{n\times n} \ni C := (c_{ij})$ where c_{ij} represents the cost

[1]Supported in part by NSF grant EIA-98-72509 and USAF grant F08635-92-C-0032

G. Di Pillo and F. Giannessi (eds.), Nonlinear Optimization and Related Topics, 235–256.

of placing facility i to location j. The objective is to find an optimal assignment of facilities to locations such that the total placement cost is minimized. Using the decision variables $x_{ij} \in \{0,1\}$ defined by

$$x_{ij} = \begin{cases} 1 & \text{if facility } i \text{ is assigned to location } j, \\ 0 & \text{otherwise,} \end{cases}$$

we can formulate the LAP as an integer $0-1$ program:

$$\begin{aligned} \min \quad & \sum_{i,j=1}^{n} c_{ij}x_{ij} \\ \text{s.t.} \quad & \sum_{i=1}^{n} x_{ij} = 1, \quad j = 1,2,\ldots,n, \\ & \sum_{j=1}^{n} x_{ij} = 1, \quad i = 1,2,\ldots,n, \\ & x_{ij} \in \{0,1\}, \quad i,j = 1,2,\ldots,n. \end{aligned}$$

Equivalently using permutations we can formulate the LAP as

$$\min_{p\in\Pi_N} \sum_{i=1}^{n} c_{ip(i)}, \tag{1}$$

where Π_N is the set of all permutations $p : N \to N$, and $N = \{1,2,\ldots,n\}$. The LAP is a classical optimization problem and can be solved very efficiently [1]. Nonlinear assignment problems are extensions of the LAP, and they can be divided into two major classes, M-adic assignment problems introduced by Lawler [17], and Multidimensional Assignment Problems (MAP's) introduced by Pierskalla [28].

In the M-adic assignment problem class, the dimension of the cost array in (1) becomes $2M, M \geq 2$, and we permute over M indices of the cost array C in the objective function. For example the 3-adic assignment problem can be formulated as

$$\min_{p\in\Pi_N} \sum_{i=1}^{n}\sum_{j=1}^{n}\sum_{k=1}^{n} c_{ijkp(i)p(j)p(k)}, \tag{2}$$

where $C := (c_{ijklms}) \in \mathbb{R}^{n^{2M}}, M = 3$. In a similar way we can define the quadratic ($M = 2$), and the biquadratic ($M = 4$) assignment problems. In the MAP class, the dimension of the cost array in (1) becomes $M \geq 3$, and we sum over n values of C. However C is permuted over $M-1$ indices by different permutations. For example the three dimensional assignment problem can be formulated as

$$\min_{p,q\in\Pi_N} \sum_{i=1}^{n} c_{ip(i)q(i)}, \tag{3}$$

where $C := (c_{ijk}) \in \mathbb{R}^{n^M}, M = 3$.

In section 2 we will present two representatives of the M-adic assignment problem class, specifically the Quadratic Assignment Problem (QAP) and the BiQuadratic Assignment Problem (BiQAP). Formulations and applications for both problems will be mentioned, and a recent application of the QAP which involves the balancing of turbine engines will be described in detail. In section 3 the general formulation of the multidimensional assignment problem and its special case the three-dimensional assignment problem are given. A major application of the multidimensional assignment problem in target tracking will be described in detail, with preliminary computational results.

2 The Quadratic Assignment Problem

Consider the 2-adic assignment problem, where given a four-dimensional cost array $C = (c_{ijkl})$ the problem is

$$\min_{p \in \Pi_N} \sum_{i=1}^{n} \sum_{j=1}^{n} c_{ijp(i)p(j)}. \tag{4}$$

The problem in (4) is the general formulation of the QAP introduced by Lawler [17]. However the above formulation is rarely used since it has few applications. The QAP is mostly known by its Koopmans and Beckman [14] formulation which is a special case of (4). Consider the problem of allocating a set of n facilities to a set of n locations, with the cost being a function of the distance and flow between the facilities. The objective is to assign each facility to a location such that the total cost is minimized. Specifically, we are given two $n \times n$ input matrices with real elements $F = (f_{ij})$ and $D = (d_{kl})$, where f_{ij} is the flow between the facility i and facility j and d_{kl} is the distance between the location k and location l. The Koopmans-Beckmann version of the QAP can be formulated as follows:

$$\min_{p \in \Pi_N} \sum_{i=1}^{n} \sum_{j=1}^{n} f_{ij} d_{p(i)p(j)}, \tag{5}$$

where Π_N is the set of all permutations $p : N \to N$. Each individual product $f_{ij} d_{p(i)p(j)}$ is the cost of assigning facility i to location $p(i)$ and facility j to location $p(j)$. In the context of facility location the matrices F and D are symmetric with zeros in the diagonal, and both matrices are nonnegative. We will refer to the Koopmans-Beckman formulation as the *standard* QAP. Clearly, a standard QAP can be formulated as a general QAP by setting $c_{ijkl} := f_{ij} d_{kl}$ for all i, j, k, l. Although extensive research has been done for more than three decades, the QAP, in contrast with its linear counterpart the LAP, remains one of the hardest optimization problems and no exact algorithm can solve problems of size $n > 20$ in reasonable computational time. In fact, Sahni and Gonzalez [32] have shown that the QAP is NP-hard and that even finding an approximate solution within some constant factor from the optimal solution cannot be done in polynomial time unless P=NP.

In addition to facility layout problems, the QAP appears in applications such as backboard wiring, computer manufacturing, scheduling, process communications, turbine balancing, and many others. Applications in other areas including data association problems and multiprocessing can be found in the books by Hubert [13] and Bokhari [4]. For a detailed treatment of the QAP the reader is referred to the recent survey by Burkard, Çela, Pardalos and Pitsoulis [6], and to the book by Pardalos, Rendl and Wolkowicz [26].

2.1 Different Formulations

In this section we will present several formulations of the QAP for the purpose of providing an insight to the nature of the problem. Almost all of the formulations presented are the result of the effort of different researchers to view the structural characteristics of the problem from a different angle.

In the formulation (5), we can associate with each $p \in \Pi_N$ a 0-1 matrix $X = (x_{ij})_{n \times n}$, such that

$$x_{ij} = \begin{cases} 1 & \text{if } p(i) = j, \\ 0 & \text{otherwise.} \end{cases}$$

In other words the elements of X must satisfy the following *assignment constraints*

$$\begin{aligned} \sum_{i=1}^{n} x_{ij} &= 1, & j &= 1, 2, \dots, n, \\ \sum_{j=1}^{n} x_{ij} &= 1, & i &= 1, 2, \dots, n, \\ x_{ij} &\in \{0, 1\}, & i, j &= 1, 2, \dots, n, \end{aligned}$$

which define the vertices of the well known assignment polytope (other names include: the Birkhoff polytope, the transportation polytope, the perfect matching polytope of $K_{n,n}$ etc.). The matrix X is also called a permutation matrix, and we denote the set of all such permutation matrices by $\mathbf{X}_n \subseteq \{0,1\}^{n \times n}$. Observe that there is one-to-one correspondence between Π_N and $\mathbf{X}_n$, and the cardinality of both sets in $n!$.

Koopmans and Beckmann introduced the standard QAP as a *quadratic 0-1 integer* program in [14]. A solution to a QAP with input matrices F and D is represented by a permutation matrix $X = (x_{ij})$ where

$$x_{ij} = \begin{cases} 1 & \text{if facility } i \text{ is assigned to location } j, \\ 0 & \text{otherwise.} \end{cases}$$

The problem can be represented then as an integer program with a quadratic objective function (hence the name quadratic assignment problem) as follows

$$\min \quad \sum_{i=1}^{n}\sum_{j=1}^{n}\sum_{k=1}^{n}\sum_{l=1}^{n} f_{ij}d_{kl}x_{ik}x_{jl} \tag{6}$$

$$\text{s.t.} \quad \sum_{i=1}^{n} x_{ij} = 1, \quad j = 1, 2, \ldots, n, \tag{7}$$

$$\sum_{j=1}^{n} x_{ij} = 1, \quad i = 1, 2, \ldots, n, \tag{8}$$

$$x_{ij} \in \{0, 1\}, \quad i, j = 1, 2, \ldots, n. \tag{9}$$

The standard QAP can be equivalently formulated in a more compact form if we define an *inner product*. More specifically, let the inner product between matrices $A, B \in \mathbb{R}^{m\times n}$ be defined as

$$\langle A, B\rangle := \sum_{i=1}^{m}\sum_{j=1}^{n} a_{ij}b_{ij}.$$

Note that given some $A \in \mathbb{R}^{n\times n}$, a permutation $p \in \Pi_N$ and the associated permutation matrix $X \in \mathbf{X}_n$, then

$$\left.\begin{array}{rl} AX^T: & \text{permutes the columns of } A \\ XA: & \text{permutes the rows of } A \end{array}\right\} \Rightarrow XAX^T = (a_{p(i)p(j)})_{n\times n}. \tag{10}$$

We can then formulate the standard QAP alternatively as

$$\begin{array}{ll} \min & \langle F, XDX^T\rangle \\ \text{s.t.} & X \in \mathbf{X}_n. \end{array} \tag{11}$$

Another formulation of the QAP which reflects the combinatorial nature of the problem utilizes the *Kronecker Product*. Let $A \in \mathbb{R}^{m\times n}$ and $B \in \mathbb{R}^{p\times q}$. Then the Kronecker Product of the matrices A and B is defined as

$$A \otimes B := \begin{pmatrix} a_{11}B & a_{12}B & \cdots & a_{1n}B \\ \vdots & \vdots & \ddots & \vdots \\ a_{m1}B & a_{m2}B & \cdots & a_{mn}B \end{pmatrix}.$$

That is, the $mp \times nq$ matrix formed from all possible pairwise element products of A and B. If we let $vec(X) \in \mathbb{R}^{n^2}$ be the vector formed by the columns of a permutation matrix X, the standard QAP can be formulated as

$$\begin{array}{ll} \min & vec(X)^T(F \otimes D)vec(X) \\ \text{s.t.} & X \in \mathbf{X}_n. \end{array} \tag{12}$$

Although operations using the Kronecker product have been studied in detail [12], the above formulation is rarely used for studying the QAP. Using the Kronecker product, Lawler in [17], provided an alternative formulation for the general QAP as a generalized $n^2 \times n^2$ LAP. An $n^2 \times n^2$ matrix C is constructed from the n^4 costs c_{ijkl}, such that the $(ijkl)^{th}$ element corresponds to the $((i-1)n+k, (j-1)n+l)^{th}$ element of C. The QAP then is equivalent to a generalized LAP of size n^2 with C as the cost matrix, and with the additional constraint that the $n^2 \times n^2$ permutation matrix which defines a feasible solution, must be the Kronecker product of two permutation matrices of size n. In other words the QAP is equivalent to

$$\begin{aligned} \min \quad & \langle C, Y\rangle \\ \text{s.t.} \quad & Y = X \otimes X, \\ & X \in \mathbf{X}_n. \end{aligned} \tag{13}$$

The resulting LAP however cannot be solved efficiently because Y, although it is an $n^2 \times n^2$ permutation matrix, is constrained to have a special structure.

2.2 The BiQuadratic Assignment Problem

A generalization of the QAP is the Biquadratic Assignment Problem (BiQAP), which is essentially a special case of a 4-adic assignment problem with cost coefficients formed by the products of two four-dimensional arrays. More specifically, the 4-adic assignment problem is

$$\min_{p\in\Pi_N} \sum_{i=1}^{n}\sum_{j=1}^{n}\sum_{k=1}^{n}\sum_{l=1}^{n} c_{ijklp(i)p(j)p(k)p(l)}, \tag{14}$$

where $C := (c_{ijklmsqt}) \in \mathbb{R}^{n^{2M}}$, $M = 4$. Suppose that there exist two $n^4 \times n^4$ arrays $F = (f_{ijkl})$ and $D = (d_{msqt})$, such that $c_{ijklmsqt} = f_{ijkl}d_{msqt}$ for all i, j, k, l, m, s, q and t. Then we have the BiQAP which can be stated as

$$\min_{p\in\Pi_N} \sum_{i=1}^{n}\sum_{j=1}^{n}\sum_{k=1}^{n}\sum_{l=1}^{n} f_{ijkl}d_{p(i)p(j)p(k)p(l)},$$

where Π_N denotes the set of all permutations of $N = \{1, 2, \ldots, n\}$ as before. The major application of the BiQAP arises in Very Large Scale Integrated (VLSI) circuit design. The majority of VLSI circuits are sequential circuits, and their design process is composed of first translating the circuit specifications into a state transition table, modeling the system using finite state machines, and finally trying to find an encoding of the states such that the actual implementation is of minimum size [5]. The BiQAP can be equivalently stated as an 0-1 integer program:

$$\begin{aligned}
\min \quad & \sum_{i=1}^{n}\sum_{j=1}^{n}\sum_{k=1}^{n}\sum_{l=1}^{n}\sum_{m=1}^{n}\sum_{p=1}^{n}\sum_{s=1}^{n}\sum_{t=1}^{n} f_{ijkl}d_{mpst}x_{im}x_{jp}x_{ks}x_{lt} \\
\text{s.t.} \quad & \sum_{i=1}^{n} x_{ij} = 1, \quad j = 1,2,\ldots,n, \\
& \sum_{j=1}^{n} x_{ij} = 1, \quad i = 1,2,\ldots,n, \\
& x_{ij} \in \{0,1\}, \quad i,j = 1,2,\ldots,n.
\end{aligned}$$

Several heuristics have been developed for the BiQAP by Burkard and Çela in [5], and in particular deterministic improvement methods and variants of simulated annealing and tabu search. Computational experiments on test problems of size up to $n = 32$, with known optimal solutions, suggest that one version of simulated annealing is best among those tested. Mavridou, Pardalos, Pitsoulis and Resende develop a GRASP heuristic for solving the BiQAP in [19], which finds the optimal solution for all the test problems presented in [5].

2.3 The Turbine Balancing Problem

In this section we present the Turbine Balancing Problem (TBP), and show how this problem can be formulated as a QAP. Computational results of a proposed algorithm in [29] are also presented. Turbine engines, either hydraulic, gas or steam powered, are composed of "fans" which rotate at high speeds around the central axis of the turbine. Each fan (see Figure 1) is composed of a number of blades placed around the periphery of a disk at equally spaced points. The masses of the blades are not identical due to errors in the manufacturing process and wear through use. These masses can vary as much as $\pm 5\%$ ([20]). Therefore, the center of mass of the assembled fan depends on the positioning of the blades around its periphery. When the center of gravity of a fan does not coincide with its central axis of rotation then static unbalance occurs. The deviation of the center of gravity of the fan from the axis of rotation is referred as the unbalance of the turbine. Static unbalance is not desirable since it results in vibrations and shortens the life of the equipment.

The current practice in balancing turbines is to find a "good" arrangement of the blades based on a trial and error process and intuition, and correct any residual unbalance by adding weights. This process is time consuming, and it may take even weeks depending on the type of the turbine and the number of fans it contains. The process of balancing a turbine has to be performed in the assembly line, where the turbine is manufactured, and thereafter every time it is maintained. Therefore an efficient way to balance a turbine fast is considered to be very beneficial.

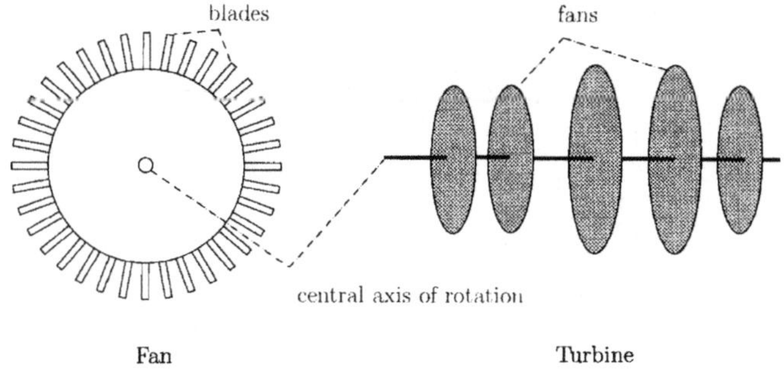

Figure 1: A turbine engine

The TBP was first addressed as a discrete combinatorial optimization problem by Mosevich [20] in 1986. In [20] a Monte Carlo approach was used for finding approximate solutions to the TBP. Korenjak and Batagelj [15] present a local search algorithm for solving the TBP, which performed a 2-opt neighborhood search on randomly generated initial feasible solutions. Laporte and Mercure [16] formulated the TBP as a QAP, and presented a series of heuristics for solving it. Their heuristics are based on an interchange algorithm by Or [23] for solving the traveling salesman problem, and the computational results indicate an improvement from the Monte Carlo approach used by Mosevich. Fathi and Ginjupalli [10] proposed two types of heuristics for solving the QAP formulation of the TBP. Mason and Rönnqvist [18] model the TBP as a variation of the standard QAP, and propose a number of local search algorithms which use as starting points solutions obtained from a Lagrangean dual scheme.

2.3.1 The TBP as a standard QAP

In this section it will be shown how the TBP can be formulated as a minimization problem which is equivalent to a standard QAP. Before we formulate the TBP as a minimization problem certain assumptions regarding the physical characteristics of the blades and the fan as a whole need to be made. Specifically:

- The weights of the blades are known. There are various methods used in the industry to measure the blade weights.
- The center of gravity for each blade lies at a distance r from the geometrical center of the fan and in the same plane when the blade is placed on the periphery of the fan.

- The center of gravity of the fan (without the blades) lies in its geometrical center and it has zero weight.

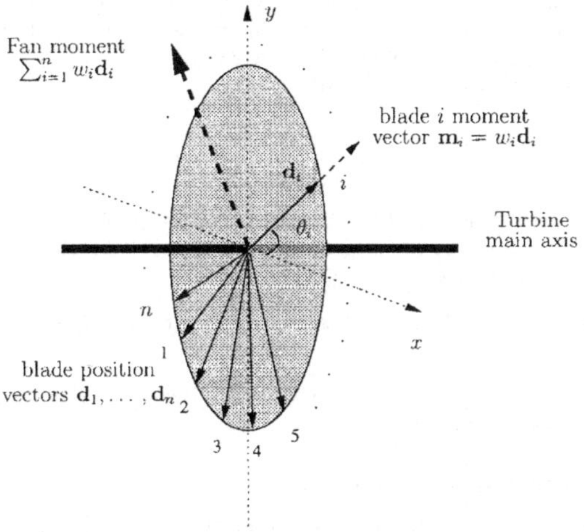

Figure 2: Fan moment vector with n blades

The above assumptions allow us to consider a two-dimensional model of the fan, where the blades are treated as point masses. It is noted that the above assumptions are in accordance with the current method used in the industry to balance turbines.

Consider that we have n blades (see Figure 2), each with a weight $w_i, i = 1, \ldots, n$, to be placed in n positions where each position is described by a vector $\mathbf{d}_i = \binom{r\cos\theta_i}{r\sin\theta_i}$, $\theta_i = 2\pi(i-1)/n, i = 1, \ldots, n$. Let Π_N be the set of all permutations of the set of integers $N := \{1, 2, \ldots, n\}$. In the TBP we want to find a permutation that minimizes the Euclidean norm of the moment of the fan about its geometrical center

$$\min_{p \in \Pi_N} \left\| \sum_{i=1}^{n} w_i \mathbf{d}_{p(i)} \right\|, \tag{15}$$

where for a given $\mathbf{x} \in \mathbb{R}^n$ the Euclidean norm is defined as $\|\mathbf{x}\| := \sqrt{\sum_{i=1}^{n} x_i^2}$. Each blade position i in Figure 2 is represented by a position vector $\mathbf{d}_i$, and each blade has a given mass w_i. Therefore, the moment vector generated by placing blade with mass w_i in the position described by the vector $\mathbf{d}_i$ is given by $\mathbf{m}_i = w_i \mathbf{d}_i$. The moment vector of the whole disk containing the blades, in other words the fan itself, would be the sum of the individual moment vectors of the blades. The problem is to assign the blades in such a way such that the magnitude of the sum of the individual moment vectors is minimized. A large fan moment vector translates into an off-balance turbine which results in vibrations when the turbine rotates about its central axis.

Recall that given two symmetric matrices in $\mathbb{R}^{n \times n}$, $F = (f_{ij})$ and $D = (d_{kl})$ with nonzero diagonal, the standard QAP is

$$\min_{p \in \Pi_N} \sum_{i=1}^{n} \sum_{j=1}^{n} f_{ij} d_{p(i)p(j)}. \tag{16}$$

The TBP can be formulated as a standard QAP where the matrices $F = (f_{ij})$ and $D = (d_{kl})$ are formed by the weights and the angles of the blade positions, as shown in the following result [29]

Remark 2.1 *The problem in (15) is equivalent to (16) with* $d_{ij} = w_i w_j$, $f_{ij} = \cos(\theta_i - \theta_j)$ *for* $i, j = 1, 2, \ldots, n$.

The turbine balancing problem is NP-Hard. This is shown by reduction from the even-odd partition problem [7].

2.3.2 Computational Results

A Greedy Randomized Adaptive Search Procedure (GRASP) is developed in [29] for solving the QAP that results from the TBP. Two sets of problems were used for testing the performance of the proposed algorithm. The first set was provided by Pratt & Whitney (P&W) which is a major turbine engine manufacturer. This problem set contains data for the NASA Space Shuttle Main Engine Advanced Turbopump High Pressure Oxygen rotor. Each turbine contains 5 discs (fans) each carrying $n = 54$ blades, and each disk has $r = 4.5286, 4.2725, 4.5640, 4.2725, 4.5927$ respectively.

The results are shown in Table 1. The first column contains the name of the turbine. The second column contains the disc number for each turbine, and the fourth column the radius r of the disc. The solution obtained from the current method employed in P&W, and that obtained by GRASP are shown in columns 5 and 7 respectively. For turbine P022-02 the solution obtained by PWBAL, a commercial turbine balancing software distributed by P&W, is given in column 6. In column 8 the number of iterations needed by GRASP to reach the given solution is reported. The maximum number of iterations that GRASP was allowed to run for all the problems was 1000, which required approximately 11 CPU minutes on an Intel Pentium 200MHz processor with 64MBytes of memory. A PC was used for performing the computational experiments, since the balancing of turbines usually occurs in the assembly floor where PC's are available. Comparison between the solution values of GRASP and the current method used at P&W demonstrates that GRASP outperforms by far the current method. On average the solutions provided by GRASP are 3 orders of magnitude smaller than those provided by the current method. In addition to the solution quality, GRASP is also much faster since it requires approximately one hour for the whole turbine, while the current method might require a week to complete the balancing process. Therefore an implementation of GRASP in the assembly floor would translate in large savings in production time, and increased quality. Moreover, GRASP also compares favorably with the commercial package PWBAL. All solutions provided by GRASP for the P022-02 turbine are less than the ones provided by PWBAL. The second set of problems used, comes from the literature and do not

Table 1: Statistics summarizing GRASP runs for the P&W problem set

Turbine	disc	current	PWBAL	GRASP	iter.
P022-02	1	0.0059	6.40×10^{-6}	5.16×10^{-6}	874
	2	0.0057	9.80×10^{-6}	3.50×10^{-6}	91
	3	0.0165	2.93×10^{-5}	3.67×10^{-6}	416
	4	0.0022	9.80×10^{-6}	4.69×10^{-6}	276
	5	0.0040	2.32×10^{-5}	4.94×10^{-6}	40
P022-01	1	0.0006	N/A	6.29×10^{-6}	116
	2	0.0005	N/A	7.07×10^{-6}	652
	3	0.0164	N/A	1.08×10^{-5}	616
	4	0.0044	N/A	8.65×10^{-6}	925
	5	0.0064	N/A	1.07×10^{-5}	103
P024-01	1	0.0086	N/A	1.75×10^{-5}	763
	2	0.0043	N/A	5.28×10^{-6}	175
	3	0.0048	N/A	9.09×10^{-6}	97
	4	0.0005	N/A	9.97×10^{-6}	201
	5	0.0064	N/A	1.15×10^{-5}	904
P030-01	1	0.0039	N/A	5.95×10^{-6}	995
	2	0.0002	N/A	1.30×10^{-6}	165
	3	0.0043	N/A	1.00×10^{-5}	185
	4	0.0022	N/A	1.35×10^{-6}	395
	5	0.0121	N/A	4.94×10^{-6}	773

correspond to actual data of turbine engines. Introduced by Korenjak and Batagelj [15] it is called the integer turbine balancing problem (ITBP) problem set, and we choose to include it in our computational results since it will illustrate how GRASP compares to other algorithms for solving the TBP. An instance of ITBP of size $n \geq 2$ is generated by setting $w_i = i, i = 1, \ldots, n$. Note that by the definition of the TBP, the angles $\theta_i, i = 1, \ldots, n$ are the same for every instance of size n since the blades are placed at equally spaced points. The computational results for the ITBP problem set are shown in table 2. The first column corresponds to the size n of the problem, while the radius of the disc for every problem was set to $r = 1000$. The second column corresponds to the solution provided by the algorithm of Korenjak and Batagelj [15] which we call KBH. Columns 3 through 8 report the solution given by the algorithms by Fathi and Ginjupalli [10]. They propose six heuristic methods based on two general schemes, the placement heuristic (PHk, $k = 1, 2, 3$) and the rotational heuristic (RHk, $k = 2, 3, 4$). The solution found by GRASP and the number of iterations needed are given in columns 9 and 10 respectively. The maximum number of iterations was set at 10000. In column 11 the CPU seconds needed for GRASP to complete 10000 iterations, or find a solution which results in 0 unbalance (the theoretical lower bound) is given. A zero value of CPU seconds means that the CPU time was less than a second. The rotational heuristic RHk only works for ITBPs with n divisible by k, while Korenjak and Batagelj [15] report solutions for ITPB instances with $n \leq 20$. From table 2 we can see that GRASP clearly outperforms the heuristics PHk, $k = 1, 2, 3$. The KBH heuristic produces the same solutions as GRASP for instances with $n \leq 12$, but as n gets larger its solutions deteriorate with respect to the ones given by GRASP. In comparing the RHk heuristics with GRASP we have to mention the following result regarding a combinatorial characteristic of the ITBP. Specifically Fathi and Ginjupalli [10] have shown that an ITBP of size n, has a perfect configuration if n is divisible by 2 but not divisible by 4, or if n is divisible by 3 but not divisible by 9. Furthermore they conjecture that an ITBP of size n has a perfect configuration if n is divisible by k but not divisible by k^2, for any prime number $k \geq 2$. A perfect configuration is an assignment of blades into positions that results in 0 unbalance. The RHk heuristics justify the above conjecture, but perform poorly for values of n that do not satisfy the above condition. GRASP finds a perfect configuration for all the values of n that satisfy the above conditions, except for $n = 15, 20$. Overall GRASP performs better than all of the heuristics mentioned and it also provides best known values for the ITBP instances with $n = 13, 17, 19, 23$. The RHk heuristic seems to be the only competitor, but it can only be applied to ITBPs with n divisible by k. Furthermore the existing algorithm can be modified to solve sparse TBP instances, and it can also be parallelized [24, 25].

3 Multidimensional Assignment Problems

A close relative to the class of M-adic assignment problems is that of the multidimensional (referred sometimes as multi-index) assignment problems (MAPs), that also arise as natural extensions from the LAP, and the most famous representative of this class is the 3-dimensional assignment problem (3-DAP).

Consider three disjoint sets I, J, K, where without loss of generality we can assume that $|I| = |J| = |K| = n$, and n^3 cost coefficients c_{ijk}. The 3-DAP can then be stated

Table 2: Statistics summarizing GRASP runs for the ITBP problem set

		Fathi and Ginjupalli						GRASP		
n	KBH	PH1	PH2	PH3	RH2	RH3	RH4	soln	iter.	CPU sec.
5	29.93	29.93	29.93	29.93				29.93	1	1
6	0.00	0.00	0.00	0.00	0.00	0.00		0.00	3	0
7	2.73	2.73	2.73	2.73				2.73	2	7
8	3.64	30.06	3.64	3.64	30.06		30.06	3.64	24	10
9	0.63	13.94	0.63	0.63		20.48		0.63	6	16
10	0.00	0.00	0.00	0.00	0.00			0.00	1	0
11	0.02	2.09	2.09	0.89				0.02	1335	35
12	0.00	0.00	0.00	0.00	0.00	0.00	0.00	0.00	46	9
13	0.04	8.00	2.74	1.79				0.03	3438	58
14	0.00	0.00	0.00	0.00	0.00			0.00	1019	40
15	0.26	15.68	0.84	0.38		0.00		0.01	8431	107
16	0.02	7.49	0.54	0.23	0.37		3.10	0.02	3375	143
17	0.05	5.03	0.14	0.11				0.03	3575	154
18	0.03	4.06	0.99	0.92	0.00	0.00		0.00	4037	77
19	0.05	0.73	0.72	0.40				0.01	1799	257
20	0.04	1.94	0.30	0.00	0.00		0.00	0.01	1765	296
21		3.05	0.82	0.75		0.00		0.00	624	20
22		2.95	0.74	0.56	0.00			0.00	3427	141
23		1.63	0.31	0.13				0.00	9848	465
24		2.46	0.87	0.49	0.00	0.00	0.00	0.00	203	11

as

$$\begin{aligned}
\min \quad & \sum_{i\in I}\sum_{j\in J}\sum_{k\in K} c_{ijk}x_{ijk} \\
\text{s.t.} \quad & \sum_{j\in J}\sum_{k\in K} x_{ijk} = 1, \quad \forall i \in I, \\
& \sum_{i\in I}\sum_{k\in K} x_{ijk} = 1, \quad \forall j \in J, \\
& \sum_{i\in I}\sum_{j\in J} x_{ijk} = 1, \quad \forall k \in K, \\
& x_{ijk} \in \{0,1\} \ \forall \ i,j,k.
\end{aligned}$$

A feasible solution to the above problem will be a three dimensional *permutation array*, that is, a 0-1 array that satisfies the above assignment constraints. The assumption $|I| = |J| = |K| = n$ does not pose a restriction theoretically, since we can add extra cost coefficients with large enough values such that they will never be included in the optimal solution.

Pierskalla [28, 27] was the first to make an in-depth description of the 3-DAP, showing that most of the results obtained for the 3-DAP can be extended to the more general MAP. Various applications of the 3-DAP are presented in [28, 27, 11].

Many algorithms derived for the 3-DAP are of branch and cut type using the facial structure of the problem, which has been the subject of study in several articles. Balas and Saltzman [3] examine the facial structure of the three-index assignment polytope and identify two facets belonging to the polytope. $O(n^4)$ separation algorithms are given for these. Balas and Qi [2] continue the above work and present $O(n^3)$ separation algorithms for earlier identified classes of facets as well as for a new class of facets. Qi, Balas and Gwan [31] base their work upon linear-time separation algorithms for different facets and construct a polyhedral procedure for solving the 3-DAP.

A generalization of the 3-DAP is the M-dimensional assignment problem. Let $n_1, n_2, \ldots, n_M$ be integers which define the ranges in each dimension. The input data consists of one cost array $C \in \mathbb{R}^{n_1 x n_2 x \cdots x n_M}$, where without loss of generality assume that $n_1 \leq n_2 \leq \cdots \leq n_M$. The integer programming formulation of the M-dimensional assignment problem is the following:

$$\min \quad \sum_{i_1=1}^{n_1} \cdots \sum_{i_M=1}^{n_M} C_{i_1 \cdots i_M} x_{i_1 \cdots i_M} \tag{17}$$

$$\text{s.t.} \quad \sum_{i_2=1}^{n_2} \cdots \sum_{i_M=1}^{n_N} x_{i_1 \cdots i_M} = 1, \quad i_1 = 1, \ldots, n_1, \tag{18}$$

$$\sum_{i_1=1}^{n_1} \cdots \sum_{i_{k-1}=1}^{n_{k-1}} \sum_{i_{k+1}=1}^{n_{k+1}} \cdots \sum_{i_M=1}^{n_M} x_{i_1 \cdots i_M} = 1, \tag{19}$$

$$\text{for } i_k = 1, \ldots, n_k \text{ and } k = 2, \ldots, M-1 \tag{20}$$

$$\sum_{i_1=1}^{n_1} \cdots \sum_{i_{M-1}=1}^{n_{M-1}} x_{i_1 \cdots i_M} = 1, \quad i_M = 1, \ldots, n_M, \tag{21}$$

$$x_{i_1 \cdots i_M} \in \{0, 1\} \text{ for all } i_1 \cdots i_M. \tag{22}$$

Multidimensional assignment problems in their general form have found many applications as a means of solving data association problems. More specifically, the central problem in any multitarget tracking and multisensor surveillance is the data association problem of partitioning the observations into tracks and false alarms. General classes of these problems can be formulated as multidimensional assignment problems. For a detailed description on the application of MAPs for multiple target tracking applications, as well as for a fast heuristic algorithm for solving the resulting problems see [21, 22]. Various applications are also contributed to special cases of the MAP. Specifically, the five-dimensional assignment problem has been successfully used for tracking elementary particles. By solving a five-dimensional assignment problem, physicists reconstruct tracks generated by charged elementary particles produced by the large electron-positron collider (LEP) at the CERN institute [30].

3.1 Problem Characteristics

In this section a different formulation of the MAP will be presented using permutations, and various characteristics of the index structure of the multidimensional input array C are examined. Let $\mathbf{X}^M$ be the set of all arrays $X = (x_{i_1 \cdots i_M})_{n_1 \times \cdots \times n_M}$ such that their elements satisfy the constraints (18) through (22). The above IP formulation of the MAP however is not very convenient with respect to computational issues. Define the sets of integers $N_i := \{1, 2, \ldots, n_i\}, i = 1, 2, \ldots, M$, and let Π_{N_i} be the set of all n_1-permutations $p_i : N_1 \to N_i$. Note that $|\Pi_{N_i}| = \frac{n_i!}{(n_i - n_1)!}$, for each $i = 1, \ldots, M$. Given $p_1, \ldots, p_M$ let $P = (p_1, \ldots, p_M)$, that is P can be regarded as a $n_1 \times M$ matrix with columns the n_1-permutations $p_1, \ldots, p_M$,

$$P := \begin{pmatrix} p_1(1) & p_2(1) & \cdots & p_M(1) \\ p_1(2) & p_2(2) & \cdots & p_M(2) \\ \vdots & \vdots & \ddots & \vdots \\ p_1(n_1) & p_2(n_1) & \cdots & p_M(n_1) \end{pmatrix}. \tag{23}$$

Now given $n_1 \leq n_2 \leq \cdots \leq n_M$, and denoting by $e_{N_1} = (1, 2, \ldots, n_1)$ the identity permutation of N_1, let $\mathbf{P}^M$ be the following set

$$\mathbf{P}^M := \{P = (p_1 \ldots, p_M) \mid p_1 = e_{N_1}, p_i \in \Pi_{N_i}, i = 2, \ldots, M\}.$$

It can be shown that $|\mathbf{P}^M| = \prod_{i=2}^{M} \frac{n_i!}{(n_i - n_1)!}$. Note that there is a one-to-one correspondence between the elements of $\mathbf{X}^M$ and $\mathbf{P}^M$ for given $n_1 \leq n_2 \leq \cdots \leq n_M$, since given any $P \in \mathbf{P}^M$ we can construct an $X \in \mathbf{X}^M$ as

$$x_{i_1 \cdots i_M} = \begin{cases} 1 & \text{if } p_j(i_1) = i_j, j = 2, \ldots, M, \\ 0 & \text{otherwise,} \end{cases} \tag{24}$$

for $i_1 = 1, \ldots, n_1$, and vice versa. Considering the above discussion we can now make the following statement:

Remark 3.1 *The MAP as stated in (17) through (22) is equivalent to:*

$$\begin{aligned} \min \quad & \sum_{i=1}^{n_1} C_{i p_2(i) \cdots p_{M-1}(i) p_M(i)} \\ s.t. \quad & p_i \in \Pi_{N_i}, \quad i = 2, \ldots, M. \end{aligned} \tag{25}$$

A solution to the MAP using the above formulation will be any $P \in \mathbf{P}^M$ such that the above objective function is minimized.

In the discussion that follows we will try to answer the following question:

> Given that an element with index $(i_1, i_2, \ldots, i_M)$ of the array C is included in the objective function of the MAP, determine which elements become infeasible.

In order to answer this question, we will first represent the array C in a more conventional format. Let $\pi_l := \prod_{i=1}^{l} n_i$ for $l = 1, \ldots, M$ with $\pi_M := \pi$. The array $C \in \mathbb{R}^{n_1 \times n_2 \times \cdots \times n_M}$ is stored as an one-dimensional array $c \in \mathbb{R}^{\pi}$, in column format. In addition to the sets of integers N_i, $i = 1, \ldots, M$ defined previously define also the set $N := \{1, 2, \ldots, \pi\}$. The set $N_1 \times \cdots \times N_M$ can be regarded as the index set of C while N the index set of c. The transformation between an index $(i_1, i_2, \ldots, i_M)$ of C to an index k of c is given by the following function $f : N_1 \times \cdots \times N_M \to N$

$$f(i_1, i_2, \ldots, i_M) := i_1 + \sum_{j=2}^{M} \pi_{j-1}(i_j - 1) = k. \tag{26}$$

The inverse (from an index k of c to an index $(i_1, i_2, \ldots, i_M)$ of C) can be obtained in a similar fashion from the function $f^{-1} : N \to N_1 \times \cdots \times N_M$, the inverse of f, which is defined by

$$f^{-1}(k) := \begin{pmatrix} k - \sum_{l=1}^{M-1} \pi_{M-l}(i_{M-l+1} - 1) \\ \vdots \\ \left\lfloor \frac{k-1-\sum_{l=1}^{M-m} \pi_{M-l}(i_{M-l+1}-1)}{\pi_{m-1}} \right\rfloor + 1 \\ \vdots \\ \left\lfloor \frac{k-1}{\pi_{M-1}} \right\rfloor + 1 \end{pmatrix}^T = \begin{pmatrix} i_1 \\ \vdots \\ i_m \\ \vdots \\ i_M \end{pmatrix}^T.$$

Remark 3.2 *The function f is a bijection.*

From f and f^{-1} it is clear that we can construct the c array as $c(i) = C(f^{-1}(i)), i = 1, \ldots, \pi$.

Assume that we are given an element $(\bar{i_1}, \bar{i_2}, \ldots, \bar{i_M})$ of C, and for a given k we wish to generate the set

$$N \supset S_{i_k} := \{i \in N \mid f^{-1}(i) = (i_1, \ldots, \bar{i_k}, \ldots, i_M), i_j = 1, \ldots, n_j, j = 1, \ldots, M, j \neq k\},$$

that is the set of all elements in N such that their corresponding element in $N_1 \times \cdots \times N_M$ contains $\bar{i_k}$ in the kth position. A direct approach would be to generate the set $S'_{i_k} \subset N_1 \times \cdots \times N_M$ defined as

$$S'_{i_k} := \{(i_1, \ldots, \bar{i_k}, \ldots, i_M) \mid i_j = 1, \ldots, n_j, j = 1, \ldots, M, j \neq k\},$$

and then take the image of S'_{i_k} under f, $S_{i_k} = f(S'_{i_k})$, a process which will require $\mathcal{O}(M\frac{\pi}{n_k})$ time. The way that c is constructed it follows that its index structure, that is the relationship between the indices of c and those of C, is well defined. Let us define an ordered π-tuple which corresponds to the index set N of c, as $I := (1, 2, \ldots, \pi - 1, \pi)$. For every $i \in I$ there is a corresponding element $(i_1, i_2, \ldots, i_M)$

in the index set of C. Define the following for every dimension $k = 1, 2, \ldots, M$ and element $i_k \in N_k$:

$$\sigma_{i_k} := (i_k - 1)\pi_{k-1} + 1,$$

$$\nu_k := \frac{\pi}{\pi_k}.$$

According to the above values, for any i_k we can identify the following π_{k-1}-sub-tuples contained in I,

$$I = (1, 2, \ldots, \underbrace{B_1^{i_k}, \ldots,}_{\pi_k} \underbrace{B_2^{i_k}, \ldots,}_{\pi_k} \underbrace{B_3^{i_k}, \ldots,}_{(\nu_k-3)\pi_k} B_{\nu_k}^{i_k}, \ldots, \pi - 1, \pi),$$

where $B_l^{i_k} = ((l-1)\pi_k + \sigma_{i_k}, (l-1)\pi_k + \sigma_{i_k} + 1, \ldots, (l-1)\pi_k + \sigma_{i_k} + \pi_{k-1} - 1)$ for $l = 1, \ldots, \nu_k$. So for every i_k, I contains ν_k sub-tuples of size π_{k-1}, starting from the element σ_{i_k} and occurring every π_k elements.

Remark 3.3 $S_{i_k} = \bigcup_{l=1}^{\nu_k} B_l^{i_k}$.

It follows from remark 3.3 that we can generate $\bigcup_{l=1}^{\nu_k} B_l^{i_k}$ and therefore S_{i_k} in $\mathcal{O}(\frac{\pi}{n_k})$ time. Hence given an element with index $(i_1, i_2, \ldots, i_M)$ of the array C to be included in the objective function, the elements contained in $\bigcup_{k=1}^{M} S_{i_k}$ become infeasible.

3.2 The Multi-Target Multi-Sensor Problem

One of the major applications of the MAP is its equivalence with the data association problem. Data association is a very difficult combinatorial problem often encountered in sensor fusion problems, one example being the MultiTarget MultiSensor Tracking (MTMST) problem. When tracking a single target with a single sensor, noisy measurements are made at discrete, periodic points in time from the sensor regarding the moving target. The goal is to establish a time continuous estimate of the target's trajectory. For the MTMST problem, noisy measurements are made from an arbitrary number of spatially diverse sensors regarding an arbitrary number of targets with the goal of estimating the trajectories of all of the targets present. Furthermore, the number of targets may change (e.g. by wandering out of and into detection range) during the problem time interval and the sensors themselves may be in motion. For example in Figure 3, two sensors observe three targets, thereby we have three measurements from each sensor. The MTMST problem is to partition all six measurements from the sensors into three subsets that correspond to targets.

Target state estimates for single target single sensor tracking are obtained satisfactorily by applying one of several variants of the Kalman filter. However, no method known can consistently obtain good trajectory estimates for the more general MTMST problem.

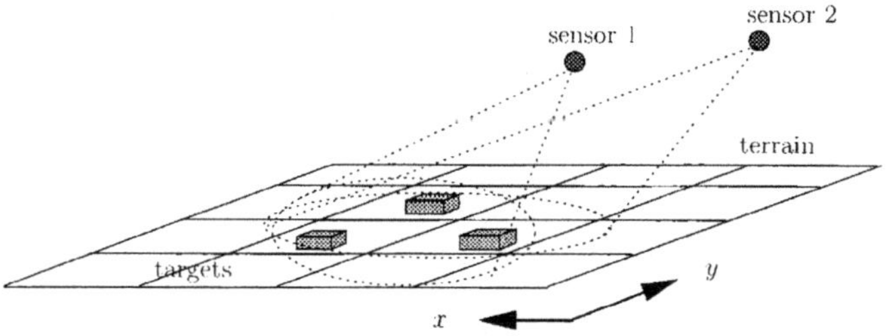

Figure 3: Sensor fusion

The MTMST has only recently been addressed [8], [9]. In these and other papers the MTMST problem has generally been divided into 2 subproblems: state estimation/propagation and data association. The combinatorial nature of the MTMST problem results from the data association subproblem; that is, given that each sensor m, $m = 1 \ldots M$, makes n_m measurements, $n_i \neq n_j, \forall i \neq j$, how do we correctly partition the measurements into all of the targets, ensuring that all of the measurements are used exactly once? Total enumeration of all possible partitions is obviously unrealistic.

In MTMST a data association problem is presented at discrete, periodic moments in time. We expect that by obtaining the correct partition or association, the data from *all* of the sensors may be combined in some manner to make estimates of target positions that are more accurate than that obtained from a single measurement.

A detailed description of how the MTMST problem can be formulated as an MAP is given in the papers by Murphey, Pardalos and Pitsoulis [21, 22]. In these papers an algorithm is also provided for solving the resulting MAP in real time utilizing multiple processors. In Table 3 the performance of the proposed algorithm in solving instances provided by the US Air Force is shown. Columns 2 through 4 in Table 3 contain respectively, the best solution found by GRASP, the CPU time needed to complete 100 iterations, and the iteration number where the best solution was attained. The last column contains the cost of the true partition of each instance. In all of the instances GRASP found a lower cost solution, than that of the true partition. This is to be expected since the maximum likelihood cost function is a probabilistic function which is subject to variability. However although the most likely partition might not (possibly) correspond to the optimal partition, the optimal partition provides a very good estimate of the true partition. This is illustrated for example in the partitioning provided by the solution of GRASP for the instance `Inst03` as illustrated in Figure 4. In `Inst03` we have 5 sensors observing 5 targets, and z_{ij} corresponds to the measurement of sensor i for some target j. For target 5

Table 3: GRASP computational results

Instance	GRASP Sol.	CPU time	iterations	True Partition
Inst01	-52.2	0.465	1	-52.1
Inst02	-63.5	0.476	1	-58.0
Inst03	-100.5	3.64	6	-48.5
Inst04	-44.5	0.16	3	-36.0
Inst05	-126.7	2.21	2	-101.8
Inst06	-131.7	2.19	1	-107.1
Inst07	-288.6	18.69	1	-163.9

for example we can see that the true partition should be $(z_{15}, z_{35}, z_{55}, z_{45})$ but the partition $(z_{15}, z_{35}, z_{55}, z_{46})$ obviously has a lower cost.

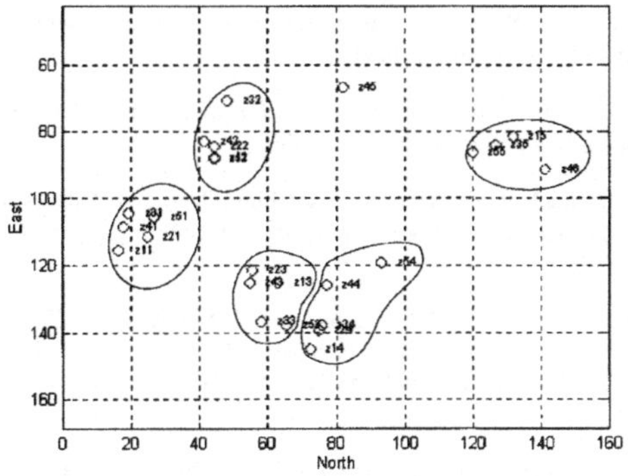

Figure 4: Inst03 GRASP Partitioning

4 Concluding Remarks

Nonlinear assignment problems are combinatorial optimization problems which have applications in many diverse fields. They are natural extensions of the linear assignment problem, and they are characterized by the large input data and their resistance to be solved to optimality efficiently.

In this paper, along with an introduction to nonlinear assignment problems, we present two relatively recent applications, the turbine balancing problem, and the multitarget multisensor tracking problem. The fact that both problems have to be solved in real time, and their input data is inaccurate, makes exact algorithms inappropriate. Computational results with heuristic algorithms that provide good quality solutions in almost real time are presented.

References

[1] R. K. Ahuja, T. Magnanti, and J. B. Orlin. *Network Flows: Theory, Algorithms, and Applications.* Prentice Hall, Englewood-Cliffs, New Jersey, 1993.

[2] E. Balas and L. Qi. Linear-time separation algorithms for the three-index assignment polytope. *Discrete Applied Mathematics*, pages 1–12, 1993.

[3] E. Balas and M. J. Saltzman. Facets of the three-index assignment polytope. *Discrete Applied Mathematics*, pages 201–229, 1989.

[4] S. H. Bokhari. *Assignment Problems in Parallel and Distributed Computing.* Kluwer Academic Publishers, Boston, 1987.

[5] R.E. Burkard and E. Çela. Heuristics for biquadratic assignment problems and their computational comparison. *European Journal of Operational Research*, 83:283–300, 1995.

[6] R.E. Burkard, E. Çela, P.M. Pardalos, and L.S. Pitsoulis. The quadratic assignment problem. In Ding-Zhu Du and P.M. Pardalos, editors, *Handbook of Combinatorial Optimization*, volume 3, pages 241–337. Kluwer Academic Publishers, 1998.

[7] E. Çela. *The Quadratic Assignment Problem: Theory and Algorithms.* Kluwer Academic Publishers, Dordrecht,The Netherlands, 1998.

[8] Y. Bar-Shalom (Editor). *Multitarget, Multisensor Tracking: Advanced Applications.* Artech House, Dedham, MA, 1990.

[9] Y. Bar-Shalom (Editor). *Multitarget, Multisensor Tracking: Applications and Advances.* Artech House, Dedham, MA, 1992.

[10] Y. Fathi and K. K. Ginjupalli. A mathematical model and a heuristic for the turbine balancing problem. *European Journal of Operations Research*, 63:336–342, 1993.

[11] A. M. Frieze. A bilinear programming formulation of the 3-dimensional assignment problem. *Mathematical Programming*, 7:376–379, 1974.

[12] A. Graham. *Kronecker Products and Matrix Calculus: with Applications*. Halsted Press, Toronto, 1981.

[13] L. J. Hubert. *Assignment Methods in Combinatorial Data Analysis*. Marcel Dekker, Inc., New York, NY 10016, 1987.

[14] T. C. Koopmans and M. J. Beckmann. Assignment problems and the location of economic activities. *Econometrica*, 25:53–76, 1957.

[15] S. Korenjak and V. Batagelj. Turbine balancing problem. Technical report, Department of Mathematics, University of Edvard Kardelj of Eljubljana, Yugoslavia, 1987.

[16] G. Laporte and H. Mercure. Balancing hydraulic turbine runners: A quadratic assignment problem. *European Journal of Operations Research*, 35:378–381, 1988.

[17] E. L. Lawler. The quadratic assignment problem. *Management Science*, 9:586–599, 1963.

[18] A. Mason and M. Rönnqvist. Solution methods for the balancing of jet turbines. *Computers and Operations Research*, 24(2):153–167, 1997.

[19] T. Mavridou, P. M. Pardalos, L. S. Pitsoulis, and M. G. C. Resende. A GRASP for the biquadratic assignment problem. *European Journal of Operations Research*, 105(3):613–621, 1998.

[20] J. Mosevich. Balancing hydraulic turbine runners: A discrete combinatorial optimization problem. *European Journal of Operations Research*, 26:202–204, 1986.

[21] R. A. Murphey, P. M. Pardalos, and L. Pitsoulis. A greedy randomized adaptive search procedure for the multitarget multisensor tracking problem. In P. M. Pardalos and Ding-Zhu Du, editors, *Network Design: Connectivity and Facility Location*, volume 40 of *DIMACS Series on Discrete Mathematics and Theoretical Computer Science*, pages 277–302. American Mathematical Society, 1997.

[22] R. A. Murphey, P. M. Pardalos, and L. Pitsoulis. A parallel grasp for the data association multidimensional assignment problem. In *Parallel Processing of Discrete Problems*, volume 106 of *IMA Volumes in Mathematics and its Applications*, pages 159–180. Springer-Verlag, 1998.

[23] I. Or. *Travelling salesman-type combinatorial problems and their relation to the logistics of blood banking.* PhD thesis, Northwestern University, Evaston, IL, 1976.

[24] P. M. Pardalos, L. S. Pitsoulis, and M. G. C. Resende. A parallel GRASP implementation for solving the quadratic assignment problem. In A. Ferreira and José D.P. Rolim, editors, *Parallel Algorithms for Irregular Problems: State of the Art*, pages 115–133. Kluwer Academic Publishers, 1995.

[25] P. M. Pardalos, L. S. Pitsoulis, and M. G. C. Resende. Algorithm 769: FORTRAN subroutines for approximate solution of sparse quadratic assignment problems. *ACM Transactions on Mathematical Software*, 23:196–208, 1997.

[26] P. M. Pardalos, F. Rendl, and H. Wolkowicz. The quadratic assignment problem: A survey and recent developments. In P.M. Pardalos and H. Wolkowicz, editors, *Quadratic assignment and related problems*, volume 16 of *DIMACS Series on Discrete Mathematics and Theoretical Computer Science*, pages 1–42. American Mathematical Society, 1994.

[27] W. P. Pierskalla. The tri-substitution method for the three-multidimensional assignment problem. *CORS J.*, 5:71–81, 1967.

[28] W. P. Pierskalla. The multidimensional assignment problem. *Operations Research*, 16:422–431, 1968.

[29] L. S. Pitsoulis. *Algorithms for Nonlinear Assignment Problems.* PhD thesis, Department of Industrial and Systems Engineering, University of Florida, Gainesville, USA, 1998.

[30] J. Pusztaszeri, P.E. Rensing, and T.M. Liebling. Tracking elementary particles near their primary vertex: a combinatorial approach. *Journal of Global Optimization*, 16:422–431, 1995.

[31] L. Qi, E. Balas, and G. Gwan. A new facet class and a polyhedral method for the three-index assignment problem. In D.-Z. Du, editor, *Advances in Optimization*, pages 256–274. Kluwer Academic, 1994.

[32] S. Sahni and T. Gonzalez. P-complete approximation problems. *Journal of the Association of Computing Machinery*, 23:555–565, 1976.

A new merit function and an SQP method for non-strictly monotone variational inequalities

Michael Patriksson (mipat@math.chalmers.se)
Department of Mathematics
Chalmers University of Technology
S-412 96 Gothenburg, Sweden

Abstract

Merit functions utilized to monitor the convergence of sequential quadratic programming (SQP) methods for nonlinear programs and variational inequality problems have in common that they include a penalty function for the explicit constraints, the value of the penalty parameter for which is subject to the requirement of being large enough compared to estimates of the optimal Lagrange multipliers. In existing applications of the SQP algorithm to variational inequality problems, the possible choices of merit functions used in descent algorithms are also limited by conditions which require the knowledge of problem parameters.

This paper introduces a merit function for use in SQP methods for possibly non-strictly monotone variational inequality problems which does not include an explicit penalty function, and whose application in descent algorithms does not require the knowledge of any problem parameters. This function is an extension of the merit function of Marcotte and Zhu (1995) to primal–dual variational inequality problems. Among its other properties, it is directionally differentiable, and exact for strongly monotone problems. Based on the new merit function, we establish the convergence of a general algorithm for primal–dual variational inequalities which includes an adaptive SQP algorithm as a special case.

Keywords: Variational inequalities, gap function, descent algorithm, sequential quadratic programming.

G. Di Pillo and F. Giannessi (eds.), Nonlinear Optimization and Related Topics, 257–275.

1 Introduction

Consider the following instance of the variational inequality problem:

$$F(x^*) + N_{X\cap G}(x^*) \ni 0^n, \tag{1}$$

where $X \subset \Re^n$ is a nonempty, closed, convex and bounded set, $G := \{ x \in \Re^n \mid g_i(x) \leq 0, \quad i = 1, \ldots, m \}$, where each function $g_i : \Re^n \mapsto \Re$ is convex and in C^2. We assume that $X \cap G \neq \emptyset$ and that a Slater constraint qualification holds, that is, that there exists a point $x \in X$ with $g_i(x) < 0$, $i = 1, \ldots, m$. Further, we assume that $F : X \mapsto \Re^n$ is monotone, Lipschitz continuous and in C^1 on X. $N_S(x)$ denotes the normal cone to the closed and convex set S at x, that is,

$$N_S(x) = \begin{cases} \{ z \in \Re^n \mid z^{\mathrm{T}}(y - x) \leq 0, \quad \forall y \in S \}, & x \in S, \\ \emptyset, & x \notin S. \end{cases} \tag{2}$$

With these assumptions, it is clear that the solution set, $\mathrm{SOL}(F, X \cap G)$, of (1), is nonempty, closed, convex and bounded.

We consider applications of (1) in which the ground set X is simple, for example a polytope, while G is defined by complicating constraints, such as when some of the functions g_i are nonlinear. In the development of a computational method for (1), we consider a relaxation approach wherein the functions g_i are iteratively linearized, thus iteratively replacing G by polyhedral approximations. Examples of such a method are sequential linear programming (SLP) and sequential quadratic programming (SQP). (We may also consider the case where the set X is not included in the original problem formulation, but is an a priori given trust region for the algorithm, although we have chosen not to describe the methodology in the framework of trust region methods.)

We introduce the following equivalent primal–dual formulation of (1). Introducing multipliers $\lambda_i \in \Re_+$, $i = 1, \ldots, m$, for the constraints defining G results in

$$\begin{pmatrix} F(x^*) + \nabla g(x^*)^{\mathrm{T}} \lambda^* \\ -g(x^*) \end{pmatrix} + N_{X \times \Re^m_+}(x^*, \lambda^*) \ni 0^{n+m}. \tag{3}$$

This variational inequality in the variables (x, λ) describes the Karush–Kuhn–Tucker (KKT) conditions for (1); under the given Slater constraint qualification, the solutions in x coincide ([Tob86, HaP90]).

In the SQP algorithm for the solution of (3) the search-direction finding subproblem defined at a given iterate, $(x^t, \lambda^t) \in X \times \Re^m_+$, is the linearization of (3) at (x^t, λ^t): find $y^t \in X \cap T(x^t)$ such that

$$[F(x^t) + H(x^t, \lambda^t)(y^t - x^t)]^{\mathrm{T}}(z - y^t) \geq 0, \qquad \forall z \in X \cap T(x^t), \tag{4}$$

where

$$H(x, \lambda) := \nabla_x [F(x) + \nabla g(x)^{\mathrm{T}} \lambda] \tag{5}$$

is the Jacobian of the Lagrangean of (1) with respect to x, and

$$T(x) := \{\, y \in \Re^n \mid g(x) + \nabla g(x)(y - x) \leq 0^m \,\}, \qquad x \in X \tag{6}$$

is the linearization of the set G defined at x.

A new iterate is usually defined by means of an inexact line search in the primal space, along the direction of $d^t := y^t - x^t$ in a penalty merit function for (1) [equivalently for (3)]. The penalty merit function utilized is the sum of a merit function for (1) and a penalty function for the set G. So, for nonlinear programming, which is obtained from (1) when $F \equiv \nabla f$ for some real-valued (objective) function $f : X \mapsto \Re$, the function most often considered is (e.g., [Han77, Fle87, Psh94, Ber95])

$$f(x) + r \sum_{i=1}^{m} \max\{0, g_i(x)\}, \qquad x \in X,$$

where r is a positive penalty parameter, while in applications to variational inequality problems, the objective function f must be replaced by a merit (or, gap) function for (1). Taji and Fukushima [TaF95] use the merit function

$$\psi_{\mathrm{TF}}(x) := \underset{y \in X}{\operatorname{maximum}} \;\; F(x)^{\mathrm{T}}(x - y) - (1/2)(y - x)^{\mathrm{T}} Q (y - x), \tag{7a}$$

$$\text{subject to} \;\; g(x) + \nabla g(x)(y - x) \leq 0^m, \tag{7b}$$

defined on X, where $Q \in \Re^{n \times n}$ is a symmetric and positive definite matrix.

Common to the existing analyses in applications of SQP algorithms in nonlinear programming and to variational inequality problems is that the value of the penalty parameter r needs to be sufficiently large in order for the penalty merit function to be exact, and (conversely) for the SQP subproblem to yield descent at every non-stationary point. The value of r is related to predictions on some norm of a vector λ^* of Lagrange multipliers for the explicit constraints; for example, the following condition is present in [Han77, Psh94, TaF95]:

$$r \geq \sup_t \{\, \|\lambda\|_\infty \mid \lambda \in \Lambda(x^t) \,\}, \tag{8}$$

where $\Lambda(x)$ denotes the set of Lagrange multipliers for the linearized constraints (6) in the solution to (4) defined at x, and $\|\cdot\|_\infty$ denotes the maximum norm.

In the case of variational inequality problems, the analysis in [TaF95] introduces the further requirement on the problem that F is strongly monotone (recall that F is strongly monotone on X if, for some $m_F > 0$, $[F(x) - F(y)]^{\mathrm{T}}(x - y) \geq m_F \|x - y\|^2$ holds for all $x, y \in X$), and the possible choices of matrices Q in the merit function (7) for (1) are restricted to those for which $\|Q\| < 2m_F$ holds.

Since an estimate of neither the size of a multiplier vector λ^* nor the modulus m_F of strong monotonicity is available a priori, estimates must be constructed either through experiments or adaptively in the algorithm.

Our approach to solving the problem (3) introduces some novelties in SQP methods. First, we introduce a merit (or, gap) function for (1) which is based on the same linearization of the constraints defining the set G as the SQP subproblem. This merit function extends the primal gap function introduced by Marcotte and Zhu [MaZ95] to the primal–dual setting, as well as the merit function (7) utilized in [TaF95, TaF96] in descent algorithms for (3). The analysis of this merit function (which differs substantially from that in [TaF95, TaF96] and from previous analyses of exact penalty functions) allows for the first time the penalty parameter r to take on any non-negative value. The explanation for this possibility is that the merit function introduced has an inherent strive for feasibility (which, for example, the function f in the case of nonlinear programming has not).

Another novelty is the introduction of a scaling parameter in the merit function. (This was earlier proposed in [MaZ95] in primal applications, but it is here developed further and put to an effective use in the primal–dual setting.) This scaling parameter serves the purpose of adaptively adjusting the merit function so that at any point outside the solution set $\mathrm{SOL}(F, X \cap G)$ of (3), the SQP subproblem solution eventually yields descent with respect to the adjusted merit function. In so doing, we remove the condition on the a priori choice of the merit function present in the analysis in [TaF95]. In particular, the scaling enables us to execute the algorithm without the knowledge (or estimation) of any problem parameters. (In the case where F is strongly monotone, the rescaling will only be made a finite number of times, resulting in an exact penalty merit function satisfying the above mentioned requirement on the merit function in [TaF95, TaF96]; cf. Corollary 3.4.) We stress that the scaling parameter does not have an interpretation as a penalty parameter, since (in the strongly monotone case) its final value mentioned above is only related to the problem data in the sense of the modulus m_F of strong monotonicity of F but not to any optimal vector λ^* of Lagrange multipliers for the explicit constraints.

The new merit function and the new analysis allows us to establish convergence of the SQP algorithm for non-strictly monotone operators F. We note however that, as in [TaF95], the SQP subproblem has an unorthodox feature, in utilizing as a multiplier estimate λ in the Jacobian mapping (5) of the SQP subproblem (4) the multiplier solution that is supplied in the evaluation of the merit function, in place of a multiplier estimate obtained from a previous solution of (4).

The rest of the paper is organized as follows. In the next section, we introduce a class of merit functions for (1), and study its most important properties. In Section 3, we introduce the iterative algorithm, which includes an adaptive SQP algorithm as a special case. The algorithm is analyzed in Section 4, and Section 5 compares the analysis to the ones in [TaF95, TaF96] for a projection algorithm and an SQP algorithm for (3). The paper concludes with some observations and remarks.

2 A merit function

The class of merit functions used defines an extension to primal–dual variational inequality problems of the differentiable gap functions for primal variational inequality problems which have a long history (see, e.g., [Auc89, Fuk92, Pat93a, Pat93b, WFM93, ZhM93, LaP94, Pat94, ZhM94, Fuk96, MaZ95, Pat97, Pat98]).

Let $\varphi : X \times X \mapsto \Re^n$ be a C^1 function on $X \times X$ of the form $\varphi(y, x)$, strictly convex in y for each fixed $x \in X$, and with the additional property that $\varphi(x, x) = 0$ and $\nabla_y \varphi(x, x) = 0^n$ for all $x \in X$. Let $\alpha > 0$, and consider the function

$$\psi_\alpha(x) := \underset{y \in X}{\text{maximum}} \quad L_\alpha(y, x) := F(x)^{\mathrm{T}}(x - y) - (1/\alpha)\varphi(y, x), \tag{9a}$$

$$\text{subject to} \quad g(x) + \nabla g(x)(y - x) \leq 0^m, \tag{9b}$$

defined on X.

We first establish a characterization of $\mathrm{SOL}(F, X \cap G)$ in terms of the fixed points of the problem (9).

Proposition 2.1 (A fixed point characterization of $\mathrm{SOL}(F, X \cap G)$). *$x \in \mathrm{SOL}(F, X \cap G)$ if and only if x solves (9).*

Proof. Let $x \in \mathrm{SOL}(F, X \cap G)$. Then, it is easily seen that $x \in \mathrm{SOL}(F, X \cap T(x))$. Let y solve (9). Then, $[F(x) + (1/\alpha)\nabla_y \varphi(y, x)]^{\mathrm{T}}(z - y) \geq 0$ holds for all $z \in X \cap T(x)$. Since $y \in X \cap T(x)$, $F(x)^{\mathrm{T}}(y - x) \geq 0$ holds, which implies that $(1/\alpha)\nabla_y \varphi(y, x)^{\mathrm{T}}(x - y) \geq 0$. But the strict convexity of $\varphi(\cdot, x)$ then implies that $y = x$ must hold, since otherwise we reach a contradiction from the strict monotonicity of $\nabla_y \varphi(\cdot, x)$ and the assumptions on φ.

Conversely, assume that x solves (9). Then, $x \in X \cap G$ must hold, and by its variational inequality characterization,

$$F(x)^{\mathrm{T}}(z - x) \geq 0, \qquad \forall z \in X \cap T(x) \supseteq X \cap G, \tag{10}$$

and hence, $x \in \mathrm{SOL}(F, X \cap G)$. □

Proposition 2.2 (ψ_α is a merit function for (1)). *$\psi_\alpha(x) \geq 0$ holds for all $x \in X \cap G$. Further, $x \in X \cap G$ and $\psi_\alpha(x) = 0$ hold if and only if $x \in \mathrm{SOL}(F, X \cap G)$.*

Proof. Let $x \in X \cap G$. It is clear from the choice $y = x$ in (9) that $\psi_\alpha(x) \geq 0$ holds.

[$\Longrightarrow$] Let $x \in X \cap G$ and $\psi_\alpha(x) = 0$ hold. That $x \in X \cap T(x)$ implies that $[F(x) + (1/\alpha)\nabla_y \varphi(y, x)]^{\mathrm{T}}(x - y) \geq 0$. But

$$\begin{aligned} 0 = \psi_\alpha(x) &= F(x)^{\mathrm{T}}(x - y) - (1/\alpha)\varphi(y, x) \\ &\geq -(1/\alpha)[\nabla_y \varphi(y, x)^{\mathrm{T}}(x - y) + \varphi(y, x)] \geq (1/\alpha)\varphi(x, x) = 0, \end{aligned}$$

where the last inequality is strict by the strict convexity of $\varphi(\cdot, x)$, unless $y = x$ holds. Proposition 2.1 then implies that $x \in \mathrm{SOL}(F, X \cap G)$.

[$\Longleftarrow$] Let x solve (9). Then, $\psi_\alpha(x) = 0$ immediately follows. Further, since $y = x$ holds, (9b) implies that $x \in X \cap G$. □

We next investigate the continuity and differentiability of the merit function ψ_α.

Proposition 2.3 (Continuity of ψ_α). *The merit function ψ_α is continuous on X.*

Proof. The mapping $x \mapsto T(x)$ defined in (6) is continuous on X ([TaF96, Prop. 2.5]), and ψ_α is finite on X, whence the result follows from the continuity and boundedness of L_α on $X \times X$. □

The analysis of the differentiability properties of ψ_α requires the explicit use of the KKT conditions for (9), which, given $x \in X$, are:

$$F(x) + (1/\alpha)\nabla_y \varphi(y,x) + \nabla g(x)^{\mathrm{T}} p + N_X(y) \ni 0^n, \tag{11a}$$

$$p \geq 0^m, \tag{11b}$$

$$p^{\mathrm{T}}[g(x) + \nabla g(x)(y - x)] = 0, \tag{11c}$$

$$g(x) + \nabla g(x)(y - x) \leq 0^m. \tag{11d}$$

We let $P(x)$ denote the set of solutions in p to (11), and further introduce the notation

$$H(x,p) := \nabla_x[F(x) + \nabla g(x)^{\mathrm{T}} p], \qquad p \in P(x). \tag{12}$$

Proposition 2.4 (Directional differentiability of ψ_α). *The merit function ψ_α is directionally differentiable on X in any direction $d \in \Re^n$, with*

$$\psi'_\alpha(x;d) = \min_{p \in P(x)} \{[F(x) - H(x,p)(y - x) - (1/\alpha)\nabla_x \varphi(y,x)]^{\mathrm{T}} d\},$$

where y is the unique solution to (9).

Proof. The result follows from [Hog73, Thm. 2]. □

Although the merit function ψ_α has an inherent strive for feasibility with respect to G, we consider augmenting ψ_α with a penalty function, as follows:

$$\psi_{\alpha,r}(x) := \psi_\alpha(x) + r \sum_{i=1}^{m} \max\{0, g_i(x)\}, \qquad r \geq 0, \qquad x \in X.$$

Clearly, $\psi_{\alpha,r}$ is a continuous merit function for (1). The following result establishes that the directional derivative of $\psi_{\alpha,r}$ in the direction of $y - x$ is strictly monotonically decreasing in $r \geq 0$. While $r = 0$ will be shown to be sufficient for establishing the validity of the algorithm, a positive value of r may still be beneficial, through the addition of another force directed towards feasibility.

Proposition 2.5 ($\psi'_{\alpha,r}(x; y-x)$ is strictly monotonically decreasing in $r \geq 0$). *It holds that $\psi'_{\alpha,r}(x; y-x)$ is strictly monotonically decreasing in r. In particular, then,*

$$\psi'_{\alpha,r}(x; y-x) \leq \psi'_{\alpha}(x; y-x), \qquad r \geq 0. \tag{13}$$

Proof. The additional terms in the expression of $\psi'_{\alpha,r}(x; d)$ are

$$r \sum_{i \in \mathcal{I}_+(x)} \nabla g_i(x)^{\mathrm{T}} d + r \sum_{i \in \mathcal{I}_0(x)} \max\{0, \nabla g_i(x)^{\mathrm{T}} d\}, \tag{14}$$

where

$$\mathcal{I}_+(x) := \{\, i = 1, \ldots, m \mid g_i(x) > 0 \,\}; \qquad \mathcal{I}_0(x) := \{\, i = 1, \ldots, m \mid g_i(x) = 0 \,\}.$$

By (11d), $\nabla g_i(x)^{\mathrm{T}} d \leq -g_i(x) = 0$, $i \in \mathcal{I}_0(x)$, so the second term in (14) is zero. Further, by the same relation, $\nabla g_i(x)^{\mathrm{T}} d \leq -g_i(x) < 0$, $i \in \mathcal{I}_+(x)$, so the first term in (14) is non-positive, and strictly monotonically decreasing in r. The result follows. □

3 A modified descent algorithm for variational inequalities

We introduce a class of search direction subproblems in the class of cost approximation methods ([Pat93b]) for the solution (3) which extend the search direction finding subproblems of the algorithms of [Pat93a, Pat93b, WFM93, ZhM93, Pat94, ZhM94, MaZ95] to the explicit treatment of constraints and to the present primal–dual setting.

For any $\alpha > 0$, let $\Phi_\alpha : X \times X \times \Re^m_+ \mapsto \Re^n$ be a continuous mapping on $X \times X \times \Re^m_+$ of the form $\Phi_\alpha(y, x, p)$, monotone and in C^1 on X with respect to y. Given $(x, p) \in X \times \Re^m_+$, we consider replacing F by $\Phi_\alpha(\cdot, x, p)$; the error made is subtracted off through a fixed error term. The search-direction finding subproblem thus is to find $y \in X \cap T(x)$ such that

$$[\Phi_\alpha(y, x, p) + F(x) - \Phi_\alpha(x, x, p)]^{\mathrm{T}}(z - y) \geq 0, \qquad \forall z \in X \cap T(x). \tag{15}$$

If

$$\Phi_\alpha(\cdot, x, p) = (1/\alpha)\nabla_y \varphi(\cdot, x), \qquad x \in X, \tag{16}$$

then the problems (9) and (15) coincide, whence the search direction is defined by the point y that defines the current value of the merit function; the possible choice of Φ_α is however more general. In our convergence analysis, we will assume that either (16) holds, or the choice of Φ_α corresponds to an (approximate) Newton subproblem; the latter includes the SQP subproblem as a special case.

We next establish the descent properties of the search directions obtained from the solution of (15). We then need the following definition, which extends the one in [MaZ95] to the primal–dual setting.

Definition 3.1 (γ-strong descent). *The vector $d \in \Re^n$ defines a γ-*strong descent *direction for the merit function $\psi_{\alpha,r}$ at x if*

$$\psi'_{\alpha,r}(x;d) \leq -\gamma\psi_\alpha(x), \qquad \gamma \in (0,1). \tag{17}$$

The next result extends one from [MaZ95] to the primal–dual setting.

Proposition 3.2 (Descent properties). *Let $\alpha > 0$. Let either Φ_α be chosen such that*

$$\|\nabla_y \Phi_\alpha(y,x,p) - H(x,p)\| \leq \delta(\alpha), \qquad x,y \in Y, \quad p \in P(x), \tag{18}$$

for some function δ depending on problem data only, or (16) *hold. Let $x \in X \setminus \mathrm{SOL}(F, X \cap G)$, y solve* (15), *and $d := y - x$. Then, either d defines a γ-strong descent direction for $\psi_{\alpha,r}$ at x, for any $r \geq 0$, or*

$$\psi_\alpha(x) \leq 1/(1-\gamma)[\delta(\alpha) + 1/\alpha]b, \tag{19}$$

for some $b > 0$.

Proof. We consider the first case for Φ_α. We first note that the direction $d := y - x$ also satisfies the conclusion (13) of Proposition 2.5, arguments for which are very similar, based on the KKT conditions for (15). Let w denote the solution to (9). Then, by Proposition 2.4,

$$\begin{aligned}
\psi'_{\alpha,r}(x;d) &\leq [F(x) - H(x,p)(w-x) - (1/\alpha)\nabla_x\varphi(w,x)]^{\mathrm{T}} d \\
&\leq -\psi_\alpha(x) + F(x)^{\mathrm{T}}(x-w) - (1/\alpha)\varphi(w,x) \\
&\quad + [F(x) - H(x,p)(w-x) - (1/\alpha)\nabla_x\varphi(w,x)]^{\mathrm{T}} d \\
&\quad + [F(x) + \Phi_\alpha(y,x,p) - \Phi_\alpha(x,x,p)]^{\mathrm{T}}(w-y) \\
&= -\psi_\alpha(x) - d^{\mathrm{T}} H(x,p) d \\
&\quad + [\Phi_\alpha(y,x,p) - \Phi_\alpha(x,x,p) - H(x,p)d]^{\mathrm{T}}(w-y) \\
&\quad - (1/\alpha)[\varphi(w,x) + \nabla_x\varphi(w,x)^{\mathrm{T}} d].
\end{aligned}$$

The second row in the last equality is written as

$$[\Phi_\alpha(y,x,p) - \Phi_\alpha(x,x,p) - H(x,p)d]^{\mathrm{T}}([x-y] + [w-x]);$$

the first term is less than $d^{\mathrm{T}}H(x,p)d$, by the monotonicity of $\Phi_\alpha(\cdot,x,p)$, while the second term can be shown to be less than $(1/2)\delta(\alpha)[\|d\|^2 + \|w-x\|^2]$ by the assumption (18). We conclude that, with

$$\theta(x,\alpha) := (1/2)\delta(\alpha)[\|d\|^2 + \|w-x\|^2] - (1/\alpha)[\varphi(w,x) + \nabla_x\varphi(w,x)^{\mathrm{T}} d], \tag{20}$$

for some $b > 0$,

$$\psi'_{\alpha,r}(x;d) \leq -\psi_\alpha(x) + \theta(x,\alpha) \leq -\psi_\alpha(x) + [\delta(\alpha) + 1/\alpha]b$$

holds, since X is bounded and φ and $\nabla_x\varphi$ are continuous. If $x \notin \mathrm{SOL}(F, X \cap G)$, then it follows that either d defines a γ-strong descent direction for $\psi_{\alpha,r}$ at x or that (19) holds.

In the second case for Φ_α, $w = y$ holds; the second row thus cancels out, and the conclusion holds with $\delta(\alpha) = 0$. □

Instance 3.3 (**An SQP algorithm**). Let $\Phi_\alpha(y, x, p) := H(x, p)y$, $p \in P(x)$. Then, the CA subproblem is to find a $y \in X \cap T(x)$ such that

$$[F(x) + H(x,p)(y - x)]^{\mathrm{T}} (z - y) \geq 0, \qquad \forall z \in X \cap T(x). \tag{21}$$

This is the subproblem of an SQP algorithm in which a subset of the constraints are linearized, and where the multiplier estimate p is taken from the calculation of $\psi_\alpha(x)$.

We note that in an SQP algorithm that uses $\psi_{\alpha,r}$ as a merit function, the above result is applicable to monotone mappings F and for arbitrary choices of the penalty parameter $r \geq 0$, because, by construction, (18) holds with $\delta(\alpha) = 0$. □

Corollary 3.4 (**Descent under strong monotonicity**). *In addition to the assumptions of Proposition* 3.2, *assume that F is strongly monotone on X. Let either the search direction d be given by the SQP subproblem described above or let* (16) *hold.*

(a) *For every $x \in X \setminus \mathrm{SOL}(F, X \cap G)$, there exists an $\overline{\alpha} > 0$ such that*

$$\alpha \geq \overline{\alpha} \qquad \Longrightarrow \qquad \psi'_{\alpha,r}(x; d) \leq -\psi_\alpha(x).$$

(b) *Let further $\varphi(y, x) := (1/2)(y - x)^{\mathrm{T}} Q (y - x)$, for some symmetric and positive definite matrix Q. Then, for every $x \in X \setminus \mathrm{SOL}(F, X \cap G)$,*

$$\alpha \geq \|Q\|/(2m_F) \qquad \Longrightarrow \qquad \psi'_{\alpha,r}(x; d) \leq -\psi_\alpha(x).$$

Proof.

(a) The proof of Proposition 3.2 shows that

$$\psi'_{\alpha,r}(x; d) \leq -\psi_\alpha(x) - d^{\mathrm{T}} \nabla F(x) d - (1/\alpha)[\varphi(y, x) + \nabla_x \varphi(y, x)^{\mathrm{T}} d],$$

whence the result follows by the boundedness of X, and the continuity of φ and $\nabla_x\varphi$.

(b) The above estimate reduces to

$$\psi'_{\alpha,r}(x; d) \leq -\psi_\alpha(x) - d^{\mathrm{T}} [\nabla F(x) - 1/(2\alpha) Q] d,$$

and the result follows. □

The result establishes that the merit function constructed is exact for large enough values of α whenever F is strongly monotone, in the sense that

$$\psi'_{\alpha,r}(x; d) \geq 0 \text{ and } x \in X \cap G \qquad \Longleftrightarrow \qquad x \in \text{SOL}(F, X \cap G).$$

The requirement in (b) on the value of α (albeit with strict inequality) is present in the convergence conditions for the SQP algorithm of Taji and Fukushima [TaF95] based on the same merit function, but their convergence analysis is different, and requires in addition that the value of r fulfills a condition of the form (8).

We are now ready to state the algorithm; it is given in Table 1.

Table 1: A modified descent algorithm

---

0. (*Initialization*): Choose an initial point $x^0 \in X$, let $\alpha_0 > 0$, $\Delta\alpha > 0$, $\gamma \in (0,1)$, $\beta \in (0,1)$, $\eta \in (0,1)$, $s \in (0,1]$, and $r \geq 0$. Set $t := 0$.
1. (*Search direction generation*): Let

$$\psi_{\alpha_t}(x^t) := \underset{y \in X \cap T(x^t)}{\text{maximum}} \; L_{\alpha_t}(y, x^t) := F(x^t)^{\mathrm{T}}(x^t - y) - (1/\alpha_t)\varphi(y, x^t). \quad (22)$$

If the second alternative for Φ_α is chosen, then let y^t be the solution to (22), otherwise, let $y^t \in X \cap T(x^t)$ be a vector such that

$$[\Phi_{\alpha_t}(y^t, x^t, p^t) + F(x^t) - \Phi_{\alpha_t}(x^t, x^t, p^t)]^{\mathrm{T}}(z - y^t) \geq 0, \quad \forall z \in X \cap T(x^t), \quad (23)$$

where $p^t \in P(x^t)$ is arbitrary.
The resulting search direction is $d^t := y^t - x^t$.

2. (*Termination criterion*): If x^t solves (23) or (22) $\to$ Stop [(x^t, p^t) solves (3)]. Otherwise, continue.
3. (*Armijo line search or null step*): If $\psi_{\alpha_t}(x^t) \leq \theta(x^t, \alpha_t)/(1-\gamma)$, where $\theta(x^t, \alpha_t)$ is defined by (20), then let $\alpha_{t+1} := \alpha_t + \Delta\alpha$ and $x^{t+1} := x^t$; otherwise, let $\alpha_{t+1} := \alpha_t$ and $x^{t+1} := x^t + \ell_t d^t$, where $\ell_t := s\beta^i$ for the smallest non-negative integer i such that

$$\psi_{\alpha_t,r}(x^t + s\beta^i d^t) - \psi_{\alpha_t,r}(x^t) \leq \eta s \beta^i \psi'_{\alpha_t,r}(x^t; d^t). \quad (24)$$

4. (*Termination criterion*): If x^t is acceptable $\to$ Stop. Otherwise, go to Step 1 with $t := t + 1$.

Remark 3.5 (Observations on the algorithm). (a) The validity of the termination criterion in Step 2 is clear from the relation (10) [the case (23)] and Proposition 2.1 [the case (22)].

(b) The rationale for the updating of α_t is clear from the proof of Proposition 3.2; if either (18) holds for a function δ with the property that $\{\delta(\alpha_t)\} \to 0$ when $\{\alpha_t\} \to \infty$ or if (16) holds, then the last equality in that proof shows that whenever x^t is not a solution to (1), by increasing the value of α_t, the directional derivative $\psi'_{\alpha_t,r}(x^t; d^t)$ will eventually be negative.

(c) The updating rule for α_t can also be motivated from the point of view of obtaining stationary points to the correct merit function: if a given iterate x^t is a stationary point for the current merit function used but x^t does not solve VIP$(F, X \cap G)$, then the updating rule is so designed that it adjusts the merit function so that eventually x^t is no longer stationary.

(4) If iteration $t-1$ included an Armijo line search, then, at iteration t, the solution to (22), the value of p^t used in (23), and the value of $\psi_{\alpha_t,r}(x^t)$ all are supplied automatically from the use of the Armijo rule at iteration $t-1$. □

4 Convergence analysis

Before we turn to the convergence of the proposed algorithm, we need the following technical lemmas.

Lemma 4.1 (Boundedness of $P(x)$). *For each $x \in X$, the set $P(x)$ is nonempty, closed and bounded.*

Proof. The first two properties follow from the Slater CQ and (11). Let y be the solution to (9b), $p \in P(x)$, and $\nu \in N_X(y)$ satisfy (11a). Let $\overline{x} \in X$ be such that $g_i(\overline{x}) < 0$ for all $i = 1, \ldots, m$. Then,

$$\begin{aligned} 0 &\leq [F(x) + (1/\alpha)\nabla_y\varphi(y,x) + \nabla g(x)^{\mathrm{T}}p + \nu]^{\mathrm{T}}(\overline{x} - y) \\ &\leq [F(x) + (1/\alpha)\nabla_y\varphi(y,x)]^{\mathrm{T}}(\overline{x} - y) \\ &\quad + \sum_{i=1}^{m} p_i[\nabla g_i(x)^{\mathrm{T}}(x - y) + g_i(\overline{x}) - g_i(x)] \\ &= [F(x) + (1/\alpha)\nabla_y\varphi(y,x)]^{\mathrm{T}}(\overline{x} - y) + \sum_{i=1}^{m} p_i g_i(\overline{x}), \end{aligned}$$

where we have used (11b) and (11c), the definition (2) of the normal cone to X at y, and the convexity of g_i, $i = 1, \ldots, m$. It therefore follows that

$$0 \leq \sum_{i=1}^{m} p_i \leq \frac{[F(x) + (1/\alpha)\nabla_y\varphi(y,x)]^{\mathrm{T}}(\overline{x} - y)}{\min_{i\in\{1,\ldots,m\}}\{-g_i(\overline{x})\}},$$

and the result follows. □

Lemma 4.2 (Continuity of $\psi_{\cdot}(\cdot)$). *Assume that $X \supset \{x^t\} \to x \in X$ and $\Re_{++} \supset \{\alpha_t\} \to +\infty$. Then,*

$$\{\psi_{\alpha_t}(x^t)\} \to \psi_\infty(x) := \underset{y \in X \cap T(x)}{\text{maximum}} \; F(x)^{\mathrm{T}}(x-y).$$

Proof. Let $\varepsilon > 0$ be arbitrary. First, we note that ψ_∞ is continuous on X (see the proof of Proposition 2.3), and thus

$$|\psi_\infty(x^t) - \psi_\infty(x)| \le \varepsilon/2, \qquad t \ge \tau_1.$$

Second, $\psi_{\cdot}(x)$ is continuous as a function of α for any fixed value of $x \in X$ (by arguments similar to those for $\psi_\alpha(\cdot)$ in Proposition 2.3, and hence,

$$|\psi_{\alpha_t}(x) - \psi_\infty(x)| \le \varepsilon/2, \qquad t \ge \tau_2.$$

By the above two inequalities,

$$|\psi_{\alpha_t}(x^t) - \psi_\infty(x)| \le \varepsilon, \qquad t \ge \max\{\tau_1, \tau_2\},$$

and the result follows. □

Theorem 4.3 (Global convergence). *In addition to the assumptions of Proposition 3.2, let the function δ be such that $\{\delta(\alpha)\} \to 0$ when $\alpha \to \infty$. Let $\{(x^t, p^t)\}$ be a sequence of primal and dual iterates generated by the algorithm of Table 1. Then, either the sequence $\{(x^t, p^t)\}$ is finite, whence the last iterate (x^t, p^t) is a solution to (3), or the sequence $\{(x^t, p^t)\}$ is infinite, whence it is bounded, any accumulation point is a solution to (3), and*

$$\left\{ \underset{(x,\lambda) \text{ solves } (3)}{\text{minimum}} \; \|(x^t, p^t) - (x, \lambda)\| \right\} \to 0. \tag{25}$$

Proof. The conclusion in the case of finite termination is clear from the construction of (23) and (22).

Clearly, the sequence $\{x^t\}$ is bounded, since X is bounded. The same is true for the sequence $\{p^t\}$, by the boundedness of $\{x^t\}$ and by Lemma 4.1.

We next distinguish two cases.

($\{\alpha_t\} \to \infty$) Let x^∞ be any accumulation point of $\{x^t\}$, corresponding to some subsequence $\mathcal{T}$. This assumption, together with the proof of Proposition 3.2, establish that then, $\psi_{\alpha_t}(x^t) \le [\delta(\alpha_t) + 1/\alpha_t]b/(1-\gamma)$ must hold for infinitely many $t \in \mathcal{T}$. Since $\{\delta(\alpha_t)\} \to 0$ holds by Step 3, $\{\psi_{\alpha_t}(x^t)\}_{t \in \mathcal{T}} \to 0$. Lemma 4.2 establishes that $\psi_\infty(x^\infty) = 0$. In a manner similar to that in the proof of Proposition 2.2, we conclude that x^∞ solves (9) at x^∞ for $\alpha = +\infty$, and that $x^\infty \in G$ therefore holds. Hence,

$x^\infty \in \mathrm{SOL}(F, X \cap G)$ follows. Turning to the proof for the sequence $\{p^t\}$, let p^∞ be any of its accumulation points. The fact that x^∞ solves (9) at x^∞ for $\alpha = +\infty$ then implies that (11) is satisfied with $x = y = x^\infty$, $p = p^\infty$ and $\alpha = +\infty$, which implies that (x^∞, p^∞) is a solution to (3).

($\alpha_t = \overline{\alpha}$, $t \geq \tau$) By the continuity of $\psi_{\overline{\alpha},r}$, the Armijo rule (24) is well-defined. We will show that $\{d^t\} \to 0^n$. The proof proceeds by contradiction. If $\{d^t\} \not\to 0^n$, then as $\{\psi_{\overline{\alpha},r}(x^{t+1}) - \psi_{\overline{\alpha},r}(x^t)\} \to 0$ must hold, there must be a subsequence $\mathcal{T}$ such that $\{\ell_t\}_{t \in \mathcal{T}} \to 0$ holds. There must then be a $\overline{\tau} \geq \tau$ such that

$$\psi_{\overline{\alpha},r}(x^t + [\ell_t/\beta]d^t) - \psi_{\overline{\alpha},r}(x^t) > -\eta[\ell_t/\beta]\psi'_{\overline{\alpha},r}(x^t; d^t), \qquad t \geq \overline{\tau}, \quad t \in \mathcal{T}. \tag{26}$$

Let d^∞ be an accumulation point of $\{d^t\}_{t \in \mathcal{T}}$; the existence of an accumulation point follows since $\{d^t\}$ is bounded. Dividing (26) by ℓ_t/β and taking the limit of $\mathcal{T}$, we obtain by the directional differentiability of $\psi_{\overline{\alpha},r}$ that $\psi'_{\overline{\alpha},r}(x^\infty; d^\infty) \geq 0$. In the limit of $\mathcal{T}$ of the inequality (17), we obtain that $\psi'_{\overline{\alpha},r}(x^\infty; d^\infty) \leq -\gamma\psi_{\overline{\alpha},r}(x^\infty)$. If the right-hand side of this inequality is zero, then it follows that x^∞ solves (9) at x^∞, and therefore $x^\infty \in \mathrm{SOL}(F, X \cap G)$ holds. If $\psi_{\overline{\alpha},r}(x^\infty) > 0$ holds, then we have reached a contradiction. In this case, we conclude that $\{d^t\} \to 0^n$, and we again conclude that $x^\infty \in \mathrm{SOL}(F, X \cap G)$ holds, this time by Proposition 2.1. Turning to the proof for the sequence $\{p^t\}$, we let p^∞ be any accumulation point thereof, and note that $\{d^t\} \to 0^n$ implies from (11), together with the fact that $\nabla_y \varphi(x^\infty, x^\infty) = 0^n$, that (x^∞, p^∞) is a solution to (3). □

Remark 4.4 (On step length selection rules). We observe that the proof establishes that any step length rule that provides a decrease in $\psi_{\alpha_t,r}$ that is at least as large as that of the Armijo rule (24) is a valid choice. This facilitates a plentiful of alternatives for more efficient choices of step length rules. For example, the line search can be replaced by a more general multi-dimensional search along the lines of simplicial decomposition methods; this approach is analyzed further in [Pat98, Chapter 9]. □

The following result shows that the modified descent algorithm finitely turns into a descent approach when F is strongly monotone. [The SQP method under case (b) may be viewed as an adaptive version of the SQP method of [TaF95], albeit under milder convergence conditions.]

Corollary 4.5 (Finite reduction to a descent algorithm). *In addition to the assumptions of Theorem 4.3, assume that F is strongly monotone on X. Let either* (16) *hold or* (18) *hold with $\delta(\alpha) = 0$.*

(a) *There exists a finite integer τ such that $\alpha_t = \overline{\alpha}$, for all $t \geq \tau$.*

(b) *Assume further that $\varphi(y, x) := (1/2)(y - x)^{\mathrm{T}} Q(y - x)$, for some symmetric and positive definite matrix Q. Then, $\overline{\alpha} \leq \max\{\alpha_0, \|Q\|/(2m_F) + \Delta\alpha\}$.*

Proof. The result follows from Theorem 4.3 and Corollary 3.4. □

The following result validates a more active involvement of the penalty parameter r in the algorithm.

Corollary 4.6 (Adaptive penalty parameter). *Let the assumptions of Theorem* 4.3 *hold. In the algorithm of Table* 1, *consider allowing the parameter r to be updated arbitrarily within a bounded interval* $0 \leq r \leq \overline{r}$, *whenever a null step is taken. Then, the conclusion of Theorem* 4.3 *holds for this modified algorithm.*

Proof. In the first case ($\{\alpha_t\} \to \infty$), the value of r is immaterial to the proof. In the second case ($\alpha_t = \overline{\alpha}$, $t \geq \tau$), the value of r is not updated after iteration τ, whence the proof reduces to the original one. □

Remark 4.7 (On the convergence rate of the SQP method). The merit function ψ_α is differentiable at some point $x \in X$ whenever the set $P(x)$ of multipliers for (9b) is a singleton. (This condition holds if the normals $\nabla g_i(x)$ of the constraints active at y are linearly independent and strict complementarity holds there; see [FiM68, Thm. 6].) If this condition holds uniformly over $x \in X$, then it is possible to establish a linear convergence rate in the primal space of the algorithm whenever F is strongly monotone, along the lines of existing analyses of primal SQP methods (e.g., [Pat98]).

Under these assumptions, the SQP method based on the subproblem (4) is locally superlinearly convergent, and quadratically convergent under additional assumptions; see [GaM76]. It is not yet clear if such results are precluded for the algorithm of this paper by the fact that the multiplier estimate p^t is instead chosen in the set $P(x^t)$ associated with the merit function. As remarked in [TaF95], however, under the above-mentioned additional assumptions, the multiplier vector λ^* is unique, in which case the two sequences converge to the same point; if we can guarantee that the unit step is accepted by the Armijo rule close to the solution, then at least a superlinear convergence rate is achieved. □

5 Two applications

We discuss two applications of Theorem 4.3, to a projection algorithm and to an SQP algorithm.

5.1 A modified projection algorithm

Let $\varphi(y, x) := (1/2)(y-x)^{\mathrm{T}}Q(y-x)$, for some symmetric and positive definite matrix Q, and let (16) hold. Then, for any $r \geq 0$, Theorem 4.3 applies. The algorithm thus constructed is a modified projection algorithm, wherein the strictly convex, quadratic subproblems are solved over the set $X \cap T(x^t)$, and the matrix Q is adaptively scaled by the parameter α_t whenever the direction obtained from (22) does not yield descent.

A related projection algorithm is given by Taji and Fukushima [TaF96]. Their problem is obtained from (1) by letting $X := \Re^n$, and we identify their algorithm in this case by setting $\alpha_t := 1$ [cf. (7)]. Their convergence result relies on the monotone decrease in their merit function; in order to guarantee this descent property, the following assumptions must be introduced in their analysis:

1. $\nabla F(x)$ is positive definite on $\Re^n$; and
2. r is chosen so that $r \geq \sup_t\{ \|p\|_\infty \mid p \in P(x^t) \}$, where $P(x^t)$ is the set of multipliers for the linearized constraints (7b).

They also include the assumption that the generated sequence $\{x^t\}$ is bounded. (In the present paper, we have instead assumed that the ground set X is bounded.)

5.2 An SQP algorithm

Our second, and most interesting application, is to SQP methods for (1). Given $x^t \in X$, the SQP method proceeds as follows. We construct a prediction of the multiplier vector λ^* of (3) by solving (22) and extracting an arbitrary multiplier vector $p^t \in P(x^t)$. The search-direction finding subproblem of the SQP scheme then is to find $y^t \in X \cap T(x^t)$ such that (4) holds, where λ^t is replaced by p^t.

We then note that (4) is obtained from (23) by choosing $\Phi_\alpha(y, x, p) := H(x, p)(y - x)$, $p \in P(x)$. Further, (18) holds with $\delta(\alpha) = 0$, and therefore Theorem 4.3 applies. The analysis provided in Theorem 4.3 shows that this SQP method for (1) is globally convergent for non-strictly monotone problem mappings F, and further that the penalty parameter r can be chosen arbitrarily in $\Re_+$.

We also note that in the case where $F \equiv \nabla f$ for some real-valued function $f : X \mapsto \Re$, the above adaptive SQP method is applicable to the nonlinear programming problem whenever f is convex.

This analysis improves that of the SQP method in [TaF95]. The problem is obtained by letting $X := \Re^n$, and their algorithm by choosing $\alpha_t := 1$ and moreover the merit function as the projection-based measure (7). The progress of the algorithm relies on the descent properties of the merit function, which makes it necessary to introduce the following assumptions in the analysis:

1. F is strongly monotone on $\Re^n$;
2. r is chosen so that $r \geq \sup_t\{ \|p\|_\infty \mid p \in P(x^t) \}$; and
3. $\|Q\| < 2m_F$.

Convergence is established under the additional presumption that the sequence $\{x^t\}$ is bounded.

6 Observations and final remarks

The convergence analyses of [TaF95, TaF96] for the above two algorithms do not cover the convergence of the sequence $\{p^t\}$ of multiplier estimates.

The conditions 2. and 3. above are difficult to observe a priori. We note also again in passing that condition 2. is present in some analyses of SQP algorithms applied to nonlinear optimization problems (e.g., [Han77, Psh94]). Further, as is illustrated in Corollary 4.5, the requirement 3. present in the analysis of [TaF95] is not present here, since the function φ is automatically scaled in order to yield descent. By doing so, we gain the additional advantage of not requiring the knowledge or estimation of any problem parameters, instead fulfilling the requirement 3. eventually in the case of strongly monotone problems (cf. Corollary 3.4).

The present SQP algorithm and its convergence analysis extends and improves, respectively, that of primal Newton methods in [MaD87, TFI93], and primal–dual Newton methods in [MaD89].

Recent articles on iterative methods for variational inequality problems, for example [ZhM93, MaZ95, Kon97], and in particular on modified projection methods ([Ius94, SoT96, Sun96, He97]), show that one may reach convergence results under milder conditions when one introduces adaptivity. The above analysis serves as a further illustration that this route of development is worthy of further study also for other algorithms.

We finally note that in the context of nonlinear optimization, recent experience with modifications of SQP algorithms that do not utilize a penalty function, but rather a filter using a dominance concept adapted from multiobjective optimization, shows promise, albeit no convergence analysis for these methods is available at present ([Zop95, FlL97]). A convergence result is however available for the more basic approach of SLP in combination with trust region approaches ([FLT98]), which would suggest the possibility of applying more general non-penalty methods.

Acknowledgement

The presentation of this article benefitted from discussions with Professor Torbjörn Larsson.

References

[Auc89] G. Auchmuty (1989), "Variational principles for variational inequalities," *Numerical Functional Analysis and Optimization*, vol. 10, pp. 863–874.

[Ber95] D. P. Bertsekas. *Nonlinear Programming*, Athena Scientific, Belmont, MA, 1995.

[FiM68] A. V. Fiacco and G. McCormick (1968). *Nonlinear Programming: Sequential Unconstrained Minimization Techniques*, John Wiley & Sons,, New York, NY. Also published as volume 4 in the Classics in Applied Mathematics Series, SIAM, Philadelphia, PA, 1990.

[Fle87] R. Fletcher (1987). *Practical Methods of Optimization*, John Wiley & Sons, Chichester, U.K., second ed.

[FlL97] R. Fletcher and S. Leyffer (1997), "Nonlinear programming without a penalty function," Numerical Analysis Report NA/171, Department of Mathematics, University of Dundee, Dundee, Scotland, U.K.

[FLT98] R. Fletcher, S. Leyffer and Ph. L. Toint (1998), "On the global convergence of an SLP-filter algorithm," Numerical Analysis Report NA/183, Department of Mathematics, University of Dundee, Dundee, Scotland, U.K.

[Fuk92] M. Fukushima (1992), "Equivalent differentiable optimization problems and descent methods for asymmetric variational inequality problems," *Mathematical Programming*, vol. 53, pp. 99–110.

[Fuk96] M. Fukushima (1996), "Merit functions for variational inequality and complementarity problems," In: *Nonlinear Optimization and Applications, Proceedings of the International School of Mathematics "G. Stampacchia" 21st Workshop on Nonlinear Optimization and Applications*, held June 13–21, 1995, in Erice, Italy, G. Di Pillo and F. Giannessi (eds.), Plenum Press, New York, NY, pp. 155–170.

[GaM76] U. M. Garcia Palomares and O. L. Mangasarian (1976), "Superlinearly convergent quasi-Newton algorithms for nonlinearly constrained optimization problems," *Mathematical Programming*, vol. 11, pp. 1–13.

[Han77] S. P. Han (1977), "A globally convergent method for nonlinear programming," *Journal of Optimization Theory and Applications*, vol. 22, pp. 297–309.

[HaP90] P. T. Harker and J.-S. Pang (1990), "Finite-dimensional variational inequality and nonlinear complementarity problems: a survey of theory, algorithms and applications," *Mathematical Programming*, vol. 48, pp. 161–220.

[He97] B. He (1997), "A class of projection and contraction methods for monotone variational inequalities," *Applied Mathematics & Optimization*, vol. 35, pp. 69–76.

[Hog73] W. W. Hogan (1973), "Directional derivatives for extremal-value functions with applications to the completely convex case," *Operations Research*, vol. 21, pp. 188–209.

[Ius94] A. I. Iusem (1994), "An iterative algorithm for the variational inequality problem," *Computational and Applied Mathematics*, vol. 13, pp. 103–114.

[Kon97] I. V. Konnov (1997), "A class of combined iterative methods for solving variational inequalities," *Journal of Optimization Theory and Applications*, vol. 94, pp. 677–693.

[LaP94] T. Larsson and M. Patriksson (1994), "A class of gap functions for variational inequalities," *Mathematical Programming*, vol. 64, pp. 53–79.

[MaD87] P. Marcotte and J.-P. Dussault (1987), "A note on a globally convergent Ne"wton method for solving monotone variational inequalities, *Operations Research Letters*, vol. 6, pp. 35–42.

[MaD89] P. Marcotte and J.-P. Dussault (1989), "A sequential linear programming algorithm for solving monotone variational inequalities," *SIAM Journal on Control and Optimization*, vol. 27, pp. 1260–1278.

[MaZ95] P. Marcotte and D. Zhu (1995), "Global convergence of descent processes for solving non strictly monotone variational inequalities," *Computational Optimization and Applications*, vol. 4, pp. 127–138.

[Pat93a] M. Patriksson (1993), "A unified description of iterative algorithms for traffic equilibria," *European Journal of Operational Research*, vol. 71, pp. 154–176.

[Pat93b] M. Patriksson (1993), *A unified framework of descent algorithms for nonlinear programs and variational inequalities*, PhD thesis, Department of Mathematics, Linköping Institute of Technology, Linköping, Sweden.

[Pat94] M. Patriksson (1994), "On the convergence of descent methods for monotone variational inequalities," *Operations Research Letters*, vol. 16, pp. 265–269.

[Pat97] M. Patriksson (1997), "Merit functions and descent algorithms for a class of variational inequality problems," *Optimization*, vol. 41, pp. 37–55.

[Pat98] M. Patriksson (1998). *Nonlinear Programming and Variational Inequality Problems: A Unified Approach*, Kluwer Academic Publishers, Dordrecht, The Netherlands.

[Psh94] B. N. Pshenichnyj (1994). *The Linearization Method for Constrained Optimization*, vol. 22 of Springer Series in Computational Mathematics, Springer-Verlag, Berlin. Translated from the Russian by S. S. Wilson.

[SoT96] M. V. Solodov and P. Tseng (1996), "Modified projection-type method for monotone variational inequalities," *SIAM Journal on Control and Optimization*, vol. 34, pp. 1814–1830.

[Sun96] D. Sun (1996), "A class of iterative methods for solving nonlinear projection equations," *Journal of Optimization Theory and Applications*, vol. 91, pp. 123–140.

[TaF95] K. Taji and M. Fukushima (1995), "A globally convergent Newton method for solving variational inequality problems with inequality constraints," in *Recent Advances in Nonsmooth Optimization*, D.-Z. Du, L. Qi, and R. S. Womersley, eds., World Scientific Publishing Co, Singapore, pp. 405–417.

[TaF96] K. Taji and M. Fukushima (1996), "A new merit function and a successive quadratic programming algorithm for variational inequality problems," *SIAM Journal on Optimization*, vol. 6, pp. 704–713.

[TFI93] K. Taji, M. Fukushima, and T. Ibaraki (1993), "A globally convergent Newton method for solving strongly monotone variational inequalities," *Mathematical Programming*, vol. 58, pp. 369–383.

[Tob86] R. L. Tobin (1986), "Sensitivity analysis for variational inequalities," *Journal of Optimization Theory and Applications*, vol. 48, pp. 191–204.

[WFM93] J. H. Wu, M. Florian, and P. Marcotte (1993), "A general descent framework for the monotone variational inequality problem," *Mathematical Programming*, vol. 61, pp. 281–300.

[ZhM93] D. L. Zhu and P. Marcotte (1993), "Modified descent methods for solving the monotone variational inequality problem," *Operations Research Letters*, vol. 14, pp. 111–120.

[ZhM94] D. L. Zhu and P. Marcotte (1994), "An extended descent framework for variational inequalities," *Journal of Optimization Theory and Applications*, vol. 80, pp. 349–366.

[Zop95] C. Zoppke-Donaldson (1995), *A tolerance-tube approach to sequential quadratic programming with applications*, PhD thesis, Department of Mathematics and Computer Science, University of Dundee, Dundee, Scotland, U.K.

A Logarithmic Barrier Approach to Fischer Function

Jiming Peng (j.pengits.tudelft.nl)

Cornelis Roos (c.roos@its.tudelft.nl)

Tamás Terlaky (t.terlaky@its.tudelft.nl)

Faculty of Technical Mathematics and Informatics, Delft University of Technology, Mekelweg 4, 2628 CD, Delft, The Netherlands

Abstract

Recently, reformulation of variational inequalities and complementarity problems as a system of equations or an optimization problem has become a hot topic in the field of mathematical programming. Various merit functions play important roles in such reformulations. In this paper, among others, we consider the well-known Fischer function and explore its relationship with other merit functions. It is shown that the Fischer function can be derived from a logarithmic barrier penalty method for constrained minimization. Based on the logarithmic barrier approach, some new merit functions are given and their properties are explored.

Keywords: Variational inequality, Nonlinear complementarity problem, CP-functions, Merit functions.

1. Introduction

Let X be a nonempty closed convex subset in $\Re^n$. The variational inequality problem (VIP) is to find $x^* \in X$ such that

$$F(x^*)^T(y - x^*) \geq 0, \quad \forall\, y \in X, \tag{1.1}$$

G. Di Pillo and F. Giannessi (eds.), Nonlinear Optimization and Related Topics, 277–297.

where $F(x)$ is a mapping from $\Re^n$ into itself. If the constraint set X is the nonnegative orthant $\Re^n_+$, then VIP reduces to a complementarity problem (CP) which is defined as

$$x \geq 0, \ F(x) \geq 0, \ F(x)^T x = 0. \tag{1.2}$$

VIPs and CPs have been widely studied in various fields such as mathematical programming, game theory, economics, and engineering; see e.g. [5, 11] and the references therein.

Recently, much attention has been paid on reformulating VIPs and CPs as an optimization problem or a system of equations [9], and great progress has been made in this direction. In the reformulation of VIPs or CPs, the so-called *Merit Functions* (or *CP functions* for CP) play an important role. A function $M(x)$ is a *merit function* for the VIP (1.1) if the set of global minimizers of $M(x)$ in $\Re^n$ (or on the constrained set X) coincides with the solution set of the VIP. Particularly, if $M(x) \geq 0$ for all $x \in \Re^n$ (or $x \in X$) and it equals to zero if and only x solves problem (1.1), then we can use $M(x)$ as a merit function for the VIP and solve problem (1.1) by globally minimizing $M(x)$. For complementarity problems, we often consider the so-called *CP-functions*:

Definition 1.1 *A function $\phi : \Re^2 \to \Re$ is called an CP-function if*

$$\phi(a, b) = 0 \Leftrightarrow ab = 0, \ a \geq 0, b \geq 0.$$

Via such an CP-function ϕ, one can reformulate CPs as the following equation system

$$\Phi_i(x) = \phi(x_i, F_i(x)) = 0, \ i = 1, \dots, n.$$

Consequently, the function

$$M(x) = \frac{1}{2}\|\Phi(x)\|^2$$

acts as a merit function. One can also construct a merit function $M(x) : \Re^n \to \Re$ for CP by

$$M(x) = \sum_{i=1}^{n} \phi(x_i, F_i(x)),$$

if ϕ is nonnegative on $\Re^2$. In this way we can solve CPs by locating a global minimizer of a merit function. Many CP-functions have been proposed and studied in recent years [9]. In this paper, among others we consider the so-called minimum function ϕ_{NR} and Fischer function ϕ_F defined by

(I) $\phi_{NR}(a, b) = \min\{a, b\}$.

(II) $\phi_F(a,b) = \sqrt{a^2+b^2} - a - b.$

For general VIPs, a simple but well known reformulation is to cast it as the following nonsmooth equation system [11]

$$x - H_\alpha(x) = 0,$$

where $H_\alpha(x) = P_X(x - \alpha F(x))$ with $P_X(x)$ being the projection operator onto X and α is a positive constant. Specializing to the CP case (1.2) with $\alpha = 1$, the above equation system reduces to

$$\min\{x_i, F_i(x)\} = \phi_{NR}(x_i, F_i) = 0, \; i = 1, \ldots, n.$$

In other words, the minimum function ϕ_{NR} is no more than a variance of the projection mapping.

Closely related to the nonsmooth mapping $H_\alpha(x)$ is the regularized gap function introduced by Fukushima [8]

$$\begin{aligned} f_\alpha(x) &= \max_{y \in X}\{F(x)^T(x-y) - \frac{1}{2\alpha}\|y-x\|^2\} \\ &= F(x)^T(x - H_\alpha(x)) - \frac{1}{2\alpha}\|H_\alpha(x) - x\|^2. \end{aligned}$$

An equivalent form of the regularized gap function $f_\alpha(x)$ for CPs is the restricted implicit Lagrangian

$$\phi_{Reg}(a,b) = ab + \frac{1}{2\alpha}(\max\{0, a - \alpha b\}^2 - a^2), \tag{1.3}$$

introduced by Mangasarian and Solodov [17]. VIPs and CPs can be reformulated as constrained optimization problems via the functions $f_\alpha(x)$ and $\phi_{Reg}(a,b)$.

Observe that by using the square of the Fischer function ϕ_F^2, one can also reformulate CP as an optimization problem. Many results about the above mentioned merit functions have been reported, see [6, 10, 13, 14] and the references therein. It is of interest to note that not only some similarities exist between the functions ϕ_{NR} (or $H_1(x)$) and ϕ_F, but there are also some similar attractive properties shared by the functions ϕ_F^2 and ϕ_{Reg} as well [10, 13]. Usually the function ϕ_F^2 possesses these properties under relatively weaker conditions. This motivates us to consider the following question:

Whether the Fischer function can be derived from a model similar to the regularized gap function (or projection mapping)?

In [25], Wu, Florian and Marcotte considered the following generalization of the regularized gap function

$$\begin{aligned} \hat{f}_\alpha(x) &= \max_{y\in X}\{F(x)^T(x-y) - \alpha\psi(x,y)\} \\ &= -\min_{y\in X}\{F(x)^T(y-x) + \alpha\psi(x,y)\} \end{aligned} \tag{1.4}$$

where $\psi(x,y)$ might be any function from $\Re^{2n}$ into $\Re$ satisfying the following conditions:

C.1 ψ is continuously differentiable on $\Re^{2n}$;

C.2 $\psi(x,\cdot)$ is uniformly strongly convex in x, i.e., there exists a constant $\lambda_1 > 0$ such that, for any $x, y_1, y_2 \in \Re^n$, one has

$$\psi(x,y_1) - \psi(x,y_2) \geq [\nabla_y\psi(x,y_2)]^T(y_1 - y_2) + \lambda_1\|y_1 - y_2\|^2;$$

C.3 $\psi(x,x) = 0$ for all $x \in X$ and $\nabla_x\psi(x,y)$ is uniformly Lipschitz continuous in X, i.e. there is a constant λ_2 such that

$$\|\nabla_x\psi(x^1,y) - \nabla_x\psi(x^2,y)\| \leq \lambda_2\|x^1 - x^2\|, \quad \forall x^1, x^2, y \in X;$$

C.4 ψ is nonnegative on $\Re^{2n}$.

An algorithm based on $\hat{f}_\alpha(x)$ was proposed and some theoretical results were reported in [25]. Related results can also be found in [15]. However, since for general non-quadratic ψ it is usually not easy to get an explicit form of $\hat{f}_\alpha(x)$, most of these results have only a theoretical interest.

The main goal of this paper is to further elaborate on merit functions for VIPs and CPs based on $\hat{f}_\alpha(x)$. Since the function $\hat{f}_\alpha(x)$ is a generalization of $f_\alpha(x)$, we naturally hope to get some merit functions based on (1.4) that behave better than $f_\alpha(x)$ (or the projection operator). In this paper we will show that for a special choice of ψ, the merit function based on $\hat{f}_\alpha(x)$ leads to the Fischer function. This is an interesting discovery, since the Fischer function and the merit functions based on it are widely used in the study of VIPs and CPs [2, 3, 4, 16, 21], and are also recognized among the most efficient approaches for solving VIPs and CPs.

The paper is organized as follows. In Section 2 some definitions, assumptions and preliminary results are presented. In Section 3 CP functions based on the model (1.4) are considered. It is shown that for a specific choice of ϕ, the Fischer function is obtained. A new merit function for CPs is considered in Section 4 and its properties are explored as well. Based on this new merit function a globally convergent projection-type method for monotone CPs is presented. Finally, some remarks are made.

2. Preliminaries

First we give some definitions and assumptions that will be used in our analysis.

Let Ω be a nonempty convex subset of $\Re^n$. The mapping F is said to be *monotone on* Ω if

$$[F(x) - F(y)]^T(x - y) \geq 0, \quad \forall\, x, y \in \Omega;$$

it is *strictly monotone on* Ω if

$$[F(x) - F(y)]^T(x - y) > 0, \quad \forall\, x, y \in \Omega\ (x \neq y);$$

and it is *strongly monotone on* Ω *with modulus* $\mu > 0$ if

$$[F(x) - F(y)]^T(x - y) \geq \mu\|x - y\|^2, \quad \forall\, x, y \in \Omega.$$

As is well known [19], when the mapping F is differentiable, F is monotone on Ω if and only if $\nabla F(x)$ is positive semidefinite for all $x \in \Omega$, F is strictly monotone on Ω if $\nabla F(x)$ is positive definite for all $x \in \Omega$, and F is strongly monotone with modulus μ on Ω if and only if $\nabla F(x)$ is uniformly positive definite with constant μ on Ω, i.e., if

$$(x - y)^T \nabla F(x)(x - y)\rangle \geq \mu\|x - y\|^2, \quad \forall\, x, y \in \Omega.$$

The mapping F is said to be *a* P_0*-function* if

$$\max_{1 \leq i \leq n, x_i \neq y_i} (x_i - y_i)(F_i(x) - F_i(y)) \geq 0 \quad \forall x, y \in \Re^n, x \neq y;$$

a P-function if

$$\max_{1 \leq i \leq n} (x_i - y_i)(F_i(x) - F_i(y)) > 0 \quad \forall x, y \in \Re^n, x \neq y;$$

and *a uniform P-function (with modulus* μ*)* if

$$\max_{1 \leq i \leq n} (x_i - y_i)(F_i(x) - F_i(y)) \geq \mu\|x - y\|^2, \quad \forall x, y \in \Re^n, x \neq y.$$

A $n \times n$ matrix M is a P_0-matrix if

$$\max_{1 \leq i \leq n, x_i \neq 0} x_i(Mx)_i \geq 0, \quad \forall 0 \neq x \in \Re^n;$$

and is a P-matrix if

$$\max_{1\le i\le n} x_i(Mx)_i > 0, \quad \forall 0 \neq x \in \Re^n.$$

It is known ([19], Theorem 5.8) that if a P_0(uniformly P)-function is differentiable, then its Jacobian matrices are $P_0(P)$-matrices.

Let $G : \Re^n \to \Re^n$ be a locally Lipschitzian mapping. Then by Rademacher's Theorem [1], G is differentiable almost everywhere. Let D_G denote the set of points where G is differentiable. Clarke's generalized Jacobian of G is defined by

$$\partial G(x) = co\{H|\exists\{x^k\} \subset D_G : x^k \to x, \lim_{k\to\infty} \nabla G(x^k)^T = H\},$$

where co denotes the convex hull of a set. The concept of semismoothness was first introduced for real-valued functions by Mifflin [18]. Later, Qi, Qi and Sun [23, 24] extended it to vector-valued mappings. A mapping G is said to be semismooth at x if for any $d \to 0$ and any $H \in \partial G(x+d)$,

$$Hd - G'(x;d) = o(\|d\|),$$

where $G'(x;d)$ is the directional derivative of G at x in the direction d. Further, G is said to be strongly semismooth at x if for any $d \to 0$ and $H \in \partial G(x+d)$, it holds that

$$Hd - G'(x;d) = O(\|d\|^2).$$

The concept of semismoothness plays an important role in the convergence analysis of methods for nonsmooth equations, particularly in some reformulations of VIPs and CPs. For example, the projection operator is semismooth whenever the constraint set X is polyhedral [20]. This implies that we can reformulate a VIP on a polyhedral set as semismooth equations. Since the Fischer function is strongly semismooth [6], we can reformulate CP as a system of strongly semismooth equations provided that the mapping $F(x)$ is also strongly semismooth.

As in [25], we define the mapping

$$\Gamma(x) = \arg\min_{y\in X}\{F(x)^T(y-x) + \alpha\psi(x,y)\}. \tag{2.1}$$

When the function $\psi(x,y)$ satisfies conditions **C.1-C.4**, then Propositions 3.2 and 3.3 in [25] give

Proposition 2.1 *Suppose that $\psi(x,y)$ satisfies conditions* **C.1-4**. *Then $\Gamma(x)$ is a continuous mapping whenever $\alpha > 0$ and x solves the VIP (1.1) if and only if it is a fixed point of the mapping Γ, i.e., $x \in \Gamma(x)$.*

Note that in the definition of $\Gamma(x)$, we consider a constrained optimization subproblem where $\psi(\cdot, y)$ acts as a penalty function. A usual way to deal with constrained optimization is to reformulate the problem by using some barrier functions. Barrier functions have also been used successfully in other fields such as interior point methods. A barrier function is defined only for the strictly feasible points, and it goes to infinity as we approach to the boundary. Let $\bar{X}$ be the boundary of X. Now we consider (2.1) with a barrier function $\psi(x, y)$ satisfying the following conditions

$\mathcal{C}.1$ ψ is continuously differentiable on $\Re^n \times \text{int}(X)$. For any $y \in \bar{X}$

$$\psi(x, y) = \lim_{z \in \text{int}(X), z \to y} \psi(x, z);$$

$\mathcal{C}.2$ $\psi(x, \cdot)$ is uniformly strongly convex in x, i.e., there exists a constant $\lambda > 0$ such that, for any $x \in \Re^n$ and any $y_1, y_2 \in \text{int}(X)$ one has

$$\psi(x, y_1) - \psi(x, y_2) \geq [\nabla_y \psi(x, y_2)]^T (y_1 - y_2) + \lambda \|y_1 - y_2\|^2;$$

$\mathcal{C}.3$ for any $x \in X$, it holds

$$\lim_{y \to x, y \in X} \frac{\psi(x, y) - \psi(x, x)}{\|x - y\|} = 0;$$

$\mathcal{C}.4$ for any $x \in X$, $\psi(x, y)$ is nonnegative and $\psi(x, x) = 0$.

Remark. From condition $\mathcal{C}.1$ we know that the function $\psi(x, y)$ may take an infinite value at the boundary of $\bar{X}$. Moreover, by conditions $\mathcal{C}.2$ and $\mathcal{C}.3$, we know that there always exist solutions of problem (1.4). Particularly, if x does not belong to $\bar{X}$, then the solution $\Gamma(x)$ is unique and $\Gamma(x) \in \text{int}(X)$. One can also easily verify that our new conditions $\mathcal{C}.3$-4 are weaker than the conditions **C.3-4** in [25].

Now we state our main result in this section.

Proposition 2.2 *Let $\Gamma(x)$ be defined by (2.1) with $\alpha > 0$ and let $\psi(x, y)$ satisfy $\mathcal{C}.1$-4. Then x solves the VIP (1.1) if and only if it is a fixed point of the mapping Γ, i.e., $x \in \Gamma(x)$.*

Proof It is easy to see that if x solves the considered VIP, then $x \in \Gamma(x)$. Now we assume that $x \in \Gamma(x)$, then it follows immediately that $\hat{f}_\alpha(x) = 0$. Suppose the statement of Proposition 2.2 is not true, namely there exists a point $\tilde{x} \in X$ such that

$$F(x)^T (\tilde{x} - x) < 0. \tag{2.2}$$

Since both x and $\tilde{x}$ belong to X, the line segment

$$\{x(t) : x(t) = x + t(\tilde{x} - x), 0 \leq t \leq 1\}$$

is also in X. Let us define

$$h(x,t) = F(x)^T(x(t) - x) + \alpha\psi(x, x(t)) - \alpha\psi(x, x).$$

Since α is a constant, from condition $\mathcal{C}.3$ and inequality (2.2) we derive

$$\lim_{t\to 0^+} \frac{h(x,t)}{\|x(t) - x\|} = \frac{F(x)^T(\tilde{x} - x)}{\|\tilde{x} - x\|} + \alpha \lim_{t\to 0^+} \frac{\psi(x, x(t)) - \psi(x, x)}{\|x(t) - x\|} < 0.$$

The above fact implies that $h(x, t) < 0$ for sufficiently small $t > 0$. Now by the definition of $\hat{f}_\alpha(x)$, the inequality $\hat{f}_\alpha(x) > 0$ must hold. This contradicts our assumption $x \in \Gamma(x)$. The proof is finished. □

Remark. Let $\psi_1(x, y)$ be any function satisfying $\mathcal{C}.1$-3. If $x \in X$ is the unique solution of the problem

$$\min_{y\in X} \psi_1(x, y),$$

then conditions $\mathcal{C}.1$-4 are true for the function $\psi(x, y) = \psi_1(x, y) - \psi_1(x, x)$. Since the function $\psi_1(x, x)$ always acts as a constant in problem (2.1), in the rest of this paper we will use this fact to simplify some of the analysis and consider only the part of $\psi(x, y)$ containing variable y.

3. Logarithmic Barrier Approaches

In this section, we consider the reformulation of CPs given in (1.2) by using models (1.4) and (2.1). Since for general CPs, i.e. $X = \Re^n_+$ which means the complementarity condition holds componentwise, hence without loss of generality we can simply consider a complementarity problem in $\Re$ given as

$$a \geq 0, \ b \geq 0, \ ab = 0. \tag{3.1}$$

Similarly to (1.4), we consider the following simple bounded optimization problem

$$\min_{v\geq 0}\{bv + \xi(a, v)\}.$$

If we choose $\xi(u, v)$ to be a barrier function, then we can equivalently cast the above constrained problem as an unconstrained optimization problem

$$\min\{bv + \xi(a, v)\}. \tag{3.2}$$

We now introduce some conditions which can be viewed as a refinement of the conditions $\mathcal{C}.1$-4 specialized the case that $n = 1$ and $X = [0, \infty)$.

A.1 ξ is continuously differentiable on $\Re \times (0, \infty)$. Define

$$\xi(u, 0) = \lim_{z \to 0+} \xi(u, z);$$

A.2 $\xi(u, \cdot)$ is uniformly strongly convex in u, i.e., there exists a constant $\lambda_3 > 0$ such that, for any $u \in \Re$ and any $v_1, v_2 > 0$ one has

$$\xi(u, v_1) - \xi(u, v_2) \geq \nabla_v \xi(u, v_2)(v_1 - v_2) + \lambda_3 (v_1 - v_2)^2;$$

A.3 for any $u \geq 0$, it holds

$$\lim_{v \to u, v \geq 0} \frac{\xi(u, v) - \xi(u, u)}{|u - v|} = 0;$$

A.4 for any $u, v \geq 0$, it holds $\xi(u, v) \geq 0$ and $\xi(u, u) = 0$.

Let us first choose the following *Logarithmic barrier function*

$$\xi(u, v) = \frac{1}{4}(v^2 - u^2) - \frac{1}{2} u^2 \log \frac{v}{u}. \tag{3.3}$$

This function $\xi(u, v)$ is only defined for $v \in (0, +\infty)$ and it is strongly convex in this domain. For the special case $v = 0$, we choose the value $\xi(u, 0)$ (may be infinity) as defined in Condition **A.1**. At the points $u = 0$, we use the limitation value

$$\xi(0, v) = \lim_{z \to v, z > 0} \lim_{u \to 0} \xi(u, z) = \frac{1}{4} v^2.$$

For $\xi(u, v)$ given by (3.3), via simple calculus we get the unique solution of (3.2) as follows

$$\begin{array}{ll} v = \sqrt{a^2 + b^2} - b \geq 0, & \text{if} \quad a \neq 0; \\ v = 0, & \text{if} \quad a = 0, b \geq 0; \\ v = -2b & \text{if} \quad a = 0, b < 0. \end{array}$$

With a closer look, one can see that $\xi(u, v)$ also satisfies the conditions **A.1-4**. Hence we can reformulate CP as a system of equations

$$\sqrt{a^2 + a^2} - b - a = 0, \tag{3.4}$$

which is precisely the Fischer function ϕ_F. The above relations show that some useful CP-functions can be constructed via (1.4) and (2.1).

The function ϕ_F possess some desirable properties [4, 6, 10, 13]. In what follows we will explore the mathematical background and give some explanations. For simplicity we only consider the case that the unique solution $\gamma(a,b)$ of (3.2) is nonzero. Hence it holds $\gamma(a,b) > 0$. On the other hand, from the optimality condition of problem (3.2) we get

$$b + \left.\frac{\partial \xi}{\partial v}\right|_{(a,\gamma)} = 0. \tag{3.5}$$

Since $\nabla_y^2 \xi(u,v)$ is positive, the Implicit Function Theorem implies that $\gamma(a,b)$ is also differentiable with respect to a and b. Now by (3.5), we have

$$1 + \frac{\partial^2 \xi}{\partial^2 v} \cdot \frac{\partial \gamma}{\partial b} = 0,$$

which implies $\partial\gamma/\partial b < 0$ whenever $\gamma \neq 0$. Similarly, we have

$$\frac{\partial^2 \xi}{\partial u \partial v} + \frac{\partial \gamma}{\partial a} \cdot \frac{\partial^2 \xi}{\partial^2 v} = 0,$$

from which it follows that

$$\frac{\partial \gamma(a,b)}{\partial a} = -\frac{\frac{\partial^2 \xi}{\partial u \partial v}}{\frac{\partial^2 \xi}{\partial^2 v}}.$$

Obviously if

$$\left|\frac{\partial^2 \xi}{\partial u \partial v}\right| \leq \frac{\partial^2 \xi}{\partial^2 v},$$

then

$$\left|\frac{\partial \gamma}{\partial a}\right| \leq 1.$$

For the case that $\gamma = 0$, $a = 0$ and $b \geq 0$ must hold.

Note that our above discussion covers not only the Fischer function ϕ_F, but also all the functions satisfying the conditions **A.1-4**. Now we put two additional conditions on the barrier function ξ.

A.5 For any $u \in \Re$ and $v > 0$, it holds

$$\frac{\partial^2 \xi(u,v)}{\partial u \partial v} \leq \frac{\partial^2 \xi(u,v)}{\partial^2 v}$$

where the equality holds if and only if $u = v$.

A.6 For any $u \in \Re$ and $v > 0$, it holds

$$\left.\frac{\partial \xi}{\partial v}\right|_{(tu,tv)} = t\left.\frac{\partial \xi}{\partial v}\right|_{(u,v)}$$

The following results can be viewed as a generalization of the Fischer function.

Theorem 3.1 *Suppose $\xi(u,v)$ satisfies the conditions* **A.1-6**. *Let $\gamma(a,b)$ be the unique solution of (3.2) and let $\Phi(a,b) = \frac{1}{2}(\gamma(a,b)-a)^2$. Then we have*

$$\Phi(a,b) = 0 \Leftrightarrow \frac{\partial \Phi(a,b)}{\partial a} = 0 \Leftrightarrow \frac{\partial \Phi(a,b)}{\partial b} = 0,$$

further

$$\frac{\partial \Phi(a,b)}{\partial a} \cdot \frac{\partial \Phi(a,b)}{\partial b} \geq 0,$$

where the equality holds if and only if (a,b) is complementarity. Moreover $\gamma(a,b)$ is strongly semismooth in $\Re^2$.

Proof Since $\gamma(a,b)$ is differentiable if $(a,b) \neq 0$, we need only to study the case $(a,b) = (0,0)$. By condition **A.6**, we have $\gamma(ta,tb) = t\gamma(a,b)$, thus $\Phi(a,b)$ is also differentiable at $(0,0)$.

Further, since $\gamma(ta,tb) = t\gamma(a,b)$, at the points where γ is differentiable with respect to a and b, we have

$$\left.\frac{\partial \gamma}{\partial a}\right|_{(ta,tb)} = t\left.\frac{\partial \gamma}{\partial a}\right|_{(a,b)}, \quad \left.\frac{\partial \gamma}{\partial b}\right|_{(ta,tb)} = t\left.\frac{\partial \gamma}{\partial b}\right|_{(a,b)},$$

which implies that $\gamma(a,b)$ is strongly semismooth at $(0,0)^T$. Since γ is differentiable at any other points except for the origin, γ is strongly semismooth in $\Re^2$. □

By the above theorem and Proposition 2.2, we can cast CPs as the following system of strongly semismooth equations

$$\phi(x) = \begin{pmatrix} \gamma(x_1, F_1(x)) - x_1 \\ \vdots \\ \gamma(x_n, F_n(x)) - x_n \end{pmatrix} = 0.$$

We give an example of a family of functions that satisfies conditions **A.1-6**.

Example 3.1 *Consider the functions*

$$\xi_\alpha(u,v) = \frac{1}{4}(v^2-u^2) - \frac{1}{2}u^2 \log\frac{v}{u} + \frac{\alpha}{2}(v-u)^2, \ \alpha \geq 0.$$

In this situation, $\gamma(a,b)$ takes the following form

$$\gamma(a,b) = \frac{\sqrt{(b-\alpha a)^2 + (2\alpha+1)a^2} - b + \alpha a}{1+2\alpha}.$$

Hence

$$(1+2\alpha)(\gamma(a,b) - a) = \sqrt{(1+\alpha)^2 a^2 + b^2 - 2\alpha ab} - b - (1+\alpha)a.$$

We note that apart from a constant scaling, this class of CP-functions is just the same as given in a recent work of Kanzow and Kleinmichel [14].

Remark. Based on any function $\xi(u,v)$ satisfing the conditions **A.1-6**, one can easily construct similar barrier function for problem (1.4) as follows

$$\psi(x,y) = \sum_{i=1}^{n} \xi(x_i, y_i), \quad x \in \Re^n, y \in \Re^n_+.$$

4 A New Reformulation for CP

In this section, we study some merit functions based on the logarithmic barrier approach. The section consists of two parts. In the first subsection, we consider a new CP function and study its properties. In the second part, CP is reformulated as a simple constrained optimization problem. Then some properties of the optimization problem are explored, and a globally convergent projection-type method for the monotone CP is proposed.

4.1. A New CP Function and Its Properties

For the CP (3.1), we consider the optimization problem given by (3.2) with $\xi(u,v)$ given by (3.3)[1]. This leads to the following optimization problem

$$\phi_R(a,b) = \max\{b(a-v) - \frac{1}{4}v^2 + \frac{1}{2}a^2 \log v + \frac{1}{4}a^2 - \frac{1}{2}a^2 \log a\} \tag{4.6}$$

Equivalently we have

$$\phi_R(a,b) = \begin{cases} \frac{1}{2}b^2 + ab - \frac{1}{2}b\sqrt{a^2+b^2} - \frac{1}{2}a^2\log a + \frac{1}{2}a^2 \log(\sqrt{a^2+b^2} - b) & \text{if } a > 0; \\ \frac{1}{2}b^2 + ab - \frac{1}{2}b\sqrt{a^2+b^2} & \text{if } a = 0. \end{cases}$$

[1]It is possible to consider any function $\xi(u,v)$ satisfying **A.1-6**. However in this section we study only one special choice. The interested reader can explore such possibilities in a similar way as we present here.

Note that the function ϕ_R is only defined on $\Re_+ \times \Re$. For future use we recall the following well known inequality

$$-t+1 \leq -\log t, \quad \text{for all } t > 0. \tag{4.7}$$

Now we have

Lemma 4.1 *$\phi_R(a,b) \geq 0$ and $\phi_R(a,b) = 0$ if and only if (a,b) is complementary.*

Proof We first assume $a > 0$. It follows from the definition of ϕ_R and (4.7) that

$$\begin{aligned} \phi_R(a,b) &\geq \frac{1}{2}b^2 + ab - \frac{1}{2}b\sqrt{a^2+b^2} - \frac{1}{2}a^2\left(\frac{a}{\sqrt{a^2+b^2}-b} - 1\right) \\ &= \frac{1}{2}(b^2 + ab - b\sqrt{a^2+b^2} - a\sqrt{a^2+b^2} + a^2) \\ &= \frac{1}{4}(a+b-\sqrt{a^2+b^2})^2 \geq 0. \end{aligned} \tag{4.8}$$

It is easy to verify that (4.8) still holds if $a = 0$. The second half of the lemma follows from direct algebraic calculus. This completes the proof of the lemma. □

One can easily show that $\phi_R(a,b)$ is once differentiable if $a > 0$, and twice differentiable except for the points that a and b are complementary. Further, at the point $a = 0$, it holds

$$\left.\frac{\partial \phi_R}{\partial a}\right|_{(0^+,b)} = \lim_{a \to 0^+} \frac{\phi_R(a,b) - \phi_R(0,b)}{a} = b.$$

At the points with $a = 0$, this right derivative will be denoted by $\frac{\partial \phi_R}{\partial a}$ in the rest of this paper.

The next lemma gives some results about the gradients of $\phi_R(a,b)$.

Lemma 4.2 *Let $\phi_R(a,b)$ be defined by (4.6). Then*

$$\phi_R(a,b) = 0 \Leftrightarrow \frac{\partial \phi_R}{\partial b} = 0 \Leftrightarrow \frac{\partial \phi_R}{\partial a} \cdot \frac{\partial \phi_R}{\partial b} = 0, \tag{4.9}$$

and

$$\frac{\partial \phi_R}{\partial a}\frac{\partial \phi_R}{\partial b} \geq 0 \quad \textit{for all } (a,b)^T \in \Re_+ \times \Re. \tag{4.10}$$

Proof We first assume $a > 0$. By the definition of ϕ_R, we have

$$\frac{\partial \phi_R}{\partial a} = b + a\log\frac{\sqrt{a^2+b^2}-b}{a}, \tag{4.11}$$

and

$$\frac{\partial \phi_R}{\partial b} = a + b - \sqrt{a^2 + b^2}. \tag{4.12}$$

It follows that $\phi_R(a,b) = 0$ if and only if $\frac{\partial \phi_R(a,b)}{\partial b} = 0$. Furthermore, it is easy to see that

$$a + b - \sqrt{a^2 + b^2} \begin{cases} > 0, & \text{if } a > 0, \ b > 0; \\ = 0, & \text{if } a > 0, \ b = 0; \\ < 0, & \text{if } a > 0, \ b < 0. \end{cases} \tag{4.13}$$

On the other hand, we have

$$\frac{\partial^2 \phi_R}{\partial a \partial b} = 1 - \frac{a}{\sqrt{a^2 + b^2}} > 0, \quad \text{for all } b \neq 0,$$

which implies that $\frac{\partial \phi_R}{\partial a}$ is a strictly increasing function of b, since $\frac{\partial \phi_R}{\partial a}|_{b=0} = 0$. It follows that

$$b + a \log \frac{\sqrt{a^2 + b^2} - b}{a} \begin{cases} > 0, & \text{if } a > 0, \ b > 0; \\ = 0, & \text{if } a > 0, \ b = 0; \\ < 0, & \text{if } a > 0, \ b < 0. \end{cases} \tag{4.14}$$

The inequalities (4.13) and (4.14) prove the statements (4.9) and (4.10) if $a > 0$. When $a = 0$, it holds $\frac{\partial \phi_R}{\partial a} = b$, thus the inequalities (4.14) are still true. This completes the proof of the lemma. □

We next study the growth behavior of the CP function $\phi_R(a,b)$. By (4.8) and [10, Lemma 3.7], we have

$$\lim_{a \to +\infty, |b| \to \infty} \phi_R(a,b) = +\infty, \tag{4.15}$$

and that

$$\lim_{b \to -\infty} \phi_R(a,b) = +\infty. \tag{4.16}$$

For the Fischer function ϕ_F, it is easy to see that

$$\lim_{a \to +\infty} \phi_F(a,b) = -b,$$

for any fixed $b \in \Re$, and that

$$\lim_{b \to +\infty} \phi_F(a,b) = -a,$$

for any fixed $a \in \Re$. For the CP-function $\phi_R(a,b)$, one has

Lemma 4.3 *Let $\phi_R(a,b)$ be defined by (4.6). Then*

$$\lim_{|b| \to \infty} \phi_R(a,b) = \infty, \tag{4.17}$$

for any fixed $a > 0$, and that

$$\lim_{a\to\infty} \phi_R(a,b) = \frac{b^2}{2}, \tag{4.18}$$

for any fixed $b \in \Re$.

Proof We first show that (4.17) is true. If $b \to -\infty$, then (4.17) follows from (4.8). We next consider the case that $b \to +\infty$. If $b \geq a > 0$, we have

$$\begin{aligned}
\phi_R(a,b) &= \frac{1}{2}b^2 + ab - \frac{1}{2}b\sqrt{a^2+b^2} - \frac{1}{2}a^2\log a + \frac{1}{2}a^2 \log(\sqrt{a^2+b^2} - b) \\
&\geq \frac{1}{2}b^2 + ab - \frac{1}{2}b(b + (\sqrt{2}-1)a) - \frac{1}{2}a^2\log a + \frac{1}{2}a^2\log(\sqrt{2}-1)\frac{a^2}{b} \\
&= \left(\frac{3}{2} - \frac{\sqrt{2}}{2} - \frac{\sqrt{2}+1}{2}[(\sqrt{2}-1)\frac{a}{b}]\log[(\sqrt{2}-1)\frac{a}{b}]\right) ab,
\end{aligned}$$

where the inequality follows from the fact that $b+(\sqrt{2}-1)\frac{a^2}{b} \leq \sqrt{a^2+b^2} \leq b+(\sqrt{2}-1)a$ whenever $b \geq a > 0$. When $b \to +\infty$ and $a > 0$ is fixed, the right hand side of the above inequality goes to infinity. This proves (4.17).

If $b = 0$, then (4.18) holds trivially. If $b \neq 0$, then

$$\begin{aligned}
\phi_R(a,b) - \frac{1}{2}b^2 &= ab - \frac{1}{2}b\sqrt{a^2+b^2} - \frac{1}{2}a^2\log a + \frac{1}{2}a^2\log(\sqrt{a^2+b^2} - b) \\
&= b^2\left(\frac{1}{t} - \frac{1}{2}\sqrt{1+\frac{1}{t^2}} + \frac{1}{2t^2}\log(\sqrt{1+t^2} - t)\right),
\end{aligned}$$

where $t = \frac{b}{a}$. For any fixed $b \neq 0$, one has $\lim_{a\to+\infty} t = 0$. Since

$$\begin{aligned}
\lim_{t\to 0}\left\{\frac{1}{t} - \frac{1}{2}\sqrt{1+\frac{1}{t^2}} + \frac{1}{2t^2}\log(\sqrt{1+t^2} - t)\right\} &= \lim_{t\to 0}\frac{t - \frac{t\sqrt{1+t^2}}{2} + \log(\sqrt{1+t^2} - t)}{2t^2} \\
&= \lim_{t\to 0}\frac{1-\sqrt{1+t^2}}{4t} = 0.
\end{aligned}$$

It follows that

$$\lim_{a\to+\infty} \phi(a,b) - \frac{1}{2}b^2 = 0.$$

The proof of the lemma is completed. □

4.2. A New Reformulation for CP

Now we return to problem (1.2). By Lemma 4.1, we can reformulate (1.2) as the following simple constrained minimization problem

$$\min_{x\geq 0} M_R(x) \tag{4.19}$$

with $M_R(x) = \sum_{i=1}^n \phi_R(x_i, F_i(x))$. We first consider the level sets defined by

$$L(c) := \{x \in \Re^n_+ | M_R(x) \leq c\}. \tag{4.20}$$

By (4.8) and using some known results about the level sets of the merit function based on the Fischer function (see [4, 10, 13]), one easily derives the following results.

Theorem 4.1 *Assume that $F(x)$ is a continuous uniformly P-function on $\Re^n_+$. Let the level set $L(c)$ be defined by (4.20). Then $L(c)$ is bounded for all $c \geq 0$.*

Theorem 4.2 *If F is a Lipschitz-continuous uniformly P-function on $\Re^n_+$, then there exists a constant $\eta > 0$ such that*

$$\|x - x^*\|^2 \leq \eta M_R(x)$$

for all $x \in \Re^n_+$, where x^* denotes the unique solution of CP.

Now we begin to study the stationary points of the merit function. Since the merit function might be nondifferentiable at some points where $x_i = F_i = 0$ for some $1 \leq i \leq n$, we need to define what is a *feasible direction.* Let us define the index set

$$I_0(x) = \{i : x_i = 0, 1 \leq i \leq n\}.$$

Definition 4.4 *Suppose that $x \in \Re^n_+$. A direction d is said to be a feasible direction for problem (4.19) at the current point x if*

$$d_i \geq 0, \quad \forall i \in I_0(x).$$

¿From the differentiability properties of ϕ_R, we conclude that $M_R(x)$ is directionally differentiable for any feasible direction d. Now we recall the definition of a stationary point

Definition 4.5 *Suppose that $x \in \Re^n_+$. Then x is said to be a stationary point of problem (4.19) if for any feasible direction d the directional derivative $M'_R(x; d) \geq 0$.*

Remark. If $M_R(x)$ is differentiable at the current point, then the above definition reduces to the standard stationary point condition such that

$$[\nabla M_R(x)]_i \geq 0, \quad x_i \geq 0, \quad x_i[\nabla M(x)]_i = 0, \quad \forall\, i = 1, \ldots, n.$$

Theorem 4.3 *Assume that $x^* \in \Re^n_+$ is a stationary point of the simple bounded optimization problem (4.19) such that $\nabla F^T(x^*)$ is a P_0-matrix. Then x^* is a solution of CP.*

Proof For any feasible direction d, the directional derivative

$$M_R'(x;d) = d^T(d_a + \nabla F^T(x^*)d_b),$$

where d_a and d_b denote the vectors whose i-th components are

$$\left.\frac{\partial \phi_R}{\partial a}\right|_{(x_i^*, F_i(x^*))}, \quad \text{and} \quad \left.\frac{\partial \phi_R}{\partial b}\right|_{(x_i^*, F_i(x^*))},$$

respectively. If x^* is not a solution, then there exists a nonempty index set $I_b(x^*)$ such that

$$\left.\frac{\partial \phi_R}{\partial b}\right|_{(x_i^*, F_i(x^*))} \neq 0, \quad \forall\, i \in I_b(x^*).$$

It follows from Lemma 4.2 that

$$\frac{\partial \phi_R}{\partial a} \cdot \frac{\partial \phi_R}{\partial b} > 0, \quad \forall\, i \in I_b(x^*).$$

Since $\nabla F^T(x^*)$ is a P_0-matrix, there exists an index $i_0 \in I_b(x^*)$ such that

$$(d_b)_{i_0}[\nabla F^T(x^*)d_b]_{i_0} \geq 0.$$

Now let us choose d be the direction whose i_0-th component is $-(d_b)_{i_0}$ and others components are zero. It is easy to see that d is a feasible direction. On the other hand, by the choices of the direction d and the index i_0 we get

$$M_R'(x^*;d) = -(d_a)_{i_0}(d_b)_{i_0} - (d_b)_{i_0}[\nabla F^T(x^*)d_b]_{i_0} < 0.$$

This contradicts the assumption that x^* is a stationary point for (4.19). The proof of the theorem is completed. □

Remark. Comparing the merit function $M_R(x)$ with the regularized gap function $f_1(x)$, one can easily see that they share some similar properties. But there exist also some minor differences. For instance, $f_1(x)$ is always differentiable while $M_R(x)$ is only directionally differentiable for feasible directions in the nonnegative orthant. On the other hand, a stationary point of the regularized gap function $f_1(x)$ solves the considered CP if the mapping $F(x)$ is a uniformly P-function (or in other words the Jacobian $\nabla F^T(x)$ is a P-matrix) while for $M_R(x)$ we need only to assume that $F(x)$ is a P_0-function.

Based on various motivations, several authors have suggested to reformulate CPs as simple constrained optimization problems, see [7, 12, 17, 21, 22]. In what follows we also propose a projection-type method for (4.19) in the case that the considered mapping $F(x)$ is monotone. For any given $x^k \in \Re^n_+$, let $\frac{\partial \phi_R}{\partial a}(x^k, F(x^k))$ and $\frac{\partial \phi_R}{\partial b}(x^k, F(x^k))$ denote the vectors whose i-th components are $\frac{\partial \phi_R}{\partial a}(x_i^k, F_i(x^k))$ and $\frac{\partial \phi_R}{\partial b}(x_i^k, F_i(x^k))$ respectively. Let

$$d^k := -\left.\frac{\partial \phi_R}{\partial b}\right|_{(x^k, F(x^k))} \tag{4.21}$$

Lemma 4.6 *Let $x^k \in \Re^n_+$ and $F(x)$ be a continuously differentiable and monotone function in $\Re^n_+$. Then the direction d^k defined by (4.21) is a descent direction of $M_R(x)$ at the current iterate x^k if x^k is not a solution of CP.*

Proof The proof of the lemma is analogous to the proof of Lemma 4.1 in [10], so it is omitted here. □

Now we can state the algorithm.

Algorithm

Step 0: Choose $x^0 \in \Re^n_+$, $\epsilon \geq 0$, $\delta \in (0,1)$ and $\beta \in (0,1)$. Set $k := 0$.

Step 1: If $M_R(x^k) \leq \epsilon$, then x^k is an approximate solution of CP; Otherwise go to Step 2.

Step 2: Compute d^k by (4.21).

Step 3: Compute a steplength $t_k = \beta^{m_k}$, where m_k is the smallest nonnegative integer m satisfying the Armijo-type condition

$$M_R(x^k + \beta^m d^k) \leq M_R(x^k) - \beta^{2m}\delta \|d^k\|^2.$$

Step 4: Set $x^{k+1} = x^k + t_k d^k$, $k := k+1$ and go to Step 1.

By using analogous arguments as in the proof of Lemma 3.1 in [22], we have:

Lemma 4.7 *Let the sequence $\{x^k\}$ be generated by the algorithm stated above. Then for each integral $k \geq 1$, $x^k \in \Re^m_+$.*

Now, following the proofs of Proposition 11 in Jiang [12] and Theorem 3.3 in [22], and by setting the constant ϵ in the algorithm equal to zero, we get:

Theorem 4.4 *Let $F(x)$ be smooth and monotone in $\Re^n_+$. Then the algorithm stated above is well-defined for any initial point $x^0 \in \Re^n_+$. Furthermore, if x^* is an accumulation point of the sequence $\{x^k\}$ generated by the algorithm, then x^* is a solution of CP.*

5 Concluding Remarks

In this paper, we discussed a class of merit functions $\hat{f}_\alpha(x)$ for CPs given by (1.4) and its relationship with the well-known Fischer function. This observation leads to some new reformulations for CPs. A globally convergent projection-type method for monotone CPs was proposed. Using the model (1.4), one can easily construct new merit functions similar to the Fischer function and various types of gap function.

It is interesting to note that our merit function is based on the logarithmic barrier function. Nowadays, barrier functions are widely used in different areas, such as penalty methods and interior point methods for many mathematical programming problems. Although we were not able to generate CP functions that are more effective than the Fischer function, we still hope our results can lead to some new discoveries in the study of VIP and CPs. For instance, let us consider the box constrained VIP with a simple box $X = [0, 1]$. Similarly to (3.2), we can consider the problem

$$\min\left\{by + \frac{y^2}{4} - \frac{a^2}{2}\log y + \frac{(1-y)^2}{4} - \frac{(1-a)^2}{2}\log(1-y)\right\}.$$

Let us denote the unique solution of the above problem by $\rho(a, b)$. The reader might be interested what properties may this function possess? What is the relationships between this function and other known reformulation schemes for the box constrained VIP? These questions deserve further investigations.

References

[1] F.H. Clarke, *Optimization and Nonsmooth Analysis*, Wiley, New York, 1983.

[2] F. Facchinei, A. Fischer and C. Kanzow, A semismooth Newton method for variational inequalities: The case of box constraints , In: M. Ferris and J.S. Pang (eds.), *Complementarity and Variational Problems: State of the Art*, SIAM, Philadelphia, 1997, 76–90.

[3] F. Facchinei, A. Fischer, C. Kanzow and J.M. Peng, A simply constrained optimization reformulation of KKT systems arising from variational inequalities, To appear in: *Applied Mathematics and Optimization.*

[4] F. Facchinei, J. Soares, A new merit function for nonlinear complementarity problems and a related algorithm, *SIAM J. Optimization*, 7(1997), 225–247.

[5] M. Ferris, J.S. Pang Engineering and economic applications of complementarity problems, *SIAM Review*, 39(1997), 669–713.

[6] A. Fischer, Solution of monotone complementarity problems with locally Lipschitzian functions, *Mathematical Programming*, 76(1997), 513–532.

[7] A. Fischer, A new constrained optimization reformulation for complementarity problems, *J. Optimization Theory and Applications*, 97(1998), 105–117.

[8] M. Fukushima, Equivalent differentiable optimization problems and descent methods for asymmetric variational inequality problems, *Mathematical Programming*, 53(1992), 99–110.

[9] M. Fukushima, Merit functions for variational inequality and complementarity problems, In: G. Di Pillo and F. Giannessi (eds.), *Nonlinear Optimization and Applications*, Plenum Press, New York, NY, 1996, 155–170.

[10] C. Geiger, C. Kanzow, On the resolution of monotone complementarity problems, *Computational Optimization and Applications*, 5(1996), 155–173.

[11] P.T. Harker, J.S. Pang, Finite dimensional variational inequality and nonlinear complementarity problems: A survey of theory , algorithms and applications, *Mathematical Programming*, 48(1990), 161–220.

[12] H.Y. Jiang, Unconstrained minimization approaches to nonlinear complementarity problems, *J. of Global Optimization*, 9(1996), 169–181.

[13] C. Kanzow, Nonlinear complementarity as unconstrained optimization, *J. of Optimization Theory and Applications*, 88(1996), 139–155.

[14] C. Kanzow, H. Kleinmichel, A new class of semismooth Newton-type methods for nonlinear complementarity problems, *Computational Optimization and Applications*, 11(1998), 227–251.

[15] T. Larsson, M. Patriksson, A class of gap functions for variational inequalities, *Mathematical Programming*, 64(1994), 53–79.

[16] Z.Q. Luo, P. Tseng, A new class of merit functions for the nonlinear complementarity problem, In: M. Ferris and J.S. Pang (eds.), *Complementarity and Variational Problems: State of the Art*, SIAM,Philadelphia, 1997, 204–225.

[17] O.L. Mangasarian, M.V. Solodov, Nonlinear complementarity as unconstrained and constrained minimization, *Mathematical Programming*, 62(1993), 277–297.

[18] R. Mifflin, Semismooth and semiconvex functions in constrained optimization, *SIAM J. Control and Optimization*, 15(1977), 959–972.

[19] J.J. Moré, W.C. Rheinboldt, On P- and S-Functions and related classes of n-dimensional nonlinear mappings, *Linear Algebra and Its Application*, 6(1973), 45–68.

[20] J.S. Pang, L. Qi, Nonsmooth equations: Motivation and algorithms, *SIAM J. Optimization*, 3(1993), 443–465.

[21] J.M. Peng, Global method for monotone variational inequality problems with inequality constraints, *J. of Optimization Theory and Applications*, 95(1997), 419–430.

[22] J.M. Peng, Derivative-free Methods for Monotone Variational Inequality and Complementarity Problems, *J. of Optimization Theory and Applications:* 99(1998), 235–252.

[23] L. Qi, Convergence analysis of some algorithms for solving nonsmooth equations, *Mathematics of Operations Research*, 18(1993), 227–244.

[24] L. Qi, J. Sun, A nonsmooth version of Newton's method, *Mathematical Programming*, 58(1993), 353–367.

[25] J. H. Wu, M. Florian and P. Marcotte, A general descent framework for the monotone variational inequality problem, *Mathematical Programming*, 61(1993), 281–300.

On an Approach to Optimization Problems with a Probabilistic Cost and or Constraints

Elijah Polak (polak@eecs.berkeley.edu)
Department of Electrical Engineering and Computer Sciences, University of California, Berkeley, CA 94720, USA

Roger J-B Wets (wets@ucdavis.edu)
Department of Mathematics, University of California, Davis, CA 95616, USA

Armen der Kiureghian (adk@ce.berkeley.edu)
Department of Civil and Environmental Engineering, University of California, Berkeley, CA 94720, USA

Abstract

We present a new approach to a class of probability constrained optimization problems that arise in the context of optimal engineering design. These problems are characterized by the fact that the probability of failure of one or several components either must be minimized or must not exceed a preassigned threshold. Our approach is interactive: it consists of replacing the original optimal design problem in which either the cost function or a constraint are expressed in terms of a probability of failure, by a constrained minimax problem. Once the minimax problem is solved, the actual probability of failure is computed. Depending on the outcome of this computation, we provide heuristic rules for modifying the minimax problem and repeating this process a couple of times. An important feature of our new approach is that it decouples optimization and probability of failure calculations. This decoupling allows independent selection of methods for the solution of the optimization and the reliability subproblems. We present an example to demonstrate the effectiveness of our approach.

Keywords: probability constraints, chance constraints, reliability cost/constraints.

G. Di Pillo and F. Giannessi (eds.), Nonlinear Optimization and Related Topics, 299–315.

1 Introduction

Quite often, engineering design must take into account the uncertainties present in quantities such as material properties, loads, effective dimensions, and boundary conditions. Worst case design can be overly conservative and hence it is common to use probabilistic methods to ensure satisfactory performance. In the important area of earthquake resistant structural design, satisfactory performance equates with a sufficiently low probability of failure during an earthquake.

Depending on design requirements, various formulations of a probabilistic optimal structural design problem can be envisioned. For example, *(i)* minimize the probability of failure of the structure (i.e., maximize reliability), subject to a number of deterministic design constraints or *(ii)* minimize the cost of the structure subject to a constraint on the probability of failure (minimum reliability) as well as a number of deterministic design constraints. The concept of failure can be simple, e.g., if any member or substructure fails the structure is considered to have failed, or it can be complex, e.g., it takes a combination of member/substructure failures to result in a failure of the structure. In this paper we will deal with the former case which is relevant in situations where a minimum reliability threshold is dictated by a code or specification, e.g., for the purpose of safeguarding against the loss of life or other intangibles. Most previous work in probabilistic optimal design deals with this class of problems, see [11, 8].

The type of structural reliability problems that we are interested in are characterized by a cost or constraints that are formulated in terms of real-valued limit-state functions $G^k(x, v)$, $k = 1, ..., p$, defined on $\mathbb{R}^n \times \mathbb{R}^m$, which depend on a random vector $v \in \mathbb{R}^m$ that represents uncertain structural quantities, such as material property constants, load magnitudes, member dimensions, and boundary variables, and a deterministic design vector $x = (x^1, ..., x^n) \in \mathbb{R}^n$. The limit-state functions are formulated so that an outcome v satisfying the inequality $G^k(x, v) \geq 0$[1] corresponds to a failure of the structure. The deterministic design vector x can represent either physical quantities, such as member sizes, or parameters of the probability distribution of the random variable v, e.g., mean values of strength variables, or both. To simplify computations it is common to use smooth, nonlinear, invertible transformation of variables $u = H(v)$, which has the property that the new random variables u has standard normal probability density $\phi(u) = (2\pi)^{-n/2}\exp(-\|u\|^2/2)$ (see [4] and [7] for the forms of this transformation). In the new coordinate system, one obtains limit-state functions $g^k(x, u) \stackrel{\Delta}{=} G^k(H^{-1}(u), x)$, with the inequality $g^k(x, u) \geq 0$ now indicating a failure event. Thus the probability of failure in the kth mode is given by

$$F^k(x) \stackrel{\Delta}{=} \int_{u \in S_k(x)} \phi(u) du, \qquad (1)$$

[1] In reliability engineering, the opposite convention is used: $G^k(x, v) \leq 0$ denotes a failure of the structure.

where

$$S_k(x) \stackrel{\Delta}{=} \{u \in \mathbb{R}^m \mid g^k(x,u) \geq 0\} \ . \tag{2}$$

Hence a typical optimal design problem can have one of the following forms:

$$\min\{\Psi(x) \mid f^j(x) \leq 0,\ j = 1,...,q\} \tag{3}$$

where

$$\Psi(x) \stackrel{\Delta}{=} \max_{k \in \mathbf{p}} \ F^k(x), \qquad (4),$$

with $\mathbf{p} \stackrel{\Delta}{=} \{1,...,p\}$, and the functions $f^j : \mathbb{R}^n \to \mathbb{R}$ are smooth and represent the deterministic constraints; or, alternatively,

$$\min\{f^0(x) \mid \Psi(x) - \alpha \leq 0,\ f^j(x) \leq 0,\ j = 1,...,q\}, \tag{5}$$

where $f^0 : \mathbb{R}^n \to \mathbb{R}$ is a smooth cost function, the $f^j(\cdot)$, $j = 1,...,q$ and $\Psi(\cdot)$ are as above, and $\alpha > 0$ is the maximum allowable failure probability. As we will see in the next section, the functions $F^k(\cdot)$ are continuous under essentially almost no assumptions on the limit-state functions $g^k(\cdot,\cdot)$ except continuity. Hence both problems (3) and (5) are well defined. However, as a practical matter, one cannot evaluate these functions F^k and consequently these optimization problems are considered to be 'unsolvable' and must be dealt with using some approximation or transformation technique. The technique currently in favor with the structural optimization community (see, [11]) is that of linearization of the limit-state surfaces, which is justified as follows.

The limit-state surfaces $\{u \in \mathbb{R}^m \mid g^k(u,x) = 0\}$ are frequently fairly flat and smooth. Hence one can take advantage of the rotational symmetry and exponential decay of the probability density in the standard normal space, to obtain a first-order approximation to the k-th failure probability by linearizing the limit-state surface $\{u \in \mathbb{R}^m | g^k(u,x) = 0\}$ around the point $u_k^*(x) \stackrel{\Delta}{=} \arg\min\{\|u\| \mid g^k(u,x) = 0\}$, i.e., nearest point to the origin. This is the most likely failure point in the standard normal space; its neighborhood makes a dominant contribution to the failure probability. The corresponding first-order approximation to $F^k(x)$ is given by $\tilde{M}^k(x) \stackrel{\Delta}{=} \Phi(-\|u_k^*(x)\|)$, where $\Phi(\cdot)$ is the standard normal cumulative distribution function. This leads to FORM (First Order Reliability Method) which replaces the functions $F^k(\cdot)$ by the functions $\tilde{M}^k(\cdot)$ in (4) (see, e.g., [3]).

The most obvious objections to the use of the FORM approach stem from the fact that there are no known error bounds relating $\tilde{M}^k(x)$ and $F^k(x)$ and the fact that the functions $u_k^*(x)$ may be set-valued and therefore their norm may not be differentiable. Hence the smooth optimization algorithms may jam up at non-solution points. One way to get around this is by reformulation as a bilevel problem (see [5]) in which the calculation of $u_k^*(x)$ is replaced with the corresponding KKT conditions; however this leads to first-order algorithms requiring second-order derivatives as well as to some ambiguities. A much sounder approach was presented in [5] where it was shown that an inequality of the form $\tilde{M}^k(x) \leq \alpha$ is equivalent to a simple semi-infinite inequality, and when one replaced the inequalities $\tilde{M}^k(x) \leq \alpha$ in (5) with these semi-infinite inequalities one obtained a "nice" semi-infinite optimization problem that can be solved efficiently using methods of outer approximations, such as those described in [2], [5], for example. In this paper, we begin by establishing the continuity properties of the functions $F^k(\cdot)$, in Section 2. Then, in Section 3, we present a conceptual extension of the ideas in [5] to obtain two parameterized semi-infinite optimization problems that must be iteratively adjusted in a process that yields good bounds on the solutions of the problems (3) and (5). In Section 4 we present a new semi-infinite optimization algorithm specially designed for the solution of these semi-infinite optimization problems. In Section 5 we present some numerical results, and in Section 6 some concluding remarks.

2 Continuity Properties of the Functions $F^k(\cdot)$.

We will now establish conditions under which the functions $F^k(\cdot)$ are continuous under the following assumption.

Assumption 1. *For $k = 1, ..., p$,*

(a) the functions $g^k(\cdot, \cdot)$ are continuous,

(b) the boundary of the sets $S_k(x)$, $\text{bdry}\, S_k(x) = \{u \in \mathbb{R}^m \mid g^k(x, u) = 0\}$*, has (Lebesgue) measure 0.*

We make use of the following convergence notion of sets.

Definition 1. *Let $\{S_i\}_{i=1}^{\infty}, S$ be subsets in $\mathbb{R}^m$, and for any i, let*

$$S_i \Delta S \triangleq (S_i \setminus S) \cup (S \setminus S_i) \ . \tag{6}$$

The sets S_i converge roughly *to a set S if*

$$\limsup_{i \to \infty} (S_i \Delta S) \subset \text{bdry}\, S \ , \tag{7}$$

where lim sup *is the* outer *limit,* cf. [10, p.109], for example.

Lemma 1. *Suppose that Assumption 1 is satisfied. Let $\{x_i\}_{i=1}^{\infty}$ be a sequence in $\mathbb{R}^n$. If $x_i \to x^*$, as $i \to \infty$, then, for $k = 1, ..., p$, $S_k(x_i) \to S_k(x^*)$ roughly.*

Proof. Note that for any i,

$$S_k(x_i) \Delta S_k(x^*) = \{u \mid g(x_i, u) \geq 0, g(x^*, u) < 0\} \cup \{u \mid g(x^*, u) \geq 0, g(x_i, u) < 0\} \ , \tag{8}$$

(with $u \in \mathbb{R}^m$). Suppose $x_i \to x^*$ and, without loss of generality, suppose that $\{u_i\}_{i=1}^{\infty} \subset \mathbb{R}^m$ is a sequence such that $u_i \in \{u \mid g^k(x_i, u) \geq 0, g^k(x^*, u) < 0\}$ and that $u_i \to u^*$, as $i \to \infty$. Then by continuity we must have that $g^k(x^*, u^*) \geq 0$ and $g^k(x^*, u^*) \leq 0$, i.e., $g^k(x^*, u^*) = 0$ which implies that $u^* \in \text{bdry}\, S_k(x^*)$, which completes the proof. □

We note that under a somewhat stronger assumption one actually can obtain continuity in the (usual) Painlevé-Kuratowski sense, i.e.,

$\limsup_{i\to\infty} S_k(x_i) = \liminf_{i\to\infty} S_k(x_i) = S_k(x)$ when $x_i \to x$.

Lemma 2. *Suppose that Assumption 1 is satisfied and that for $x \in \mathbb{R}^n$ and $u \in S_k(x)$, there exists a sequence $\{u_i\}_{i=1}^{\infty}$ in $S_k(x)$ such that $u_i \to u$ and $g^k(x, u_i) > 0$ for all i.* For $k = 1, ..., p$, if $x_i \to x^*$, as $i \to \infty$, then $S_k(x_i) \to S_k(x^*)$, as $i \to \infty$.

Proof. Suppose that $x_i \to x$, as $i \to \infty$ and $k \in \{1, ..., p\}$.

(a) For each i, let $u_i \in S_k(x_i)$ and, without loss of generality, suppose that $u_i \to u^*$, as $i \to \infty$. Then $g^k(x_i, u_i) \geq 0$ for all i and by continuity, $g^k(x_i, u_i) \to g^k(x^*, u^*) \geq 0$. Hence $u^* \in S_k(x^*)$, which proves that

$$\limsup_{i\to\infty}\ S_k(x_i) \subset S_k(x*)\ , \tag{9}$$

where lim sup denotes the outer limit.

(b) Next, suppose that $u^* \in S_k(x^*)$. Then, by Assumption 2, there exists an infinite sequence $\{u_j\}_{j=0}^{\infty}$, such that $g^k(x^*, u_j) > 0$ for all j. Next, for each j, there exists an i_j, such that $g^k(u_j, x_i) \geq 0$ for all $i \geq i_j$. For $i_j \leq i \leq i_{j+1}$, let $u_i = u_j$. Then $u_i \to u^*$, as $i \to \infty$, and $g^k(x_i, u_i) \geq 0$ for all i, i.e., $u_i \in S_k(x_i)$ for all i, which implies that

$$\liminf_{i\to\infty}\ S_k(x_i) \supset S_k(x*)\ , \tag{10}$$

where lim inf denotes the inner limit. Combining (9) and (10), we conclude that $S_k(x_i) \to S_k(x^*)$, as $i \to \infty$. □

Returning to our the functions $F^k(\cdot)$. For any subset S of $\mathbb{R}^m$, the measure $M(S)$ induced by ϕ is

$$M(S) \triangleq \int_S \phi(u) du\ . \tag{11}$$

Theorem 1. *Let $\{S_i\}_{i=1}^{\infty}$ be a sequence of sets in $\mathbb{R}^m$ and let $S \subset \mathbb{R}^m$. If $S_i \to S$ roughly and $M(\text{bdry}\, S) = 0$, then $M(S_i) \to M(S)$, as $i \to \infty$.*

Proof. First,

$$S_i \to S\ \textit{roughly} \Leftrightarrow \begin{cases} S \supset \limsup\limits_{i\to\infty} S_i \\ \\ S^c \supset \limsup\limits_{i\to\infty} S_i^c \end{cases} \tag{12}$$

Next, for all $\epsilon > 0$,

$$S \supset \limsup S_i \Leftrightarrow \forall\epsilon > 0,\quad S_i \setminus \epsilon^o S \to \emptyset\ , \tag{13}$$

where

$$\epsilon^o S = \{u \in \mathbb{R}^m \mid d(u, S) < \epsilon\} = S + \text{int}\,\epsilon\mathbf{B} \ , \tag{14}$$

where $\mathbf{B} \triangleq \{u \in \mathbb{R}^m \mid \|u\| \leq 1\}$ is the unit ball. Next,

$$\limsup_{i\to\infty} \int_{S_i} \phi(u)du \leq \limsup_{i\to\infty} \int_{S_i \setminus \epsilon^o S} \phi(u)du + \int_{\text{cl}]\,\epsilon^o S} \phi(u)du. \tag{15}$$

Letting $\epsilon \to 0$, it now follows from (13) that

$$\limsup_{i\to\infty} M(S_i) = \limsup_{i\to\infty} \int_{S_i} \phi(u)du \leq \int_S \phi(u)du = M(S) \ . \tag{16}$$

Replacing S_i and S by their complements in (16), we obtain

$$\limsup_{i\to\infty} M(S_i^c) = \limsup_{i\to\infty} \int_{S_i^c} \phi(u)du \leq \int_{S^c} \phi(u)du = M(S^c) \ , \tag{17}$$

Since $M(S_i^c) = 1 - M(S_i)$ and $M(S^c) = 1 - M(S)$, we obtain from (17) that

$$\liminf_{i\to\infty} M(S_i) = \liminf_{i\to\infty} \int_{S_i} \phi(u)du \geq \int_S \phi(u)du \geq M(S) \ ,$$

which proves that $M(S_i) \to M(S)$, as $i \to \infty$. □

Corollary 1. *For $k = 1, ..., p$, the functions $F^k(\cdot)$ are continuous.*

3 Design Problem Reformulation

Despite the fact the the function $\Psi(\cdot)$ is continuous, because the evaluation of $\Psi(x)$ and $\nabla\Psi(x)$, when it exists, requires the use of Monte Carlo methods, problems (3) and (5) are currently intractable. To be more precise, it is essentially impossible to evaluate the integrals defining $\Psi(\cdot)$ without recourse to Monte Carlo techniques when the random quantity v is an m-vector with $m \geq 2$. And this is almost always the case for structural engineering design problems. Note that there are some formulas for the gradients of the functions $F^k(\cdot)$, but these are derived under quite restrictive conditions and, in any event, require evaluation of a (complicated) multivariate integral, cf. [12].

A reasonable approach, based on approximations, is suggested by the following considerations. First suppose that the limit-state functions $g^k(x, u)$ are linear in u, i.e., for $k = 1, ..., p$, the functions $g^k(x, u)$ are of the form

$$g^k(x, u) = a_k(x) + \langle n_k(x), u \rangle \ . \tag{18}$$

For $k = 1, ..., p$, let

$$\beta^k(x) \triangleq \min_{u \in \mathbb{R}^m} \{ \|u\| \mid g^k(x, u) = 0 \} . \tag{19}$$

One has that $F^k(x) = \Phi(-\beta^k(x))$ where $\Phi : \mathbb{R} \to \mathbb{R}$ is the standard normal cumulative probability function. In this case problem (3), i.e.,

$$\min\{ \Psi(x) \mid f^j(x) \leq 0, \ j = 1, ..., q \} , \tag{20}$$

can be replaced by

$$\max\{ \min_{k \in \mathbf{p}} \beta^k(x) \mid f^j(x) \leq 0 \ , \ j = 1, ..., q \}, \tag{21}$$

where $\mathbf{p} \triangleq \{1, ..., p\}$, and problem (5), i.e.,

$$\min\{ f^0(x) \mid \Psi(x) \leq \alpha, \ f^j(x) \leq 0, \ j = 1, ..., q \}, \tag{22}$$

can be replaced by

$$\min\{ f^0(x) \mid \min_{k \in \mathbf{p}} \beta^k(x) \geq \beta^*, \ f^j(x) \leq 0, \ j = 1, ..., q \}, \tag{23}$$

where

$$\beta^* = \Phi^{-1}(-\alpha). \tag{24}$$

Now, for $k = 1, ..., p$ and any $\rho > 0$, let

$$B(\rho) \triangleq \{ u \in \mathbb{R}^m \mid \|u\| \leq \rho \} , \tag{25}$$

and, for $k = 1, ..., p$ and $\rho > 0$, let the functions $\psi^k_\rho : \mathbb{R} \to \mathbb{R}^n$ be defined by

$$\psi^k_\rho(x) \triangleq \max_{u \in B(\rho)} g^k(x, u) . \tag{26}$$

Finally, let the function $\psi_\rho : \mathbb{R} \to \mathbb{R}^n$ be defined by

$$\psi_\rho(x) \triangleq \max_{k \in \mathbf{p}} \psi^k_\rho(x) . \tag{27}$$

Consider the following substitute problem: and

$$\min\{ f^0(x) \mid \psi_{\beta^*}(x) \leq 0, \ f^j(x) \leq 0, \ j = 1, ..., q \} . \tag{28}$$

Because $\psi^k_\rho(x) < 0$ if and only if $S_k(x) \cap B(\rho) = \emptyset$, we obtain the following relations between the problems (22) and (28) when the functions $g^k(x, u)$ are linear in u. For a geometric interpretation in the general case, see Fig. 1.

Theorem 2. *Suppose that for $k = 1, ..., p$, the functions $g^k(\cdot, \cdot)$ are continuous and linear in the second argument. Then x^* is a solution of (22) if and only if x^* is a solution of (28).*

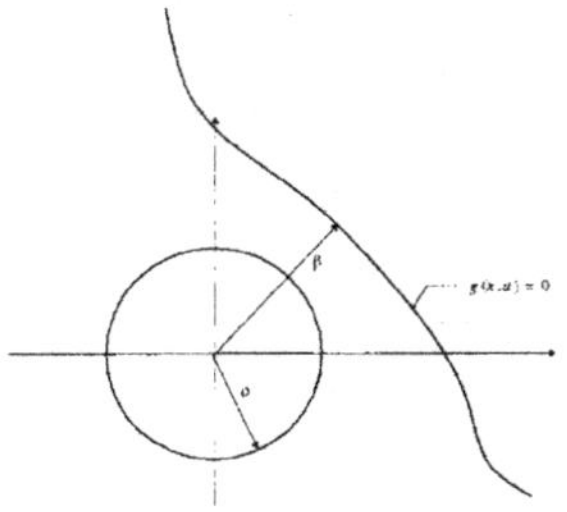

Figure 1: Geometry in the sample space

Proof. Because $\psi^k_{\beta^*}(x) \leq 0$ if and only if $S_k(x) \cap B(\beta^*)$ is empty or consists of the nearest point to the origin in $S_k(x)$, it follows that $\beta^k \geq \beta^*$ if and only if $\psi^k_{\beta^*}(x) \leq 0$. Hence the constraints $\psi_{\beta^*}(x) \leq 0$ and $\min_{k=1,\ldots,p} \beta^k \geq \beta^*$ are equivalent. Since the cost function and all the other constraints in the problems (22) and (28) are the same, the desired result follows. □

Obtaining a meaningful substitute problem for problem (21) is less straightforward. Since for any $\alpha(x) > 0$,

$$\{u \in \mathbb{R}^m \mid g^k(x,u) \geq 0\} = \{u \in \mathbb{R}^m \mid g^k(x,u)/\alpha(x) \geq 0\} , \tag{29}$$

we find it desirable to replace our limit-state functions by their normalized versions defined by

$$\tilde{g}^k(x,u) \triangleq g^k(x,u)/g^k(x,0) , \tag{30}$$

which requires the additional assumption that $g^k(x,0) < 0$ for all $x \in \mathbb{R}^n$. In view of this, we will assume that $a_k(x) \equiv -1$ in (18). Now let $\rho > 0$ and consider the following substitute problem:

$$\min\{\psi_\rho(x) \mid f^j(x) \leq 0,\ j = 1,\ldots,q\}, \quad \rho > 0, \tag{31}$$

Theorem 3. *Suppose that for $k = 1,\ldots,p$, the functions $g^k(\cdot,\cdot)$ are continuous and defined as in (18) with $a_k(x) \equiv -1$.*

(a) *If x^* is a solution of (31) for some $\rho = \rho^* > 0$, then x^* is a solution of (31) for any $\rho > 0$.*

(b) *The vector x^* is a solution of (21) if and only if x^* is a solution of (31) for any $\rho > 0$.*

Proof. *(a)* It is easy to see that for any $x \in \mathbb{R}^n$,

$$\psi^k(x) = -1 + \rho\|n_k(x)\|, \ k = 1, ..., p \ . \tag{32}$$

Hence

$$\psi_\rho(x) = -1 + \rho \max_{k \in \mathbf{p}} \|n_k(x)\| \tag{33}$$

It follows by inspection from (33) that the minimizer of $\psi_\rho(\cdot)$ does not depend on ρ. *(b)* First we note that when the $g^k(\cdot,\cdot)$ are defined as in (18), with $a_k(x) \equiv -1$, then for any $x \in \mathbb{R}^n$ and $k = 1, ..., p$,

$$\beta^k(x) = 1/\|n_k(x)\| \ . \tag{34}$$

Let

$$X \triangleq \{x \in \mathbb{R}^n \mid f^j(x) \le 0, \ j = 1, ..., q\} \ . \tag{35}$$

Then, in view of (33), we see that x^* is a solution of (31) if and only if it is a solution of

$$\min\{\max_{k \in \mathbf{p}} \|n_k(x)\| \mid x \in X\} \ , \tag{36}$$

and in view of (34), (21) assumes the form

$$\max\{\min_{k \in \mathbf{p}} \frac{1}{\|n_k(x)\|}\| \mid x \in X\} \ . \tag{37}$$

Now, x^* is a solution of (37) if and only if it is a solution of

$$\left.\begin{aligned} \max\{\min_{k \in \mathbf{p}} \ln \frac{1}{\|n_k(x)\|}\| \mid x \in X\} &= \max\{\min_{k \in \mathbf{p}} -\ln \|n_k(x)\| \mid x \in X\} \\ &= \max\{-\max_{k \in \mathbf{p}} \ln \|n_k(x)\| \mid x \in X\} \\ &= -\min\{\max_{k \in \mathbf{p}} \ln \|n_k(x)\| \mid x \in X\} \ . \end{aligned}\right\} \tag{38}$$

Since x^* solves (38) if and only if it is a solution of (36), our proof is complete. □

For the general case of the functions $g^k(\cdot,\cdot)$, it seems desirable to strengthen Assumption 1 as follows:

Assumption 2. *For $k = 1, ..., p$,*

(a) The functions $g^k(\cdot,\cdot)$ are continuous,

(b) the boundary of the sets $S_k(x)$, defined in (2), is given by bdry $S_k(x) = \{u \in \mathbb{R}^m \mid g^k(x, u) = 0\}$ *and for every $x \in \mathbb{R}^n$ and $u \in S(x)$, there exists a sequence $\{u_i\}_{i=0}^\infty$ in $S(x)$ and converging to u, such that $g^k(x, u_i) > 0$ for all i.*

It is easy to see that the last part of Assumption 2 implies that bdry $S_k(x)$ is a set of measure zero, and hence that Assumption 2 implies Assumption 1. Note also that these are the assumptions used in Lemma 2.

Theorem 4. *Suppose that Assumption 2 is satisfied. Then for any $k \in \{1, ..., p\}$ and any $x \in \mathbb{R}^n$,*

$$S^k(x) \cap \text{int}B(\rho) = \emptyset \Leftrightarrow \psi^k(x) \leq 0 \; . \tag{39}$$

Proof. First suppose that $S^k(x) \cap \text{int}B(\rho) = \emptyset$ but $\psi^k(x) > 0$. Then there exists a $u \in \text{bdry}\, B(\rho)$ such that $g^k(x, u) > 0$. Let $\{u_i\}_{i=0}^{\infty}$ in $\text{int}B(\rho)$ be any sequence such that $u_i \to u$. Then, by continuity, there exists an i' such that $g^k(x, u_{i'}) \geq 0$, i.e., $u_{i'} \in S^k(x) \cap \text{int}B(\rho)$, which is a contradiction.

Now suppose that $\psi^k(x) \leq 0$, but $S^k(x) \cap \text{int}B(\rho) \neq \emptyset$. Then there exists a $u \in \text{int}B(\rho)$ such that $g^k(x, u) \geq 0$. By Assumption 2, there exists a sequence $\{u_i\}_{i=0}^{\infty}$ in $S(x)$, converging to u, such that $g^k(x, u_i) > 0$ for all i. Since $\text{int}B(\rho)$ is an open set, it follows that there exists an i'' such that $u_{i''} \in B(\rho)$, which implies that $\psi^k(x) > 0$, a contradiction, which completes our proof. □

It follows directly from Theorem 4 that for every $x \in \mathbb{R}^n$ and $\rho > 0$ such that $\psi^k(x) \leq 0$, we have that

$$F^k(x) \leq 1 - M(B(\rho_\alpha)) \; . \tag{40}$$

Hence there is the temptation to proceed as follows with regard to problem (5). Compute a ρ_α such that $1 - M(B(\rho_\alpha)) = \alpha$, and replace in (5) the constraint

$$\Psi(x) - \alpha \leq 0 \tag{41}$$

by the more conservative, but more tractable constraint

$$\psi_{\rho_\alpha} \leq 0 \; . \tag{42}$$

Unfortunately, this approach is much too conservative, particularly when m is large. In our experience, it is much better to proceed as follows:

Heuristic For Problem (5).
To obtain a reasonable approximation to a solution of Problem (5), repeat the procedure below until a satisfactory result has been obtained.

1. Compute $\beta^* = \Phi^{-1}(-\alpha)$.

2. Solve the problem (28).

3. If x^* is the solution of (28) and $\Psi(x^*) \neq \alpha$, in (28), replace β^* by $\beta^* = [\Phi^{-1}(-\alpha)(x^*)]^2/\Phi^{-1}(-\Psi(x^*))$, and solve the problem (28) again. □

Heuristic for Problem (3).
For the general case of problem (3), it is not clear whether normalizing the limit-state functions by dividing them by $g^k(x,0)$, as in (30), is a good idea or not. Hence we propose to solve problem (31) for a range of values of ρ, which results in solutions x_ρ^*, compute the corresponding values of $\Psi(x_\rho^*)$ and accept the x_ρ^* that results in the lowest value of $\Psi(x_\rho^*)$ as a reasonable approximation to a solution of (3) □

Empirical Observation. referring to the example in Section 5, we see that, at least in this case, the choice of ρ had little effect on the solution of (31), even though the limit-state function was not normalized as in (30), and is clearly nonlinear. □

4 An Algorithm for Minimizing the Probability of Failure

We will now present an implementable phase I-phase II feasible directions algorithm for solving problem (31). To simplify notation, we will assume that $p = 1$ and hence drop the superscript on the $g^k(\cdot,\cdot)$, so that

$$\psi_\rho(x) = \max_{\|u\|\le\rho} g(x,u) \,. \tag{43}$$

Throughout this section, we will assume that the function $g(\cdot,\cdot)$ is twice continuously differentiable and that the the constraint functions $f^j : \mathbb{R}^n \to \mathbb{R}$, $j = 1,...,q$, in (31), are continuously differentiable. Let

$$\psi_X(x) \triangleq \max_{j\in\mathbf{q}} f^j(x), \tag{44}$$

and let $\psi_X(x)_+ \triangleq \max\{0, \psi_X(x)\}$. Next, let $\omega > 0$ and let

$$\bar{G}(x) \triangleq \mathrm{co}\left\{ \begin{pmatrix} \psi_\rho(x) - g(x,u) + \omega\psi_X(x)_+ \\ \nabla_x g(x,u) \end{pmatrix}, \begin{pmatrix} \psi_X(x)_+ - f^j(x) \\ \nabla f^j(x) \end{pmatrix} \right\}_{\|u\|\le\rho, j\in\mathbf{q}} \tag{45}$$

Referring to [9](Theorem 3.2.4) we find the following generalization of the Karush-Kuhn-Tucker conditions to semi-infinite optimization problems:

Theorem 5. *If x^* is a local minimizer for (31), then $0 \in \bar{G}(x^*)$.*

We can associate with the set $\bar{G}(x)$ an optimality function, as shown in Theorem 3.2.8 in [9].

Theorem 6. *Let*

$$\theta(x) \triangleq -\min_{\bar\xi\in\bar{G}(x)} \xi^0 + \tfrac{1}{2}\|\xi\|^2, \tag{46}$$

and let $\bar\xi^*(x) = (\xi^{0*}(x), \xi^*(x))$ be the solution of (46). Then

(a) *the functions* $\theta(\cdot)$ *and* $\bar{\xi}^*(\cdot)$ *are a continuous,*
(b) $0 \in G(x^*)$ *if and only if* $\theta(x^*) = 0$.
(c) *the function*

$$h(x) \triangleq -\xi^*(x) \tag{47}$$

defines a continuous search direction for problem (31). □

Next, to define the line search of our algorithm, we introduce a *unified* cost function $F : \mathbb{R}^n \times \mathbb{R}^n \to \mathbb{R}$ which is defined as follows. Let $\omega > 0$ be as in (45), then

$$F(z,x) \triangleq \max\{\psi_\rho(x) - \psi_\rho(z) - \omega\psi_X(z)_+, \psi_X(x) - \psi_X(z)_+\} \ . \tag{48}$$

The algorithm below is a straightforward extension of Algorithm 2.6.1 in [9] to the semi-infinite case. In general this algorithm would have to be considered only conceptual, but, due to the special structure of problem (31), we will see that it is in fact an implementable algorithm.

Polak-He Algorithm for problem (31).
Parameters. $\omega > 0$, $\alpha \in (0,1]$, $\beta \in (0,1)$.
Data. $x_0 \in \mathbb{R}^n$.
Step 0. Set $i = 0$.
Step 1. Compute the optimality function value $\theta_i = \theta(x_i)$, according to (46) and the corresponding solution $\xi^*(x_i)$. Set $h_i = -\xi(x_i)$.
Step 2. Compute the step-size λ_i:

$$\lambda_i = \max_{k \in \mathbb{N}}\{\beta^k \mid F(x_i, x_i + \beta^k h_i) \leq \beta^k \alpha \theta_i\} \ . \tag{49}$$

Step 3. Set $x_i + 1 = x_i + \lambda_i h_i$, replace i by $i+1$, and go to Step 1. □

Theorem 7. *If* $\{x_i\}_{i=0}^{\infty}$ *is a sequence constructed by the Polak-He Algorithm in solving problem (31), then every accumulation point* x^* *of this sequence satisfies* $\theta(x^*) = 0$.

Next we must address some of the details of implementation of the Polak-He Algorithm for problem (31). First we note that the function $\psi_\rho(x)$, defined in (43) is conveniently evaluated by a projection method, such as a stabilized form of the Levitin-Polyak Projected Newton method [6], which at each iteration solves a simple, quadratically constrained problem of the form:

$$\min\{\langle\nabla_u g(x,u), u - u\rangle + \tfrac{1}{2}\langle u - u, Q(u - u)\rangle \mid \|u\| \leq \rho\} \ . \tag{50}$$

Next we turn to the evaluation of $\theta(x)$ and of $\bar{\xi}^*(x)$. We begin by showing that the problem (46) can be converted into an ordinary nonlinear programming problem by making use of a generalization of the Caratheodory Theorem due to Bonnesen and Fenchel [1] which states that any point in the convex hull of a *connected* set S in $\mathbb{R}^n$ can be expressed as a convex combination of at most n points in S. When applied to the set $\bar{G}(x)$ this leads to the following result:

Proposition 1. *Any point $\bar{\xi} = (\xi^0, \xi)$, on the boundary of the set $\bar{G}(x)$, defined in (45) can be represented in the form*

$$\bar{\xi} = \sum_{j=1}^{n+1} \mu^j \begin{pmatrix} \psi_\rho^0(x) - g(x, u_j) + \omega\psi_X(x)_+ \\ \nabla_x g(x, u_j) \end{pmatrix} + \sum_{j=1}^{q} \mu^{n+1+j} \begin{pmatrix} \psi_X(x)_+ - f^j(x) \\ \nabla f^j(x) \end{pmatrix} \tag{51}$$

where all $\|u_j\| \leq \rho$, and $\mu \in \Sigma_{n+1+q} \triangleq \{\mu \in \mathbb{R}^{n+1+q} \mid \sum_{j=1}^{n+1+q} \mu^j = 1\}$.

Corollary 2. *For every $x \in \mathbb{R}^n$,*

$$\left.\begin{aligned} \theta(x) &= -\min_{\bar{\xi} \in \bar{G}(x)} \{\xi^0 + \tfrac{1}{2}\|\xi\|^2\} \\ &= -\min_{\substack{\mu \in \Sigma_{n+1+q} \\ \|u_j\| \leq \rho}} \sum_{j=1}^{n+1} \mu^j [\psi_\rho^0(x) - g(x, u_j) + \omega\psi_X(x)_+] \\ &\quad + \sum_{j=1}^{q} \mu^{n+1+j} [\psi_X(x)_+ - f^j(x)] \\ &\quad + \tfrac{1}{2}\|\sum_{j=1}^{n+1} \mu^j \nabla g(x, u_j) + \sum_{j=1}^{q} \mu^{n+1+j} \nabla f^j(x)\|^2 . \end{aligned}\right\} \tag{52}$$

When we replace the formula (46) for $\theta(x)$ by (52), we obtain the following implementable form of the Polak-He Algorithm:

Implementable Polak-He Algorithm

Parameters. $\omega > 0$, $\alpha \in (0, 1]$, $\beta \in (0, 1)$.

Data. $x_0 \in \mathbb{R}^n$.

Step 0. Set $i = 0$.

Step 1. Compute the *optimality function* value $\theta_i = \theta(x_i)$ by solving (52) and the corresponding solution vectors μ_i, and $u_{j,i}$, $j = 1, ..., n + 1$. Set

$$h_i = -\{\sum_{j=1}^{n+1} \mu_i^j \nabla g(x_i, u_{j,i}) + \sum_{j=1}^{q} \mu_i^{n+1+j} \nabla f^j(x_i)\} . \tag{53}$$

Step 2 Compute the *step-size* λ_i:

$$\lambda_i = \max_{k \in \mathbb{N}} \{\beta^k \mid F(x_i, x_i + \beta^k h_i) \leq \beta^k \alpha \theta_i\} . \tag{54}$$

Step 3 Set $x_{i+1} = x_i + \lambda_i h_i$, replace i by $i + 1$, and go to Step 1. □

The nonlinear programming search direction finding problem (52) may have non-unique solutions, even though the search direction $h(x)$ is uniquely defined by (54). Because of this, (52) can be ill-conditioned with respect to SQP type algorithms. In our experience, it is easy to solve it using a gradient projection algorithm that makes use of the following decomposition result:

Proposition 2. *Let* $z = (\mu, u_1, ..., u_{n+1})$, *with* $\mu \in \mathbb{R}^{n+1+q}$ *and all* $u_j \in \mathbb{R}^m$. *Let*

$$\left.\begin{array}{rcl} \kappa(z) & \triangleq & \sum_{j=1}^{n+1} \mu^j[\psi_\rho(x_i) - g(x_i, u_j) + \omega\psi_X(x_i)_+] \\ & + & \sum_{j=1}^{q} \mu^{n+1+j}[\psi_X(x_i)_+ - f^j(x_i)] \\ & + & \frac{1}{2}\| \sum_{j=1}^{n+1} \mu^j \nabla g(x_i, u_j) + \sum_{j=1}^{q} \mu^{n+1+j} \nabla f^j(x_i) \|^2 \end{array}\right\} \tag{55}$$

let $z^* = z' - \nabla\kappa(z')$, *and consider the problem*

$$\min\{\|z - z^*\|^2 \mid z \in \Sigma_{n+1+q} \times B(\rho) \times ... \times B(\rho)\} \ , \tag{56}$$

where $B(\rho) = \{u \in \mathbb{R}^m \mid \|u\| \leq \rho\}$. *Then (56) decomposes into the set of problems*

$$\min\{\|\mu - \mu^*\|^2 \mid \mu \in \Sigma_{n+1+q}\} \ , \tag{57}$$

$$\min\{\|u_j - u_j^*\|^2 \mid u_j \in B(\rho)\} \ \} \ j = 1, ..., n+1 \ . \tag{58}$$

Remark. The problem (57) is a very simple QP. The problems (58) have the following solutions: if $\|u_j^*\| \leq \rho$, then $\hat{u}_j = u_j^*$ solves (58). Else, $\hat{u}_j = u_j^*/\|u_j^*\|$ solves (58). Hence a gradient projection algorithm is very efficient in solving (58). □

5 A Numerical Example

Consider the design of a short column, as shown in Fig. 2. In terms of the random variables $v = (M_1, M_2, P, Y)$ and the design vector $x = (b, h)$, the limit-state function is given by

$$G(x, v) = -\left[1 - \frac{4M_1}{bh^2Y} - \frac{4M_2}{b^2hY} - \frac{P^2}{(bhY)^2}\right] \ , \tag{59}$$

i.e.,

$$G(x, v) = -\left[1 - \frac{4v^1}{x^1(x^2)^2v^4} - \frac{4v^2}{x^2(x^1)^2v^4} - \frac{v^3}{(x^1x^2v^4)^2}\right] \tag{60}$$

It can be assumed that the random variable (P, M_1, M_2, Y) are uncorrelated and have the following characteristics:

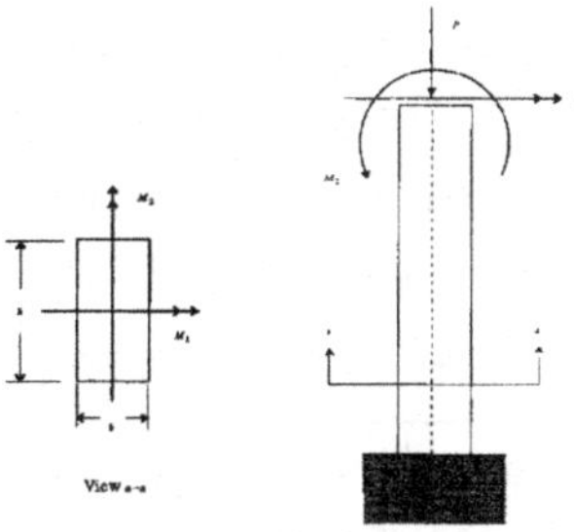

Figure 2: A Short Column

variable	distribution	mean	std dev.	λ_i	ζ_i
M_1	lognormal	2000	600	7.558	0.294
M_2	lognormal	1000	300	6.865	0.294
P	lognormal	500	100	6.195	0.198
Y	lognormal	5	0.5	1.604	0.100

where the coefficients λ_i, ζ_i can be used to obtain standard normal variables using the transformation:

$$u^i = (\ln v^i - \lambda_i)/\zeta_i, \; i = 1, 2, 3, 4 \; . \tag{61}$$

We now define the limit-state function in terms of these standard normal variables by

$$g(x, u) \triangleq G(x, v(u)) \; . \tag{62}$$

We can design the short column in at least 2 ways. The first is to minimize the probability of failure subject to a constraint on the area, e.g., $A = bh \leq 300$, a positivity constraint, $0 \leq b, h$, and an aspect ratio constraint, e.g., $0.5 \leq b/h \leq 2$. This leads to a reformulated optimization problem of the form

$$\min\{\psi_p(x) \mid x \in X_a\} \; , \tag{63}$$

where

$$X_a = \{x \in \mathbb{R}^2 \mid x^1 x^2 \leq 300, x^1 - 2x^2 \leq 0, \; 0.5x^2 - x^1 \leq 0, x^1 \geq 0, x^2 \geq 0\} \; . \tag{64}$$

Alternatively, we can attempt to minimize the area $A = bh$ subject to a constraint on the probability of failure in the form $\Psi(x) \leq \Phi(-3)$ (where $\beta(x)$ was defined in (19)), together with a positivity constraint, and an aspect ratio constraint, as above. This leads to a reformulated optimization problem of the form

$$\min\{f^0(x) \mid \psi_\rho(x) \leq 0,\ x \in X_b\}\ , \tag{65}$$

where $f^0(x) = x^1 x^2$ is a differentiable cost function and the set X_b is given by

$$X_b = \{x \in \mathbb{R}^2 \mid x^1 - 2x^2 \leq 0,\ 0.5x^2 - x^1 \leq 0, x^1 \geq 0, x^2 \geq 0\}\ . \tag{66}$$

Numerical results for problem (63) are given in the table below:

ρ	1.0	1.5	2.0	2.5	3.0	3.5	4.0
$\hat{x}^1_\rho$	12.363	12.414	12.336	12.352	12.326	12.266	12.403
$\hat{x}^2_\rho$	24.266	24.167	24.318	24.288	24.338	24.458	24.188
$\psi_\rho(\hat{x}_\rho)$	-0.343	-0.267	-0.181	-0.085	-0.022	0.143	0.280
β_{FORM}	2.903	2.903	2.902	2.903	2.903	2.903	2.903
β_{SORM}	2.701	2.701	2.701	2.701	2.701	2.701	2.701
β_{MC}	2.703	2.704	2.704	2.704	2.704	2.705	2.703

In the above table, $\hat{x}_\rho$ denotes the solution of (63) for the value of ρ indicated, β_{FORM} is the approximation to $\beta(\hat{x}_\rho)$ (defined in (19)) obtained using the first-order reliability method, β_{SORM} is the approximation to $\beta(\hat{x}_\rho)$ obtained using the second-order reliability method (using quadratic approximations to the surface defined by $g(\hat{x}_\rho, u) = 0$) and β_{MC} is an accurate evaluation of $\beta(\hat{x}_\rho)$ using a Monte Carlo method. Note how little the solution $\hat{x}_\rho$ changes with ρ, even though the limit-state function is nonlinear and not normalized. For the problem of minimum-area design (65), we obtained the following results using the heuristic procedure outlined in Section 2: 1. First cut: we set $\rho = 3$. The solution was $\hat{x} = (13.828, 22.104)$, $\beta_{FORM}(\hat{x}) = 3.001$, $\beta_{MC}(\hat{x}) = 2.820$, area $= 305.654$. 2. Second cut: we set $\rho = 3^2/2.820$. The solution was $\hat{x} = (13.344, 23.434)$, $\beta_{FORM}(\hat{x}) = 3.192$, $\beta_{MC}(\hat{x}) = 3.001$, area $= 312.703$.

6 Conclusion

References

[1] T. Bonnesen and W. Fenchel (1934), "Theorie der konvexen Körper," *Ergebnisse der Mathematik*, 3, Springer, Berlin.

[2] C. Gonzaga and E. Polak (1979), "On Constraint Dropping Schemes and Optimality Functions for a Class of Outer Approximations Algorithms," *SIAM J. Control and Optimization*, Vol. 17, No. 4, pp. 477-493.

[3] O. Ditlevsen and H. O. Madsen (1996), *Structural Reliability Methods*, J. Wiley and Sons, New York.

[4] M. Hohenbichler and R. Rackwitz (1981), "Non-Normal Dependent Vectors in Structural Safety," *J. Engineering Mechanics ASCE*, Vol. 107, No. 6, pp. 1227-1238.

[5] C. Kirjner-Neto, E. Polak and A. Der Kiureghian (1998), "An outer approximations approach to reliability-based optimal design of structures," *JOTA*, Vol. 98, No.1, pp. 1-17.

[6] E. S. Levitin, and B. T. Polyak (1966), "Constrained Minimization Methods," *U.S.S.R. Computational Mathematics and Mathematical Physics*, Vol. 6, No. 5, pp. 1-50.

[7] P.-L. Liu, and A. Der Kiureghian (1986), "Multivariate Distribution Models with Prescribed Marginals and Covariances," *Probabilistic Engineering Mechanics*, Vol. 1, pp. 105-112.

[8] K. Marti (1994), "Approximations and derivatives of probabilities in structural reliability and design," in *Structural Safety & Reliability*, Shinozuka and Yao (eds), Balkema, Rotterdam. pp. 1295-1299.

[9] E. Polak (1997), *Optimization: Algorithms and Consistent Approximations*, Springer-Verlag, New York.

[10] R.T. Rockafellar and R.J-B Wets (1998), *Variational Analysis*, Springer, New York.

[11] P. Thoft-Christensen (1991), "151 References in Reliability-Based Structural Optimization," *Proc. IFIP WG 7.5 Working Conference on Reliability and Optimization of Structural Systems*, Munich, September 1991.

[12] S. Uryasev (1994), Derivatives of probability functions and integrals over sets given by inequalities, *Journal of Computational and Applied Mathematics*, Vol. 56, No. 1-2, pp. 197-223.

Semiderivative Functions and Reformulation Methods for Solving Complementarity and Variational Inequality Problems [1]

Liqun Qi (L.Qi@unsw.edu.au)
School of Mathematics, The University of New South Wales,
Sydney, New South Wales 2052, Australia

Danny Ralph (danny@ms.unimelb.edu.au)
Department of Mathematics and Statistics, The University of Melbourne,
Parkville, Victoria 3052, Australia

Guanglu Zhou [2] (zhou@maths.unsw.edu.au)
School of Mathematics, The University of New South Wales,
Sydney, New South Wales 2052, Australia

Abstract

Two major reformulation methods, the nonsmooth method and the smoothing method, for solving nonlinear complementarity problems and variational inequality problems, have been rapidly developed in recent years. Superlinear convergence of these methods is linked with semismoothness, which is based on generalized Jacobians of locally Lipschitz functions. However, the definition of generalized Jacobians relies on the Rademacher theorem, and the exact calculus rules do not hold for generalized Jacobians. These pose some restrictions and difficulties for these methods. In this paper, we define a semiderivative function G for a continuous function F and show that this concept is indeed an extension of the concept of the derivative function. The semiderivative function G is single-valued and satisfies exact calculus rules. Hence it is relatively easy to calculate. Several common nonsmooth equation reformulations of nonlinear complementarity problems and variational inequality problems can be

[1]This research is supported by the Australian Research Council.

[2]This author's permanent address: Institute of Operations Research, Qufu Normal University, Qufu 273165, Shandong, P. R. China.

G. Di Pillo and F. Giannessi (eds.), Nonlinear Optimization and Related Topics, 317–350.

regarded as componentwise compositions of some smooth functions and some simple generalized semismooth functions, such as the plus function, the Fischer-Burmeister function and the median function. The derivatives of several well-known smoothing functions, such as the Chen-Harker-Kanzow-Smale function, the Chen-Mangasarian function and the Gabrial-Moré function, converge to semiderivative functions of the corresponding nonsmooth reformulation functions as the smoothing parameter goes to zero. Based upon this new concept, superlinear convergence conditions for both the nonsmooth method and the smoothing method are established.

Keywords: nonsmooth reformulation, generalized Newton methods, superlinear convergence, generalized semismooth functions.

1 Introduction

We consider the box constrained variational inequality problem (BVIP for short): find $x \in X := \{x \in \Re^n : l \leq x \leq u\}$, where $l \in \{\Re \cup \{-\infty\}\}^n, u \in \{\Re \cup \{+\infty\}\}^n$ and $l < u$, such that for all $y \in X$,

$$(y - x)^T \Phi(x) \geq 0, \tag{1.1}$$

where $\Phi : \Re^n \to \Re^n$ is a smooth vector function. This problem is also called the mixed complementarity problem, see [1, 2], where the reader may also find a number of interesting applications of this problem. If all bounds are infinite, then BVIP reduces to the problem of finding a solution of the system of equations $\Phi(x) = 0$, whereas, if $l_i = 0$, and $u_i = +\infty$ for all $i = 1, 2, \cdots, n$, then BVIP is just the well-known complementarity problem

$$x \geq 0, \quad \Phi(x) \geq 0, \quad x^T \Phi(x) = 0. \tag{1.2}$$

We denote this problem by NCP. Let

$$N_0 := \{(a, 0) : a \geq 0\} \cup \{(0, b) : b \geq 0\}.$$

Definition 1.1 *A function $\psi : \Re^2 \to \Re$ is called an NCP function if $\psi(a, b) = 0$ if and only if $(a, b) \in N_0$.*

Two most famous NCP functions are the *min function* [24], defined by

$$\psi(a, b) := \min\{a, b\} \tag{1.3}$$

and the *Fischer-Burmeister function* [14], defined by

$$\psi(a, b) := \sqrt{a^2 + b^2} - a - b. \tag{1.4}$$

With an NCP function ψ, NCP can be reformulated as a system of nonsmooth equations

$$F(x) = 0, \tag{1.5}$$

where $F : \Re^n \to \Re^n$ is defined by

$$F_i(x) = \psi(x_i, \Phi_i(x)) \tag{1.6}$$

for each i.

Let $\alpha \in \Re \cup \{-\infty\}$, $\beta \in \Re \cup \{+\infty\}$ and $\alpha \leq \beta$. Denote

$$B_0 := \{(\alpha, b) : b \geq 0\} \cup \{(a, 0) : \alpha \leq a \leq \beta\} \cup \{(\beta, b) : b \leq 0\}.$$

Definition 1.2 *We say $\psi_{(\alpha,\beta)}(a, b) : \Re^2 \to \Re$ is a BVIP function if $\psi_{(\alpha,\beta)}(a, b) = 0$ if and only if $(a, b) \in B_0$.*

Some BVIP functions have been presented in recent years, e.g., the Billups BVIP function [1] and the Sun-Womersley BVIP function [38], for details see [31]. Let $\psi = \psi_{(\alpha,\beta)}$ be a BVIP function. If $\psi_i = \psi_{(l_i,u_i)}$, then solving BVIP is equivalent to solving $F(x) = 0$, where F is defined by

$$F_i(x) = \psi_i(x_i, \Phi_i(x)) \tag{1.7}$$

for each i.

In recent years, based on NCP and BVIP functions, two major reformulation methods, the nonsmooth method and the smoothing method, for solving nonlinear complementarity problems and variational inequality problems, have been rapidly developed, e.g., see [1, 4, 5, 7, 8, 9, 10, 11, 12, 19, 20, 24, 25, 33, 36, 42] and references therein. In particular, based on the *min function*, Pang and Gabriel [25] proposed an algorithm for NCP called the NE/SQP method. Jiang and Qi [20] and De Luca et al. [7] proposed globally and superlinearly (quadratically) convergent nonsmooth Newton methods for NCP based on the *Fischer-Burmeister* function, which only require solving a system of linear equations to determine a search direction at each iteration.

To deal with the nonsmoothness of $F(x)$, smoothing methods for NCP and BVIP were proposed, e.g., by Chen and Harker [3], Chen and Mangasarian [4], Chen, Qi and Sun [5], Gabriel and Moré [15] and Kanzow [22]. The basic idea of the smoothing methods is to approximate $F(x)$ by a smooth function $F_t(x)$, where $t > 0$ denotes the smoothing parameter, and then to solve a sequence of problems

$$F_t(x) = 0 \tag{1.8}$$

and to force t to go to zero. In particular, a globally and superlinearly (quadratically) convergent smoothing Newton method was proposed by Chen, Qi and Sun in [5], where the authors exploited a Jacobian consistence property and applied this property to an infinite sequence of smoothing approximation functions to get high-order convergent methods.

Recently Qi, Sun and Zhou [33] and Jiang [19] proposed smoothing methods for NCP and BVIP by treating the smoothing parameter t as a variable, in which the smoothing parameter t is driven to zero automatically and no additional procedure for adjusting the smoothing parameter is necessary. Some regularized versions of the method in [33] were proposed in [28, 42] for BVIP, and the global convergence of the methods in [28, 42] were proved under the assumption that Φ is a P_0-function on X and the solution set of the BVIP is nonempty and bounded.

In these reformulation methods, some generalized Jacobian or directional derivative concepts have been used, e.g.,Clarke's generalized Jacobians [6], B-subdifferentials [29] and C-differentials [30, 27, 40]. However, the exact calculus rules do not hold for Clarke's generalized Jacobians, B-subdifferentials and the approximate Jacobians [18]; Clarke's generalized Jacobians, B-subdifferentials, C-differentials and the approximate Jocabians are all set-valued. Such properties can cause difficulties in analysing and implementing these methods.

In this paper, we define a semiderivative function G for a continuous function F, and show that this concept is indeed an extension of the concept of the derivative function. The semiderivative function G is single-valued and satisfies exact calculus rules. Hence it is relatively easy to calculate. Several common nonsmooth equation reformulations of nonlinear complementarity problems and variational inequality problems can be regarded as componentwise compositions of some smooth functions and some simple semidifferentiable functions, such as the plus function, the Fischer-Burmeister function and.the median function. The derivatives of several well-known smoothing functions, such as the Chen-Harker-Kanzow-Smale function, the Chen-Mangasarian function and the Gabrial-Moŕe function, converge to semiderivative functions of the corresponding nonsmooth reformulation functions as the smoothing parameter goes to zero. Based upon this new concept, superlinear convergence conditions for both the nonsmooth method and the smoothing method are established.

The organization of the paper is as follows. In section 2, we present basic definitions and results needed in the paper. In section 3, we give four examples. These examples motivate our study of semiderivative function in this paper. In section 4, after giving the definition of the semiderivative function, we study its relation with the derivative function when the latter exists, and its calculus rules. We also define (strongly) generalized semismooth functions there and show that this concept is indeed a generalization of the concept of (strongly) semismooth functions. Some applications of the semiderivative function on the nonsmooth approach and the smoothing approach for BVIP are given in sections 5-8 and some conclusions are drawn in section 9.

To ease our discussion, we introduce some notation here: For a continuously differentiable function $\Phi : \Re^m \to \Re^m$, we denote the Jacobian of Φ at $x \in \Re^m$ by $\Phi'(x)$, whereas the transposed Jacobian as $\nabla\Phi(x)$, and define $\partial_j\Phi_i(x) = \partial\Phi_i(x)/\partial x_j$.

$\|\cdot\|$ denotes the Euclidean norm. If M is an $m \times m$ matrix with entries M_{jk}, $j, k = 1, \ldots, m$, and $\mathcal{J}$ and $\mathcal{K}$ are nonempty index sets such that $\mathcal{J}, \mathcal{K} \subseteq \{1, \ldots, m\}$, we denote by $M_{\mathcal{J}\mathcal{K}}$ the $|\mathcal{J}| \times |\mathcal{K}|$ sub-matrix of M consisting of entries M_{jk}, $j \in \mathcal{J}$, $k \in \mathcal{K}$. If $M_{\mathcal{J}\mathcal{J}}$ is nonsingular, we denote by $M/M_{\mathcal{J}\mathcal{J}}$ the Schur-complement of $M_{\mathcal{J}\mathcal{J}}$ in M, i.e., $M/M_{\mathcal{J}\mathcal{J}} := M_{\mathcal{K}\mathcal{K}} - M_{\mathcal{K}\mathcal{J}} M_{\mathcal{J}\mathcal{J}}^{-1} M_{\mathcal{J}\mathcal{K}}$, where $\mathcal{K} = \{1, \ldots, m\} \backslash \mathcal{J}$. If $\mathcal{J} = \{1, \ldots, m\}$ then define $M/M_{\mathcal{J}\mathcal{J}} = M^{-1}$; if $\mathcal{J} = \emptyset$, define $M/M_{\mathcal{J}\mathcal{J}} = M$.

Let $F : \Re^n \to \Re^m$ be a locally Lipschitzian vector function. By Rademacher's theorem, F is differentiable almost everywhere. Let Ω_F denote the set of points where F is differentiable. Then the B-subdifferential of F at $x \in \Re^n$ is defined to be

$$\partial_B F(x) = \{ \lim_{\substack{x^k \to x \\ x^k \in \Omega_F}} \nabla F(x_k)^T \}, \tag{1.9}$$

while Clarke's generalized Jacobian of F at x is defined to be

$$\partial F(x) = \text{conv} \partial_B F(x), \tag{1.10}$$

(see [29]). The B-subdifferential was introduced in [29]. See [26, 16, 35] for some of its other applications besides its application in the generalized Newton method for solving NCP and BVIP. Sun and Han [37] introduced a generalized Jacobian ∂_C, defined by

$$\partial_C F(x) = \partial F_1(x) \times \cdots \times \partial F_m(x). \tag{1.11}$$

By (1.9-1.11), for any x,

$$\partial_B F(x) \subseteq \partial F(x) \subseteq \partial_C F(x). \tag{1.12}$$

F is called semismooth at x if F is directionally differentiable at x and for all $V \in \partial F(x+d)$ and $d \to 0$,

$$F'(x; d) = Vd + o(\|d\|); \tag{1.13}$$

F is called strongly semismooth at x if F is semismooth at x and for all $V \in \partial F(x+d)$ and $d \to 0$,

$$F'(x; d) = Vd + O(\|d\|^2); \tag{1.14}$$

F is called a (strongly) semismooth function if it is (strongly) semismooth everywhere. Here, $o(\|d\|)$ stands a vector function of d, satisfying

$$\lim_{d \to 0} \frac{o(\|d\|)}{\|d\|} = 0,$$

while $O(\|d\|^2)$ stands a vector function of d, satisfying

$$\|O(\|d\|^2)\| \leq M\|d\|^2$$

for all d satisfying $\|d\| \leq \delta$, and some $M > 0$ and $\delta > 0$.

2 Some Preliminaries

We first restate some definitions.

Definition 2.1 *A matrix $M \in \Re^{n\times n}$ is called a*

(i) P_0-matrix if each of its principal minors is nonnegative;

(ii) P-matrix if each of its principal minors is positive.

It is obvious that a positive semidefinite matrix is a P_0-matrix, a positive definite matrix is a P-matrix. We shall need the following characterization of P_0 and P matrices [7].

Proposition 2.1 *A matrix $M \in \Re^{n\times n}$ is a*

(i) P_0-matrix iff, for every $x \in \Re^n$ with $x \neq 0$, there is an index $i_0 = i_0(x)$ with

$$x_{i_0} \neq 0 \quad \text{and} \quad x_{i_0}[Mx]_{i_0} \geq 0;$$

(ii) P-matrix iff, for every $x \in \Re^n$ with $x \neq 0$, it holds that

$$\max_i x_i[Mx]_i > 0.$$

Definition 2.2 *A function $F : \Re^n \to \Re^n$ is called a*

(i) P_0-function if, for all $x, y \in \Re^n$ with $x \neq y$, there is an index $i_0 = i_0(x, y)$ with

$$x_{i_0} \neq y_{i_0} \quad \text{and} \quad (x_{i_0} - y_{i_0})[F_{i_0}(x) - F_{i_0}(y)] \geq 0;$$

(ii) P-function if, for all $x, y \in \Re^n$ with $x \neq y$, it holds that

$$\max_i (x_i - y_i)[F_i(x) - F_i(y)] > 0;$$

(iii) uniform P-function if there is a constant $\mu > 0$ such that

$$\max_i (x_i - y_i)[F_i(x) - F_i(y)] \geq \mu\|x - y\|^2$$

holds for all $x, y \in \Re^n$.

Definition 2.3 *A function $F : \Re^n \to \Re^n$ is called a*

(i) monotone function if, for all $x, y \in \Re^n$ with $x \neq y$,

$$(x - y)^T[F(x) - F(y)] \geq 0;$$

(ii) strictly monotone function if, for all $x, y \in \Re^n$ with $x \neq y$,

$$(x-y)^T[F(x)-F(y)] > 0.$$

It is known that every uniform P-function is P-function and every P-function is a P_0-function. Moreover, the Jacobian of a continuously differentiable P_0-function (uniform P-function) at a point is a P_0-matrix (P-matrix).

Associated to a solution x^* of BVIP we define three index sets:

$$\mathcal{I} := \{i \in N : l_i < x_i^* < u_i, \Phi_i(x^*) = 0\},$$

$$\mathcal{J} := \{i \in N : x_i^* = l_i, \Phi_i(x^*) = 0\} \cup \{i \in N : x_i^* = u_i, \Phi_i(x^*) = 0\}$$

and

$$\mathcal{K} := \{i \in N : x_i^* = l_i, \Phi_i(x^*) > 0\} \cup \{i \in N : x_i^* = u_i, \Phi_i(x^*) < 0\}.$$

The solution x^* is said to be nondegenerate if $\mathcal{J} = \emptyset$.

The followings are two regularity concepts which will be used later.

Definition 2.4 *A solution x^* of BVIP is called*

(i) b-regular if the principal submatrices $\Phi'(x^*)_{\mathcal{I}\cup\mathcal{L},\mathcal{I}\cup\mathcal{L}}$ are nonsingular for all subsets $\mathcal{L}$ such that $\emptyset \subseteq \mathcal{L} \subseteq \mathcal{J}$;

(ii) R-regular if the submatrix $\Phi'(x^*)_{\mathcal{I}\mathcal{I}}$ is nonsingular and the Schur-complement of this matrix in

$$\begin{pmatrix} \Phi'(x^*)_{\mathcal{I}\mathcal{I}} & \Phi'(x^*)_{\mathcal{I}\mathcal{J}} \\ \Phi'(x^*)_{\mathcal{J}\mathcal{I}} & \Phi'(x^*)_{\mathcal{J}\mathcal{J}} \end{pmatrix}$$

is a P-matrix.

We recall that the above mentioned Schur-complement is defined by

$$\Phi'(x^*)_{\mathcal{J}\mathcal{J}} - \Phi'(x^*)_{\mathcal{J}\mathcal{I}}\Phi'(x^*)_{\mathcal{I}\mathcal{I}}^{-1}\Phi'(x^*)_{\mathcal{I}\mathcal{J}} \in \Re^{|\mathcal{J}|\times|\mathcal{J}|}. \tag{2.1}$$

Obviously, there is no difference between b- and R-regularity at a nondegenerate solution of BVIP. The following theorem on the relationship between b- and R-regularity is a generalization of Theorem 2.5 in [8]. So we omit its proof.

Theorem 2.1 *Let x^* be a solution of BVIP. Then the following statements are equivalent:*

(i) x^* is an R-regular solution of BVIP.

(ii) x^* is a b-regular solution of BVIP and the Schur-complement (2.1) is a P-matrix.

(iii) x^* is a b-regular solution of BVIP and the Schur-complement (2.1) is a P_0-matrix.

3 Motivation: Nonsmooth Reformulation and Smooth Functions

In some common nonsmooth equation reformulation methods for solving nonlinear complementarity and variational inequality problems, the problem is reformulated as a system of nonsmooth equations, $F(x) = 0$, where each component of F, F_i is a composition of a simple nonsmooth function ψ, such as the plus function, the Fischer-Burmeister function or the median function, and the corresponding component function Φ_i, of the smooth function Φ in the original problem. In the nonsmooth approach, the generalized Jacobian or the directional derivative of F is used to solve $F(x) = 0$ to get a solution of the original problem. In the smooth approach, F is approximated by a family of smooth functions F_t, where $t > 0$ is a parameter. The smooth function F_t can also be regarded as a componentwise composite function of a family of simple smooth functions ψ_t, and the function Ψ. As $t \to 0^+$, $\psi_t \to \psi$ pointwise, hence $F_t \to F$ pointwise. Then the derivative of ψ_t, $\nabla\psi_t$, converges to a function ξ pointwise. Correspondingly, the derivative of F_t, $(\nabla F_t)^T$, converges to a matrix-valued function G pointwise. If we regard ξ as a certain generalized derivative function of ψ, then we see that G can be regarded as a generalized derivative function of F correspondingly, generated from the chain rule for composite functions. The function G plays an important role in convergence analysis of smoothing methods. The following four examples illustrate this phenomenon and motivate our study of semiderivative functions in this paper.

Example 3.1 *The nonlinear complementarity problem is to find $x \in \Re^n$ such that*

$$\Phi(x) \geq 0, \quad x \geq 0, \quad x^T\Phi(x) = 0, \tag{3.1}$$

where $\Phi : \Re^n \to \Re^n$ is a smooth function. We may reformulate this problem as a nonsmooth equation problem $F(x) = 0$, where F is defined by

$$F_i(x) = x_i - \psi(x_i - \Phi_i(x)) \tag{3.2}$$

for each i, and $\psi : \Re \to \Re$ is the plus function, defined by

$$\psi(a) \equiv a_+ := \max\{a, 0\}. \tag{3.3}$$

We may approximate ψ by the Chen-Harker-Kanzow-Smale function ψ_t, defined by

$$\psi_t(a) := \frac{1}{2}\left(a + \sqrt{a^2 + 4t^2}\right), \tag{3.4}$$

where $t > 0$. Then F is approximated by F_t, defined by

$$(F_t)_i(x) = x_i - \psi_t(x_i - \Phi_i(x)) \tag{3.5}$$

for each i. For each a, we have

$$\lim_{t\downarrow 0} \psi_t(a) = \psi(a). \tag{3.6}$$

Correspondingly, for each x, we have

$$\lim_{t\downarrow 0} F_t(x) = F(x). \tag{3.7}$$

For $t > 0$, ψ_t, hence F_t, is differentiable. We have

$$\psi_t'(a) = \frac{1}{2}\left(1 + \frac{a}{\sqrt{a^2 + 4t^2}}\right). \tag{3.8}$$

Then

$$\partial_j (F_t)_i(x) = \delta_{ij} - \psi_t'(x_i - \Phi_i(x))(\delta_{ij} - \partial_j \Phi_i(x)), \tag{3.9}$$

where the Kronecker symbol $\delta_{ii} = 1$ and $\delta_{ij} = 0$ for $i \neq j$. Then

$$\lim_{t\downarrow 0} \psi_t'(a) = \xi(a) := \begin{cases} 0 & \text{if } a < 0, \\ 0.5 & \text{if } a = 0, \\ 1 & \text{if } a > 0. \end{cases} \tag{3.10}$$

We also have

$$\lim_{t\downarrow 0} (\nabla F_t(x))^T = G(x), \tag{3.11}$$

where

$$G_{ij}(x) = \delta_{ij} - \xi(x_i - \Phi_i(x))(\delta_{ij} - \partial_j \Phi_i(x)). \tag{3.12}$$

So, if we regard ξ as a certain generalized derivative function of ψ, then we see that G can be regarded as a generalized derivative function of F correspondingly, generated from the chain rule for composite functions.

Example 3.2 *In Example 3.1, a more general form of ψ_t is the Chen-Mangasarian function, see [4], defined by*

$$\psi_t(a) := \int_{-\infty}^{a} \int_{-\infty}^{r} \frac{1}{t} \rho(\frac{s}{t}) ds dr, \tag{3.13}$$

where ρ is a probability density function. If we let

$$\rho(s) = \frac{2}{s^2 + 4},$$

then ψ_t is the Chen-Harker-Kanzow-Smale function. If we let

$$\rho(s) = \frac{e^{-s}}{(1 + e^{-s})^2},$$

then ψ_t is the neural network smooth (sigmoid) function. If we let

$$\rho(s) = \frac{e^{-s^2/2}}{\sqrt{2\pi}},$$

then ψ_t is the normal smooth function. Replacing (3.4) by (3.13), we have

$$\psi_t'(a) = \int_{-\infty}^{a/t} \rho(s)ds. \tag{3.14}$$

Then

$$\lim_{t\downarrow 0} \psi_t'(a) = \xi(a) := \begin{cases} 0 & \text{if } a < 0, \\ \int_{-\infty}^{0} \rho(s)ds & \text{if } a = 0, \\ 1 & \text{if } a > 0. \end{cases} \tag{3.15}$$

We see that all the other arguments and formulas in Example 3.1 are still valid if we replace (3.4), (3.8) and (3.10) by (3.13-3.15).

Example 3.3 *The nonlinear complementarity problem (3.1) may also be reformulated as a nonsmooth equation problem $F(x) = 0$, where F is defined by*

$$F_i(x) = \psi(x_i, \Phi_i(x)) \tag{3.16}$$

for each i, and $\psi : \Re^2 \to \Re$ is the Fischer-Burmeister function, defined by

$$\psi(a, b) = \sqrt{a^2 + b^2} - a - b. \tag{3.17}$$

We may approximate ψ by the Kanzow-Fischer function ψ_t, defined by

$$\psi_t(a, b) = \sqrt{a^2 + b^2 + t} - a - b. \tag{3.18}$$

where $t > 0$. Then F is approximated by F_t, defined by

$$(F_t)_i(x) = \psi_t(x_i, \Phi_i(x)) \tag{3.19}$$

for each i. For each (a, b), we have

$$\lim_{t\downarrow 0} \psi_t(a, b) = \psi(a, b). \tag{3.20}$$

Correspondingly, for each x, we still have (3.7). Again, for $t > 0$, ψ_t, hence F_t, is differentiable. We have

$$\nabla \psi_t(a, b) = \begin{pmatrix} \frac{a}{\sqrt{a^2+b^2+t}} - 1 \\ \frac{b}{\sqrt{a^2+b^2+t}} - 1 \end{pmatrix}. \tag{3.21}$$

Then

$$\partial_j (F_t)_i(x) = \partial_1 \psi_t(x_i, \Phi_i(x))\delta_{ij} + \partial_2 \psi_t(x_i, \Phi_i(x))\partial_j \Phi_i(x). \tag{3.22}$$

We have

$$\lim_{t\downarrow 0} \nabla\psi_t(a,b) = \xi(a,b) := \begin{cases} \binom{-1}{-1} & \text{if } a=b=0, \\ \begin{pmatrix} \frac{a}{\sqrt{a^2+b^2}}-1 \\ \frac{b}{\sqrt{a^2+b^2}}-1 \end{pmatrix} & \text{othersise.} \end{cases} \tag{3.23}$$

We still have

$$\lim_{t\downarrow 0}(\nabla F_t(x))^T = G(x), \tag{3.24}$$

where

$$G_{ij}(x) = \xi_1(x_i, \Phi_i(x))\delta_{ij} + \xi_2(x_i, \Phi_i(x))\partial_j\Phi_i(x). \tag{3.25}$$

Again, if we regard ξ as a certain generalized derivative function of ψ, then we see that G can be regarded as a generalized derivative function of F correspondingly, generated from the chain rule for composite functions.

For any three numbers $\alpha \in \Re \cup \{-\infty\}$, $\beta \in \Re \cup \{\infty\}$ with $\alpha \le \beta$ and $a \in \Re$, the median function $\phi_{(\alpha,\beta)}(\cdot)$ is defined by

$$\phi_{(\alpha,\beta)}(a) = \begin{cases} \alpha & \text{if } a < \alpha \\ a & \text{if } \alpha \le \ a \le \beta \\ \beta & \text{if } \beta < a \end{cases}. \tag{3.26}$$

Then the Gabriel-Moré smoothing function $\phi_{t\alpha\beta}(\cdot)$ for $\phi_{(\alpha,\beta)}(\cdot)$ [15] is defined by

$$\phi_{t\alpha\beta}(a) = \int_{-\infty}^{\infty} \phi_{(\alpha,\beta)}(a-ts)\rho(s)ds, \quad (t,a) \in \Re_{++} \times \Re, \tag{3.27}$$

where ρ is a probability density function.

Example 3.4 *The box constrained variational inequality problem is to find $x \in X := \{x \in \Re^n : l \le x \le u\}$, where $l \in \{\Re \cup \{-\infty\}\}^n, u \in \{\Re \cup \{+\infty\}\}^n$ and $l < u$, such that for all $y \in X$,*

$$(y-x)^T\Phi(x) \ge 0, \tag{3.28}$$

where $\Phi : \Re^n \to \Re^n$ is a smooth function. We may reformulate this problem as a nonsmooth equation problem $F(x) = 0$, where F is defined by

$$F_i(x) = x_i - \psi_i(x_i - \Phi_i(x)), \tag{3.29}$$

for each i, and $\psi_i = \phi_{(l_i,u_i)}$, where $\phi_{(l_i,u_i)}$ is defined in (3.26). Let $\psi_{i,t}(a) = \phi_{tl_iu_i}(a)$, where $\phi_{tl_iu_i}$ is defined in (3.27), then F is approximated by F_t, defined by

$$(F_t)_i(x) = x_i - \psi_{i,t}(x_i - \Phi_i(x)) \tag{3.30}$$

for each i. For each i and a, we have

$$\lim_{t\downarrow 0} \psi_{i,t}(a) = \psi_i(a). \tag{3.31}$$

Correspondingly, for each x, we still have (3.7). For $t > 0$, $\psi_{i,t}$, hence F_t, is differentiable. We have

$$\psi'_{i,t}(a) = \int_{(a-u_i)/t}^{(a-l_i)/t} \rho(s)ds. \tag{3.32}$$

Then

$$\partial_j(F_t)_i(x) = \delta_{ij} - \psi'_{i,t}(x_i - \Phi_i(x))(\delta_{ij} - \partial_j\Phi_i(x)). \tag{3.33}$$

Then

$$\lim_{t\downarrow 0} \psi'_{i,t}(a) = \xi_i(a) := \begin{cases} 0 & \text{if } a < l_i, \\ \int_{-\infty}^{0} \rho(s)ds & \text{if } a = l_i, \\ 1 & \text{if } l_i < a < u_i, \\ \int_0^{\infty} \rho(s)ds & \text{if } a = u_i, \\ 0 & \text{if } a > u_i. \end{cases} \tag{3.34}$$

We also have

$$\lim_{t\downarrow 0}(\nabla F_t(x))^T = G(x), \tag{3.35}$$

where

$$G_{ij}(x) = \delta_{ij} - \xi_i(x_i - \Phi_i(x))(\delta_{ij} - \partial_j\Phi_i(x)), \tag{3.36}$$

Again, if we regard ξ_i as a certain generalized derivative function of ψ_i for each i, then we see that G can be regarded as a generalized derivative function of F correspondingly, generated from the chain rule for composite functions.

Through these four examples, we see a similar pattern. We find a certain generalized derivative function ξ (ξ_i) of ψ (ψ_i), and then determine G as a certain generalized derivative function of F, which can be calculated by the chain rule. Unlike the generalized Jacobian function, which is set-valued and does not satisfy exact calculus rules, this generalized derivative function is single-valued and satisfies exact calculus rules (here the chain rule). These stimulate us to explore an adequate definition of this generalized derivative function, to analyse its properties and its applications in both smoothing Newton methods and nonsmooth Newton methods in the next few sections.

4 Semiderivative Functions: Definition and Calculus Rules

In this section, we first define a semiderivative function G for a continuous function F. Theorem 4.1 will show that this concept is indeed an extension of the concept of the derivative function.

Recall that if $F : D \subseteq \Re^n \rightarrow \Re^m$, where D is an open set, is continuously differentiable at $x \in D$, then for $d \rightarrow 0$,

$$F(x + d) = F(x) + F'(x + d)d + o(\|d\|). \tag{4.1}$$

If $F : D \subseteq \Re^n \rightarrow \Re^m$ is semismooth at $x \in D$, then for all $V \in \partial F(x+d)$ and $d \rightarrow 0$,

$$F(x+d) = F(x) + Vd + o(\|d\|). \tag{4.2}$$

A function $H : D \subseteq \Re^n \rightarrow \Re^p$ is called locally bounded at $x \in D$ if there is a neighborhood of x, $N_\epsilon(x) \subset D$, such that H is bounded in this neighborhood. We say that H is locally bounded if it is locally bounded at any $x \in D$. Note that ξ, ξ_i and G in the four examples in the last section are all locally bounded. Furthermore, a relation between F and G in the four examples in Section 3 is that for any fixed x and $d \rightarrow 0$,

$$F(x+d) = F(x) + G(x+d)d + o(\|d\|).$$

Actually, this relation also holds between ψ and ξ in the four examples in Section 3, and this property is retained under composition. This motivates us to give the following definition. Also notice that, for any subset $\mathcal{A}$ of $\Re^{m \times n}$, Sweeter [39] defines

$$plen\mathcal{A} = \{A \in \Re^{m \times n} : Ab \in \mathcal{A}b, \ \forall\, b \in \Re^n\}. \tag{4.3}$$

Definition 4.1 *Suppose that $F : D \subseteq \Re^n \rightarrow \Re^m$ is a continuous function, where D is an open set. We say that $G : D \subseteq \Re^n \rightarrow \Re^{m \times n}$ is a semiderivative function of F at x if G is locally bounded at x, and for $d \rightarrow 0$,*

$$F(x+d) = F(x) + G(x+d)d + o(\|d\|). \tag{4.4}$$

If furthermore,

$$F(x+d) = F(x) + G(x+d)d + O(\|d\|^2), \tag{4.5}$$

then we say that G is a strong semiderivative function of F at x. We call $F : D \subseteq \Re^n \rightarrow \Re^m$ (strongly) semidifferentiable at $x \in D$ if F has a (strong) semiderivative function at x. Let

$$\partial_S F(x) = plenAG(x), \tag{4.6}$$

where

$$AG(x) = \{\lim G(x^k) : x^k \rightarrow x, x^k \neq x\}. \tag{4.7}$$

We call $\partial_S F(x)$ the semidifferential of F at x, associated with the semiderivative function G.

Theorem 4.1 *Suppose that $F : D \subseteq \Re^n \rightarrow \Re^m$ is a continuous function, where D is an open set, and $G : D \subseteq \Re^n \rightarrow \Re^{m \times n}$ is a semiderivative function of F at x.*

- *(i) If G is continuous at x, then F is F-differentiable at x with $F'(x) = G(x)$, and $\partial_S F(x) = \{G(x)\}$;*
- *(ii) If F is G-differentiable at x, then $F'(x) \in \partial_S F(x)$.*

Proof. From (4.4) and the continuity of G at x, (i) holds. If F is G-differentiable at x, then for any $d \in \Re^n$,

$$\begin{aligned} F'(x)d &= F'(x;d) \\ &= \lim_{t\downarrow 0} G(x+td)d, \text{ by } (4.4) \\ &= \left(\lim_{t\downarrow 0} G(x+td)\right) d \in AG(x)d. \end{aligned}$$

So (ii) holds. □

We now introduce the concept "generalized semismooth functions" and show in the next two theorems that this concept is indeed a generalization of the concept of semismooth functions: we will show that a (strongly) semismooth function is a (strongly) generalized semismooth function, while a generalized semismooth function is always locally Lipschitz.

Definition 4.2 *Suppose that $F : D \subseteq \Re^n \to \Re^m$ is a continuous function, where D is an open set. If there exists a function $G : D \subseteq \Re^n \to \Re^{m\times n}$ such that G is a (strong) semiderivative function of F at any $x \in D$, then F is said to be (strongly) generalized semismooth on D.*

Theorem 4.2 *Suppose that $F : D \subseteq \Re^n \to \Re^m$ is a continuous function, where D is an open set. If F is generalized semismooth on D, then F is locally Lipschitz on D.*

Proof. For any $x \in D$, since the semiderivative function G of F is locally bounded there exists a closed neighborhood $B_\epsilon(x)$ of x, $x \in B_\epsilon(x) \subset D$, and a constant $L > 0$ such that for any $y \in intB_\epsilon(x)$, $\|G(y)\| \leq \frac{L}{2}$. For any $y \in intB_\epsilon(x)$, from (4.4) there exists an open neighborhood $B_{\epsilon(y)}(y) \subset B_\epsilon(x)$ of y such that when $y + d \in B_{\epsilon(y)}(y)$,

$$\|F(y+d) - F(x)\| \leq L\|d\|. \tag{4.8}$$

For any $z, z+d \in intB_\epsilon(x)$,

$$[z, z+d] \subset \bigcup_{y\in[z,z+d]} B_{\epsilon(y)}(y). \tag{4.9}$$

Then there exist $y^k \in [z, z+d]$, $k = 1, \cdots, p$, where p is a finite number, such that $y^k \in [y^{k-1}, y^{k+1}]$, where $y^0 = z$ and $y^{p+1} = z + d$, and

$$[z, z+d] \subset \bigcup_k B_{\epsilon(y^k)}(y^k). \tag{4.10}$$

In fact we can find $p-1$ points z^k, $z^k = z + t_k d$, $0 < t_1 < t_2 < \cdots < t_{p-1} < 1$, such that each pair of successive points z^{k-1} and z^k are in the same $B_{\epsilon(y^k)}(y^k)$ and

$y^k \in [z^{k-1}, z^k]$, where $z^0 = z$, $z^p = z + d$, $t_0 = 0$ and $t_p = 1$. Hence by (4.8),

$$\begin{aligned} \|F(z+d) - F(x)\| &\leq \sum_{k=0}^{p-1} \left\{ \|F(z^{k+1}) - F(y^k)\| + \|F(y^k) - F(z^k)\| \right\} \\ &\leq \sum_{k=0}^{p-1} L(t_{k+1} - t_k)\|d\| = L\|d\|. \end{aligned}$$

This shows F is locally Lipschitz on D. □

Theorem 4.3 *If $F : \Re^n \rightarrow \Re^m$ is (strongly) semismooth, then F is (strongly) generalized semismooth.*

Proof. By [32, 29], with $G(x) \in \partial F(x)$ everywhere, (4.4) holds at any x if F is semismooth while (4.5) holds at any x if F is strongly semismooth. Also, if $G(x) \in \partial F(x)$, then G is locally bounded. □

However, the converse of this theorem may not be true since we need directional differentiability for semismoothness.

In the four examples in the last section, instead of using definition, we may also use this theorem and the facts that ψ and ψ_i are strongly semismooth functions as well as $\xi(x) \in \partial\psi(x)$ and $\xi_i(x) \in \partial\psi_i(x)$ for all x, to verify that ξ and ξ_i are strong semiderivative functions of ψ and ψ_i respectively.

We have exact calculus rules for semiderivative functions in the sense of the following propositions and theorem.

Proposition 4.1 *Suppose that $F : \Re^n \rightarrow \Re^m$ and $\hat{F} : \Re^n \rightarrow \Re^m$ are two continuous functions and α is a real number. If G and $\hat{G}$ are (strong) semiderivative functions of F and $\hat{F}$ at $x \in \Re^n$ respectively, then αG and $G + \hat{G}$ are (strong) semiderivative functions of αF and $F + \hat{F}$ at x respectively.*

Proposition 4.2 *Suppose that $F : \Re^n \rightarrow \Re^m$ is a continuous function with components $F_1,..., F_m$. If $G_1, ..., G_m$ are (strong) semiderivative functions of $F_1, ..., F_m$ at $x \in \Re^n$ respectively, then $G = G_1 \times G_2 \times \cdots G_m$ is a (strong) semiderivative function of F at x.*

Proposition 4.3 *Suppose that $f : \Re^n \rightarrow \Re$ and $\hat{f} : \Re^n \rightarrow \Re$ are two continuous functions. If g and $\hat{g}$ are (strong) semiderivative functions of f and $\hat{f}$ at $x \in \Re^n$ respectively, then $g + \hat{g}$, $g - \hat{g}$ and $f\hat{g} + g\hat{f}$ are (strong) semiderivative functions of $f + \hat{f}$, $f - \hat{f}$ and $f\hat{f}$ at x respectively. Furthermore, if $\hat{f}(x) \neq 0$ for all $x \in \Re^n$, then $\frac{\hat{f}g - f\hat{g}}{\hat{f}^2}$ is a (strong) semiderivative function of $f/\hat{f}$ at x.*

The proofs of these three propositions are direct from the definition. They may also be regarded as corollaries of the following theorem.

Theorem 4.4 *Suppose that $F : \Re^n \to \Re^m$ and $S : \Re^m \to \Re^p$ are two continuous vector functions. If G is a (strong) semiderivative function of F at $x \in \Re^n$ and T is a (strong) semiderivative function of S at $F(x) \in \Re^m$ respectively, then $R : \Re^n \to \Re^{p\times n}$, defined by*

$$R(\cdot) = T(F(\cdot))G(\cdot)$$

is a (strong) semiderivative function of $S \circ F : \Re^n \to \Re^p$ at x.

Proof. Clearly, R is locally bounded. For any $d \in \Re^n$, let $y = F(x)$ and $y_d = F(x+d)$. Then we have

$$F(x+d) = F(x) + G(x+d)d + o(\|d\|), \tag{4.11}$$

which implies that

$$\|y_d - y\| = O(\|d\|) \tag{4.12}$$

by local boundedness of G at x. Thus, we have

$$\begin{aligned} & S(F(x+d)) \\ = \; & S(y_d) \\ = \; & S(y) + T(y_d)(y_d - y) + o(\|y_d - y\|) \\ = \; & S(y) + T(F(x+d))(G(x+d)d + o(\|d\|)) + o(\|d\|) \quad \text{(by (4.11) and (4.12))} \\ = \; & S(F(x)) + R(x+d)d + o(\|d\|) \text{ (by local boundedness of } T \text{ at } x). \end{aligned}$$

Hence R is a semiderivative function of $S \circ F$. Similarly, we can prove that R is a strong semiderivative function of $S \circ F$. □

Since ξ and ξ_i are strong semiderivative functions of ψ and ψ_i respectively in the four examples in the last section, by this theorem, G is a semiderivative function of F is these examples. If furthermore Φ has a locally Lipschitz derivative function, then G is a strong semiderivative function of F is these examples. In reformulation methods for solving complementarity and variational problems, most nonsmooth functions are composite functions of some simple nonsmooth functions and some smooth functions. Hence, by this theorem, we may calculate their semiderivative functions easily. Another example of using this theorem is as follows.

Proposition 4.4 *Suppose that $f : \Re^n \to \Re$ and $\hat{f} : \Re^n \to \Re$ are two continuous functions. Let $h : \Re^n \to \Re$ be defined by*

$$h(x) = \min\{f(x), \hat{f}(x)\}.$$

Suppose that g and $\hat{g}$ are (strong) semiderivative functions of f and $\hat{f}$ respectively. Define

$$r(x) = \begin{cases} \hat{g}(x) & \text{if } \hat{f}(x) < f(x), \\ g(x) & \text{if } f(x) < \hat{f}(x), \\ \frac{1}{2}(g(x) + \hat{g}(x)) & \text{if } f(x) = \hat{f}(x). \end{cases}$$

Then r is a (strong) semiderivative function of h.

Proof. We can use the definition or Theorem 4.3 to verify that $\psi : \Re^2 \to \Re$, defined by

$$\psi(a, b) = \min\{a, b\},$$

has a strong semiderivative function ξ, defined by

$$\xi(a, b) = \begin{cases} \binom{1}{0} & \text{if } a < b, \\ \binom{0}{1} & \text{if } b < a, \\ \binom{0.5}{0.5} & \text{if } a = b. \end{cases}$$

Now, applying Theorem 4.4, we have this proposition. □

Applying Theorem 4.3 and Proposition 4.2, we have the following proposition.

Proposition 4.5 *Suppose that $F : \Re^n \to \Re^m$ is locally Lipschitz and $G(x) \in \partial_C F(x)$ for all x. Then G is a (strong) semiderivative function of F if F is a (strongly) semismooth function.*

By (1.12), this proposition is stronger than Theorem 4.3.

5 Semismooth Newton Methods

Definition 5.1 *Let $\psi : \Re^2 \to \Re$ be a strongly semismooth function. Denote by P_ψ the set of points that ψ maps to zero. Let E_ψ be the extreme point set of P_ψ. We say ψ is a pseudo-smooth function if it is smooth everywhere in $\Re^2 \setminus E_\psi$.*

Let $N := \{1, 2, \cdots, n\}$. For each $i \in N$, let $\psi_i = \psi_{(l_i, u_i)}$ be a pseudo-smooth BVIP function. Then solving $F(x) = 0$, where $F_i(x)$ is defined in (1.6), is equivalent to solving the BVIP. For each $i \in N$, define $\xi_i : \Re^2 \to \Re^2$ by

$$\begin{cases} \xi_i(a, b) = \nabla\psi_i(a, b) & \text{if } (a, b) \notin E_{\psi_i} \\ \xi_i(a, b) \in \partial_B\psi_i(a, b) & \text{if } (a, b) \in E_{\psi_i}. \end{cases} \tag{5.1}$$

From Theorem 4.3, we have that ξ_i is a strong semiderivative function of ψ_i. From Theorem 4.4, we have that $G : \Re^n \to \Re^{n \times n}$, defined by

$$G(x) = D(x) + E(x)\nabla\Phi(x)^T, \tag{5.2}$$

is a semiderivative function of F, where $D(x)$ is a diagonal matrix whose ith diagonal element is the first element of $\xi_i(x_i, \Phi_i(x))$ for each $i \in N$, and $E(x)$ is a diagonal matrix whose ith diagonal element is the second element of $\xi_i(x_i, \Phi_i(x))$ for each $i \in N$. If $\nabla\Phi(x)$ is locally Lipschitz, then G is a strong semiderivative function of F.

Let $\theta : \Re^n \to \Re$ be the norm function of $F(x)$, defined by

$$\theta(x) = \frac{1}{2}F(x)^T F(x).$$

Theorem 5.1 *The norm function θ is continuously differentiable at $\Re^n$ with*

$$\nabla\theta(x)^T = F(x)^T G(x) \tag{5.3}$$

for all $x \in \Re^n$. If $F(x) \neq 0$ and

$$F(x) + G(x)d(x) = 0$$

has a solution $d(x)$, then $d(x)$ is a descent direction of θ.

We omit the proof of this theorem since it is similar to that of Theorem 3.1 of [31].

Proposition 5.1 *[31, Proposition 3.1] Let $A(x^*)$ be the set of all $n \times n$ matrices G such that*

$$G = D(x^*) + E(x^*)\nabla\Phi(x^*)^T,$$

where $D(x^*)$ is a diagonal matrix whose ith diagonal element is the first element of a vector s^i in $\partial_B\psi_i(x_i^*, \Phi_i(x^*))$ for $i \in N$, and $E(x^*)$ is a diagonal matrix whose ith diagonal element is the second element of a vector s^i in $\partial_B\psi_i(x_i^*, \Phi_i(x^*))$ for $i \in N$. If all the matrices in $A(x^*)$ are nonsingular, then there are a positive number c and a neighborhood $N(x^*)$ of x^* such that for all $x \in N(x^*)$, $G(x)$ is nonsingular and

$$\|G(x)^{-1}\| \leq c.$$

Definition 5.2 *We say that F is A-regular at x^* if all the matrices in $A(x^*)$ are nonsingular.*

Algorithm 5.1

Step 1. *Let $x^0 \in \Re^n$, $\sigma, \rho \in (0,1), \eta > 0, p > 2$ and $k := 0$.*

Step 2. *If $F(x^k) = 0$ then stop. Otherwise, let d^k be a solution of*

$$F(x^k) + G(x^k)d(x^k) = 0. \tag{5.4}$$

If (5.4) is not solvable, or if $\nabla\theta(x^k)^T d^k > -\eta\|d^k\|^p$, set $d^k = -\nabla\theta(x^k)$.

Step 3. *Let $\alpha^k = \rho^{j_k}$, where j_k be the smallest nonnegative integer j satisfying*

$$\theta(x^k + \rho^j d^k) - \theta(x^k) \le \sigma\rho^j \nabla\theta(x^k)^T d^k, \tag{5.5}$$

where ρ^j means the jth power of ρ.

Step 4. *Let $x^{k+1} = x^k + \alpha^k d^k$ and $k := k+1$. Go to Step 2.*

This algorithm is a generalization of the algorithms in [7, 11, 31]. We have the following convergence theorem for this algorithm. We omit its proof since it is similar to the proof of Theorem 7.1 of [31].

Theorem 5.2 *Assume that x^* is an accumulation point of $\{x^k\}$ generated by Algorithm 5.1. If F is A-regular at x^*, then x^* is a solution of the pseudo-smooth equation $F(x) = 0$, and the whole sequence converges to x^* superlinearly. If furthermore $\nabla\Phi$ is locally Lipschitz around x^*, then the sequence converges to x^* quadratically.*

Now we study under what conditions F is A-regular at a solution point x^* of $F(x) = 0$. We have the following lemma.

Lemma 5.1 *If $n \times n$ matrix M is a P-matrix and D and E are two diagonal matrices satisfying that $D_{ii}E_{ii} \ge 0$ and $D_{ii} + E_{ii} \ne 0$ for each $i \in N$, then*

$$D + EM \text{ and } D + ME$$

are nonsingular.

Proof. Define the following index set associated with D and E:

$$\mathcal{A} = \{i \in N : D_{ii} \ne 0, E_{ii} = 0\},$$

$$\mathcal{B} = \{i \in N : D_{ii} \ne 0, E_{ii} \ne 0\}$$

and

$$\mathcal{C} = \{i \in N : D_{ii} = 0, E_{ii} \ne 0\}.$$

Let $(D + EM)q = 0$ for some $q \in \Re^n$. Then we have that $q_i = 0$ for $i \in \mathcal{A}$, $q_i = -\frac{E_{ii}}{D_{ii}}(Mq)_i$ for $i \in \mathcal{B}$ and $(Mq)_i = 0$ for $i \in \mathcal{C}$. If there exists an $i \in \mathcal{B}$ such that $q_i \ne 0$, then we have that $\max_{i\in N} q_i(Mq)_i \le 0$. This is a contradiction. Thus $q_i = 0$ for all $i \in \mathcal{B}$. Similarly we can prove that $q_i = 0$ for all $i \in \mathcal{C}$. So $D + EM$ is nonsingular. Because $(D + ME)^T = D + EM^T$ and M^T is a P-matrix, $D + EM^T$ is nonsingular. Then $D + ME$ is nonsingular. □

Lemma 5.2 *If $n \times n$ matrix M is a P_0-matrix and D and E are two diagonal matrices satisfying that $D_{ii}E_{ii} \geq 0$ and $D_{ii} \neq 0$ for each $i \in N$, then*

$$D + EM \text{ and } D + ME$$

are nonsingular.

The proof of this lemma is similar to that of Lemma 6.1 in [31].

Definition 5.3 *Let ψ be a BVIP function. We say that ψ is P-regular at $(a, b) \in \Re^2$ if for all $(\xi_1, \xi_2)^T \in \partial_B \psi(a, b)$, $\xi_1 \xi_2 \geq 0$ and $\xi_1 + \xi_2 \neq 0$.*

We can verify that the Billups BVIP function [1] and the Sun-Womersley BVIP function [38] are P-regular at any $(a, b) \in \Re^2$.

Form Lemma 5.1 we have the following theorem.

Theorem 5.3 *Assume that $\nabla\Phi(x^*)^T$ is a P-matrix and ψ_i is P-regular at $(x_i^*, \Phi_i(x^*))$ for each $i \in N$. Then F is A-regular at x^*.*

The condition in Theorem 5.3 that $\nabla\Phi(x^*)^T$ is a P-matrix can be weakened to the condition that x^* is a R-regular solution of BVIP for some special BVIP functions. For instance, we assume that ψ_i is a regular pseudo-smooth BVIP function for each $i \in N$, see [31]. Let

$$\xi_i(a, b) := \begin{cases} \begin{pmatrix} 0 \\ 1 \end{pmatrix} & \text{if } (a, b) = (l_i, 0) \text{ or } (a, b) = (u_i, 0), \\ \nabla\psi_i(a, b) & \text{otherwise.} \end{cases} \tag{5.6}$$

Then we have the following theorem for regular pseudo-smooth BVIP functions.

Theorem 5.4 *Let $M = \nabla\Phi(x^*)^T$. Assume that x^* is a R-regular solution of BVIP. Then F is A-regular at x^*.*

We omit its proof since it is similar to the proof of Theorem 6.2 of [31].

6 Smoothing Newton Methods

Let $\psi_{(\alpha,\beta)} : \Re^2 \to \Re$ be a strongly semismooth BVIP function.

Definition 6.1 *We call $\psi_t : \Re^2 \times \Re_{++} \to \Re$ a smoothing approximation of $\psi_{(\alpha,\beta)}$ if ψ_t satisfies the following conditions:*

(i) ψ_t is continuously differentiable for any $t > 0$ and there are two constants $\mu > 0$ and $c > 0$ such that for any $(a, b) \in \Re^2$ and $t \in (0, c]$,

$$\|\psi_t(a, b) - \psi_{(\alpha,\beta)}(a, b)\| \leq \mu t; \tag{6.1}$$

(ii) $\lim_{t\downarrow 0}\psi_t(a,b) = \psi_{(\alpha,\beta)}(a,b)$, for each $(a,b) \in \Re^2$;

(iii) $\lim_{t\downarrow 0}\nabla\psi_t(a,b) = \xi(a,b)$, for each $(a,b) \in \Re^2$;

(iv) $\xi(a,b) \in \partial_B\psi_{(\alpha,\beta)}(a,b)$, for each $(a,b) \in \Re^2$.

For example, let

$$\psi_{(\alpha,\beta)}(a,b) = a - \phi_{(\alpha,\beta)}(a-b), \tag{6.2}$$

where $\phi_{(\alpha,\beta)}$ is defined in (3.26). Then $\psi_{(\alpha,\beta)}$ defined in (6.2) is a strongly semismooth BVIP function. Let

$$\psi_t(a,b) = a - \phi_{t\alpha\beta}(a-b), \tag{6.3}$$

where $\phi_{t\alpha\beta}(\cdot)$ is defined in (3.27). From Lemma 2.3 [33] we can easily verify that ψ_t defined in (6.3) is a smoothing approximation of $\psi_{(\alpha,\beta)}$ defined in (6.2).

For each $i \in N$, let $\psi_i = \psi_{(l_i,u_i)}$ be a BVIP function and $\psi_{i,t}$ be a smoothing approximation BVIP function of ψ_i. Define $F : \Re^n \to \Re^n$ and $F_t : \Re^n \times \Re_{++} \to \Re^n$ by

$$F_i(x) = \psi_i(x_i, \Phi_i(x)) \tag{6.4}$$

and

$$F_{t,i}(x) = \psi_{i,t}(x_i, \Phi_i(x)) \tag{6.5}$$

for each $i \in N$. Then we have that F_t is continuously differentiable for any $t > 0$ and $\lim_{t\downarrow 0} F_t(x) = F(x)$ for each $x \in \Re^n$. From Theorem 4.4, we have that $G : \Re^n \to \Re^{n\times n}$, defined by

$$G(x) = D(x) + E(x)\nabla\Phi(x)^T, \tag{6.6}$$

is a semiderivative function of F, where $D(x)$ is a diagonal matrix whose ith diagonal element is the first element of $\xi_i(x_i, \Phi_i(x))$ for each $i \in N$, and $E(x)$ is a diagonal matrix whose ith diagonal element is the second element of $\xi_i(x_i, \Phi_i(x))$ for each $i \in N$. If $\nabla\Phi(x)$ is locally Lipschitz, then G is a strong semiderivative function of F.

Let $\nabla F_t(x)$ be the derivative of F_t, then

$$\lim_{t\downarrow 0}(\nabla F_t(x))^T = G(x) \tag{6.7}$$

for each $x \in \Re^n$.

We denote

$$\theta(x) = \frac{1}{2}\|F(x)\|^2$$

and

$$\theta_t(x) = \frac{1}{2}\|F_t(x)\|^2.$$

Algorithm 6.1

Step 1. *Given $\rho, \alpha, \eta \in (0,1)$, $\gamma \in (0, +\infty)$ and a starting point $x^0 \in \Re^n$. Choose $\sigma \in (0, \frac{1}{2}(1-\alpha))$, and $\mu > 0$ satisfying (6.1). Let $\beta_0 = \|F(x_0)\|$, $t_0 = \frac{\alpha}{2\mu}\beta_0$ and $k := 0$.*

Step 2. *If $F(x^k) = 0$ then stop. Otherwise, let d^k be a solution of*

$$F(x^k) + (\nabla F_{t^k}(x^k))^T d(x^k) = 0. \tag{6.8}$$

Step 3. *Let j_k be the smallest nonnegative integer j satisfying*

$$\theta_{t^k}(x^k + \rho^j d^k) - \theta_{t^k}(x^k) \leq -2\sigma\rho^j\theta(x^k). \tag{6.9}$$

where ρ^j means the jth power of ρ. Set $x^{k+1} = x^k + \rho^{j_k} d^k$.

Step 4. *If*

$$\|F(x^{k+1})\| \leq \max\{\eta\beta_k, \alpha^{-1}\|F(x^{k+1}) - F_{t^k}(x^{k+1}\|\}, \tag{6.10}$$

we let

$$\beta_{k+1} = \|F(x^{k+1})\|$$

and choose a t^{k+1} satisfying

$$0 < t^{k+1} \leq \min\{\frac{\alpha}{2\mu}\beta_{k+1}, \frac{t^k}{2}\} \tag{6.11}$$

and

$$\|(\nabla F_{t^{k+1}}(x^{k+1}))^T - G(x^{k+1})\| \leq \gamma\beta_{k+1}. \tag{6.12}$$

If (6.10) does not hold, we let $\beta_{k+1} = \beta_k$ and $t^{k+1} = t^k$. Set $k := k+1$, and go to step 2.

Assumption 6.1 *The level set*

$$L = \{x \in \Re^n : \theta(x) \leq (1+\alpha)^2\theta(x^0)\}$$

is bounded.

Assumption 6.2 *For any $t > 0$ and $x \in L$, $(\nabla F_t(x))^T$ is nonsingular.*

Lemma 6.1 *Let $\{V_k\}$ and $\{W_k\}$ be two $n \times n$ matrix sequences. Suppose that W_k is nonsingular and there exists a constant number $c > 0$ such that $\|W_k^{-1}\| \leq c$ for all sufficient large k. If $\lim_{k\to\infty} \|V_k - W_k\| = 0$, then V_k is nonsingular and*

$$\|V_k^{-1}\| \leq 2c$$

for all sufficient large k.

Proof. Because $\lim_{k\to\infty} \|V_k - W_k\| = 0$, we have

$$\|V_k - W_k\| \leq \frac{1}{2c}$$

for all k large enough. Then

$$\|W_k^{-1}(W_k - V_k)\| \leq c \cdot \frac{1}{2c} = \frac{1}{2}$$

for all k large enough. From Theorem 2.3.2 in [23], we obtain that V_k is nonsingular for all k large enough with

$$\|V_k^{-1}\| \leq \frac{\|W_k^{-1}\|}{1 - \|W_k^{-1}(W_k - V_k)\|} \leq 2c. \quad \square$$

Algorithm 6.1 is a generalization of the algorithm in [5]. We have the following global and quadratic convergence theorem for this algorithm. We omit its proof since it is similar to the proof of Theorem 3.2 in [5], by using Lemma 6.1.

Theorem 6.1 *Suppose that Assumption 6.1 and 6.2 hold. Assume that x^* is an accumulation point of $\{x^k\}$ generated by Algorithm 6.1. If F defined in (6.4) is A-regular at x^*, then x^* is a solution of the equation $F(x) = 0$, and the whole sequence converges to x^* superlinearly. If furthermore $\nabla\Phi$ is locally Lipschitz around x^*, then the sequence converges to x^* quadratically.*

We have the following theorem for BVIP function $\psi_{(\alpha,\beta)}$ defined in (6.2) and its smoothing approximation $\psi_t(a, b)$ defined in (6.3). We omit its proof since it is similar to the proof of Proposition 3.2 in [13].

Theorem 6.2 *Let $M = \nabla\Phi(x^*)^T$. Assume that x^* is a R-regular solution of BVIP. Then F defined in (6.4) is A-regular at x^*.*

7 Augmented Smoothing Newton Methods

Definition 7.1 *We call $\phi_{(\alpha,\beta)} : \Re^3 \to \Re$ an augmented BVIP function if $\phi_{(\alpha,\beta)}$ satisfies the following conditions:*

(i) $\phi_{(\alpha,\beta)}$ is a continuous function and $\phi_{(\alpha,\beta)}$ is continuously differentiable for any $(t, a, b) \in \Re^3$ with $t \neq 0$;

(ii) Define $\psi_{(\alpha,\beta)} : \Re^2 \to \Re$ by

$$\psi_{(\alpha,\beta)}(a, b) = \phi_{(\alpha,\beta)}(0, a, b) \tag{7.1}$$

and $\psi_{(\alpha,\beta)}$ is a BVIP function;

(iii) $\psi^2_{(\alpha,\beta)}$ *is continuously differentiable for any* $(a,b) \in \Re^2$.

For example, define $\phi_{(\alpha,\beta)} : \Re^3 \to \Re$ by

$$\phi_{(\alpha,\beta)}(t,a,b) = \psi_{FB}(t, a-\alpha, \psi_{FB}(t, \beta - a, b)), \tag{7.2}$$

where

$$\psi_{FB}(t,a,b) = \sqrt{t^2 + a^2 + b^2} - (a+b). \tag{7.3}$$

The function in (7.3) is a smoothed form of Fischer-Burmeister function and was first defined by Kanzow [21]. We can verify that $\phi_{(\alpha,\beta)}(t,a,b)$ defined in (7.2) is an augmented BVIP function.

For each $i \in N$, let $\phi_i = \phi_{(l_i,u_i)}$ be a strongly semismooth augmented BVIP function and $\psi_i = \psi_{(l_i,u_i)}$, where $\psi_{(l_i,u_i)}$ is defined in (7.1). Then ψ_i is a strongly semismooth BVIP function. For any given $x \in \Re^n$ and $t \in \Re$, define $F : \Re^n \to \Re^n$ by

$$F_i(x) = \psi_i(x_i, \Phi_i(x)) \tag{7.4}$$

for each $i \in N$, $\hat{H} : \Re^{n+1} \to \Re^n$ by

$$\hat{H}_i(t,x) = \phi_i(t, x_i, \Phi_i(x)) \tag{7.5}$$

for each $i \in N$ and $H : \Re^{n+1} \to \Re^{n+1}$ by

$$H(t,x) = \begin{pmatrix} \kappa(t) \\ \hat{H}(t,x) \end{pmatrix}, \tag{7.6}$$

where κ satisfies the following conditions: (i) $\kappa : \Re \to \Re$ is continuously differentiable with $\kappa'(t) > 0$ for any t; (ii) $\kappa(t) = 0$ implies that $t = 0$; (iii) $dt = -\frac{\kappa(t)}{\kappa'(t)} \in (-t, 0)$ for any $t > 0$. For example, let $\kappa(t) = e^t - 1$ where e is the Euler constant. We can show that this function satisfies the above conditions; see [19] for details. Consequently, we may define two systems of equations in the following

$$F(x) = 0 \tag{7.7}$$

and

$$H(t,x) = 0. \tag{7.8}$$

We call (7.8) an augmented equation of (7.7). The main difference between (7.7) and (7.8) is that (7.8) has one more variable t and one more equation than (7.7). $H(t,x)$ is continuously differentiable at all $(t,x) \in \Re^{n+1}$ with $t \neq 0$. If $H(t^*, x^*) = 0$, then $t^* = 0$ and $F(x^*) = 0$.

For each $i \in N$, define $\xi_i : \Re^3 \to \Re^3$ by

$$\begin{cases} \xi_i(t,a,b) = \nabla\phi_i(t,a,b) & \text{if } t \neq 0 \\ \xi_i(t,a,b) \in \partial_B\phi_i(t,a,b) & \text{if } t = 0. \end{cases} \tag{7.9}$$

From Theorem 4.3, we have that ξ_i is a strong semiderivative function of ϕ_i. From Theorem 4.4, we have that $G : \Re^{n+1} \rightarrow \Re^{(n+1)\times(n+1)}$, defined by

$$G(t,x) = \begin{pmatrix} \kappa'(t) & 0 \\ C(t,x) & D(t,x) + E(t,x)\nabla\Phi(x)^T \end{pmatrix}, \tag{7.10}$$

is a semiderivative function of H, where $C(t,x)$ is a vector in $\Re^n$ whose ith element is the first element of $\xi_i(t,x_i,\Phi_i(x))$ for each $i \in N$, $D(t,x)$ is a diagonal matrix whose ith diagonal element is the second element of $\xi_i(t,x_i,\Phi_i(x))$ for each $i \in N$, and $E(t,x)$ is a diagonal matrix whose ith diagonal element is the third element of $\xi_i(t,x_i,\Phi_i(x))$ for each $i \in N$. If $\nabla\Phi(x)$ is locally Lipschitz, then G is a strong semiderivative function of H. Because $\kappa'(t) > 0$ for any t, for any $(t,x) \in \Re^{n+1}$, $G(t,x)$ is nonsingular if and only if $D(t,x) + E(t,x)\nabla\Phi(x)^T$ is nonsingular.

Proposition 7.1 *Let $A(t^*,x^*)$ be the set of all $(n+1)\times(n+1)$ matrices G such that*

$$G = \begin{pmatrix} \kappa'(t^*) & 0 \\ C(t^*,x^*) & D(t^*,x^*) + E(t^*,x^*)\nabla\Phi(x^*)^T \end{pmatrix} \tag{7.11}$$

and $B(t^*,x^*)$ be the set of all $n \times n$ matrices $\bar{G}$ such that

$$\bar{G} = D(t^*,x^*) + E(t^*,x^*)\nabla\Phi(x^*)^T,$$

where $C(t^*,x^*)$ is a vector in $\Re^n$ whose ith element is the first element of $\xi_i(t_i^*, x_i^*, \Phi_i(x^*))$ for each $i \in N$, $D(t^*,x^*)$ is a diagonal matrix whose ith diagonal element is the second element of $\xi_i(t^*,x_i^*,\Phi_i(x^*))$ for each $i \in N$, and $E(t^*,x^*)$ is a diagonal matrix whose ith diagonal element is the third element of $\xi_i(t^*,x_i^*,\Phi_i(x^*))$ for each $i \in N$.

(i) If all the matrices in $A(t^*,x^*)$ are nonsingular, then there are a positive number c and a neighborhood $N(t^*,x^*)$ of (t^*,x^*) such that for all $(t,x) \in N(t^*,x^*)$, $G(t,x)$ is nonsingular and

$$\|G(t,x)^{-1}\| \le c.$$

(ii) All the matrices in $A(t^*,x^*)$ are nonsingular if and only if all the matrices in $B(t^*,x^*)$ are nonsingular.

We omit the proof of this proposition since it is similar to that of Proposition 3.1 of [31].

Let $\theta : \Re^{n+1} \rightarrow \Re$ be the norm function of H, defined by

$$\theta(t,x) = \frac{1}{2}\|H(t,x)\|^2.$$

Algorithm 7.1

Step 1. *Let* $z^0 = (t^0, x^0) \in \Re^{n+1}$ *where* $t^0 > 0$, $\sigma \in (0, \frac{1}{2})$, $\rho \in (0,1)$ *and* $k := 0$.

Step 2. *If* $H(z^k) = 0$ *then stop. Otherwise, let* d^k *be a solution of*

$$H(z^k) + G(z^k)d(z^k) = 0. \tag{7.12}$$

Step 3. *Let* $\alpha^k = \rho^{j_k}$, *where* j_k *be the smallest nonnegative integer* j *satisfying*

$$\theta(z^k + \rho^j d^k) - \theta(z^k) \le \sigma\rho^j \nabla\theta(z^k)^T d^k, \tag{7.13}$$

where ρ^j *means the* j*th power of* ρ.

Step 4. *Let* $z^{k+1} = z^k + \alpha^k d^k$ *and* $k := k + 1$. *Go to Step 2.*

This algorithm is a generalization of the algorithm in [19]. We have the following global and quadratic convergence theorem for this algorithm. We omit its proof since it is similar to the proof of Theorem 4.1 in [19].

Theorem 7.1 *Suppose (7.12) in Step 2 is solvable for each* k. *Assume that* $z^* = (t^*, x^*)$ *is an accumulation point of* $\{z^k\}$ *generated by Algorithm 7.1. If all the matrices in* $A(t^*, x^*)$ *defined in (7.11) are nonsingular, then* $t^* = 0$, x^* *is a solution of the BVIP, and* $\{z^k\}$ *converges* $(0, x^*)$ *superlinearly. The convergence rate is quadratic if* $\nabla\Phi$ *is locally Lipschitz around* x^*.

Now we study under what conditions all the matrices in $A(t, x)$ defined in (7.11) are nonsingular at a solution point (t^*, x^*) of $H(t, x) = 0$. We have the following theorem for the augmented BVIP functions defined in (7.2). We omit its proof since it is similar to the proof of Proposition 3.2 in [13].

Theorem 7.2 *Let* $M = \nabla\Phi(x^*)^T$. *Assume that* $t^* = 0$ *and* x^* *is a R-regular solution of BVIP. Then all the matrices in* $A(t^*, x^*)$ *defined in (7.11) are nonsingular.*

8 Squared Smoothing Newton Methods

In section 5-7 we always assume that Φ is defined on the whole space $\Re^n$. However, for some BVIP problems, Φ is not defined on the whole space $\Re^n$. In this section, we will give a squared smoothing Newton method for BVIP which only requires Φ to be defined on X.

It is well known that solving BVIP is equivalent to solving Robinson's normal equation,

$$F(x) := \Phi(\Pi_X(x)) + x - \Pi_X(x) = 0 \tag{8.1}$$

in the sense that if $x^* \in \Re^n$ is a solution of (8.1) then $\Pi_X(x^*)$ is a solution of BVIP, and conversely if $x^* \in X$ is a solution of BVIP then $x^* - \Phi(x^*)$ is a solution of (8.1)

[34], where $\Pi_X(x)$ is the Euclidean projection of x on X and $(\Pi_X(x))_i = \phi_{(l_i,u_i)}(x_i)$ for each $i \in N$, where $\phi_{(l_i,u_i)}$ is defined in (3.26).

We have the following relationship between the solution set of BVIP and the solution set of (8.1).

Lemma 8.1 *The solution set of BVIP is nonempty and bounded if and only if the solution set of (8.1) nonempty and bounded.*

Proof. Let $\mathcal{S}$ be the solution set of BVIP and $\mathcal{S}'$ be the solution set of (8.1). Suppose that $\mathcal{S}$ is nonempty and bounded, then $\mathcal{S}'$ is nonempty. For any $y \in \mathcal{S}'$, from (8.1) we have that $y = \Pi_X(y) - \Phi(\Pi_X(y))$. Let $x = \Pi_X(y)$. Then $x \in \mathcal{S}$. So $y = x - \Phi(x)$. Because $\mathcal{S}$ is bounded, there exists a $c > 0$ such that $\|x - \Phi(x)\| \le c$ for all $x \in \mathcal{S}$. Hence $\|y\| \le c$, i.e., $\mathcal{S}'$ is bounded.

Suppose that $\mathcal{S}'$ is nonempty and bounded. Then $\mathcal{S}$ is nonempty. For any $x \in \mathcal{S}$, from (1.1) we have that $x = \Pi_X(x - \Phi(x))$. Let $y = x - \Phi(x)$. Then $y \in \mathcal{S}'$. So $\mathcal{S} \subseteq \Pi_X(\mathcal{S}')$. Hence $\mathcal{S}$ is bounded. □

For $\phi_{(\alpha,\beta)}$ defined in (3.26), we give the following definition.

Definition 8.1 *We call $\psi_{(\alpha,\beta)} : \Re^2 \to \Re$ an augmented function of $\phi_{(\alpha,\beta)}$ if $\psi_{(\alpha,\beta)}$ satisfies the following conditions:*

(i) $\psi_{(\alpha,\beta)}$ is continuously differentiable on any $(t,a) \in \Re^2$ with $t \neq 0$;

(ii) $\psi_{(\alpha,\beta)}$ is a strongly semismooth function;

(iii) $\psi_{(\alpha,\beta)}(0,a) = \phi_{(\alpha,\beta)}(a)$ for any $a \in \Re$;

(iv) $\psi_{(\alpha,\beta)}(t,a) \in [\alpha,\beta] \cap \Re$ for any $(t,a) \in \Re^2$.

For example, define $\psi_{(\alpha,\beta)} : \Re^2 \to \Re$ by

$$\psi_{(\alpha,\beta)}(t,a) = \begin{cases} \phi_{t\alpha\beta}(a) & \text{if } t > 0 \\ \phi_{(-t)\alpha\beta}(a) & \text{if } t < 0 \\ \phi_{(\alpha,\beta)}(a) & \text{if } t = 0 \end{cases}, \tag{8.2}$$

where $\phi_{t\alpha\beta}$ is defined in (3.27). If $\phi_{t\alpha\beta}$ is one of the three most often used Gabriel-Moré smoothing functions, i.e., the neural networks smoothing function, the Chen-Harker-Kanzow-Smale smoothing function and the uniform smoothing function, see [33], then from Proposition 3.1-3.3 [33] we have that $\psi_{(\alpha,\beta)}$ defined in (8.2) is an augmented function of $\phi_{(\alpha,\beta)}$.

For each $i \in N$, let $\psi_i = \psi_{(l_i,u_i)}$ be an augmented function of $\phi_{(l_i,u_i)}(a)$. Define $p(t,x) : \Re^{n+1} \to \Re^n$ by

$$p_i(t,x) = \psi_i(t,x_i) \tag{8.3}$$

for each $i \in N$. From (iv) of Definition 8.1 we have that $p(t,x) \in X$ for any $(t,x) \in \Re^{n+1}$.

Define $\hat{H}(t,x) : \Re^{n+1} \to \Re^n$ by

$$\hat{H}(t,x) = \Phi(p(t,x)) + x - p(t,x), \tag{8.4}$$

$\tilde{H}(t,x) : \Re^{n+1} \to \Re^n$ by

$$\tilde{H}(t,x) = \hat{H}(t,x) + \lambda t p(t,x), \tag{8.5}$$

where $\lambda > 0$, and $H(t,x) : \Re^{n+1} \to \Re^{n+1}$ by

$$H(t,x) = \begin{pmatrix} t \\ \tilde{H}(t,x) \end{pmatrix}. \tag{8.6}$$

$H(t,x)$ is continuously differentiable at all $(t,x) \in \Re^{n+1}$ with $t \neq 0$ and $H(t^*,x^*) = 0$ if and only if $t^* = 0$ and $F(x^*) = 0$.

For each $i \in N$, define $\xi_i : \Re^2 \to \Re^2$ by

$$\begin{cases} \xi_i(t,a) = \nabla\psi_i(t,a) & \text{if } t \neq 0 \\ \xi_i(t,a) \in \partial_B\psi_i(t,a) & \text{if } t = 0. \end{cases} \tag{8.7}$$

From Theorem 4.3, we have that ξ_i is a strong semiderivative function of ψ_i. From Theorem 4.4, we have that $G : \Re^{n+1} \to \Re^{(n+1)\times(n+1)}$, defined by

$$G(t,x) = \begin{pmatrix} 1 & 0 \\ \hat{G}(t,x) & \tilde{G}(t,x) \end{pmatrix}, \tag{8.8}$$

where

$$\hat{G}(t,x) = \left(\nabla\Phi(p(t,x))^T + (\lambda t - 1)I\right) D(t,x) + p(t,x) \tag{8.9}$$

and

$$\tilde{G}(t,x) = \left(\nabla\Phi(p(t,x))^T + \lambda t I\right) E(t,x) + I - E(t,x), \tag{8.10}$$

is a semiderivative function of H, where $D(t,x)$ is a vector in $\Re^n$ whose ith element is the first element of $\xi_i(t,x_i)$ for each $i \in N$, $E(t,x)$ is a diagonal matrix whose ith diagonal element is the second element of $\xi_i(t,x_i)$ for each $i \in N$. If $\nabla\Phi(x)$ is locally Lipschitz, then G is a strong semiderivative function of H. From (8.8) we have that $G(t,x)$ is nonsingular if and only if $\tilde{G}(t,x)$ is nonsingular.

Proposition 8.1 *Let $A(t^*,x^*)$ be the set of all $(n+1)\times(n+1)$ matrices G such that*

$$G = \begin{pmatrix} 1 & 0 \\ \hat{G}(t^*,x^*) & \tilde{G}(t^*,x^*) \end{pmatrix} \tag{8.11}$$

where $\hat{G}(t^*, x^*)$ *and* $\tilde{G}(t^*, x^*)$ *are defined in (8.9) and (8.10), and* $B(t^*, x^*)$ *be the set of all* $n \times n$ *matrices* $\bar{G}$ *such that*

$$\bar{G} = \left(\nabla\Phi(p(t^*, x^*))^T + \lambda t^* I\right) E(t^*, x^*) + I - E(t^*, x^*),$$

where $D(t^*, x^*)$ *is a vector in* $\Re^n$ *whose ith element is the first element of* $\xi_i(t^*, x_i^*)$ *for each* $i \in N$, $E(t^*, x^*)$ *is a diagonal matrix whose ith diagonal element is the second element of* $\xi_i(t^*, x_i^*)$ *for each* $i \in N$.

(i) If all the matrices in $A(t^*, x^*)$ *are nonsingular, then there are a positive number* c *and a neighborhood* $N(t^*, x^*)$ *of* (t^*, x^*) *such that for all* $(t, x) \in N(t^*, x^*)$, $G(t, x)$ *is nonsingular and*

$$\|G(t, x)^{-1}\| \leq c.$$

(ii) All the matrices in $A(t^*, x^*)$ *are nonsingular if and only if all the matrices in* $B(t^*, x^*)$ *are nonsingular.*

The proof of this proposition is similar to that of Proposition 3.1 of [31].

Now we present the squared smoothing method for BVIP. Choose $\bar{t} \in \Re_{++}$ and $\gamma \in (0, 1)$ such that $\gamma\bar{t} < 1$. Let $\bar{z} := (\bar{t}, 0) \in \Re \times \Re^n$. Define the merit function $\theta : \Re^{n+1} \to \Re$ by

$$\theta(z) := \|H(z)\|^2$$

and define $\beta : \Re^{n+1} \to \Re$ by

$$\beta(z) := \gamma \min\{1, \theta(z)\}.$$

Algorithm 8.1

Step 0. *Choose constants* $\rho \in (0, 1)$ *and* $\sigma \in (0, \frac{1}{2})$. *Let* $t^0 := \bar{t}$, $x^0 \in \Re^n$ *be an arbitrary point and* $k := 0$.

Step 1. *If* $H(z^k) = 0$ *then stop. Otherwise, let* $\beta_k := \beta(z^k)$.

Step 2. *Compute* $\Delta z^k := (\Delta t^k, \Delta x^k) \in \Re \times \Re^n$ *by*

$$H(z^k) + G(z^k)\Delta z^k = \beta_k \bar{z}. \tag{8.12}$$

Step 3. *Let* j_k *be the smallest nonnegative integer* j *satisfying*

$$\theta(z^k + \rho^j \Delta z^k) \leq [1 - 2\sigma(1 - \gamma\bar{t})\rho^j]\theta(z^k). \tag{8.13}$$

Define $z^{k+1} := z^k + \rho^{j_k}\Delta z^k$.

Step 4. *Replace* k *by* $k + 1$ *and go to Step 1.*

This algorithm is a regularized version of the algorithm in [33]. We have the following convergence theorem for this algorithm. We omit its proof since it is similar to the proof of Theorem 3.1 and 3.2 in [42].

Theorem 8.1 *Assume that (8.12) in step 2 is solvable for each k. Suppose that $z^* = (t^*, x^*)$ is an accumulation point of the infinite sequence $\{z^k\}$ generated by Algorithm 8.1 and all the matrices in $A(t^*, x^*)$ defined in (8.11) are nonsingular. Then $t^* = 0$, $\Pi_X(x^*)$ is a solution of BVIP, and the whole sequence $\{z^k\}$ converges to z^* superlinearly. The convergence rate is quadratic if $\nabla\Phi(x)$ is locally Lipschitz around x^*.*

If $\psi_{(\alpha,\beta)}$ is defined in (8.2), then we have the following theorem.

Theorem 8.2 *(i) If $\nabla\Phi(x)^T$ is a P_0-matrix for any $x \in X$, then for any $(t, x) \in \Re^{n+1}$ with $t \neq 0$, $G(t, x)$ is nonsingular.*

(ii) Let $M = \nabla\Phi(\Pi_X(x^*))^T$. Assume that $t^* = 0$ and $\Pi_X(x^*)$ is a R-regular solution of BVIP. Then all the matrices in $A(t^*, x^*)$ defined in (8.11) are nonsingular.

Proof. (i) For any $(t, x) \in \Re^{n+1}$ with $t \neq 0$, we can easily prove that $\nabla\Phi(p(t, x))^T + \lambda tI$ is a P-matrix. From (3.32) we have that $E_{ii}(t, x) \in [0, 1]$. Then from Lemma 4.1 we have that $\tilde{G}(t, x)$ is nonsingular. So $G(t, x)$ is nonsingular. The proof of (ii) is similar to the proof of Proposition 7.1 in [33]. □

We have the following global convergence theorem for this method if $\psi_{(\alpha,\beta)}$ is defined in (8.2) and $\phi_{t\alpha\beta}$ is one of the neural networks smoothing function, the Chen-Harker-Kanzow-Smale smoothing function and the uniform smoothing function, see [33]. We omit its proof since it is similar to the proof of Theorem 3.1 in [42], by using Lemma 8.1. It deserves to be pointed out that we cannot get global convergence theorems for Algorithms 5.1-7.1 under the conditions of Theorem 8.3. So the assumptions in Theorem 8.3 may be the weakest ones in the literature.

Theorem 8.3 *Suppose that Φ is a P_0-function on X. Then an infinite sequence $\{z^k\}$ is generated by Algorithm 8.1 and*

$$\lim_{k \to \infty} \theta(z^k) = 0. \tag{8.14}$$

In particular, each accumulation point z^* of $\{z^k\}$ is a solution of $H(t, x) = 0$. If we further assume that the solution set of BVIP is nonempty and bounded, then we have that the infinite sequence $\{z^k\}$ generated by Algorithm 8.1 is bounded.

9 Conclusions

In this paper, we define a semiderivative function for a continuous function, which is single-valued and satisfies exact calculus rules. Based on this concept, superlinear convergence conditions for both nonsmooth methods and smoothing methods for BVIP are established.

Algorithms 5.1-7.1 are not restricted to the BVIP. In fact, we can apply Algorithms 5.1-7.1 to general box constrained variational inequalities (GBVIP for short), i.e., finding a vector $x \in \Re^n$ such that $\Psi(x) \in X$, where X is defined in (1.1), and for all $y \in X$,

$$(y - \Psi(x))^T \Phi(x) \geq 0, \tag{9.1}$$

where $\Phi,\ \Psi : \Re^n \to \Re^n$ are two smooth vector functions. All the results in sections 5-7 hold for the GBVIP if we modify some functions in these sections.

Acknowledgement. The authors would like to thank Dr. Defeng Sun for his helpful comments.

References

[1] S. C. Billups (1995), *Algorithms for complementarity problems and generalised equation*, Ph.D. Thesis, Computer Sciences Department, University of Wisconsin, Madison, WI.

[2] S.C. Billups, S.P. Dirkse and M.C. Ferris (1997), "A comparison of algorithms for large-scale mixed complementarity problems," *Computational Optimization and Applications*, 7, 3–25.

[3] B. Chen and P.T. Harker (1993), "A non-interior-point continuation method for linear complementarity problems," *SIAM Journal on Matrix Analysis and Applications*, 14, 1168–1190.

[4] C. Chen and O.L. Mangasarian (1996), "A class of smoothing functions for nonlinear and mixed complementarity problems," *Computational Optimization and Applications*, 5, 97–138.

[5] X. Chen, L. Qi and D. Sun (1998), "Global and superlinear convergence of the smoothing Newton method and its application to general box constrained inequalities," *Mathematics of Computation*, 67, 519–540.

[6] F.H. Clarke (1983), *Optimization and Nonsmooth Analysis*, Wiley, New York.

[7] T. De Luca, F. Facchinei and C. Kanzow (1996), "A semismooth equation approach to the solution of nonlinear complementarity problems," *Mathematical Programming*, 75, 407-439.

[8] T. De Luca, F. Facchinei and C. Kanzow (1997), "A theoretical and numerical comparison of some semismooth algorithms for complementarity problems," Mathematical Programming Technical Report 97-15, Computer Sciences Department, University of Wisconsin - Madison, Madison, WI.

[9] F. Facchinei, A. Fischer and C. Kanzow (1996), "Inexact Newton methods for semismooth equations with applications to variational inequality problems," in: G. Di Pillo and F. Giannessi, eds., *Nonlinear Optimization and Applications*, Plenum Press, New York, 125-139.

[10] F. Facchinei, A. Fischer and C. Kanzow (1997), "A semismooth Newton method for variational inequalities: The case of box constraints," in: M.C. Ferris and J.S. Pang, eds., *Complementarity and Variational Problems - State of the Art*, SIAM Publications, Philadelphia, 76-90.

[11] F. Facchinei, A. Fischer and C. Kanzow (1998), "Regularity properties of a semismooth reformulation of variational inequalities," *SIAM Journal on Optimization*, 8, 850–869.

[12] F. Facchinei and C. Kanzow (1997), "A nonsmooth inexact Newton method for the solution of large-scale nonlinear complementarity problems," *Mathematical Programming*, 76, 493-512.

[13] F. Facchinei and J. Soares (1997), "A new merit function for nonlinear complementarity problems and a related algorithm," *SIAM Journal on Optimization*, 7, 225–247.

[14] A. Fischer (1995), "An NCP-function and its use for the solution of complementarity problems," in: D. Du, L. Qi and R. Womersley, eds., *Recent Advances in Nonsmopth Optimization*, World Scientific Publishers, New Jersey, 88-105.

[15] S.A. Gabriel and J.J. Moré (1997), "Smoothing of mixed complementarity problems," in: M.C. Ferris and J.S. Pang, eds., *Complementarity and Variational Problems: State of the Art*, SIAM, Philadelphia, Pennsylvania, 105–116.

[16] M.S. Gowda and R. Sznajder (1995), "On the pseudo-Lipschitzian behaviour of the inverse of a piecewise affine function," Preprint, Department of Mathematics and Statistics, University of Maryland Baltimore County, USA.

[17] J. Han and D. Sun (1997), "Newton and Quasi-Newton methods for normal maps with polyhedral sets," *J. Optimization Theory and Appl.*, 94, 659-676.

[18] V. Jeyakumar and D.T. Luc (1998), "Approximate Jacobian matrices for nonsmooth continuous maps and C^1 optimization," *SIAM Journal on Control and Optimization*, 36, 1815-1832.

[19] H. Jiang (1997), "Smoothed Fischer-Burmeister equation methods for the complementarity problem," Preprint, Department of Mathematics, The University of Melbourne, Victoria, Australia.

[20] H. Jiang and L. Qi (1997), "A new nonsmooth equations approach to nonlinear complementarity problems," *SIAM Journal on Control and Optimization*, 35, 178-193.

[21] C. Kanzow (1994), "Some equation-based methods for the nonlinear complementarity problem," *Optimization Methods and Software*, 3, 327–340.

[22] C. Kanzow (1996), "Some noninterior continuation methods for linear complementarity problems," *SIAM Journal on Matrix Analysis and Applications*, 17, 851–868.

[23] J. M. Ortega and W. C. Rheinboldt (1970), *Iterative solution of nonlinear equations in several variables*, Academic Press, New York.

[24] J.-S. Pang (1991), "A B-differentiable equation based, globally, and locally quadratically convergent algorithm for nonlinear programs, complementarity and variational inequality problems," *Mathematical Programming*, 51, 101-131.

[25] J.-S. Pang and S.A. Gabriel (1993), "NE/SQP: A robust algorithm for the nonlinear complementarity problem," *Mathematical Programming*, 60, 295-337.

[26] J.-S. Pang and D. Ralph (1996), "Piecewise smoothness, local invertibility, and parametric analysis of normal maps," *Mathematics of Operations Research*, 21, 401-426.

[27] D. Pu and J. Zhang (1998), "Inexact generalized Newton methods for second-order C-differentiable optimization," *Journal of Computational and Applied Mathematics*, 93, 107-122.

[28] H.-D. Qi (1999), "A regularized smoothing Newton method for box constrained variational inequality problems with P_0 functions," *SIAM Journal on Optimization*, to appear.

[29] L. Qi (1993), "Convergence analysis of some algorithms for solving nonsmooth equations," *Mathematics of Operations Research*, 18, 227-244.

[30] L. Qi (1996), "C-differential operators, C-differentiability and generalized Newton methods," Applied Mathematics Report AMR 96/5, School of Mathematics, The University of New South Wales, Sydney 2052, Australia.

[31] L. Qi (1999), "Regular pseudo-smooth NCP and BVIP functions and globally and quadratically convergent generalized Newton methods for complementarity and variational problems," *Mathematics of Operations Research*, 24, 440-471.

[32] L. Qi and J. Sun (1993), "A nonsmooth version of Newton's method," *Mathematical Programming*, 58, 353-368.

[33] L. Qi, D. Sun and G. Zhou (1997), "A new look at smoothing Newton methods for nonlinear complementarity problems and box constrained variational inequalities," Applied Mathematics Report AMR 97/13, School of Mathematics, The University of New South Wales, Sydney 2052, Australia.

[34] S. M. Robinson (1992), "Normal maps induced by linear transformation," *Mathematics of Operations Research*, 17, 691-714.

[35] S. Scholtes and M. Stöhr (1999), "Exact penalisation of mathematical programs with equilibrium constraints," *SIAM J. Control Optim*, 37, 617-652.

[36] D. Sun (1997), "A regularization Newton method for solving nonlinear complementarity problems," *Applied Mathematics and Optimization*, to appear.

[37] D. Sun and J. Han (1997), "Newton and quasi-Newton methods for a class of nonsmooth equations and related problems," *SIAM Journal on Optimization*, 7, 463-480.

[38] D. Sun and R.S. Womersley (1999), "A new unconstrained differentiable merit function for box constrained variational inequality problems and a damped Gauss-Newton method," *SIAM Journal on Optimization*, 9, 409-434.

[39] T.H. Sweeter (1977), "A minimal set-valued strong derivative for vector-valued Lipschitz functions," *Journal of Optimization Theory and Applications*, 23, 549-562.

[40] H. Xu (1998), "Point-based set-valued approximation, C-differential operator and application", Preprint, School of Information Technology and Mathematical Sciences, University of Ballarat, Ballarat, Victoria, Australia.

[41] N. Yamashita and M. Fukushima (1997), "Modified Newton methods for solving semismooth reformulations of monotone complementarity problems," *Mathematical Programming*, 76, 273-284.

[42] G. Zhou, D. Sun and L. Qi (1999), "Numerical experiments for a class of squared smoothing Newton methods for box constrained variational inequality problems," in: M. Fukushima and L. Qi, eds., *Reformulation: Nonsmooth, Piecewise Smooth, Semismooth and Smoothing Methods* Kluwer, Boston, 421–441.

Global Lagrange multiplier rule and smooth exact penalty functions for equality constraints[1]

Tamás Rapcsák (rapcsak@oplab.sztake.hu)
Computer and Automation Institute, Hungarian Academy of Sciences, Budapest

Abstract

Smooth exact penalty functions based on Courant's and Fletcher's ideas are reconsidered. After a short survey, the original ideas are combined with the global Lagrange multiplier rule formulated by the first and second covariant derivatives of the objective function with respect to the induced Riemannian metric of the constraint manifold. The tensor approach is described by the usual tools of nonlinear optimization giving a clearer geometric background of these methods.

Keywords: Equality constrained optimization, Lagrange multipliers, exact penalty functions, Riemannian manifolds, geodesic convexity.

1 Introduction

The Lagrange multiplier rule was introduced in 1762, and later, in "Lagrange, J.L., Mécanique analytique I-II, Paris, 1788" for minimizing a function subject to equality constraints, and thus, for finding a stable equilibrium of a mechanical system. Since then, a great many applications of wide variety have been published in the fields of theory, methodology and practice without changing its scope from the point of view of mathematics. Prékopa (1980) stated that "This ingenious method was unable to attract mathematics students and teachers for a long time. The method of presentation used nowadays by many instructors is to prove the necessary condition first in the case of inequality constraints, give a geometric meaning to this, and then refer to the case of equality constraints. This can make the students more enthusiastic about this theory." The Lagrange multiplier rule was further developed by Rapcsák (1991, 1997) who combined the optimization theory with Riemannian geometry in order to describe the geometric structure of smooth nonlinear optimization problems by tensors and to extend the local results of Lagrange to global ones.

[1]This research was supported by the Hungarian National Research Foundation, Grandt No. OTKA-T029572.

G. Di Pillo and F. Giannessi (eds.), Nonlinear Optimization and Related Topics, 351–368.

In 1970, Fletcher gave the theoretical justification of a class of exact penalty methods for solving nonlinear optimization problems with smooth nonlinear objective and equality constraint functions. The class of methods has interpretations as a Lagrangian type method, and also as a penalty function method. Yet, the latter has the advantage according to which only one well-conditioned unconstrained optimization problem has to be solved, so the method seems to unify the Lagrangian approach with a penalty function method. The method was extended to include inequality constraints [8].

Since then, a great number of investigations, both from theoretical and computational points of view, have been devoted to methods for solving nonlinear optimization problems by means of one unconstrained minimization. Following the survey paper (Di Pillo, 1994), it can be said that exact penalty methods are based on the construction of a function depending on a penalty parameter, such that the unconstrained minimum points of this function are also solutions for the constrained problem for all sufficiently large values of the parameter. Exact penalty methods can be subdivided in two classes: methods based on exact penalty functions depending only on the original variables of the problem, while the term "exact augmented Lagrangian function" is used if the unconstrained problem has to be minimized in the product space of the problem variables and the multipliers. The class of exact penalty functions consists of nondifferentiable and continuously differentiable ones.

In the paper, the original idea of Fletcher (1970) to define smooth exact penalty functions and that of Courant (1943) to use a quadratic penalty term are reconsidered and developed further by the global version of the Lagrange multiplier rule clarifying the geometric meaning of the Lagrange theorem, as well as Courant's and Fletcher's ideas.

2 Lagrange multiplier rule for the case of equality constraints

The optimization problem studied by Lagrange is to minimize a function f subject to equalities, given in the form of

$$\begin{gathered} \min f(\mathbf{x}), \qquad \mathbf{x} \in M, \\ M = \{\mathbf{x} \in R^n \mid h_j(\mathbf{x}) = 0, \quad j = 1, \ldots, n-k\}, \end{gathered} \tag{2.1}$$

where f and $\mathbf{h} = (h_1, \ldots, h_{n-k}) \in C^2(R^n, R^{n-k})$. Let us introduce the Lagrangian function

$$L(\mathbf{x}, \mu) = f(\mathbf{x}) - \sum_{j=1}^{n-k} \mu_j h_j(\mathbf{x}), \qquad \mathbf{x} \in R^n, \quad \mu \in R^{n-k}, \tag{2.2}$$

the sets

$$TM_x = \{\mathbf{v} \in R^n \mid \nabla h_j(\mathbf{x})\mathbf{v} = 0, \quad j = 1, \ldots, n-k\}, \quad \mathbf{x} \in M, \tag{2.3}$$

and a regularity condition

$$r(J\mathbf{h}(\mathbf{x})) = n - k, \qquad \mathbf{x} \in M, \tag{2.4}$$

where the row vectors ∇h_j, $j = 1, \ldots, n-k$, are the gradients of the functions h_j, $j = 1, \ldots, n-k$, $J\mathbf{h}$ is the $(n-k) \times n$ Jacobian matrix of the mapping $\mathbf{h}$, and $r(J\mathbf{h}(\mathbf{x}))$ means the rank of the Jacobian matrix at the point $\mathbf{x}$. The Lagrange method, formulated in the following statement, consists in transforming equality constrained optimization problems into unconstrained minimizations of the Lagrangian function.

LAGRANGE THEOREM

Necessary conditions :

If the point $\mathbf{x}_0$ is a local minimum of problem (2.1) and $r(J\mathbf{h}(\mathbf{x}_0)) = n-k$, then there exists a unique vector $\mu \in R^{n-k}$ (consisting of the components of the Lagrange multipliers) such that

$$\nabla L(\mathbf{x}_0, \mu) = 0 \tag{2.5}$$

and

$$\mathbf{v}^T H_x L(\mathbf{x}_0, \mu)\mathbf{v} \geq 0, \qquad \mathbf{v} \in TM_{x_0}, \tag{2.6}$$

where $H_x L(\mathbf{x}_0, \mu)$ is the Hessian matrix of the Lagrangian function at the point $\mathbf{x}_0$ (differentiated with respect to the variables $\mathbf{x}$).

Sufficient conditions :

If there exists a point $\mathbf{x}_0 \in M$ and a vector $\mu \in R^{n-k}$ such that condition (2.5) holds and

$$\mathbf{v}^T H_x L(\mathbf{x}_0, \mu)\mathbf{v} > 0, \qquad \mathbf{v} \in TM_{x_0}, \quad \mathbf{v} \neq 0, \tag{2.7}$$

then there exists a neighbourhood $U(\mathbf{x}_0) \subseteq M$ where the point $\mathbf{x}_0$ is a strict local minimum of problem (2.1).

The Lagrange theorem shows that a solution of nonlinear optimization problem (2.1) is a stationary point of the Lagrangian function but it is not necessarily a local or global minimum.

3 Lagrangian function depending only on the variables

By a new kind of penalty function approach, the starting point of Fletcher was that "iteration within iteration" is, in general, expensive, moreover, also ill-conditioning arises when terminating the inner iteration. The primary aim of Fletcher was to construct a smooth penalty function so that its minimum should be at the solution of the nonlinear optimization problem to be obtained by solving only one unconstrained optimization problem. It is possible to show that under the second-order sufficient optimality conditions, a penalty function has an unconstrained local minimum at the solution, thus a penalization makes it possible to determine a local minimum instead of a stationary point. Another advantage is that local convergence can be ensured by using an unconstrained minimization method with the local convergence property. Further desirable features should be that an exact penalty function is well-conditioned and it is easy to be used with inequality constraints. This penalty function approach permits an alternative explanation in terms of Lagrange multipliers and can be considered as an augmented Lagrangian function.

The basic idea of Fletcher was to construct a penalty function with a minimum at the stationary point of the nonlinear optimization problem, a necessary condition of which is that the gradient of the modified Lagrangian function be equal to zero.

The first-order optimality condition is

$$\nabla L(\mathbf{x}_0, \mu) = \nabla f(\mathbf{x}_0) - \sum_{j=1}^{n-k} \mu_j \nabla h_j(\mathbf{x}_0) = \nabla f(\mathbf{x}_0) - \mu^T J\mathbf{h}(\mathbf{x}_0) = 0. \tag{3.1}$$

It follows that the Lagrange multipliers can be given in the form of

$$\mu^T = \nabla f(\mathbf{x}_0) J\mathbf{h}(\mathbf{x}_0)^T (J\mathbf{h}(\mathbf{x}_0) J\mathbf{h}(\mathbf{x}_0)^T)^{-1}, \tag{3.2}$$

thus, by introducing the notations

$$\begin{aligned} J\mathbf{h}(\mathbf{x}_0)^+ &= J\mathbf{h}(\mathbf{x}_0)^T (J\mathbf{h}(\mathbf{x}_0) J\mathbf{h}(\mathbf{x}_0)^T)^{-1}, \\ P(\mathbf{x}_0) &= J\mathbf{h}(\mathbf{x}_0)^T (J\mathbf{h}(\mathbf{x}_0) J\mathbf{h}(\mathbf{x}_0)^T)^{-1} J\mathbf{h}(\mathbf{x}_0), \end{aligned} \tag{3.3}$$

we obtain that

$$\nabla L(\mathbf{x}_0, \mu) = \nabla f(\mathbf{x}_0) - \nabla f(\mathbf{x}_0) J\mathbf{h}(\mathbf{x}_0)^+ J\mathbf{h}(\mathbf{x}_0) = \nabla f(\mathbf{x}_0)(I - P(\mathbf{x}_0)) = 0, \tag{3.4}$$

where I is the $(n \times n)$ identity matrix. Since, $\nabla f(\mathbf{x})(I - P(\mathbf{x}))$, $\mathbf{x} \in M$, are the orthogonal projections of the gradients $\nabla f(\mathbf{x})$, $\mathbf{x} \in M$, to the tangent spaces of M, thus the Lagrange multipliers can be extended to a vector function as follows:

$$\mu(\mathbf{x})^T = \nabla f(\mathbf{x}) J\mathbf{h}(\mathbf{x})^+ = \nabla f(\mathbf{x}) J\mathbf{h}(\mathbf{x})^T (J\mathbf{h}(\mathbf{x}) J\mathbf{h}(\mathbf{x})^T)^{-1}, \qquad \mathbf{x} \in M. \tag{3.5}$$

Let us introduce the Lagrangian function

$$L(\mathbf{x}, \mu(\mathbf{x})) = f(\mathbf{x}) - \mu(\mathbf{x})^T \mathbf{h}(\mathbf{x}) = f(\mathbf{x}) - \nabla f(\mathbf{x}) J\mathbf{h}(\mathbf{x})^+ \mathbf{h}(\mathbf{x}), \qquad \mathbf{x} \in M, \tag{3.6}$$

where the vector function μ is given by (3.5).

Lemma 3.1 *If the objective function is quadratic and the constraints are linear in problem (2.1), then the matrices*

$$HL(\mathbf{x}, \mu(\mathbf{x})) = (I - P) Hf (I - P) - PHfP, \qquad \mathbf{x} \in M, \tag{3.7}$$

are constant where P is given by formula (3.3).

Proof: If the objective function is quadratic and the constraints are linear, then the Hessian matrix Hf and the Jacobian matrix $J\mathbf{h}$ are constant and

$$\begin{aligned} \nabla L(\mathbf{x}, \mu(\mathbf{x})) &= \nabla f(\mathbf{x}) - \mu(\mathbf{x})^T J\mathbf{h} - \mathbf{h}(\mathbf{x})^T J\mu(\mathbf{x}) = \\ & \nabla f(\mathbf{x}) - \mu(\mathbf{x})^T J\mathbf{h} - \mathbf{h}(\mathbf{x})^T (J\mathbf{h}^+)^T Hf = \\ & \nabla f(\mathbf{x})(I - P) - \mathbf{h}(\mathbf{x})^T (J\mathbf{h}^+)^T Hf, \qquad \mathbf{x} \in M, \\ HL(\mathbf{x}, \mu(\mathbf{x})) &= Hf - J\mathbf{h}^T (J\mathbf{h}^+)^T Hf - Hf J\mathbf{h}^+ J\mathbf{h} = \\ & Hf - PHf - HfP, \qquad \mathbf{x} \in M. \end{aligned} \tag{3.8}$$

Since,

$$Hf = [(I - P) + P]Hf[(I - P) + P] =$$
$$(I - P)Hf(I - P) + PHfP + (I - P)HfP + PHf(I - P) =$$
$$(I - P)Hf(I - P) - PHfP + HfP + PHf,$$

the statement is proved. ∎

Lemma 3.1 states that in the quadratic/linear case, the Hessian matrix of the Lagrangian function (3.6) is of the same character as that of the objective function for the vectors orthogonal to the gradients of the constraints, given by (2.3), and the opposite one to the objective function in the subspace spanned by the gradients of the constraints. Because this result seems to be fundamental in understanding the general problem, it is illustrated by some examples and diagrams.

Example 3.1 *Let us consider the problem*

$$\min\ f(\mathbf{x}) = x_1^2 + 4x_2^2$$
$$h(\mathbf{x}) = x_1 + 2x_2 - 1 = 0, \qquad (x_1, x_2) \in R^2,$$
$$M = \{\mathbf{x} \in R^2 |\ h(\mathbf{x}) = 0\}.$$

The solution is at $x^* = (1/2, 1/4)$. The corresponding Lagrange multiplier function is given by

$$\mu(\mathbf{x}) = \frac{2}{5}(x_1 + 8x_2), \qquad \mathbf{x} \in M,$$

and the Lagrangian function depending only on the variables is as follows:

$$L(\mathbf{x}, \mu(\mathbf{x})) = \frac{1}{5}(3x_1^2 - 12x_2^2 - 20x_1x_2 + 2x_1 + 16x_2),$$
$$\nabla(5L(\mathbf{x}, \mu(\mathbf{x}))) = (6x_1 - 20x_2 + 2, -24x_2 - 20x_1 + 16),$$
$$H(5L(\mathbf{x}, \mu(\mathbf{x}))) = \begin{bmatrix} 6 & -20 \\ -20 & -24 \end{bmatrix}, \qquad (x_1, x_2) \in M.$$

In Figure 1, certain contours of the positive quadratic function $f(x) = x_1^2+4x_2^2$, $(x_1, x_2) \in R^2$, are shown together with the linear equality constraints $x_1 + 2x_2 = 1$ and the solution of the problem. The corresponding Lagrangian function is plotted in Figure 2 which shows that along the unique linear constraint the character of L and f are the same, while along the gradient of the equality constraints it is opposite to that of the objective function. Thus, if the Hessian matrix of the objective function Hf is positive definite, then the Lagrange function (3.6) must have a saddle point at the solution $\mathbf{x}^*$.

Naturally, the Lagrangian function can have a local minimum at the stationary point if in formula (3.7), both of the terms are positive definite matrices at this point, which is illustrated by the following example.

A quadratic equality problem

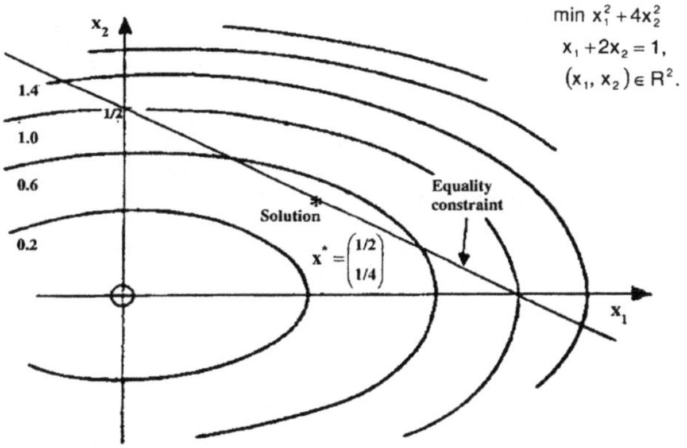

Figure 1.

Contours of the Lagrangian function

$$L(\mathbf{x}, \mu(\mathbf{x})) = \frac{1}{5}\left(3x_1^2 - 12x_2^2 - 20x_1x_2 + 2x_1 + 16x_2\right), \quad (x_1, x_2) \in R^2.$$

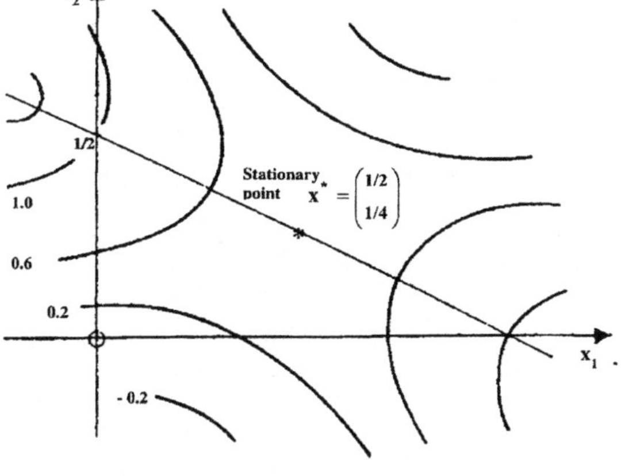

Figure 2.

Contours for the Rosenbrock's parcel problem in the plane $x_2 = x_3$

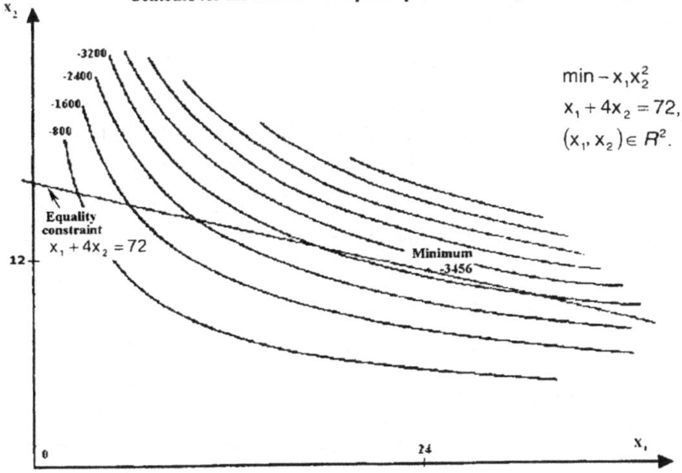

Figure 3.

Contours of the Lagrangian function

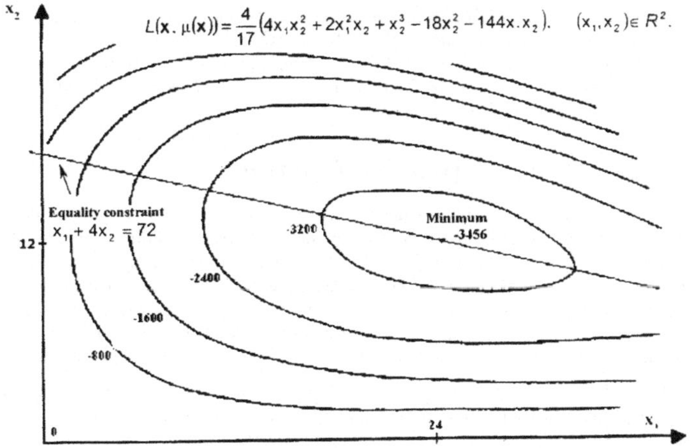

Figure 4.

Example 3.2 *The Rosenbrock's parcel problem (1960) is given in the form of*

$$\min \ -x_1x_2x_3$$

$$x_1 + 2x_2 + 2x_3 - 72 = 0, \qquad (x_1, x_2, x_3) \in R^3.$$

Let $x_2 = x_3$, *then the problem is*

$$\min \ f(\mathbf{x}) = -x_1x_2^2$$

$$h(\mathbf{x}) = x_1 + 4x_2 - 72 = 0, \qquad (x_1, x_2) \in R^2,$$

$$M = \{\mathbf{x} \in R^2 | \ h(\mathbf{x}) = 0\}.$$

The solution is at $x^* = (24, 12)$. *The corresponding Lagrange multiplier function is given by*

$$\mu(\mathbf{x}) = \frac{1}{17}(-x_2^2 - 8x_1x_2), \qquad \mathbf{x} \in M,$$

and the Lagrangian function depending only on the variables is as follows:

$$L(\mathbf{x}, \mu(\mathbf{x})) = \frac{4}{17}(4x_1x_2^2 + 2x_1^2x_2 + x_2^3 - 18x_2^2 - 144x_1x_2),$$

$$\nabla\left(\frac{17}{4}L\left(\mathbf{x}, \mu(\mathbf{x})\right)\right) = (4x_2^2 + 4x_1x_2 - 144x_2, 8x_1x_2 + 2x_1^2 + 3x_2^2 - 36x_2 - 144x_1),$$

$$H\left(\frac{17}{4}L(\mathbf{x}, \mu(\mathbf{x}))\right) = 4\begin{bmatrix} x_2 & 2x_2 + x_1 - 36 \\ 2x_2 + x_1 - 36 & 2x_1 + \frac{3}{2}x_2 - 9 \end{bmatrix}, \qquad (x_1, x_2) \in M.$$

The Rosenbrock's parcel problem in the plane $x_2 = x_3$ *is demonstrated in Figure 3. When the Lagrangian function depending only on the variables is constructed for this function (see Figure 4), then it can be seen that the Hessian matrix of the Lagrangian function is positive definite in a region of the solution. In this case, the Lagrange function would in fact be a suitable penalty function to minimize.*

4 Smooth exact penalty functions

Although, there are cases in which the Lagrange function $L(\mathbf{x}, \mu(\mathbf{x}))$, $\mathbf{x} \in M$, would have the required minimum, still in general, this will not be the case and further investigations must be made. However, equation (3.4) gives the clue how the Lagrange function L ought to be modified. In this part, we follow Fletcher's paper (1970). The directions in which the quadratic approximation of L may be wrong are along the gradients of the equality constraints. If a term is added to L with zero derivative at $\mathbf{x}^*$ but with large positive values of the quadratic approximations in the directions of the gradients and zero value elsewhere, then a suitable penalty function is obtained. Such a term can be $\frac{1}{2}\mathbf{h}^TQ\mathbf{h}$, where Q is a sufficiently large positive definite matrix. It seems to be obvious to simplify this function by choosing

$$Q = qI, \tag{4.1}$$

where q is a scalar. Let $M = \{\mathbf{x} \in R^n | \; J\mathbf{h}(\mathbf{x}) \text{ is of full rank}\}$. Thus, the simplest member of the class is

$$\mathcal{P}_1(\mathbf{x}; q) = L(\mathbf{x}, \mu(\mathbf{x})) + \frac{1}{2} q \mathbf{h}^T(\mathbf{x}) \mathbf{h}(\mathbf{x}), \qquad \mathbf{x} \in \mathcal{M}, \tag{4.2}$$

where any value of the parameter q above a certain threshold is convenient and one minimization of the function $\mathcal{P}_1$ can solve the problem. The initial value for q is increased if it becomes apparent during the minimization that q is not sufficiently large. In classical methods, the parameter must be varied to the limiting case and a minimization should be carried out for each value of the parameter. This new function is clearly superior from theoretical point of view.

A key problem of penalty function $\mathcal{P}_1$ is the choice of the parameter q. A very large value would work in theory but would leave $\mathcal{P}_1$ very ill-conditioned with much greater values along the normals than elsewhere. Alternatively, a small value q could be chosen arbitrarily and increased - say by a power of ten - if the Hessian matrix of $\mathcal{P}_1$ seems to be negative (or just positive) definite during the minimization.

For the quadratic/linear problem, $\mathcal{P}_1$ can be differentiated twice to give

$$H\mathcal{P}_1(\mathbf{x}; q) = (I - \mathcal{P})Hf(I - \mathcal{P}) - \mathcal{P}Hf\mathcal{P} + qJ\mathbf{h}^T J\mathbf{h}, \qquad \mathbf{x} \in \mathcal{M}, \tag{4.3}$$

from which $q > ||J\mathbf{h}^{+T} Hf J\mathbf{h}^{+}||_2$ can be obtained as a suitable condition for the positive definiteness of the matrices in (4.3) by using the spectral norm for matrices. This suggests that in the non-quadratic/linear case a trial value $2||J\mathbf{h}^{+T} Hf J\mathbf{h}^{+}||_2$ could be used to estimate the parameter q.

A penalty function more satisfactory than $\mathcal{P}_1$ seems to be one in which the positive definite matrix Q in $\mathcal{P}_1$ is replaced by another one as follows:

$$\mathcal{P}_2(\mathbf{x}; q) = L(\mathbf{x}, \mu(\mathbf{x})) + \frac{1}{2} q \mathbf{h}^T(\mathbf{x}) J\mathbf{h}(\mathbf{x})^{+} J\mathbf{h}(\mathbf{x})^{+T} \mathbf{h}(\mathbf{x}), \qquad \mathbf{x} \in \mathcal{M}. \tag{4.4}$$

This function takes no longer to compute than $\mathcal{P}_1$ because $\mathbf{h}^T(\mathbf{x}) J\mathbf{h}(\mathbf{x})^{+}$ is already available from the middle term. Thus, in the quadratic/linear case, the Hessian matrix of $\mathcal{P}_2$ is equal to

$$H\mathcal{P}_2(\mathbf{x}; q) = (I - P)Hf(I - P) - PHfP + qP, \qquad \mathbf{x} \in \mathcal{M}, \tag{4.5}$$

which suggests an initial choice of q as $2||Hf||_2$. The preference for $\mathcal{P}_2$ over $\mathcal{P}_1$ should arise in part from this simpler formula for q.

Another function in the class is obtained by bringing the matrix of second derivatives into the penalty function by setting

$$Q(\mathbf{x}; q) = J\mathbf{h}(\mathbf{x})^{+}(Hf + qI)J\mathbf{h}(\mathbf{x})^{+T}, \qquad \mathbf{x} \in \mathcal{M}, \tag{4.6}$$

giving

$$H\mathcal{P}_2(\mathbf{x}; q) = (I - P)Hf(I - P) + qP, \qquad \mathbf{x} \in \mathcal{M}, \tag{4.7}$$

and any value of $q > 0$ would be suitable. Requiring the matrix of second derivatives to define functions will, however, usually be a disadvantage of this formulation. The following statement guarantees the local optimality of the penalty function at the solution.

Theorem 4.1 *At a solution point of any equality constrained nonlinear optimization problem (2.1) for which the second-order sufficiency conditions fulfil, it is possible to choose a sufficiently large q so that the Hessian matrix of the function P_1 is positive definite.*

Proof: By differentiating $\mathcal{P}_1$, we obtain that

$$\mathcal{P}_1(\mathbf{x};q) = f(\mathbf{x}) - \sum_{j=1}^{n-k} \mu_j(\mathbf{x}) h_j(\mathbf{x}) + \frac{1}{2} q \sum_{j=1}^{n-k} h_j^2(\mathbf{x}), \qquad \mathbf{x} \in \mathcal{M}, \tag{4.8}$$

$$\begin{aligned} \nabla \mathcal{P}_1(\mathbf{x};q) = \nabla f(\mathbf{x}) - \sum_{j=1}^{n-k} \mu_j(\mathbf{x}) \nabla h_j(\mathbf{x}) - \sum_{j=1}^{n-k} h_j(\mathbf{x}) \nabla \mu_j(\mathbf{x}) + \\ q \sum_{j=1}^{n-k} h_j(\mathbf{x}) \nabla h_j(\mathbf{x}), \qquad \mathbf{x} \in \mathcal{M}, \end{aligned} \tag{4.9}$$

$$\begin{aligned} H\mathcal{P}_1(\mathbf{x};q) = Hf(\mathbf{x}) - \sum_{j=1}^{n-k} \mu_j(\mathbf{x}) Hh_j(\mathbf{x}) - \sum_{j=1}^{n-k} \nabla h_j(\mathbf{x})^T \nabla \mu_j(\mathbf{x}) - \\ \sum_{j=1}^{n-k} \nabla \mu_j(\mathbf{x})^T \nabla h_j(\mathbf{x}) - \sum_{j=1}^{n-k} h_j(\mathbf{x})^T H\mu_j(\mathbf{x}) + \\ q \sum_{j=1}^{n-k} \nabla h_j(\mathbf{x})^T \nabla h_j(\mathbf{x}) + q \sum_{j=1}^{n-k} h_j(\mathbf{x})^T Hh_j(\mathbf{x}), \qquad \mathbf{x} \in \mathcal{M}, \end{aligned} \tag{4.10}$$

and by considering only the feasible points of problem (2.1),

$$\begin{aligned} H\mathcal{P}_1(\mathbf{x};q) = Hf(\mathbf{x}) - \sum_{j=1}^{n-k} \mu_j(\mathbf{x}) Hh_j(\mathbf{x}) - \sum_{j=1}^{n-k} \nabla h_j(\mathbf{x})^T \nabla \mu_j(\mathbf{x}) - \\ \sum_{j=1}^{n-k} \nabla \mu_j(\mathbf{x})^T \nabla h_j(\mathbf{x}) + q \sum_{j=1}^{n-k} \nabla h_j(\mathbf{x})^T \nabla h_j(\mathbf{x}), \qquad \mathbf{x} \in M. \end{aligned} \tag{4.11}$$

By the second-order optimality conditions, the matrix

$$\begin{aligned} H\mathcal{P}_1(\mathbf{x})_{|TM} = \Bigg(Hf(\mathbf{x}) - \sum_{j=1}^{n-k} \mu_j(\mathbf{x}) Hh_j(\mathbf{x}) - \\ \sum_{j=1}^{n-k} \nabla h_j(\mathbf{x})^T \nabla \mu_j(\mathbf{x}) - \sum_{j=1}^{n-k} \nabla \mu_j(\mathbf{x})^T \nabla h_j(\mathbf{x}) \Bigg)_{|TM}, \qquad \mathbf{x} \in M, \end{aligned} \tag{4.12}$$

is positive definite at every solution point where the symbol $|TM$ means the restriction of a matrix to the corresponding tangent space of the manifold M. Thus, by Finsler lemma [6] or [1, p.75], the statement is proved. ∎

This class of exact penalty functions seems to be good from theoretical point of view, but numerical experience did not back this expectation because of a possible matrix inversion, or the solution of a linear system for evaluating the multiplier functions at the iteration points [2,4,5,9].

5 Global Lagrange multiplier rule for the case of equality constraints

It is emphasized that in the Lagrange therorem, the regularity condition and second-order conditions (2.7) are assumed to be satisfied at a fix point and the sufficient conditions ensure strict local minimality. Now, the local character of conditions (2.4) and (2.7) will be shown.

Lemma 5.1 *If $r(J\mathbf{h}(\mathbf{x}_0)) = n-k$, $\mathbf{x}_0 \in M$, and (2.7) holds, then there exists a neighbourhood $\mathcal{U}(\mathbf{x}_0) \subseteq M$ such that*

$$r(J\mathbf{h}(\mathbf{x})) = n-k, \quad \mathbf{x} \in U(\mathbf{x}_0) \quad and \tag{5.1}$$

$$\mathbf{v}^T H_x L(\mathbf{x},\mu)\mathbf{v} > 0, \quad \mathbf{v} \in TM_x, \quad \mathbf{v} \neq 0, \quad \mathbf{x} \in \mathcal{U}(\mathbf{x}_0). \tag{5.2}$$

Proof: By the rank condition of the Jacobian matrix at $\mathbf{x}_0$, their row vectors are linearly independent at this point, so they can be completed by vectors forming a basis of R^n. Since a basis of R^n consists of linearly independent vectors, the determinant of this basis is different from zero. Because the component functions of the basis vectors are continuous, the determinant will be different from zero in a neighbourhood of $\mathbf{x}_0$ from which statement (5.1) follows.

By Mann theorem (1943), condition (2.7) holds iff the bordered determinant

$$(-1)^p \det \begin{pmatrix} H_x L(\mathbf{x}_0,\mu)_p & J\mathbf{h}(\mathbf{x}_0)_p^T \\ J\mathbf{h}(\mathbf{x}_0)_p & 0 \end{pmatrix} > 0, \tag{5.3.}$$

for $p = n-k+1, \ldots, n$, where $H_x L(\mathbf{x}_0,\mu)_p$ means the matrix obtained from $H_x L(\mathbf{x}_0,\mu)$ by keeping the elements in the first p rows and columns, $J\mathbf{h}(\mathbf{x}_0)_p$ is the corresponding $(n-k)\times p$ submatrix of the Jacobian matrix and the notation of the determinant is det. Because the functions of the matrix elements are continuous, inequalities (5.3) remain valid in a neighbourhood of $\mathbf{x}_0$ from which statement (5.2) follows. ∎

The necessary and sufficient conditions of the Lagrange multiplier rule consist of properties at a given point, which properties hold also in a neighbourhood by Lemma 5.1. It follows that each of these results is of local character. As to the mathematical background, mainly the differentiability of functions in the Euclidean space and some principal theorems of linear vector spaces are used. Our idea is to obtain results of global character by combining the tools of Riemannian geometry with that of linear vector spaces. Thus, an improvement of the Lagrange sufficient conditions can be obtained under the same type of assumptions.

Consider problem (2.1) where regularity condition (2.4) holds. Since the gradients $\nabla h_j(\mathbf{x})$, $j = 1, \dots, n-k$, $\mathbf{x} \in M$, are linearly independent, the feasible set M is a Riemannian C^2 k-manifold where the Riemannian metric on M is induced by the Euclidean one of R^n. It is convenient to introduce the Lagrangian function associated with f and M, defined as

$$L(\mathbf{x}, \mu(\mathbf{x})) = f(\mathbf{x}) - \sum_{j=1}^{n-k} \mu_j(\mathbf{x}) h_j(\mathbf{x}), \qquad \mathbf{x} \in M, \tag{5.4}$$

where

$$\mu(\mathbf{x})^T = \nabla f(\mathbf{x}) J\mathbf{h}^T(\mathbf{x}) \left[J\mathbf{h}(\mathbf{x}) J\mathbf{h}(\mathbf{x})^T \right]^{-1}, \tag{5.5}$$

$$J\mathbf{h}(\mathbf{x}) = \begin{bmatrix} \nabla h_1(\mathbf{x}) \\ \vdots \\ \nabla h_{n-k}(\mathbf{x}) \end{bmatrix}.$$

Let the geodesic gradient vector and the geodesic Hessian matrix of the Lagrangian function be defined in the forms of

$$\nabla^g L(\mathbf{x}, \mu(\mathbf{x})) = \nabla f(\mathbf{x}) - \sum_{j=1}^{n-k} \mu_j(\mathbf{x}) \nabla h_j(\mathbf{x}), \quad \mathbf{x} \in M, \tag{5.6}$$

$$H^g L(\mathbf{x}, \mu(\mathbf{x})) = \left(Hf(\mathbf{x}) - \sum_{j=1}^{n-k} \mu_j(\mathbf{x}) H h_j(\mathbf{x}) \right)_{|TM}, \quad \mathbf{x} \in M. \tag{5.7}$$

In finite-dimensional problems, the Lagrangian function has constant multipliers, but the following statement points out the meaning of multipliers $\mu(\mathbf{x})$ depending on $\mathbf{x} \in M$. Let us recall a definition of geodesic convex sets where the geodesic has the classical meaning.

Definition 5.1[15] *Let M be a Riemannian C^2 k-manifold, then the set $A \subseteq M$ is geodesic convex if any two points of A are joined by a geodesic belonging to A, moreover, a singleton is geodesic convex.*

It is emphasized that every Riemannian metric generates a geodesic convexity notion. In the case of a Euclidean space, the geodesic convexity coincides with the classical one. In optimization theory related to the Lagrange multiplier rule, the induced Riemannian metric, defined as the restriction of the $n \times n$ identity matrix to all tangent spaces of the manifold M, seems to be important.

GLOBAL LAGRANGE MULTIPLIER RULE. *(Rapcsák, 1997) If the point $\mathbf{x}_0 \in M$ is a (strict) local or global minimum of problem (2.1), (2.4), then*

$$\begin{aligned} &\nabla^g L(\mathbf{x}_0, \mu(\mathbf{x}_0)) = 0, \\ &\mathbf{v}^T H^g L(\mathbf{x}_0, \mu(\mathbf{x}_0)) \mathbf{v} \geq (>) 0, \qquad \mathbf{v} \in TM_{x_0}. \end{aligned} \tag{5.8}$$

If $A \subseteq M$ *is an open geodesic convex set with respect to the induced Riemannian metric and there exists a point* $\mathbf{x}_0 \in A$ *such that*

$$\nabla^g L(\mathbf{x}_0, \mu(\mathbf{x}_0)) = 0, \tag{5.9}$$

$$\mathbf{v}^T H^g L(\mathbf{x}, \mu(\mathbf{x}))\mathbf{v} \;\geq\; (>)0, \qquad \mathbf{v} \in TM_x,\ (\mathbf{v} \neq 0), \quad \mathbf{x} \in A, \tag{5.10}$$

then $\mathbf{x}_0$ *is a (strict) global minimum of* f *on* A*. Moreover,*

$$\nabla^g L(\mathbf{x}, \mu(\mathbf{x})) \;=\; Df(\mathbf{x}), \tag{5.11}$$

$$H^g L(\mathbf{x}, \mu(\mathbf{x})) \;=\; D^2 f(\mathbf{x}), \qquad \mathbf{x} \in A, \tag{5.12}$$

where Df *and* $D^2 f$ *denotes the first and second covariant derivatives of* f *with respect to the induced Riemannian metric, respectively.*

By the global Lagrange multiplier rule, the necessary and sufficient optimality conditions are given by the same tensor formulae based on the first and second covariant derivatives, only the domains are different where the second-order formula holds.

Corollary 5.1 *In the global Lagrange multiplier rule, the first-order optimality condition of the Lagrange theorem holds at* $\mathbf{x}_0$ *and the quadratic forms in (2.7) and (5.10) are equal at* $\mathbf{x}_0 \in A$.

Proof: It is enough to show the equivalence of the first-order conditions (2.5) and (5.9) at $\mathbf{x}_0$. By using the definition of the geodesic gradient of the Lagrangian function, we have that

$$\begin{aligned}
\nabla^g L(\mathbf{x}_0, \mu(\mathbf{x}_0)) &= \nabla f(\mathbf{x}_0) - \sum_{j=1}^{n-k} \mu_j(\mathbf{x}) \nabla h_j(\mathbf{x}) \\
&= \nabla f(\mathbf{x}_0) - \nabla f(\mathbf{x}_0) J\mathbf{h}^T(\mathbf{x}_0) \left[J\mathbf{h}(\mathbf{x}_0) J\mathbf{h}(\mathbf{x}_0)^T\right]^{-1} J\mathbf{h}^T(\mathbf{x}_0) \\
&= \nabla f(\mathbf{x}_0) \left(I - J\mathbf{h}^T(\mathbf{x}_0) \left[J\mathbf{h}(\mathbf{x}_0) J\mathbf{h}(\mathbf{x}_0)^T\right]^{-1} J\mathbf{h}^T(\mathbf{x}_0)\right) = \nabla f(\mathbf{x}_0)_T,
\end{aligned} \tag{5.13}$$

where ∇f_T means the tangential component of the gradient vector with respect to the Euclidean metric. So, conditions (2.5) and (5.9) coincide at $\mathbf{x}_0$. ∎

6 Geometric background of the global Lagrange multiplier rule

First-order condition (5.9) means that the gradient of the objective function ∇f is orthogonal to the tangent space of the Riemannian manifold M at the given point $\mathbf{x}_0$, the gradients of the equality constraints ∇h_j, $j = 1, \ldots, n-k$, generate the normal space to the tangent space with respect to the Euclidean metric of R^n at $\mathbf{x}_0$, thus ∇f is a linear combination of ∇h_j, $j = 1, \ldots, n-k$, at the point $\mathbf{x}_0$. In the Lagrange theorem, the first-order and second-order conditions ensure a strict local minimum while in the global Lagrange multiplier rule, a global minimum is characterized on an open geodesic convex set. It seems that the new multiplier rule contains a new assumption, the geodesic convexity of the set A. Because of the local character of the Lagrange theorem, there exists a neighbourhood such that (2.5) and (2.7) hold locally (see 5.1 Lemma), thus the existence of a geodesic convex neighbourhood around the given point can be proved (e.g., Spivak, 1983). It follows that the geodesic convexity property fulfils locally also in the classical case, studied by Lagrange.

Definition 6.1 *If $A \subseteq M$ is a geodesic convex set, then a function $f : A \to R$ is geodesic (strictly) convex if its restrictions to all geodesic arcs belonging to A are (strictly) convex in the arc length parameter.*

The following statement shows the geometric character of the global Lagrange multiplier rule.

Theorem 6.1 *(Rapcsák, 1991) If $A \subseteq M \subseteq R^n$ is an open geodesic convex set and $f \in C^2(A, R)$, then f is geodesic (strictly) convex on A with respect to the induced Riemannian metric iff $H^g L$ is positive semidefinite (definite) on A.*

From the point of view of geometry, the existence of a constrained minimum is equivalent to the existence of a geodesic convex function with respect to the induced Riemannian metric. This class of functions is exactly described by the second-order covariant derivative. This tensor provides the sufficient optimality conditions on an open domain for problem (2.1). It follows that the Lagrange method can be considered the transformation of an equality constrained nonlinear optimization problem given in R^n into an unconstrained problem defined on the constraint submanifold with the induced Riemannian metric and immersed in R^n.

7 Global Lagrangian function with a penalty term

As we have seen, the original idea of Lagrange was to transform an equality constrained optimization problem into an unconstrained one. Since then, a large class of the methods resulted in an unconstrained optimization of a nonsmooth or a continuously differentiable function defined on a compact domain with nonemoty interior in R^n. In order to preserve the smoothness of the penalty functions in the case of smooth optimization problems, the tools of the global Lagrange multiplier rule can be used. An advantage of this approach is the possibility to approximate the smooth nonlinear optimization problems on the feasible set by the first and second covariant derivatives of the objective functions with respect to the induced Riemannian metric. These tensors can replace the terms of the Taylor's expansion of the given functions. A disadvantage of the numerical realization seems to be the determination of the second covariant tensors given as the restriction of matrices to the corresponding tangent spaces.

On the basis of the global Lagrange multiplier rule, the Lagrange function (5.4) is considered with the classical penalty term, the squared Euclidean norm of the constraint violations (usually attributed to Courant, 1943), which leads to the well-known quadratic penalty functions and the unconstrained minimization problems (see, e.g., [10, p.270])

$$\min \ \mathcal{P}_1(\mathbf{x}; q) = f(\mathbf{x}) - \sum_{j=1}^{n-k} \mu_j(\mathbf{x}) h_j(\mathbf{x}) + \frac{1}{2} q \sum_{j=1}^{n-k} h_j^2(\mathbf{x}), \qquad \mathbf{x} \in \mathcal{M}, \tag{7.1}$$

where q^* is a given positive number, $q > q^*$ and

$$M \subset \mathcal{M} = \{\mathbf{x} \in R^n | \ J\mathbf{h}(\mathbf{x}) \ \text{ is of full rank}\}.$$

Now, a basic result will be proved that seems to be stronger than Proposition 4.22 (Bertsekas, 1982) because in the following theorem, the feasible domain is not compact in statements (1) and (2); the feasible domain is not compact, the global optimal solution is not necessarily unique and not necessarily a point in the interior of the feasible domain in statement (3); moreover, in statements (2) and (3), not only one direction is proved, but a one-to-one correspondence is given between the local and global optimum points of problem (2.1), (2.4) and problem (7.1), respectively.

Theorem 7.1 *Let us consider problem (2.1) with the regularity condition (2.4) and problem (7.1). Then,*

(1) *a point belonging to M is a stationary point of P_1 on M for all $q > q^*$ iff it is a stationary point of problem (2.1);*

(2) *a point belonging to M is a local optimal point of P_1 on M for all $q > q^*$ iff it is a local optimal point of problem (2.1);*

(3) *a point belonging to M is a global optimal point of P_1 on M for all $q > q^*$ iff it is a global optimal point of problem (2.1).*

Proof:

(1) At a stationary point $\mathbf{x}_0 \in M$ of (2.1) $Df(\mathbf{x}_0) = 0$, thus,

$$\nabla \mathcal{P}_1(\mathbf{x}_0; q) = Df(\mathbf{x}_0) - \sum_{j=1}^{n-k} h_j(\mathbf{x}_0) \nabla \mu_j(\mathbf{x}_0) + q \sum_{j=1}^{n-k} h_j(\mathbf{x}_0) \nabla h_j(\mathbf{x}_0) = 0. \tag{7.2}$$

If $\nabla \mathcal{P}_1(\mathbf{x}_0; q) = 0$, then

$$Df(\mathbf{x}_0) - \sum_{j=1}^{n-k} h_j(\mathbf{x}_0) \nabla \mu_j(\mathbf{x}_0) = -q \sum_{j=1}^{n-k} h_j(\mathbf{x}_0) \nabla h_j(\mathbf{x}_0), \tag{7.3}$$

for all $q > q^*$, where q^* is a given scalar.
Let us assume that $\mathbf{x}_0$ is not a stationary point of (2.1). It follows that $\mathbf{x}_0 \notin M$, $\mathbf{x}_0 \in \mathcal{M}$, and $q \sum_{j=1}^{n-k} h_j(\mathbf{x}_0) \nabla h_j(\mathbf{x}_0) \neq 0$ for all $q > q^*$, which contradicts formula (7.3). This proves the first statement.

(2) By statement (1), a stationary point of $\mathcal{P}_1$ belongs to M. It follows that if $\mathbf{x}_0$ is local optimum point of $\mathcal{P}_1$ on $\mathcal{M}$, then $\mathbf{x}_0$ is a local optimum point of (2.1) because $M \subseteq \mathcal{M}$. Let us assume that $\mathbf{x}_0$ is a local optimum point of (2.1), but it is not

a local optimum point of $\mathcal{P}_1$ on $\mathcal{M}$. It follows that there exists a point $\hat{\mathbf{x}} \notin M$, $\hat{\mathbf{x}} \in \mathcal{M}$, in a neighbourhood of $\mathbf{x}_0$, such that

$$f(\mathbf{x}_0) > f(\hat{\mathbf{x}}) - \sum_{j=1}^{n-k} \mu_j(\hat{\mathbf{x}}) h_j(\hat{\mathbf{x}}) + q\frac{1}{2}\sum_{j=1}^{n-k} h_j(\hat{\mathbf{x}})^2, \tag{7.4}$$

for all $q > q^*$. Since, $q\frac{1}{2}\sum_{j=1}^{n-k} h_j(\hat{\mathbf{x}})^2 > 0$ for all $q > q^*$, this gives a contradiction in (7.4), which proves the second statement.

(3) Because of the property $M \subseteq \mathcal{M}$, the proof is equivalent to the preceding one. ∎

By the global Lagrange multiplier rule, a sufficient condition for the global optimality of the objective function on the constraint manifold or an open geodesic convex set is the geodesic convexity of the objective function with respect to the induced Riemannian metric. The question arises how it is possible to construct a class of smooth exact penalty functions with good algorithmical properties so that these functions should be convex in the Euclidean space, and simultanously, they should be geodesic convex with respect to the induced Riemannian metric on the constraint manifold. In the following statement, the function $\mathcal{P}_1$ is investigated from this point of view.

Theorem 7.2 *If the objective function f is strictly geodesic convex with respect to the induced Riemannian metric on an open geodesic convex set containing a compact subset A of the constraint manifold M in problem (2.1) and (2.4), then $H\mathcal{P}_1$ is positive definite on $A \subseteq M$.*

Proof: By Theorem 6.1, the strictly geodesic convexity of f on an open geodesic convex set is equivalent to the positive definiteness of $D^2 f$ on the same set. Because of the tensor character of the covariant derivatives $D^2 f$ and $H\mathcal{P}_1$ (the latter one is the second covariant derivative of $\mathcal{P}_1$ with respect to the Euclidean metric), any regular coordinate representaion is convenient. Thus,

$$\begin{aligned} H\mathcal{P}_1(\mathbf{x}; q) = Hf(\mathbf{x}) - \sum_{j=1}^{n-k} \mu_j(\mathbf{x}) Hh_j(\mathbf{x}) - \sum_{j=1}^{n-k} \nabla h_j(\mathbf{x})^T \nabla \mu_j(\mathbf{x}) - \\ \sum_{j=1}^{n-k} \nabla \mu_j(\mathbf{x})^T \nabla h_j(\mathbf{x}) + q\sum_{j=1}^{n-k} \nabla h_j(\mathbf{x})^T \nabla h_j(\mathbf{x}), \qquad \mathbf{x} \in \mathcal{A} \subseteq M. \end{aligned} \tag{7.5}$$

By using formula (5.12), Theorem 6.1 and following the proof of Theorem 4.1, we have that there exists a scalar q at every point of A such that $H\mathcal{P}_1$ is positive definite in a neighbourhood of this point. Since, the set A is compact, a finite number of covering neighbourhoods and a sufficiently large value for q^* can be chosen. ∎

8 Concluding remark

The aim of the paper is to reconsider smooth exact penalty functions based on Courant's and Fletcher's ideas. After a short survey, the original idea is combined with the global

Lagrange multiplier rule giving a clearer geometric background of this field. An interesting problem is how to construct smooth penalty functions with the property that they are convex or generalized convex in the Euclidean space, and simultaneously, geodesic convex with respect to the induced Riemannian metric on the constraint manifold. From computational point of view, the question is whether it is possible to improve the algorithmical and numerical features of the exact penalty functions by using this tensor approach.

Acknowledgement.

This is to express my thanks to F. Giannessi for his profound perceptions and his valuable advice.

References

[1] Bellmann, R., Introduction to matrix analysis, McGraw-Hill Book Company, New York, Toronto, London, 1960.

[2] Bertsekas, D.P., Constrained optimization and Lagrange multiplier methods, Academic Press, New York, London, 1982.

[3] Courant, R., Variational methods for the solution of problems of equilibrium and vibrations, Bulletin of the American Mathematical Society 49 (1943) 1-23.

[4] Di Pillo, G., Exact penalty methods, in: Algorithms for continuous optimization: the state of the art, E.Spedicato (ed.), Kluwer Academic Publishers, Boston (1994) 1-45.

[5] Di Pillo, G. and Grippo, L., Exact penalty functions in constrained optimization, SIAM Journal on Control and Optimization 27 (1989) 1333-1360.

[6] Finsler, P., Über das Vorkommen definiter und semidefiniter Formen in schären quadratischen Formen, Commentarii Mathematicii Helveticii 9 (1937) 188-192.

[7] Fletcher, R., A class of methods for nonlinear programming with termination and convergence properties, in: Integer and nonlinear programming, J. Abadie (ed.), North-Holland Publishing Company, Amsterdam, London (1970) 157-175.

[8] Fletcher, R., An exact penalty function for nonlinear programming with inequalities, Mathematical Programming 5 (1973) 129-150.

[9] Fletcher, R., Practical methods of optimization, John Wiley & Sons, 1987.

[10] Hestenes, M.R., Optimization theory, John Wiley & Sons, New York, London, Sydney, Toronto, 1975.

[11] Lagrange, J.L., Essai sur une nouvelle méthode pour determiner les maxima et minima des formules integrales indefinies, in: Miscellanea Taurinensia II (1762) 173-195.

[12] Lagrange, J.L., Mécanique Analytique I-II., Paris, 1788.

[13] Mann, H.B., Quadratic forms with linear constraints, American Mathematical Monthly 50 (1943) 430-433.

[14] Prékopa, A., On the development of optimization theory, American Mathematical Monthly 87 (1980) 527-542.

[15] Rapcsák, T., Geodesic convexity in nonlinear optimization, Journal of Optimization Theory and Applications 69 (1991) 169-183.

[16] Rapcsák, T., Smooth nonlinear optimization in R^n, Kluwer Academic Publishers, 1997.

[17] Rosenbrock, H.H., An automatic method for finding the greatest or least value of a function, Computer Journal 3 (1960) 175-184.

[18] Spivak, M., A comprehensive introduction to differential geometry I-V., Publish or Perish, Berkeley, 1979.

Structural Methods in the Solution of Variational Inequalities

Stephen M. Robinson (smrobins@facstaff.wisc.edu)
Department of Industrial Engineering, University of Wisconsin-Madison, 1513 University Avenue, Madison, WI 53706-1572, USA

Abstract

This paper shows how to apply the recently-proposed tool of composition duality, together with exploitation of features such as block structure, to simplify the solution of variational inequalities. As a byproduct of this approach we recover as special cases a number of known methods, including the Schur complement and the Sherman-Morrison-Woodbury formula for inverting perturbed matrices.

Keywords: Variational inequalities, block structure, composition duality, Schur complement, small-rank perturbation.

1 Introduction

In this paper we deal with the analysis and solution of *variational inequality* problems in finite-dimensional spaces, under the simplifying assumption that the problems have certain types of special structure. We argue that a suitable tool for taking advantage of such special structure is the recently-proposed scheme of *composition duality,* and we show how one can exploit this scheme, together with special structure, to simplify the solution process. As we show, in some cases the simplification can reduce very substantially the difficulty of solving the problems.

The remainder of this paper is organized as follows: in Section 2 we give a brief introduction to variational inequalities and to composition duality, sufficient for the work of the following sections. Those sections then present applications.

G. Di Pillo and F. Giannessi (eds.), Nonlinear Optimization and Related Topics, 369–380.

Section 3 shows how to use this methodology to solve a problem consisting of a relatively simple variational inequality subjected to a low-rank linear perturbation. We show that when restricted to the special case of linear equations and rank-1 perturbations, we recover the well known Sherman-Morrison-Woodbury formula.

Section 4 exhibits a different kind of application, in which a variational inequality with block structure is, in effect, partially solved: this yields formulas that reduce, again in the special case of linear equations, to the Schur complement, which has long been a useful device in linear algebra.

We conclude the paper in Section 5, with a brief summary and some ideas on further application of the ideas presented here. References follow.

2 Variational inequalities and composition duality

This section gives a brief introduction to variational inequalities and to composition duality. The purpose is to provide sufficient detail so that the reader can follow the work of the remaining sections of this paper. References to the literature give pointers to papers giving further detail and examples.

Suppose that C is a closed convex subset of $\mathbb{R}^n$ and f is a single-valued function from C to $\mathbb{R}^n$. The *variational inequality* in $\mathbb{R}^n$ is the problem of finding $x \in C$ such that $f(x)$ is an inward normal to C at x: that is, for each $c \in C$,

$$\langle f(x), c - x \rangle \geq 0. \tag{1}$$

The associated *generalized equation* is

$$0 \in f(x) + N_C(x), \tag{2}$$

where $N_C(x)$ is the *normal cone* of C at x:

$$N_C(x) = \begin{cases} \{x^* \mid \langle x^*, c - x \rangle \leq 0 & \text{if } x \in C, \\ \emptyset & \text{if } x \notin C \end{cases}$$

The relations (1) and (2) are equivalent ways of expressing the same fact. They include, in order of increasing generality, the following special cases:

- The system of nonlinear equations $f(x) = 0$ (take $C = \mathbb{R}^n$ so that $N_C(x)$ is simply the zero operator).

- The *complementarity problem,* in which $C = \mathbb{R}^n_+$ and thus the problem is to find $x \geq 0$ such that $f(x) \geq 0$ and $\langle x, f(x) \rangle = 0$. Here the vector inequality notation $x \geq 0$ means that each component of x must be nonnegative.

- The *box-constrained variational inequality,* in which C is restricted to be a product of nonempty closed intervals.

Variational inequalities were introduced in the 1960s: see, for example, [12, 13]. Since that time they have been widely used as a device for conveniently formulating problems in a wide variety of technical fields, as well as for gaining additional insight into the nature and properties of these problems. Ferris and Pang [3] have produced an excellent survey article giving many applications of the complementarity problem in engineering and in economics. For an earlier survey, also devoted to complementarity problems, with coverage of theory and algorithms as well as applications, see [4]. A survey of generalized equations in the more general form treated here is in [6].

We now turn to composition duality, briefly sketching the main idea and some of its consequences. For more information and proofs, see the general exposition and treatment given in [8]. An earlier paper [7] used a particular application of this method to reduce the computational burden in solving a specially-structured variational inequality. A closely related technique has been presented independently by Pennanen in [5]. In keeping with the theme of this paper, we develop the method in a restricted form in finite-dimensional spaces, though it applies in general spaces.

The method begins with a "primal" inclusion involving a composition of operators:

$$0 \in A(x) + QBP(x), \tag{3}$$

where $A : \mathbb{R}^n \to \mathbb{R}^n$, $P : \mathbb{R}^n \to \mathbb{R}^m$, $B : \mathbb{R}^m \to \mathbb{R}^m$, $Q : \mathbb{R}^m \to \mathbb{R}^n$ are arbitrary multifunctions. The key observation is the following.

Theorem 1. *Solvability of* (3) *is equivalent to solvability of*

$$0 \in (-P)A^{-1}(-Q)(y) + B^{-1}(y) \tag{4}$$

The proof of this result is very simple, and we omit it; rather, the formulation and the insight that it can provide are the real reasons for interest in this method. We point out next some aspects of the duality between (3) and (4) that are relevant to our development here.

First, (3) is in $\mathbb{R}^n$ while (4) is in $\mathbb{R}^m$. Therefore this duality framework permits the dual problem to be posed in a different space than the primal. Second, as we shall see, the formulation can be especially useful when P and Q are taken to be linear operators suited to the structure of the problem.

Finally, this structure contains all of the previously developed duality structures of which we are aware. For example, with $n = m$ and $P = I = Q$ we recover the duality of Attouch and Théra [1]. Several other examples are given in [8]. One of these, which has been particularly significant from an applications standpoint, is the conjugate duality framework of Rockafellar [9, 10]. We develop that connection in order to illustrate the generality of the composition duality framework.

Consider a closed proper convex function

$$F : \mathbb{R}^p \times \mathbb{R}^q \to (-\infty, +\infty].$$

Let $L : \mathbb{R}^p \to \mathbb{R}^{p+q}$ with $L(x) = (x, 0)$ and assume there is $\hat{x} \in \mathbb{R}^p$ with $L(\hat{x}) \in \operatorname{ri}\operatorname{dom} F$. We know that $x \in \mathbb{R}^p$ minimizes $F(\cdot, 0)$ if and only if $0 \in \partial(F \circ L)(x)$. We can rewrite this as

$$0 \in 0(x) + L^*(\partial F)L(x), \tag{5}$$

where $0(x)$ is the zero operator on $\mathbb{R}^p$. Applying composition duality, we find that the dual problem is then

$$0 \in (-L)(0^{-1})(-L^*)(y) + (\partial F)^{-1}(y), \tag{6}$$

where $0^{-1}(z) = \mathbb{R}^p$ if $z = 0$ and $\emptyset$ otherwise, and $(\partial F)^{-1} = \partial F^*$. The connection with Rockafellar's framework is now shown by the following theorem, proved in [8].

Theorem 2. *The following are equivalent:*

1. $x_0 \in \mathbb{R}^p$ *solves* (5) *and* $y_0 \in \mathbb{R}^{p+q}$ *solves* (6),
2. $y_0 = (0, -p_0^*)$ *and* $y_0 \in \partial F(x_0, 0)$,
3. $y_0 = (0, -p_0^*)$ *and* (x_0, p_0^*) *is a saddle point of* $\mathcal{L}(x, p^*) := \inf_p\{\langle p^*, p\rangle + F(x, p)\}$.

Note that freedom to have primal and dual in different spaces was vitally important here.

Composition duality is very new, and its full range of applicability is not yet well understood. In the remainder of this paper we show how to apply it to two classes of problems, the first being a "simple" variational inequality perturbed by a low-rank linear operator that appears to destroy the simplicity, and the second being the "partial solution" of a variational inequality to yield an operator reminiscent of the classical Schur complement.

3 Application: dimension reduction for perturbed variational inequalities

This section gives a simple illustration of an application of composition duality to the numerical solution of a structured variational inequality. The situation we have in mind is a variational inequality of the form

$$0 \in f(x) + UV^T x + N_C(x), \tag{7}$$

where $f : \mathbb{R}^n \to \mathbb{R}^n$, C is a closed convex subset of $\mathbb{R}^n$, and U and V are $n \times k$ matrices. We are interested in the case in which $k < n$; indeed, in the illustration we give below, $k = 1$. We also suppose that the problem

$$z \in f(x) + N_C(x) \tag{8}$$

is fairly easy to solve for different values of z.

In the particularly simple numerical illustration we give here, we have $f(x) = Dx$, where D is a diagonal matrix with positive diagonal elements, and $C = \mathbb{R}^n_+$. In this case the problem (8) can be solved by componentwise projection, whereas (7) is a linear complementarity problem with a full matrix. If n were quite large, the difference in difficulty between (8) and (7) would then seem very considerable.

However, as we shall show, by applying composition duality we can obtain an equation in $\mathbb{R}^k$, which may have various desirable properties depending on the structure of (8): for instance, in our example this is a piecewise affine equation in a single real variable. By finding a zero of this equation and then performing a simple computation involving solution of (8), we obtain a solution of (7).

To apply composition duality we take $A(x) = f(x) + N_C(x)$, $B(y) = y$ $(y \in \mathbb{R}^k)$, $P(y) = Uy$, $Q(x) = V^T x$. Then the problem (7) is

$$0 \in A(x) + PBQ(x).$$

The equivalent dual problem is then

$$0 \in (-Q)A^{-1}(-P)(y) + B^{-1}(y), \tag{9}$$

which is an equation in $\mathbb{R}^k$. We can rewrite (9) as

$$0 = y - V^T F(-Uy) =: y - \phi(y), \tag{10}$$

where $F(z) = (f + N_C)^{-1}(z)$. This F is, by our assumption, easy to compute. We now proceed to solve (10) in $\mathbb{R}^k$ to obtain $\bar{y}$ with $\bar{y} - \phi(\bar{y}) = 0$. Then the vector $\bar{x} := F(-U\bar{y})$ solves the original problem (7).

We have therefore solved the problem (7) in the space $\mathbb{R}^n$, by solving the equation (10) in $\mathbb{R}^k$, involving (possibly repeated) solutions of the "easy" problem (8). In fact, for this structure we could have developed an equivalent procedure by applying a splitting argument, but that would have led to a subsidiary equation in $\mathbb{R}^n$, not $\mathbb{R}^k$. That equation could, with some thought, be rearranged to yield the same equation as we found directly by using the duality framework. A convenient feature of using composition duality, however, is that in addition to providing a mechanical procedure for producing (10), it naturally focuses our attention on the lower-dimensional space $\mathbb{R}^k$.

To illustrate this procedure, consider the problem in $\mathbb{R}^5$ in which $C = \mathbb{R}^5_+$ and $f(x) = 10x + a$, where $a = (\ 1\ \ -2\ \ 3\ \ -4\ \ 5\ )^T$, and where

$$U = (\ 1\ \ 1\ \ 1\ \ 1\ \ 1\ )^T, \qquad V = (\ 1\ \ -1\ \ 1\ \ -1\ \ 1\ )^T.$$

Then (7) is

$$0 \in \begin{pmatrix} 11 & -1 & 1 & -1 & 1 \\ 1 & 9 & 1 & -1 & 1 \\ 1 & -1 & 11 & -1 & 1 \\ 1 & -1 & 1 & 9 & 1 \\ 1 & -1 & 1 & -1 & 11 \end{pmatrix} x + \begin{pmatrix} 1 \\ -2 \\ 3 \\ -4 \\ 5 \end{pmatrix} + N_{\mathbb{R}^5_+}(x),$$

a linear complementarity problem in $\mathbb{R}^5$. By contrast, (8) is

$$z \in \begin{pmatrix} 10 & 0 & 0 & 0 & 0 \\ 0 & 10 & 0 & 0 & 0 \\ 0 & 0 & 10 & 0 & 0 \\ 0 & 0 & 0 & 10 & 0 \\ 0 & 0 & 0 & 0 & 10 \end{pmatrix} x + \begin{pmatrix} 1 \\ -2 \\ 3 \\ -4 \\ 5 \end{pmatrix} + N_{\mathbb{R}^5_+}(x),$$

whose solution is easily seen to be a vector x whose ith component is

$$x_i = (0.1)(z_i - a_i)_+,$$

where ζ_+ is defined to be $\max\{0, \zeta\}$. Therefore in this case (8) is trivial to solve.

To apply our procedure to this problem, we set up the equation (10), which in this case becomes

$$\begin{aligned} 0 &= y - \phi(y) \\ &= y - (0.1)[(-y-1)_+ - (-y+2)_+ + (-y-3)_+ - (-y+4)_+ + (-y-5)_+]. \end{aligned} \tag{11}$$

The expression on the right of (11) is a piecewise affine function of the real variable y, and it is therefore only a matter of locating an interval on which the function is affine and a zero occurs. By ordering the breakpoints of the max functions in (11) we see that the boundaries of the intervals are -5, -3, -1, 2, and 4 On the interval $[-1, 2]$ we have $F(-Uy) = (\ 0 \quad 2-y \quad 0 \quad 4-y \quad 0\ )^T$ and $y - \phi(y) = (0.1)(8y + 6)$, so that $\bar{y} = -0.75$. Then $\bar{x} = (\ 0 \quad .275 \quad 0 \quad .475 \quad 0\ )^T$, and one can easily check that this $\bar{x}$ solves (7).

To apply this procedure it is only necessary that we be able to solve (8) easily. Such a solution can always be found for important classes of problems, such as the form in our illustration, which involved a positive definite diagonal matrix. However, the applicability of the procedure is by no means restricted to such special cases. It is worth pointing out also that even in the special case we illustrated, C could have been taken to be an arbitrary polyhedral convex set; in that case the solution of (7) would involve projection on C, which could be done using a quadratic programming code.

A very well known special case of this approach can be recovered by taking $f(x) = Ax - a$, where A is an $n \times n$ nonsingular matrix and $a \in \mathbb{R}^n$, U and V to be n-vectors with $V^T A^{-1} U \neq -1$, and C to be $\mathbb{R}^n$ (so that the normal-cone operator is identically zero). Then (7) is the problem of solving $(A + UV^T)x = a$, while (8) is that of computing $A^{-1}(a + z)$. Applying our procedure, we find that

$$y - \phi(y) = (1 + V^T A^{-1} U)y - V^T A^{-1} a,$$

and therefore

$$\bar{y} = (1 + V^T A^{-1} U)^{-1} V^T A^{-1} a, \qquad \bar{x} = [A^{-1} - A^{-1} U (1 + V^T A^{-1} U)^{-1} V^T A^{-1}] a,$$

so that we have the well known Sherman-Morrison-Woodbury formula for inverting a rank-1 perturbation of a nonsingular matrix.

To conclude the development of this section, we consider briefly the question of conditions for solvability of (10) in the general case. One can give various such conditions depending on the assumptions one is willing to place on the elements of (7). Here is one set of conditions, which covers our example in particular and which also gives a method by which (10) may be numerically solved even if $k > 1$. We recall that a monotone operator $T : \mathbb{R}^n \to \mathbb{R}^n$ is termed *strongly monotone* with modulus ρ if for each pair of points (x_1, y_1) and (x_2, y_2) in the graph of T one has $\langle x_1 - x_2, y_1 - y_2 \rangle \geq \rho \|x_1 - x_2\|^2$. An application of the Schwarz inequality then shows that the inverse operator T^{-1} is Lipschitzian with modulus ρ^{-1}. Also, an operator $S : \mathbb{R}^n \to \mathbb{R}^n$ is said to be a *strong contraction* if it is Lipschitzian with a modulus less than 1. All norms used in the remainder of this paper are Euclidean.

Proposition 1. *Suppose that $f + N_C$ is a maximal monotone operator that is strongly monotone with modulus $\rho > 0$, and that $\|U\|\|V\| < \rho$. Then ϕ is a strong contraction with modulus $\mu := \rho^{-1}\|U\|\|V\|$, and (10) has a unique solution $\bar{y}$. Further, for any $y_0 \in \mathbb{R}^n$ the sequence defined by $y_{i+1} = \phi(y_i)$ converges to $\bar{y}$ at a rate that is at least linear, with $\|y_{i+1} - \bar{y}\| \leq \mu\|y_i - \bar{y}\|$ for each i.*

Proof. We know that $(f + N_C)^{-1}$ is maximal monotone, and is Lipschitzian with modulus ρ^{-1}. Under these conditions the domain of $(f + N_C)^{-1}$ is all of $\mathbb{R}^n$. Therefore for any v and w in $\mathbb{R}^k$ we have

$$\begin{aligned} \|\phi(v) - \phi(w)\| &= \|V^T F(-Uv) - V^T F(-Uw)\| \\ &\leq \|V\|\|(f + N_C)^{-1}(-Uv) - (f + N_C)^{-1}(-Uw)\| \\ &\leq \|V\|\rho^{-1}\|U\|\|v - w\|, \end{aligned}$$

so that ϕ is a strong contraction with modulus μ. Now the statements about the existence of $\bar{y}$ and the convergence of the approximating sequence $\{y_i\}$ follow from the contraction mapping principle. □

In this section we have shown how to apply composition duality to take advantage of special structure in a perturbed variational inequality, and thereby to simplify considerably its solution. We also showed that this procedure amounts to a natural extension of the well known Sherman-Morrison-Woodbury formula. Next we turn to the question of partial solution of structured variational inequalities. There also we shall obtain an extension of a known result, in that case the Schur complement known from matrix algebra.

4 Application: Generalized Schur complements

In this section we consider how a variational inequality with a certain structure, corresponding to block structure in linear algebra, may be partially solved. This partial solution yields formulas which, in the special case of linear equations, include the well known *Schur complement* [2], which has long been a useful device in linear algebra. We extend this idea to cases in which, instead of linear equations, we have various kinds of linear variational inequalities.

We begin with a simple formula involving a completely general multifunction between product spaces, and a simple linear operator, whose inverse is here intended in the sense of multifunctions, not of linear algebra. The formula can easily be stated in general spaces, but in keeping with the setting of this paper we give it here for the finite-dimensional case.

Proposition 2. *Let A be a multifunction from $\mathbb{R}^p \times \mathbb{R}^q$ to $\mathbb{R}^p \times \mathbb{R}^q$, and let S be the linear transformation from $\mathbb{R}^p \times \mathbb{R}^q$ to $\mathbb{R}^p$ defined by $S(x, y) = x$. For any pair (x, x^*) in $\mathbb{R}^p$ we have $x^* \in (S^*)^{-1}AS^{-1}(x)$ if and only if there is some $y \in \mathbb{R}^q$ such that*

$$\begin{pmatrix} x^* \\ 0 \end{pmatrix} \in A \begin{pmatrix} x \\ y \end{pmatrix}. \tag{12}$$

Proof. The statement that $x^* \in (S^*)^{-1}AS^{-1}(x)$ is equivalent to the statement that there exist u, u^*, y, and y^* such that

$$\begin{pmatrix} u^* \\ y^* \end{pmatrix} \in A \begin{pmatrix} u \\ y \end{pmatrix}, \qquad S^*x^* = \begin{pmatrix} u^* \\ y^* \end{pmatrix}, \qquad S \begin{pmatrix} u \\ y \end{pmatrix} = x.$$

But then $u^* = x^*$, $y^* = 0$, and $u = x$, so that (12) holds. If (12) holds, reverse the argument to obtain $x^* \in (S^*)^{-1}AS^{-1}(x)$. □

Proposition 2 shows that the operator $\bar{A} = (S^*)^{-1}AS^{-1}$ yields a partial solution of an inclusion involving the operator A: when we evaluate $\bar{A}$ at x, we obtain x^* such that for some y, (12) holds. Roughly speaking, we have solved the second part of the inclusion involving A, while holding the input to the first part at x; then $x^* = \bar{A}(x)$ is the associated output of the first part of the inclusion. One can conveniently describe this operation and others like it by using the partial-inverse notation of Pennanen [5], which builds on earlier work of Spingarn [11].

The operator $\bar{A}$ arises naturally in several useful operations. One of these is the application of composition duality to "factor out" part of a structured inclusion. As an example, consider the inclusion

$$0 \in \begin{pmatrix} g(x) \\ 0 \end{pmatrix} + A \begin{pmatrix} x \\ y \end{pmatrix}, \tag{13}$$

in which g is some nonlinear function and A is a multifunction having some special properties (for example it might be the sum of a linear transformation and a normal-cone operator). If we wish to solve (13), it may be easier to do so if we can temporarily neglect the variables y, and concentrate on the variables x appearing in the nonlinear function.

Using the S of Proposition 2, we rewrite (13) as

$$0 \in (S^*)(g)(S)\begin{pmatrix} x \\ y \end{pmatrix} + A\begin{pmatrix} x \\ y \end{pmatrix},$$

then apply composition duality to obtain the equivalent inclusion (in the lower-dimensional space $\mathbb{R}^p$)

$$0 \in g^{-1}(z) + (-S)A^{-1}(-S^*)(z).$$

Finally, we apply Attouch-Théra duality [1] in $\mathbb{R}^p$ to obtain another equivalent form, namely

$$0 \in g(x) - [(-S)A^{-1}(-S^*)]^{-1}(-x) = g(x) + (S^*)^{-1}AS^{-1}(x).$$

We have thus reduced the original inclusion in $\mathbb{R}^{p+q}$ to one in $\mathbb{R}^p$ that is equivalent in the sense that one is solvable if and only if the other is. Frequently we can solve the lower-dimensional inclusion and then, from information accumulated during that solution, easily recover a solution to the original one.

We used this device in [7] to reduce the dimensionality of a Walrasian economic equilibrium problem, at a considerable saving in solution time. That suggested the desirability of investigating some other manifestations of this particular functional form. As we now show, it does indeed occur in other useful contexts.

One of these contexts is that in which the multifunction A is taken to be a partitioned linear transformation with a particular block being nonsingular. In this case we can show that the partial solution operator yields the well known Schur complement. Recall that if

$$A = \begin{pmatrix} A_{11} & A_{12} \\ A_{21} & A_{22} \end{pmatrix}, \tag{14}$$

with A_{11} and A_{22} square and A_{22} nonsingular, then the *Schur complement* of A_{22} in A is defined to be the matrix

$$(A/A_{22}) = A_{11} - A_{12}A_{22}^{-1}A_{21}.$$

For a very interesting exposition of this idea with many useful formulas and references, see [2].

Corollary 1. *Let A and S be as in Proposition 2. If A is taken to be a linear transformation with matrix partitioned as in* (14) *and with A_{22} nonsingular, then $(S^*)^{-1}AS^{-1} = (A/A_{22})$.*

Proof. Choose any $x \in \mathbb{R}^p$ and let $x^* = (A/A_{22})(x)$. Let $y = -A_{22}^{-1}A_{21}(x)$. Then

$$A\begin{pmatrix} x \\ y \end{pmatrix} = \begin{pmatrix} A_{11}(x) + A_{12}(-A_{22}^{-1}A_{21})(x) \\ A_{21}(x) + A_{22}(-A_{22}^{-1}A_{21}(x) \end{pmatrix} = \begin{pmatrix} (A/A_{22})(x) \\ 0 \end{pmatrix} = \begin{pmatrix} x^* \\ 0 \end{pmatrix}.$$

Now apply Proposition 2 to show that $x^* \in (S^*)^{-1}AS^{-1}(x)$. Conversely, if $x^* \in (S^*)^{-1}AS^{-1}(x)$ then Proposition 2 shows that for some y, (12) holds. Using the form of A, we obtain $x^* = (A/A_{22})(x)$. □

Identification of the Schur complement with this particular case of the operator $(S^*)^{-1}AS^{-1}$ sometimes permits one to simplify the proofs of various properties. As an example, consider the fact that when the matrix A of Corollary 1 is nonsingular, the inverse of the Schur complement (A/A_{22}) is the upper left block (associated with the block indices $\{1, 1\}$) of A^{-1}. To see this, just note that

$$(A/A_{22})^{-1} = [(S^*)^{-1}AS^{-1}]^{-1} = SA^{-1}S^* = (A^{-1})_{11}.$$

One can obtain generalizations of the Schur complement by combining the block structure of Corollary 1 with a linear generalized equation of more general form. For example, consider the case in which A is given by

$$A\begin{pmatrix} x \\ y \end{pmatrix} = \begin{pmatrix} A_{11} & A_{12} \\ A_{21} & A_{22} \end{pmatrix}\begin{pmatrix} x \\ y \end{pmatrix} + \begin{pmatrix} a_1 \\ a_2 \end{pmatrix} + N_P(x) \times N_Q(y), \qquad (15)$$

where P and Q are polyhedral convex sets in $\mathbb{R}^p$ and $\mathbb{R}^q$ respectively. First suppose that $Q = \mathbb{R}^q$ and A_{22} is nonsingular. Then with

$$b = a_1 - A_{12}A_{22}^{-1}a_2,$$

one can apply Proposition 2 to verify that

$$\bar{A}(x) = (A/A_{22})(x) + b + N_P(x).$$

In this case the domain of the operator is all of P, and if we set $P = \mathbb{R}^p$ and a_1 and a_2 equal to zero, we recover the Schur complement as we would expect to.

Another interesting case is that occurring in the Walrasian equilibrium problem analyzed in [7]. There one has $a_1 = m$, $a_2 = 0$, A_{11} and A_{22} both zero, $A_{12} = M$ and $A_{21} = -M^T$, so that

$$A\begin{pmatrix} x \\ y \end{pmatrix} = \begin{pmatrix} 0 & M \\ -M^T & 0 \end{pmatrix}\begin{pmatrix} x \\ y \end{pmatrix} + \begin{pmatrix} 0 \\ m \end{pmatrix} + N_P(x) \times N_Q(y).$$

In addition, Q is restricted to be a cone. In that case, as noted in [8], one has

$$(S^*)^{-1}AS^{-1} = N_{P\cap Z},$$

where Z is the polyhedral convex set defined by

$$Z = \{z \mid M^T z - m \in Q^\circ\},$$

and where Q° is the polar cone of Q.

Thus in different special cases we can obtain apparently radically different forms for the operator $(S^*)^{-1}AS^{-1}$, ranging from a Schur complement to the normal cone of an intersection of two sets. This fact, together with the fact that this operator occurs naturally in formulas obtained by composition duality, suggests that its further study might be worthwhile.

5 Summary

In this paper we have briefly reviewed the method of composition duality as applied to variational inequality problems, and have then shown that it permits us to take advantage of structural features of these problems that might otherwise be difficult to exploit effectively. We showed how this could be done in two different cases, one being the solution of a variational inequality perturbed by a low-rank linear transformation, and the other being the exploitation of block structure in the linear part of a nonlinear variational inequality. In the first case, the transformation reduced the work involved in solving the problem to the repeated solution of a simpler variational inequality combined with the solution of an equation in a space whose dimensionality equaled the rank of the perturbation. In the second, the transformation produced an equivalent variational inequality in a space of lower dimension, involving an operator that was shown to reduce in various special cases to known operators including the Schur complement.

The composition duality setup has already proved to be of use both in enhancing computation and in increasing our understanding of the role of structure in the analysis of a problem. The eventual usefulness of this technique could be considerably enhanced by a more comprehensive body of knowledge about how it interacts in particular cases with the structure of the underlying problem. We have given a few examples here of how such a body of knowledge could be applied, and we hope these may interest the reader in contributing to its further development.

Acknowledgment. The research reported here was sponsored by the U. S. Army Research Office under grant number DAAG55-97-1-0324. The U. S. Government has certain rights in this material, and is authorized to reproduce and distribute reprints for Governmental purposes notwithstanding any copyright notation thereon. The views and conclusions contained herein are those of the author and should not be interpreted as necessarily representing the official policies or endorsements, either expressed or implied, of the sponsoring agency or the U. S. Government.

References

[1] H. Attouch and M. Théra (1996), "A general duality principle for the sum of two operators," *Journal of Convex Analysis*, 3, 1–24.

[2] R. W. Cottle (1974), "Manifestations of the Schur complement," *Linear Algebra and Its Applications*, 8, 189–211.

[3] M. C. Ferris and J.-S. Pang (1997), "Engineering and economic applications of complementarity problems," *SIAM Review*, 39, 669–713.

[4] P. T. Harker and J.-S. Pang (1990), "Finite-dimensional variational inequality and nonlinear complementarity problems: A survey of theory, algorithms, and applications," *Mathematical Programming*, 48, 161–220.

[5] T. Pennanen (1998), "Dualization of generalized equations of maximal monotone type," Manuscript, Department of Mathematics, University of Washington, Seattle, WA.

[6] S. M. Robinson (1983), "Generalized equations," In A. Bachem, M. Grötschel, and B. Korte, editors, *Mathematical Programming: The State of the Art, Bonn 1982*, Springer-Verlag, Berlin, 346–367.

[7] S. M. Robinson (1998), "A reduction method for variational inequalities," *Mathematical Programming*, 80, 161–169.

[8] S. M. Robinson (1999), "Composition duality and maximal monotonicity," *Mathematical Programming*, 85, 1–13.

[9] R. T. Rockafellar (1970), *Convex Analysis*, Princeton University Press, Princeton, NJ.

[10] R. T. Rockafellar (1974), *Conjugate Duality and Optimization*, Number 16 in CBMS Regional Conference Series in Applied Mathematics, Society for Industrial and Applied Mathematics, Philadelphia, PA.

[11] J. E. Spingarn (1983), "Partial inverse of a monotone operator," *Applied Mathematics and Optimization*, 10, 247–265.

[12] G. Stampacchia (1964), "Formes bilinéaires coercitives sur les ensembles convexes," *Comptes Rendus de l'Académie des Sciences de Paris*, 258, 4413–4416.

[13] G. Stampacchia (1968), "Variational inequalities," In *Theory and Applications of Monotone Operators*, Proceedings of a NATO Advanced Study Institute held in Venice, Italy, June 17–30, 1968, Edizioni "Oderisi," Gubbio, Italy.

Extended Nonlinear Programming

R. Tyrrell Rockafellar (rtr@math.washington.edu)
Dept. of Mathematics, Box 354350
University of Washington, Seattle, WA 98195-4350

Abstract

Shortcomings of the conventional problem format in nonlinear programming suggest the need for a broader model that features composite terms. Such a model, promoting better representation of the problem structures found in applications, can be adopted with no real sacrifice in computing practicality.

Keywords: Nonlinear programming, extended linear-quadratic programming, composite optimization, optimality conditions, quadratic approximations

1 Introduction

The basic problem in nonlinear programming, and for that matter in all of finite-dimensional optimization, is usually explained as having the form

$$(\mathcal{P}_0) \qquad \text{minimize } f_0(x) \text{ subject to } f_i(x) = \begin{cases} \leq 0 & \text{for } i = 1, \ldots, s, \\ = 0 & \text{for } i = s+1, \ldots, m. \end{cases}$$

Sometimes a geometric constraint like $x \in X$ is included, where X is an orthant or more generally a box, i.e., a product of closed intervals (whether bounded or unbounded). For purposes such as the introduction of Lagrange multipliers, however, X is suppressed in favour of additional equalities or inequalities.

This description is by now so ingrained that many people accept it without question. It is thought to provide a good statement of what optimization is about, and to offer the right model for setting up applications. In that role, it has become the language of communication between the optimization community and its clients. But is this really the best we can do?

G. Di Pillo and F. Giannessi (eds.), Nonlinear Optimization and Related Topics, 381–399.

All of optimization revolves around minimizing or maximizing some function over some feasible set. The issue is what framework is most advantageous for specifying that function and set, in order to promote effective computation and analysis along with good modeling in applications. From that angle, an imbalance in $(\mathcal{P}_0)$ is apparent. The format demands thinking about feasibility in a structured way but offers no suggestions for representing structure also in the objective. A user is just expected to produce a finite family of functions and to designate which give the objective, the equations, and the inequalities.

History has a lot to do with this. The prototype in modern optimization was linear programming. In a linear programming problem, the constraint and objective functions are affine, and there is little more to be said. Problem structure in linear programming has typically been viewed in terms of the sparsity patterns of a coefficient matrix. Nonlinear programming started out as a sort of negation of linear programming, with functions merely no longer affine. In the early days no need was felt to be more detailed than that, and no other guidelines were developed.

People took for granted that, for practicality, all the functions in a problem should be smooth (i.e., continuously differentiable) to whatever degree might be helpful. This was a much more serious restriction than may have been realized, though, especially in applying the same standard to the objective as to the constraints. Certain kinds of nonsmoothness are common in expressing objective structure, and nowadays it's known how to handle them quite simply through composite terms which allow the numerical utilization of *underlying* smoothness in the problem data to go forward.

Even in a context of smoothness, it's easy to come up with applications where composite structure in the objective function provides special support in computation. In nonlinear least-squares problems, the expression to be minimized has the form

$$f_0(x) = \lambda_1 f_{01}(x)^2 + \cdots + \lambda_r f_{0r}(x)^2 \tag{1}$$

with positive weights λ_j. This function f_0 comes from composing a smooth mapping $x \mapsto (f_{01}(x), \ldots, f_{0r}(x))$ with a convex function $(u_{01}, \ldots, u_{0r}) \mapsto \lambda_1 u_{01}^2 + \cdots + \lambda_r u_{0r}^2$. The numerical methodology revolves around linearizations of the mapping.

For nonsmooth objective structures, a smooth mapping is often in the background as well, but the conventional format in $(\mathcal{P}_0)$ doesn't permit it to be accessed directly. *Nonsmoothness of functions has to be recast as nonsmoothness of sets by way of inequality constraints.* A typical example is "min-max" optimization, where the goal is to minimize an expression like

$$f_0(x) = \max\{f_{01}(x), \ldots, f_{0r}(x)\}. \tag{2}$$

If f_0 has this form in $(\mathcal{P}_0)$, an additional variable x_0 has to be introduced. The task is translated into the minimization of $\tilde{f}_0(x_0, x) = x_0$ subject to $0 \geq \tilde{f}_{0j}(x_0, x) = f_{0j}(x) - x_0$ for $j = 1, \ldots, r$ and the other constraints already in $(\mathcal{P}_0)$. Note that f_0 in (2) is obtained by composing a smooth mapping $x \mapsto (f_{01}(x), \ldots, f_{0r}(x))$ with the convex, piecewise linear function $(u_{01}, \ldots, u_{0r}) \mapsto \max\{u_{01}, \ldots, u_{0r}\}$.

A similar maneuver would be triggered by a seemingly innocent shift of squares to absolute values in (1). If the expression to be minimized in $(\mathcal{P}_0)$ is

$$f_0(x) = \lambda_1|f_{01}(x)| + \cdots + \lambda_r|f_{0r}(x)|, \tag{3}$$

one is obliged to introduce extra variables x_{0j} and minimize $\tilde{f}_0(x_{01}, \ldots, x_{0r}, x) := \lambda_1 x_{01} + \cdots + \lambda_r x_{0r}$ subject to the given constraints on x and the additional conditions $f_{0j}(x) - x_{0j} \leq 0$ and $-f_{0j}(x) - x_{0j} \leq 0$ for $j = 1, \ldots, r$. But why should such a drastic reformulation be required for the f_0 in (3), where none was needed for the f_0 in (1)? The two objective functions have almost the same form, except that in (3) the smooth mapping $x \mapsto (f_{01}(x), \ldots, f_{0r}(x))$ is composed with a convex function $(u_{01}, \ldots, u_{0r}) \mapsto \lambda_1|u_{01}| + \cdots \lambda_r|u_{0r}|$ that's piecewise linear.

Penalty expressions offer further illustration. What if we wish to modify $(\mathcal{P}_0)$ by replacing the constraint $f_1(x) \leq 0$ by a objective term, say a "linear" penalty term,

$$f_1(x) \leq 0 \qquad \longrightarrow \qquad \rho_1 \max\{0, f_1(x)\} \tag{4}$$

with penalty parameter $\rho_1 > 0$? Again, this doesn't fit the conventional format directly. One has to resort to introducing a new variable ξ_1 and passing to the problem of minimizing $\tilde{f}_0(x, \xi_1) = f_0(x) + \rho_1\xi_1$ subject to $f_1(x) - \xi_1 \leq 0$, $-\xi_1 \leq 0$, and the remaining constraints in $(\mathcal{P}_0)$ for $i = 2, \ldots, m$. Objective structure has once more been converted to constraint structure in order to conform to the model that people have gotten accustomed to taking as standard. But the penalty expression in (4) is simply the composition of a smooth function with a convex, piecewise linear function, so one might hope for an approach that's less disruptive.

Penalties need not be just "linear," of course. Expressions with nonlinear pieces often come up, and the nonsmoothness may then be just of second order. For instance, an equation $f_i(x) = 0$ in $(\mathcal{P}_0)$ could be replaced by an objective term $\theta_i(f_i(x))$ in which θ_i is a convex penalty function that's quadratic on an interval around 0 but switches to affine expressions outside of that interval—without any jumps in the first derivative, only in the second derivative. Such composite terms arise in particular in "robust statistical" modifications of least-squares problems which aim at reducing the influence of outliners. Observe, though, that in this situation it's not evident any more how to squeeze the problem into the mold of $(\mathcal{P}_0)$ with functions that are all $\mathcal{C}^2$, if the given f_i's are all $\mathcal{C}^2$.

These difficulties aren't only technical. They have important consequences for the mathematical modeling that goes into optimization. People are turned away from problem formulations that might be very good at representing the right structure and are forced instead into something less apt and perhaps inappropriate. This is an issue that deserves more attention from the optimization community than it has so far received.

The distorting effect of a narrow format is most serious perhaps in the emphasis of $(\mathcal{P}_0)$ on constraints that have to be satisfied with no room for penalty or reward.

In practice, the constraints encountered in optimization models can be divided into two categories, "hard" and "soft." In the "hard" category are constraints that record immutable requirements on the variables involved, such as nonnegativity and defining relationships. In the "soft" category are constraints that express properties deemed desirable, but which might be subject eventually to trade-off with other properties, or with costs. There are parameters in such constraints that are assigned wishful values, which might later be altered if the results of optimization aren't satisfactory. In many circumstances it might be better to replace soft constraints by objective terms that directly model a trade-off or gray areas between having an inequality simply be fulfilled or not fulfilled. The statement of $(\mathcal{P}_0)$ discourages that, however.

Stochastic programming gives a very telling example. Beginners who have grown up with $(\mathcal{P}_0)$ tend to set up a stochastic programming model with exact constraints, not realizing how costly that can be. Because of uncertainty, a constraint that's supposed to be satisfied at a future stage, but is influenced by present decisions, may be handled much better by introducing penalties for its violation than by insisting that earlier actions be foolproof.

By now, it's well understood by theorists that a modeling distinction between objective and constraints is artificial. While one can take the position that any optimization problem comes down to minimizing some function over some set, one can equally well take the position that any optimization problem in n real variables consists, in principle, just of minimizing some function f over $\mathbb{R}^n$—if f is allowed to take on ∞. There's no real difference between including the condition $f_1(x) \leq 0$ in $(\mathcal{P}_0)$ as an *explicit* constraint or representing it *implicitly* in the objective by a penalty term $\theta_1(f_1(x))$ in which $\theta_1(u_1) = 0$ for $u_1 \leq 0$ but $\theta_1(u_1) = \infty$ for $u_1 > 0$.

To promote good modeling, we ought to explain optimization to the world in a framework in which transitions between such infinite penalties and alternative finite penalties are easy and attractive. Of course, the framework should also be one that's just as conducive, or even more conducive, to capturing and utilizing smoothness as the conventional one. And, as long as a list is being made, we might ask for the framework to be better than $(\mathcal{P}_0)$ at making use of what's now known about dualization and Lagrange multipliers. For instance, it should be able to handle a box constraint $x \in X$ without conversion to linear inequality constraints. It should furnish a way around the current obstacles to dualizing quadratic programming problems.

All this is possible with surprisingly little adjustment. By accepting the notion of composite terms as a basic modeling tool, one can arrive at a problem format that's much more flexible than $(\mathcal{P}_0)$ and much richer in ways of indicating structure, yet with essentially no sacrifice in computational practicality. Through only a small investment in learning how some common composite terms are dualized, in order to associate them with Lagrange multipliers, one can achieve a new level of standardization in which problems can be inputted to software without first reformulating them. An automatic interface can be supplied for invoking numerical packages as they now exist, but the door is opened also to further algorithmic developments, designed to take full advantage of the special structures that might be represented.

2 Format with Composite Modeling

As a first idea of the kind of extension that might be made—but one which we will end up simplifying—it could be proposed to replace $(\mathcal{P}_0)$ by

$$(\mathcal{P}_1) \qquad \text{minimize } \theta_0(F_0(x)) + \theta_1(F_1(x)) + \cdots + \theta_q(F_q(x)) \text{ over } x \in X,$$

where X is a subset of $\mathbb{R}^n$ and each composite term involves a smooth mapping $F_l : \mathbb{R}^n \to \mathbb{R}^{d_l}$ and a function θ_l on $\mathbb{R}^{d_l}$ that *might take on* ∞. The effective domains $D_l := \{u_l \in \mathbb{R}^{d_l} \mid \theta_l(u_l) < \infty\}$ would give constraints that are implicit in $(\mathcal{P}_1)$: a point x is feasible if and only if $x \in X$ and $F_l(x) \in D_l$ for $l = 1, \ldots, q$. As a special case, a particular θ_l might be just the *indicator* of D_l, i.e., the function that vanishes on D_l but has the value ∞ everywhere outside of D_l.

Much could be done with this fully composite model, but it seems too great a leap into inscrutability to be sold easily to users of optimization. It's too far from the conventional model for instant comparisons, and it appears to insist on too many things being specified.

What we propose instead, therefore, is a focus on the following problem format, which we speak of as *extended nonlinear programming*, or ENLP:

$$(\mathcal{P}) \qquad \text{minimize } f_0(x) + \theta(f_1(x), \ldots, f_m(x)) \text{ over } x \in X.$$

The feasible solutions to $(\mathcal{P})$ are the vectors $x \in X$ with $(f_1(x), \ldots, f_m(x)) \in D$, where

$$D = \{u \in \mathbb{R}^m \mid \theta(u) < \infty\}. \tag{5}$$

This covers the conventional NLP format in $(\mathcal{P}_0)$ as the case of $X = \mathbb{R}^n$ and

$$\theta(f_1(x), \ldots, f_m(x)) = \theta_1(f_1(x)) + \cdots + \theta_m(f_m(x)), \tag{6}$$

with the functions $\theta_1, \ldots, \theta_s$ being the indicator of $(-\infty, 0]$ but $\theta_{s+1}, \ldots, \theta_m$ being the indicator of $\{0\}$. It also has the virtue of indicating at once how exact constraints can be replaced by expressions imposing penalties or rewards, since all that's needed is the replacement of indicator functions in (6) by other functions θ_i.

In fact, $(\mathcal{P})$ also encompasses the general composite model $(\mathcal{P}_1)$ and thus all the examples that we've been discussing. This is because the functions $f_0, f_1, \ldots, f_m$ don't *have* to be identified one-by-one with the functions having the same symbol in $(\mathcal{P}_0)$, if that's not convenient. In $(\mathcal{P})$, we could choose to take $f_0 \equiv 0$, and then the entire objective is a composite expression $\theta(f_1(x), \ldots, f_m(x))$. It's a small step from that to specializing into a sum of composite terms as in $(\mathcal{P}_1)$.

For an illustration at an intermediate level, suppose the aim is to minimize a "max" expression like the one in (2) subject to equations and inequalities like those in $(\mathcal{P}_0)$. This amounts to minimizing

$$\theta_0(f_{01}(x), \ldots, f_{0r}(x)) + \theta_1(f_1(x)) + \cdots + \theta_m(f_m(x)) \tag{7}$$

over $X = \mathbb{R}^n$ with $\theta_0(u_{01}, \ldots, u_{0r}) = \max\{u_{01}, \ldots, u_{0r}\}$ and the other θ_i's indicators or $(-\infty, 0]$ of $\{0\}$, as already seen. We then have an extended nonlinear programming problem of elementary type in which the f_0 in $(\mathcal{P})$ (as opposed to the f_0 in (2)) is just the constant function 0.

In like manner, nonlinear programming problems $(\mathcal{P}_0)$ with objective functions of the types in (1) or (3) can be handled as extended nonlinear programming problems with objective functions as in (7) but through different choices of θ_0.

We haven't yet been clear about the assumptions that should go along with the designation of $(\mathcal{P})$ as a problem of extended nonlinear programming. As in conventional nonlinear programming, we want the functions f_i to be smooth, of course, even with continuous second derivatives if needed. That's no longer a real restriction, but rather a modeling choice. The philosophy is that $f_0, f_1, \ldots, f_m$ furnish all the smoothness we wish to build into the problem for the purpose of generating Taylor expansions and other classical approximations. Anything nonsmooth is to be captured through the specification of θ.

What should be assumed, though, about θ and for that matter about the set X in $(\mathcal{P})$? Optimization theory is capable now of treating very general θ and X, but our goal here isn't in that direction. Instead, we want a workable compromise between generality and simplicity, moreover one that concentrates on a systematic and elementary way of specifying θ and X. That turns out to be much easier than might be expected, although the solution requires a bit of explaining.

Assumptions. *In an extended nonlinear programming problem $(\mathcal{P})$, it will be supposed that*

(A1) *the functions $f_0, f_1, \ldots, f_m$ are smooth,*

(A2) *the set X in $\mathbb{R}^n$ is nonempty and polyhedral (convex), and*

(A3) *the function θ on $\mathbb{R}^m$ is convex, proper and lower semicontinuous, furthermore representable in the form*

$$\theta(u) = \sup_{y \in Y}\{y{\cdot}u - k(y)\} \tag{8}$$

by means of a nonempty polyhedral set Y in $\mathbb{R}^m$ and a smooth function k that is convex on Y.

The smoothness of the functions f_i has already been addressed. The polyhedral convexity of X may seem unnecessarily limiting, but it's good enough for many purposes. It covers nonnegativity constraints, upper and lower bounds on variables, and also situations where one wishes to minimize over a linear subspace of $\mathbb{R}^n$, with $\mathbb{R}^n$ itself as a special case, since linear subspaces are polyhedral sets in particular. Also covered are cases of linear equations where certain variables have to add to 1, or where defining relationships are expressed between various quantities.

There's no controversy over how to specify a polyhedral set. Everyone knows how to do that in more than one way, as convenience dictates. Anything nonpolyhedral could of course be handled in $(\mathcal{P})$ by other means.

It's the assumption on θ that may seem mysterious. The mystery will dissipate with an appeal to elementary convex analysis.

Taking θ to be convex and having ∞ as a possible value isn't itself controversial. (An extended-real-valued function is *proper* if it doesn't take on $-\infty$ and isn't just the constant function ∞.) In all the examples brought out so far, the outer function in the composite term was convex. One can go a very long way with that. Lower semicontinuity of θ is equivalent to the epigraph of θ being closed and is a minor technical requirement. The rest of (A3) is what raises eyebrows.

An expression of θ as in assumption (A3) will be called a *dualizing representation*. It will soon be seen that, in all the examples we've encountered and a vast array of others, such an expression of θ is indeed available. Dualizing representations will play a big role with respect to Lagrange multipliers. There's much more to them than that, however.

Many functions θ of interest are only piecewise smooth and therefore difficult to describe directly, but a dualizing representation (8) furnishes an alternative, if indirect, description that's actually quite simple and easy to work with, all the more so once the implications of it are understood. In order to specify θ through a dualizing representation, all one has to specify is a polyhedral set Y and a smooth function k that's convex on Y. As it turns out, it would be enough most of the time to have k be quadratic, hence specified in terms of an $m \times m$ positive-semidefinite matrix (in some cases just the 0 matrix).

This is a crucial observation because it reveals that the structural features so essential to the nonsmooth functions θ one wants to use in modeling don't have to be an impediment in practice. *To specify $(\mathcal{P})$, all one has to do is to specify the smooth functions $f_0, f_1, \ldots, f_m$ on $\mathbb{R}^n$ and k on $\mathbb{R}^m$ along with two polyhedral sets $X \subset \mathbb{R}^n$ and $Y \subset \mathbb{R}^m$.* That's no harder than specifying a conventional nonlinear programming problem $(\mathcal{P})$, by it does require an appreciation of how θ corresponds to Y and k. The results coming next address that concern, first in general terms and then through examples.

Proposition 1. *If a function θ has a dualizing representation as in (A3), it automatically satisfies the requirements of being convex, proper and lower semicontinuous. Moreover Y and k, at least in its restriction to Y, can be recovered then from θ by*

$$Y = \{y \in \mathbb{R}^m \mid \theta^*(y) < \infty\}, \qquad k(y) = \theta^*(y) \text{ for } y \in Y,$$

where $\theta^*(y) = \sup_u \{u{\cdot}y - \theta(y)\}$. (This θ^* is the convex function conjugate to θ.)

Proof. Define $\psi(y)$ to equal $k(y)$ for $y \in Y$ but ∞ for $y \notin Y$. Then ψ is convex function on $\mathbb{R}^m$ that's proper and lower semicontinuous. According to (8), the conjugate convex function ψ^* is θ. It follows then that $\theta^* = \psi^{**} = \psi$, as is well known in convex analysis. □

On the basis of Proposition 1 there is a *one-to-one* correspondence between the functions θ admitted in assumption (A3) and the pairs (Y, k) described there, except that only the values of k on Y count. We show now how this correspondence can be broken down to a term by term description when θ is to some degree separable.

Proposition 2. *Suppose that $\theta(u) = \theta_1(u_1) + \cdots + \theta_q(u_q)$ for $u = (u_1, \ldots, u_q)$ with $u_l \in \mathbb{R}^{d_l}$. Let each θ_l have a dualizing representation in terms of a polyhedral set $Y_l \subset \mathbb{R}^{d_l}$ and a smooth function k_l that is convex on Y_l. Then θ has a dualizing representation with respect to $y = (y_1, \ldots, y_q)$ and $y_l \in \mathbb{R}^{d_l}$ in terms of*

$$Y = Y_1 \times \cdots \times Y_q, \qquad k(y) = k_1(y_1) + \cdots + k_q(y_q). \tag{9}$$

Proof. This is just of matter of recognizing that

$$\sum_{l=1}^{q} \sup_{y_l \in Y_l} \{u_l \cdot y_l - k_l(y_l)\} = \sup_{y \in Y} \{u \cdot y - k(y)\}$$

under (9). Note that the components u_l and y_l can be vectors or, in the case of dimension $d_l = 1$, merely scalars. □

Let's proceed now with some examples of dualizing representations, taking the cue from Proposition 2 that it's enough to consider individual terms. One-dimensional terms are a good place to begin.

We've seen that inequality constraints $f_i(x) \leq 0$ correspond to objective terms $\theta_i(f(x))$ in which $\theta_i(u_i)$ is 0 when $u_i \leq 0$ but ∞ when $u_i > 0$. The dualizing representation for such θ is obtained by taking $k_i \equiv 0$ on $Y_i = [0, \infty)$. For equality constraints $f_i(x) = 0$, we instead have $\theta_i(u_i)$ equal to 0 for $u_i = 0$ but ∞ everywhere else, and then $k_i \equiv 0$ on $Y_i = (-\infty, \infty)$. A linear penalty term as in (4) with $\theta_i(u_i) = \rho_i \max\{0, u_i\}$ comes out, however, as corresponding to $k_i \equiv 0$ on $Y_i = [0, \rho_i]$. Likewise, such a penalty term for an equality constraint, with $\theta_i = \rho_i |u_i|$, has $k_i \equiv 0$ on $Y_i = [-\rho_i, \rho_i]$.

It's valuable to observe that in these cases, where the function k_i doesn't really enter, the effect of replacing a classical constraint by a linear penalty term is to replace an unbounded interval Y_i by a truncated one. We'll see later that this corresponds to introducing bounds on Lagrange multipliers. In general, for $k_i \equiv 0$ on $Y_i = [\sigma_i, \rho_i]$ one gets $\theta_i(u_i) = \rho_i u_i$ for $u_i \geq 0$, and $\theta_i(u_i) = \sigma_i u_i$ for $u_i \leq 0$, regardless of the signs of σ_i and ρ_i.

What can happen with $k_i \not\equiv 0$? A simple case is a term like those in the least-squares setting of (1), with $\theta_{0j}(u_{0j}) = \lambda_j u_{0j}^2$, $\lambda_j > 0$. The dualizing representation is obtained then with $k_{0j}(y_{0j}) = (1/4\lambda_j) y_{0j}^2$ on $Y_{0j} = (-\infty, \infty)$. If we kept the same function $k_{0j}(y_{0j})$ but truncated the interval to $Y_{0j} = [-\rho_j, \rho_j]$, however, we would get the piecewise linear-quadratic function θ_{0j} that has the formula $\lambda_j u_{0j}^2$ on the interval of u_{0j} values where the derivative of this term is between $-\rho_j$ and ρ_j, i.e., where $\alpha_j \leq u_{0j} \leq \beta_j$ with $\alpha_j = -\rho_j/2\lambda_j$ and $\beta_j = \rho_j/2\lambda_j$, but is extrapolated linearly outside that interval, with formula $\lambda_j\alpha_j^2 - \rho_j[u_{0j} - \alpha_j]$ to the left and $\lambda_j\beta_j^2 + \rho_j[u_{0j} - \beta_j]$ to the right. (These affine pieces take off as tangents from the original quadratic graph.) Note that this is just the kind of function of interest in "robust statistics." The expression for θ_{0j} is readily generalized to the case of $k_{0j}(y_{0j}) = (1/4\lambda_j)y_{0j}^2$ on an arbitrary interval $Y_{0j} = [\sigma_j, \rho_j] \subset (-\infty, \infty)$. Terms of such type come up in augmented Lagrangians.

Just with these very simple, one-dimensional dualizing representations, we have already taken care of all the composite terms mentioned in earlier examples except for the "max" term in (2). That requires an appeal to higher dimensions: we get

$$\begin{aligned} &\theta(u_{01}, \ldots, u_{0r}) = \max\{u_{01}, \ldots, u_{0r}) \;\text{ for }\; k \equiv 0 \;\text{ on} \\ &Y = \{(y_{01}, \ldots, y_{0r}) \,|\, y_{0j} \geq 0,\; y_{01} + \cdots + y_{0r} = 1\}. \end{aligned} \tag{10}$$

The set Y is polyhedral, as stipulated in (A3). In another example, if Y is any polyhedral cone in $I\!R^m$ and $k(y) = \frac{1}{2}|y|^2$ for the Euclidean norm $|y|$, then one has $\theta(u) = \frac{1}{2}d(u, K)^2$ for the polar cone $K = Y^*$, with $d(u, K)$ denoting the distance of u from K.

All the examples so far fit the pattern of k being a purely quadratic convex function, perhaps identically 0. As seen from Proposition 2, if that holds for the individual terms in θ, whatever they might be, it also holds for θ as a whole. This case is particularly deserving of attention, and we give it special notation:

$$\theta = \theta_{YQ} \quad\text{when}\quad \theta(u) = \sup_{y \in Y}\{u{\cdot}y - \tfrac{1}{2}y{\cdot}Qy\} \tag{11}$$

for a polyhedral set $Y \subset I\!R^m$ and a symmetric, positive-semidefinite matrix $Q \in I\!R^{m\times m}$ (possibly $Q = 0$). When Y is a box and Q is diagonal, one has a decomposition of θ_{YQ} into one-dimensional terms.

Note from Proposition 1 that although Y and the values of the quadratic form $y{\cdot}Qy$ for $y \in Y$ can be recovered from θ in (11), this wouldn't be enough to pin down Q uniquely unless Y has nonempty interior. Situations where Y has empty interior do arise, as for instance in (10).

Proposition 3. *Any function θ of the form θ_{YQ} in (11) is piecewise linear-quadratic, in the sense that its effective domain D in (5) is polyhedral (in particular closed and convex) and can be partitioned into a finite collection of polyhedral subsets, with respect to each of which the formula for θ is a polynomial of degree at most 2.*

Indeed, D is a polyhedral cone which is polar to $\{y \,|\, Qy = 0,\; Y + y \subset Y\}$. For D to be the whole space (so that θ is finite everywhere), it is necessary and sufficient therefore that the latter cone (likewise polyhedral) contain no $y \neq 0$.

Proof. For the first assertion, see Theorem 11.14(b) of [1]. The rest was proved in Proposition 2.4 of [2]. □

It follows for instance that when Y is bounded or Q is positive-definite, the feasible set in $(\mathcal{P})$ is the polyhedral set X. Otherwise the condition $(f_1(x), \ldots, f_m(x)) \in D$ could come into play in determining feasibility. Because D is a polyhedral cone, this condition can in theory be expressed by constraints $0 \geq \sum_{i=1}^{m} \alpha_{ki} f_i(x) = g_k(x)$ for a collection of coefficient vectors $(\alpha_{k1}, \ldots, \alpha_{km})$ (chosen to generate the cone polar to D in Proposition 3). This sheds light on the nature of the feasible set in an extended nonlinear programming problem $(\mathcal{P})$, revealing that it's neither more nor less general in principle than the kind of feasible set in a conventional problem $(\mathcal{P}_0)$—but that's also somewhat beside the point. The focus isn't merely on constraints any more, and because of the possibility of D having to be broken down into many pieces to get a direct description of θ, one may have to rely anyway on the dual description of θ furnished by Y and Q.

If we concentrate on functions θ of form θ_{YQ} in (11) and at the same time restrict f_0 to be quadratic and $f_1, \ldots, f_m$ to be affine in $(\mathcal{P})$ we get *extended linear-quadratic programming*, or ELQP for short:

$$(\mathcal{Q}) \qquad \text{minimize } c{\cdot}x + \tfrac{1}{2}x{\cdot}Px + \theta_{YQ}(b - Ax) \text{ over } x \in X,$$

where $A \in I\!R^{m \times n}$ and $P \in I\!R^{n \times n}$ (symmetric). This kind of model goes back to Rockafellar and Wets [3], where it was introduced for the sake of penalty modeling and algorithm development in stochastic programming. The topic was expanded in [2], where many special cases of ELQP were worked out and applications were made to continuous-time optimal control. Other aspects of ELQP methodology and applications can be found in [4], [5], [6], along with [1].

The general nature of $(\mathcal{Q})$, in contrast to conventional quadratic programming, can be perceived from the case where Y is a box and Q is a diagonal matrix, so that

$$\theta_{YQ}(b - Ax) = \theta_1(b_1 - a_1{\cdot}x) + \cdots + \theta_m(b_m - a_m{\cdot}x)$$

for the components b_i of b and the vectors a_i giving the rows of A. Our discussion of such one-dimensional terms shows that they can stand for piecewise linear-quadratic penalty expressions as well as standard linear constraints. Of course, "max" expressions can also be represented, by following the pattern in (10). When $P = 0$ and $Q = 0$ in $(\mathcal{Q})$, one has *extended linear programming*, ELP, where X can impose upper or lower bounds on the variables while θ allows for linear penalty.

An important advantage of ELQP over ordinary QP is that problem $(\mathcal{Q})$ can be dualized without difficulty to get another ELQP problem, as long as P is positive-semidefinite, so that convexity prevails in $(\mathcal{Q})$. Similarly, an ELP problem dualizes to an ELP problem. This will be explained near the end of the next section.

3 Extended Lagrangian and Multiplier Rule

The dualizing representation in assumption (A3) is the route to the *Lagrangian function* we associate with an extended nonlinear programming function $(\mathcal{P})$, namely

$$L(x,y) = f_0(x) + y_1 f_1(x) + \cdots + y_m f_m(x) - k(y) \quad \text{for } x \in X,\ y \in Y. \tag{12}$$

The sets X and Y are regarded as an integral part of the specification of L. Obviously $(\mathcal{P})$ is completely determined by its Lagrangian in this sense, because the expression being minimized over X in $(\mathcal{P})$ is

$$f(x) = \sup_{y \in Y} L(x,y) = f_0(x) + \theta(f_1(x) \ldots, f_m(x)). \tag{13}$$

The main difference between this extended Lagrangian for $(\mathcal{P})$ and the one associated with a conventional nonlinear programming problem $(\mathcal{P}_0)$ is the presence of the term $-k(y)$ (which could however be 0) and a general polyhedral set Y instead of just the special cone $\mathbb{R}^s_+ \times \mathbb{R}^{m-s}$ that classically expresses the sign requirements on Lagrange multipliers for constraints $f_i(x) \le 0$ or $f_i(x) = 0$. Here, Y will express requirements on Lagrange multipliers more broadly.

There's also a difference in the presence of a set X, where for $(\mathcal{P}_0)$ one would have $X = \mathbb{R}^n$. This is less important, though, because a set X has often been brought into discussions of Lagrangians. If one goes to the original paper of Kuhn and Tucker [7], say, one finds a treatment of how full or partial nonnegativity of x can be represented in that way, and what it means for the statement of first-order optimality conditions. That simple modification has also been common in linear and quadratic programming.

Beyond such special instances of X, little has been made of how to adapt Lagrange multiplier rules to a constraint $x \in X$, at least within the optimization community at large, although theorists have long had answers. This is true even though the case of a box X is very common in numerical optimization. Actually the adaptation is quite easy, and it's worth looking at because the same ideas are need for understanding how to adapt to multiplier vectors $y \in Y$ for nonclassical Y.

The only notion that's needed is that of the *normal cone* $N_X(\bar{x})$ to a convex set X at one of its points $\bar{x}$, as introduced in convex analysis [8], [1]:

$$v \in N_X(\bar{x}) \iff \bar{x} \in X \text{ and } v \cdot [x - \bar{x}] \le 0 \text{ for all } x \in X. \tag{13}$$

This is polar to the *tangent cone* $T_X(\bar{x})$, which for polyhedral X consists of the origin and all rays of the form $\{\tau[x - \bar{x}] \mid \tau \ge 0\}$ generated by points $x \ne \bar{x}$ in X. For a box X, the description of $N_X(\bar{x})$ is especially simple. Then X is a product of closed intervals X_j which constrain the components x_j of x, and we have

$$\begin{aligned}
&(v_1, \ldots, v_n) \in N_X(\bar{x}_1, \ldots, \bar{x}_n) \quad \text{for } X = X_1 \times \cdots \times X_n \\
&\iff \begin{cases} v_j = 0 & \text{when } \bar{x}_j \text{ is an interior point of } X_j, \\ v_j \ge 0 & \text{when } \bar{x}_j \text{ is the right endpoint (only) of } X_j, \\ v_j \le 0 & \text{when } \bar{x}_j \text{ is the left endpoint (only) of } X_j. \end{cases}
\end{aligned} \tag{14}$$

When $\bar{x}_j$ is both the right and left endpoint of X_j, i.e., X_j is a one-point interval, there's no restriction on v_j; it can be any number in $(-\infty, \infty)$. All this holds also for Y with only a change of notation. In particular, when Y is a box we have

$$(u_1, \ldots, u_m) \in N_Y(\bar{y}_1, \ldots, \bar{y}_m) \text{ for } Y = Y_1 \times \cdots \times Y_m$$
$$\iff \begin{cases} u_i = 0 & \text{when } \bar{y}_i \text{ is an interior point of } Y_i, \\ u_i \geq 0 & \text{when } \bar{y}_i \text{ is the right endpoint (only) of } Y_i, \\ u_i \leq 0 & \text{when } \bar{y}_i \text{ is the left endpoint (only) of } Y_i. \end{cases} \tag{15}$$

Theorem 1. *In an extended nonlinear programming problem* $(\mathcal{P})$, *let* $\bar{x}$ *be a locally optimal solution. Suppose that the following constraint qualification is satisfied:*

$$\nexists \bar{y} \in Y,\ \bar{y} \neq 0,\ \textit{with} \begin{cases} -[\bar{y}_1 \nabla f_1(\bar{x}) + \cdots + \bar{y}_m \nabla f_m(\bar{x})] \in N_X(\bar{x}), \\ \quad (\bar{y}_1, \ldots, \bar{y}_m) \in N_D(f_1(\bar{x}), \ldots, f_m(\bar{x})). \end{cases} \tag{16}$$

Then necessarily

$$\exists \bar{y} \in Y \textit{ with } \begin{cases} -[f_0(\bar{x}) + \bar{y}_1 \nabla f_1(\bar{x}) + \cdots + \bar{y}_m \nabla f_m(\bar{x})] \in N_X(\bar{x}), \\ (\bar{u}_1, \ldots, \bar{u}_m) \in N_Y(\bar{y}_1, \ldots, \bar{y}_m) \textit{ for } \bar{u}_i = f_i(\bar{x}) - (\partial k/\partial y_i)(\bar{y}). \end{cases} \tag{17}$$

When $(\mathcal{P})$ *is an extended linear-quadratic programming problem* $(\mathcal{Q})$, *the constraint qualification need not be invoked.*

Proof. See Rockafellar [9] for ENLP and [2] for ELQP. □

The constraint qualification in (16) is equivalent to the Mangasarian-Fromovitz constraint qualification when $(\mathcal{P})$ is a conventional problem $(\mathcal{P}_0)$. It can be viewed therefore as an apt extension of that well known condition to our context. When $X = \mathbb{R}^n$, the first normality condition in (16) reduces to $\bar{y}_1 \nabla f_1(\bar{x}) + \cdots + \bar{y}_m \nabla f_m(\bar{x}) = 0$. On the other hand, $(\mathcal{P}_0)$ has D equal to the product of intervals $D_i = (-\infty, 0]$ for $i = 1, \ldots, s$ and $D_i = \{0\}$ for $i = s+1, \ldots, m$. In that case, by (14) as applied to D, the second normality condition in (16) requires $f_i(\bar{x}) = 0$ for inequality constraints with $\bar{y}_i > 0$ but allows $f_i(\bar{x}) \leq 0$ for inequality constraints with $\bar{y}_i = 0$; for equality constraints, one must of course have $f_i(\bar{x}) = 0$. In other words, we get the usual conditions of *complementary slackness* associated with multipliers in $(\mathcal{P}_0)$.

The interpretation of the multiplier rule in (17) is similar. If $\bar{x}$ is an interior point of X, as for instance when $X = \mathbb{R}^n$, the cone $N_X(\bar{x})$ is just the zero cone $\{0\}$. In asking the gradient expression $-[\nabla f_0(\bar{x}) + \bar{y}_1 \nabla f_1(\bar{x}) + \cdots + \bar{y}_m \nabla f_m(\bar{x})]$ to belong to $N_X(\bar{x})$, one is asking it to be 0. When $\bar{x}$ is a boundary point of X, however, the gradient expression is required to have a certain relationship to $\bar{x}$. What this might be can easily be seen for instance when X is a box. Then, according to (14), the requirement is for the jth component of the gradient sum to be positive, negative or zero according to the location of $\bar{x}_j$ within the jth interval X_j.

The meaning of the requirement $(f_1(\bar{x}), \ldots, f_m(\bar{x})) \in N_Y(\bar{y}_1, \ldots, \bar{y}_m)$ in (17) is simplest when Y is a box $Y_1 \times \cdots \times Y_m$. In that case it places a sign restriction on $f_i(\bar{x})$ that's tied to the location of the multiplier $\bar{y}_i$ within Y_i, as seen from (15). Again we have a generalization of the complementary slackness conditions in a nonlinear programming problem $(\mathcal{P}_0)$. When $Y_i = [0, \infty)$, the requirement comes out as $\bar{u}_i \leq 0$ if $\bar{y}_i = 0$ but $\bar{u}_i = 0$ if $\bar{y}_i > 0$. When $Y_i = (-\infty, \infty)$, it's just $\bar{u}_i = 0$. For general intervals Y_i, (15) describes a broader version of complementary slackness.

First-order optimality conditions for $(\mathcal{P})$ as in Theorem 1 were first developed in [9]. The second normality condition in (17) can equivalently be stated in terms of subgradients of θ, namely as

$$(\bar{y}_1, \ldots, \bar{y}_m) \in \partial\theta(f_1(\bar{x}), \ldots, f_m(\bar{x})). \tag{18}$$

When θ is separable as in (6), this comes down to $\bar{y}_i \in \partial\theta_i(f_i(\bar{x}))$ for $i = 1, \ldots, m$. An advantage of the version in (17), however, is a connection with variational inequalities.

Theorem 2. *The normality conditions in the multiplier rule (17) in Theorem 1 can be equivalently expressed in the form*

$$-\nabla_x L(\bar{x}, \bar{y}) \in N_X(\bar{x}), \qquad \nabla_y L(\bar{x}, \bar{y}) \in N_Y(\bar{y}). \tag{19}$$

In terms of $z = (x, y)$, $Z = X \times Y$ and $F : (x, y) \mapsto (\nabla_x L(x, y), -\nabla_y L(x, y))$, this is the variational inequality $F(\bar{z})\cdot[z - \bar{z}] \geq 0$ for all $z \in Z$, the same as

$$F(\bar{z}) + N_Z(\bar{z}) \ni 0. \tag{20}$$

Proof. The equivalence of these various statements is immediate from the definition of L and the fact that $N_Z(\bar{z}) = N_X(\bar{x}) \times N_Y(\bar{y})$ for $\bar{z} = (\bar{x}, \bar{y})$. □

The set Z in the variational inequality of Theorem 2 is, like X and Y, polyhedral. It's a box when X and Y are boxes. For variational inequalities over polyhedral sets, unusually powerful results are available; see e.g. [10].

To explore convexity and duality in extended linear programming, we introduce now the notion of $(\mathcal{P})$ being a problem of *extended convex programming*. By that we'll mean that the data elements $f_0, f_1, \ldots, f_m$, X and Y in $(\mathcal{P})$ have the property that, for each $y \in Y$, $L(x, y)$ is convex in x relative to X.

Theorem 3. *Suppose $(\mathcal{P})$ is an extended convex programming problem in the general sense just defined. Then the expression $f_0(x) + \theta(f_1(x), \ldots, f_m(x))$ being minimized in $(\mathcal{P})$ is convex relative to X, and, the normal cone conditions in (17), or the equivalent versions of them in Theorem 2, are sufficient for $\bar{x}$ to be a globally optimal solution, without regard to the constraint qualification (16). Indeed, these conditions mean that the pair $(\bar{x}, \bar{y}) \in X \times Y$ gives a saddle point of L on $X \times Y$:*

$$L(x, \bar{y}) \geq L(\bar{x}, \bar{y}) \geq L(\bar{x}, y) \quad \text{for all } x \in X,\ y \in Y. \tag{21}$$

Furthermore, in this case the mapping F in the variational inequality in Theorem 2 is monotone on $Z = X \times Y$, i.e., one has

$$[F(z') - F(z)]\cdot[z' - z] \geq 0 \quad \text{for all } z, z' \in Z. \tag{22}$$

Proof. The convexity of the function being minimized over X follows from (13) and the assumed convexity of $L(x,y)$ in $x \in X$, since the supremum of any collection of convex functions is convex.

To say that $L(x,\bar{y}) \geq L(\bar{x},\bar{y})$ for all $x \in X$ is to say that for every choice of $x \in X$ the function $\varphi(\tau) = L((1-\tau)\bar{x} + \tau x, \bar{y})$ has $\varphi(\tau) \geq \varphi(0)$ for $\tau \in [0,1]$. The convexity of $L(\cdot,\bar{y})$ relative to X implies that φ is convex relative to $[0,1]$, so that this inequality on φ holds if and only if $\varphi'(0) \geq 0$. But $\varphi'(0) = \nabla_x L(\bar{x},\bar{y}) \cdot [x - \bar{x}]$. Hence we have $L(x,\bar{y}) \geq L(\bar{x},\bar{y})$ for all $x \in X$ if and only if $\nabla_x L(\bar{x},\bar{y}) \cdot [x - \bar{x}] \geq 0$ for all $x \in X$, or in other words, $-\nabla_x L(\bar{x},\bar{y}) \in N_X(\bar{x})$.

We always have $L(\bar{x},y)$ concave in $y \in Y$ on the basis of the formula for L in (12) and the convexity of k on Y that was assumed in (A3). By a parallel argument, therefore, we have $L(\bar{x},y) \leq L(\bar{x},\bar{y})$ for all $y \in Y$ if and only if $\nabla_y L(\bar{x},\bar{y}) \in N_Y(\bar{y})$.

A minor extension of these two arguments brings out the fact that actually, for arbitrary choices of $x, x' \in X$ and $y, y' \in Y$, we have

$$\begin{aligned}
L(x',y) &\geq L(x,y) + \nabla_x L(x,y) \cdot [x' - x], \\
L(x,y') &\leq L(x,y) + \nabla_y L(x,y) \cdot [y' - y], \\
L(x,y') &\geq L(x',y') + \nabla_x L(x',y') \cdot [x - x'], \\
L(x',y) &\leq L(x',y') + \nabla_y L(x',y') \cdot [y - y'].
\end{aligned}$$

In multiplying the first and third inequalities by -1 and then adding all four together, we get $0 \geq [\nabla_x L(x,y) - \nabla_x L(x',y')] \cdot [x' - x] - [\nabla_y L(x,y) - \nabla_y L(x',y')] \cdot [y' - y]$, which comes out as (22) for $z = (x,y)$ and $z' = (x',y')$. □

In the light of Theorem 3, the vectors $\bar{y}$ paired with $\bar{x}$ in the first-order optimality conditions for an extended *convex* programming problem $(\mathcal{P})$ solve the associated *dual problem*

$$(\mathcal{P}^*) \qquad \text{maximize } g(y) \text{ over } y \in Y, \text{ where } g(y) = \inf_{x \in X} L(x,y).$$

It's especially interesting to see how this kind of duality works out in extended linear-quadratic programming.

Theorem 4. *In the case of an extended linear-quadratic programming problem* $(\mathcal{Q})$, *the Lagrangian is*

$$L(x,y) = c \cdot x + \tfrac{1}{2} x \cdot Px + b \cdot y - \tfrac{1}{2} y \cdot Qy - y \cdot Ax \quad \text{on } X \times Y \tag{23}$$

and the first-order optimality conditions take the form

$$-c - P\bar{x} + A^T\bar{y} \in N_X(\bar{x}), \qquad b - A\bar{x} - Q\bar{y} \in N_Y(\bar{y}), \tag{24}$$

thus corresponding to a variational inequality (20) with F monotone and affine.

If P is positive-semidefinite along with Q, these conditions are equivalent to having $(\bar{x},\bar{y})$ be a saddle point of L on $X \times Y$, and they hold if and only if $\bar{x}$ is optimal in $(\mathcal{Q})$

while $\bar{y}$ is optimal in the problem dual to $(\mathcal{Q})$. Moreover, that dual problem belongs again to extended linear-quadratic programming and has the form

$$(\mathcal{Q}^*) \qquad \text{maximize } b{\cdot}y - \tfrac{1}{2}y{\cdot}Qy - \theta_{XP}(A^T y - c) \text{ over } y \in Y.$$

Proof. In $(\mathcal{Q})$ we have $\theta = \theta_{YQ}$ and therefore the Lagrangian in (23). Everything then follows from Theorem 3 except for the particular form of $(\mathcal{Q}^*)$. That emerges from the description of the general dual problem $(\mathcal{P}^*)$ through the fact that

$$\begin{aligned}\inf_{x \in X}&\{c{\cdot}x + \tfrac{1}{2}x{\cdot}Px + b{\cdot}y - \tfrac{1}{2}y{\cdot}Qy - y{\cdot}Ax\}\\ &= b{\cdot}y - \tfrac{1}{2}y{\cdot}Qy - \sup_{x \in X}\{y{\cdot}Ax - c{\cdot}x - \tfrac{1}{2}x{\cdot}Px\}.\end{aligned}$$

Here $y{\cdot}Ax = x{\cdot}A^T y$, so the supremum is $\theta_{XP}(A^T y - c)$ by the definition of the function θ_{XP} (in parallel to that of θ_{YQ} in (11)). □

For the conclusions of Theorem 4 to be valid, it's not really essential that P and Q be positive semidefinite, but just that the expressions $x{\cdot}Px$ and $y{\cdot}Qy$ be convex with respect to $x \in X$ and $y \in Y$, respectively. In that more subtle case, though, the mapping F may only be monotone relative to $Z = X \times Y$ rather than the whole space $\mathbb{R}^n \times \mathbb{R}^m$. The same extra bit of generality is available also in the following duality theorem.

Theorem 5. *In the case of extended linear-quadratic programming with both the matrix P and the matrix Q positive-semidefinite, one has*

$$\min(\mathcal{Q}) = \max(\mathcal{Q}^*) \text{ (with the existence of optimal solutions)}$$

in any of the following circumstances, which in fact are equivalent:

(a) *the optimal value in $(\mathcal{Q})$ is finite;*

(b) *the optimal value in $(\mathcal{Q}^*)$ is finite;*

(c) *feasible solutions exist for both $(\mathcal{Q})$ and $(\mathcal{Q}^*)$.*

This remains valid even if P and Q are not positive-semidefinite, as long as the expressions $x{\cdot}Px$ and $y{\cdot}Qy$ are convex with respect to $x \in X$ and $y \in Y$, respectively.

Proof. For positive-definite P and Q, this theorem was first proved in [3]. It has subsequently been presented in that mode also in [2] and [1]. The following argument confirms that it holds true also under the weaker conditions of relative convexity.

Without loss of generality it can be supposed, for the purpose at hand, that $0 \in X$ and that $0 \in Y$. This just amounts to a change of variables: for any choice of $x_0 \in X$ and $y_0 \in Y$, we can rewrite everything in terms of $x' \in X_0 = X - x_0$ and $y' \in Y_0 = Y - y_0$. The shifted Lagrangian $L_0(x', y') = L(x_0 + x', y_0 + y')$ on X_0 and Y_0 gives rise to primal and dual problems $(\mathcal{Q}_0)$ and $(\mathcal{Q}_0^*)$ in x' and y' that are equivalent to $(\mathcal{Q})$ and $(\mathcal{Q}^*)$, as readily can be checked.

Once we have $0 \in X$ and $0 \in Y$, we know that the affine hulls of X and Y are subspaces of $\mathbb{R}^n$ and $\mathbb{R}^m$, respectively. Through coordinate transformations in these spaces, we can identify these affine hulls with spaces $\mathbb{R}^{n_0}$ and $\mathbb{R}^{m_0}$ of perhaps lower dimension. Since X and Y, being convex sets, have nonempty interior relative to their affine hulls, we can in this way reduce the whole issue to the case where X and Y have nonempty interior. In that case, however, the convexity of $x{\cdot}Px$ and $y{\cdot}Qy$ with respect to $x \in X$ and $y \in Y$ implies that P and Q are positive-semidefinite. The earlier result can then be invoked, and its conclusion can be brought over. □

4 Solution Methodology

Many of the approaches to solving nonlinear programming problems can be explained in terms of their Lagrangians and approximations that can be made of them. That's true for techniques of the sequential quadratic programming variety and even steepest descent, as well as for algorithms based on successive approximations to Karush-Kuhn-Tucker conditions and the techniques for handling variational inequalities. Because the extended Lagrangian in (12) is so close to the classical one, it's tempting to think that virtually all of those approaches may work also for extended nonlinear programming with appropriate adjustments in concept and formulation.

For example, around any pair of points $\hat{x} \in X$ and $\hat{y} \in Y$, the Lagrangian L for problem $(\mathcal{P})$ has a second-order expansion

$$\begin{aligned}\widehat{L}(x,y) \;=\; & L(\hat{x},\hat{y}) + \nabla_x L(\hat{x},\hat{y}){\cdot}[x-\hat{x}] + \nabla_y L(\hat{x},\hat{y}){\cdot}[y-\hat{y}] \\ & +\tfrac{1}{2}[x-\hat{x}]{\cdot}\nabla^2_{xx}L(\hat{x},\hat{y})[x-\hat{x}] + \tfrac{1}{2}[y-\hat{y}]{\cdot}\nabla^2_{yy}L(\hat{x},\hat{y})[y-\hat{y}] \\ & +[y-\hat{y}]{\cdot}\nabla^2_{yx}L(\hat{x},\hat{y})[x-\hat{x}],\end{aligned} \tag{25}$$

which reduces algebraically to

$$\widehat{L}(x,y) = \hat{d} + \hat{c}{\cdot}x + \tfrac{1}{2}x{\cdot}\hat{P}x + \hat{b}{\cdot}y - \tfrac{1}{2}y{\cdot}\hat{Q}y - y{\cdot}\hat{A}x \tag{26}$$

for a certain choice of constant $\hat{d}$, vectors $\hat{c}$ and $\hat{b}$, and matrices $\hat{P}$, $\hat{Q}$ and $\hat{A}$, with $\hat{P}$ and $\hat{Q}$ symmetric and the expression $y{\cdot}\hat{Q}y$ convex for $y \in Y$. (The latter holds because $\hat{Q} = \nabla^2 k(\hat{y})$ and the function k is assumed to be convex relative to Y.) Associated with $\widehat{L}$ and the sets X and Y is an *extended linear-quadratic* programming problem

$$(\hat{\mathcal{Q}}) \qquad \text{minimize } \hat{c}{\cdot}x + \tfrac{1}{2}x{\cdot}\hat{P}x + \theta_{Y\hat{Q}}(\hat{b} - \hat{A}x) \text{ over } x \in X,$$

(where the constant $\hat{d}$ has been dropped as unessential). This might be exploited as an approximation of $(\mathcal{P})$ in some iterative scheme. Moreover X and Y might be replaced in $(\hat{\mathcal{Q}})$ by $\widehat{X} = X \cap X_0$ and $\widehat{Y} = Y \cap Y_0$ for polyhedral neighborhoods X_0 of $\hat{x}$ and Y_0 of $\hat{y}$, so that such an approximation would have a trust region character, primally and even dually.

Such ideas offer stimulating prospects for research that could be rewarding for optimization in practice. The extended problem format could be matched with numerical methods able to take direct advantage of the extra structure that modelers could bring out in it. Progress could be made in many directions.

Here, we can't go further than speculation about such possibilities. In the meantime, it's important to know that there's no need to hold back from using the extended format. We'll now demonstrate that in the absence of computer codes tailored to extended nonlinear programming, a technical trick can be applied to convert a problem $(\mathcal{P})$ with θ of type θ_{YQ} into a conventional nonlinear programming problem. Not only a solution $\bar{x}$ but also an associated multiplier vector $\bar{y}$ for it in $(\mathcal{P})$ can be obtained then by calling on existing software to solve the converted problem. Moreover the conversion can be achieved in terms of standardized representations of X, Y and Q so that, if desirable, it can be carried out automatically in a programming interface which the user doesn't even have to be aware of. That wouldn't be as efficient presumably as a more direct approach, but could be helpful nevertheless.

Proposition 4. *Let $\theta = \theta_{YQ}$ as in (11) with Q positive-semidefinite and consider any representations of the form*

$$Y = \{y \mid S^T y \le s\}, \qquad Q = DJ^{-1}D^T, \tag{27}$$

where J is symmetric and positive-definite (for instance $J = I$). Then

$$\theta(u) = \inf_{z \ge 0,\, w} \{s{\cdot}z + \tfrac{1}{2}w{\cdot}Jw \mid Sz + Dw = u\} \ \text{ for every } u. \tag{28}$$

Proof. For fixed u, let $(\mathcal{Q}_1)$ be the problem in (z, w) that underlies the right side of (28). This is the primal problem associated with the saddle point problem for

$$L_1(z, w; y) = s{\cdot}z + \tfrac{1}{2}w{\cdot}Jw + y{\cdot}[u - Sz - Dw]$$

with respect to minimizing in (z, w) with $z \ge 0$ but maximizing in $y \in I\!R^m$. The associated dual problem $(\mathcal{Q}_1)^*$, in the framework of extended linear-quadratic programming that's been presented here, is to maximize over all $y \in I\!R^m$ the expression

$$g(y) = \inf_{z \ge 0,\, w} L_1(z, w; y) = \inf_{z \ge 0,\, w} \{u{\cdot}y + z{\cdot}[s - S^T y] + \tfrac{1}{2}w{\cdot}Jw - w{\cdot}[D^T y]\}.$$

Since J is symmetric and positive definite, this calculates out to

$$g(y) = \begin{cases} u{\cdot}y - \tfrac{1}{2}[D^T y]{\cdot}J^{-1}[D^T y] & \text{if } y \in Y, \\ \infty & \text{if } y \notin Y. \end{cases}$$

Thus, the feasible solution set in $(\mathcal{Q}_1^*)$ is Y, which is nonempty, whereas the optimal value in $(\mathcal{Q}_1^*)$ is

$$\sup_{y \in Y}\{u{\cdot}y - \tfrac{1}{2}[D^T y]{\cdot}J^{-1}[D^T y]\} = \sup_{y \in Y}\{u{\cdot}y - \tfrac{1}{2}y{\cdot}[DJ^{-1}D^T]y\} = \theta_{YQ}(u).$$

It follows from the duality facts in Theorem 5, as specialized here, that this value and the one in (28) are equal; either both are finite, or both are ∞. □

Theorem 6. *Let $(\mathcal{P})$ be a problem of extended nonlinear programming with $\theta = \theta_{YQ}$ as in (11). Let Y and Q be represented as in (27) and express $X = \{x \mid Rx \leq r\}$ for some matrix R and vector r. Write $F(x) = (f_1(x), \dots, f_m(x))$. Then the optimal solutions $\bar{x}$ to $(\mathcal{P})$ are the $\bar{x}$ components of the optimal solutions $(\bar{x}, \bar{z}, \bar{w})$ to*

$$(\overline{\mathcal{P}}_0) \qquad \begin{array}{c} \text{minimize } \; f_0(x) + s{\cdot}z + \tfrac{1}{2}w{\cdot}Jw \; \text{ subject to} \\ Rx - r \leq 0, \;\; -z \leq 0, \;\; F(x) - Sz - Dw = 0, \end{array}$$

which belongs to conventional nonlinear programming. Moreover a multiplier vector $\bar{y}$ for the equation constraint in $(\overline{\mathcal{P}}_0)$ in the usual sense will be a multiplier vector associated with $\bar{x}$ in the sense of the extended Lagrangian for $(\mathcal{P})$.

Proof. The validity of the reformulation is immediate from the expression for θ achieved in the lemma. The claim about multiplier vectors $\bar{y}$ comes out of the duality developed in the proof of the lemma, according to which the multiplier vectors for the constraint $u - Sz - Dw = 0$ in $(\mathcal{Q}_1)$ must, on the basis of Theorem 5, be the optimal solutions to the dual problem $(\mathcal{Q}_1^*)$. Since the maximization in that dual problem expresses the conjugacy formula

$$\theta_{YQ}(u) = \sup_{y \in R^m} \{u{\cdot}y - \psi(y)\} \;\text{ for }\; \psi(y) = \begin{cases} \tfrac{1}{2}y{\cdot}Qy & \text{if } y \in Y, \\ \infty & \text{if } y \notin Y, \end{cases}$$

such vectors y are precisely the subgradients of $\theta = \theta_{YQ}$ at u. That means in the case of $\bar{u} = F(\bar{x})$ that they are the elements of $\partial\theta(F(\bar{x}))$. But those vectors, as in (18), are known to be the multiplier vectors $\bar{y}$ associated with $\bar{x}$ in the sense of the optimality condition in Theorem 1 for $(\mathcal{P})$. □

Corollary. *In a problem $(\mathcal{Q})$ of extended linear-quadratic programming in which Y and Q are furnished expressions of the kind in (27), the optimal solutions $\bar{x}$ to $(\mathcal{Q})$ are the $\bar{x}$ components of the optimal solutions $(\bar{x}, \bar{z}, \bar{w})$ to*

$$(\overline{\mathcal{Q}}_0) \qquad \begin{array}{c} \text{minimize } \; c{\cdot}x + \tfrac{1}{2}x{\cdot}Px + s{\cdot}z + \tfrac{1}{2}w{\cdot}Jw \; \text{ subject to} \\ x \in X, \;, -z \leq 0, \;\; b - Ax - Sz - Dw = 0. \end{array}$$

Moreover the multiplier vectors $\bar{y}$ for the constraint $b - Ax - Sz - Dw = 0$ in this reformulated problem are identical to the ones associated with $\bar{x}$ in $(\mathcal{Q})$. Hence, if P is positive semidefinite, they are the optimal solutions to the dual problem of extended linear-quadratic programming, $(\mathcal{Q}^*)$.

References

[1] R.T. Rockafellar and R.J-B Wets (1998), *Variational Analysis*, Springer Verlag, Berlin.

[2] R.T. Rockafellar (1987), "Linear-quadratic programming and optimal control," *SIAM J. Control Opt.* 25, 781–814.

[3] R.T. Rockafellar and R.J-B Wets (1986), "A Lagrangian finite-generation technique for solving linear-quadratic problems in stochastic programming," *Math Programming Studies* 28, 63–93.

[4] R.T. Rockafellar and R.J-B Wets (1990), "Generalized linear-quadratic problems of deterministic and stochastic optimal control in discrete time," *SIAM J. Control Opt.* 28, 810–822.

[5] R.T. Rockafellar (1987), "Computational schemes for large-scale problems in extended linear-quadratic programming," *Math. Programming* 48, 447–744.

[6] R.T. Rockafellar and C.-Y. Zhu (1987), "Primal-dual projected gradient algorithms for extended linear-quadratic programming," *SIAM J. Optimization* 3, 751–761.

[7] H.W. Kuhn and A.W. Tucker (1951), "Nonlinear programming," *Proceedings of the Second Berkeley Symposium on Mathematical Statistics and Probability* (J. Neyman, ed.), Univ. of California Press, Berkeley, California, 481–492.

[8] R.T. Rockafellar (1970), *Convex Analysis*, Princeton University Press, Princeton, New Jersey (available from 1997 also in paperback in the series Princeton Landmarks in Mathematics).

[9] R.T. Rockafellar (1993), "Lagrange multipliers and optimality," *SIAM Review* 35, 183–238.

[10] A.L. Dontchev and R.T. Rockafellar (1996), "Characterizations of strong regularity for variational inequalities over polyhedral sets," *SIAM J. Optimization* 6, 1087–1105.

On the Efficiency of Splitting and Projection Methods for Large Strictly Convex Quadratic Programs

Valeria Ruggiero (rgv@dns.unife.it)
Dipartimento di Matematica, Università di Ferrara, Ferrara, Italy.

Luca Zanni (zanniluca@unimo.it)
Dipartimento di Matematica, Università di Modena e Reggio Emilia, Modena, Italy.

Abstract

In this paper we analyse the behaviour of the classical splitting and projection methods for solving large-scale strictly convex quadratic programming problems with linear constraints. The drawbacks of these classical methods are overcome by the recent modified projection-type and variable projection methods. These new approaches have the same complexity and a similar structure: each iteration consists of a projection step followed by a correction formula. Nevertheless, on the contrary of the modified projection-type methods, the variable projection method does not require to prefix any scalar parameters and is weakly dependent on a priori scaling of the objective function. The results of a numerical experimentation permit to compare the new approaches with the classical splitting and projection methods and to evaluate the effectiveness of the variable projection method as solver of large quadratic programs.

Keywords: Strictly convex quadratic programs, large-scale optimization, splitting method, projection-type method.

1 Introduction

We consider the linearly constrained strictly convex quadratic programming (QP) problem:

$$\begin{array}{ll} \textit{minimize} & f(\boldsymbol{x}) = \frac{1}{2}\boldsymbol{x}^T G\boldsymbol{x} + \boldsymbol{q}^T\boldsymbol{x} \\ \textit{subject to} & C\boldsymbol{x} = \boldsymbol{d}, \qquad A\boldsymbol{x} \geq \boldsymbol{b} \end{array} \tag{1}$$

G. Di Pillo and F. Giannessi (eds.), Nonlinear Optimization and Related Topics, 401–413.

where G is a symmetric positive definite matrix of order n, C is an $m_e \times n$ matrix of full row rank ($m_e \leq n$) and A is an $m_i \times n$ matrix; we assume that the QP problem is feasible and that the matrices G, C and A are large, sparse and without a particular structure.
This problem assumes great importance since an efficient solver for large QP problems is the key of success of many methods, such as the SQP methods in large–scale nonlinear constrained optimization problems [16] or the iterative methods in the solution of variational inequalities (linear approximation methods [19] and descent methods based upon projective gap function [5]).
In this paper we are interested into well-known approaches, called Splitting Methods (SM) and Projection Methods (PM), which solve the original problem by solving a sequence of easier QP subproblems. In particular, these classes of iterative methods consist in to generate, starting from an arbitrary vector $\boldsymbol{x}^{(0)}$, a sequence of vectors $\{\boldsymbol{x}^{(k)}\}$ such that $\boldsymbol{x}^{(k)}$ is the solution of the following subproblem ($k = 1, 2, \ldots$):

$$\begin{array}{ll} \textit{minimize} & \frac{1}{2}\boldsymbol{x}^T \frac{D}{\rho} \boldsymbol{x} + (\boldsymbol{q} + (G - \frac{D}{\rho})\boldsymbol{x}^{(k-1)})^T \boldsymbol{x} \\ \textit{subject to} & C\boldsymbol{x} = \boldsymbol{d}, \qquad A\boldsymbol{x} \geq \boldsymbol{b} \end{array} \tag{2}$$

where D is a prefixed symmetric positive definite matrix of order n and ρ is a positive parameter. The difference between the splitting methods and the projection methods consists in the way in which the convergence of the sequence $\{\boldsymbol{x}^{(k)}\}$ is guaranteed.
The classical splitting methods are obtained by setting $\rho = 1$ in (2), (see the iterative scheme in [4], the diagonalization algorithm in [12] and the splitting algorithm in [8]; see also [1], [13], [19] for similar approaches in variational inequality and linear complementarity problems); the sufficient condition for the convergence is that $(D, G-D)$ is a P–regular splitting [18] of G. Under this hypothesis, the sequence $\{\boldsymbol{x}^{(k)}\}$ is linearly convergent to the unique solution $\boldsymbol{x}^*$ of (1) and its convergence rate is dependent on the spectral radius of the matrix $(D^{-1/2}GD^{-1/2} - I)$ [8]. Thus, the choice of the matrix D is crucial for the behaviour of a splitting method. In some applications, the structure of the Hessian matrix G suggests a "good" P–regular splitting, that is a P–regular splitting that implies a fast convergence and in which D is easily solvable (see, for example, the numerical results obtained by splitting methods for QP problems arising in the solution of constrained bivariate interpolation problems [9]). On the other hand, when G has no particular structure, the choice of a matrix D easily solvable and such that $(D, G - D)$ is a P–regular splitting can produce slow convergence.
By introducing a parameter $\rho \neq 1$, the class of the projection methods allows to select any positive definite matrix D and this feature could be used to generate easily solvable and well-conditioned subproblems. The convergence of these methods can be immediately derived from the projection schemes for the variational inequality problems; by proceeding as in [3], the convergence of $\{\boldsymbol{x}^{(k)}\}$ can be obtained for $\rho < \frac{2\,\lambda_{min}(G)}{\lambda_{max}(GD^{-1}G)}$ or, following [15], for $\rho < \frac{2\lambda_{min}(D)}{\lambda_{max}(G)}$ ($\lambda_{min}(\cdot)$ and $\lambda_{max}(\cdot)$ denote the minimum and the maximum eigenvalue of a matrix respectively). The convergence

rate is linear and it is strongly dependent on the value of ρ; in particular, in order to satisfy one of the above restrictive sufficient convergence conditions, the parameter ρ may assume a "too small" value that implies a very slow convergence (as in projection methods for variational inequalities [24]). Thus, with the classical splitting and projection methods the question is how to choose a good splitting and how to choose a good value of the parameter ρ respectively.

2 The Modified Projection–type Methods

The drawbacks of the classical splitting and projection methods are avoided in the modified projection–type methods by introducing a correction step after the solution of each QP subproblems. One of the most general Modified Projection–type Methods (MPM) is the scheme proposed by Solodov and Tseng in [22] in the context of the solution of monotone affine variational inequality problems. Starting from an arbitrary vector $\boldsymbol{x}^{(0)}$, this scheme consists in to generate a sequence $\{\boldsymbol{x}^{(k)}\}$, where $\boldsymbol{x}^{(k)}$, $k = 1, 2, ...$, is obtained by a projection step that computes the solution $\boldsymbol{y}^{(k)}$ of the following QP subproblem:

$$\begin{array}{ll} \textit{minimize} & \frac{1}{2}\boldsymbol{y}^T\boldsymbol{y} + (\boldsymbol{q} + (G - I)\boldsymbol{x}^{(k-1)})^T\boldsymbol{y} \\ \textit{subject to} & C\boldsymbol{y} = \boldsymbol{d} \qquad A\boldsymbol{y} \geq \boldsymbol{b} \end{array} \tag{3}$$

followed by the correction formula:

$$\boldsymbol{x}^{(k)} = \boldsymbol{x}^{(k-1)} + \gamma_{k-1}P^{-1}(I + G)(\boldsymbol{y}^{(k)} - \boldsymbol{x}^{(k-1)}) \tag{4}$$

with

$$\gamma_{k-1} = \theta \frac{\|\boldsymbol{y}^{(k)} - \boldsymbol{x}^{(k-1)}\|_2^2}{\|P^{-\frac{1}{2}}(I + G)(\boldsymbol{y}^{(k)} - \boldsymbol{x}^{(k-1)})\|_2^2}. \tag{5}$$

Here P is a symmetric positive definite matrix and θ is a parameter in the interval $(0, 2)$. When $\theta = 1$ and $P = I + G$, we have the Projection and Contraction method proposed in [11]. For appropriate choices of P (e.g. $P = I + G$), the computational cost introduced by the correction step is essentially the cost for a matrix–vector product. The performance of this scheme is strongly dependent on the choice of P, θ and on the scaling of G and $\boldsymbol{q}$, as emphasized also in [22].

3 The Variable Projection Method

Following the same idea of introducing a correction step, in [21] we propose another scheme, called Variable Projection Method (VPM), with the same complexity (the solution of a QP subproblem and the computation of a matrix–vector product at each iteration) of the MPM.

In the sequel of the paper, we denote by $\|G\|_F$ the Frobenius norm of the matrix G ($\|G\|_F = (\sum_i \sum_j g_{ij}^2)^{\frac{1}{2}}$) and by $\|G\|_2$ the spectral norm of G.
If we assume that D is a symmetric positive definite matrix, the VPM is stated as follows:

1. Let $\boldsymbol{x}^{(0)}$ be an arbitrary vector and $\rho_1 = \frac{\lambda_{min}(D)}{\|G\|_F}$; $k \leftarrow 1$.

2. Compute the direction $\boldsymbol{d}^{(k)} = \boldsymbol{y}^{(k)} - \boldsymbol{x}^{(k-1)}$ where $\boldsymbol{y}^{(k)}$ is the unique solution of the subproblem

$$\begin{array}{ll} minimize & \frac{1}{2}\boldsymbol{y}^T\frac{D}{\rho_k}\boldsymbol{y} + (\boldsymbol{q} + (G - \frac{D}{\rho_k})\boldsymbol{x}^{(k-1)})^T\boldsymbol{y} \\ subject\ to & C\boldsymbol{y} = \boldsymbol{d} \qquad A\boldsymbol{y} \geq \boldsymbol{b} \end{array} \tag{6}$$

3. If $\nabla f(\boldsymbol{x}^{(k-1)})^T\boldsymbol{d}^{(k)} < 0$ and $k \neq 1$, then

   compute the solution α_k of the problem $\min_{\alpha>0} f(\boldsymbol{x}^{(k-1)} + \alpha\boldsymbol{d}^{(k)})$:

$$\alpha_k = -\frac{\nabla f(\boldsymbol{x}^{(k-1)})^T\boldsymbol{d}^{(k)}}{\boldsymbol{d}^{(k)^T}G\boldsymbol{d}^{(k)}} \tag{7}$$

   and set

$$\theta_k = \min(1, \alpha_k) \tag{8}$$

   else

$$\theta_k = 1 \tag{9}$$

4. Compute

$$\boldsymbol{x}^{(k)} = \boldsymbol{x}^{(k-1)} + \theta_k\boldsymbol{d}^{(k)} \tag{10}$$

5. Terminate if $\boldsymbol{x}^{(k)}$ satisfies an appropriate stopping rule, otherwise compute

$$\rho_{k+1} = \frac{\boldsymbol{d}^{(k)^T}G\boldsymbol{d}^{(k)}}{\boldsymbol{d}^{(k)^T}GD^{-1}G\boldsymbol{d}^{(k)}} \tag{11}$$

   then $k \leftarrow k+1$ and go to step 2.

As in the MPM, any iteration of the VPM requires a projection step (that consists in to solve the QP subproblem (6)), but the projection matrix $\frac{D}{\rho_k}$ is changed at each iteration, by trying to select a convenient parameter ρ_k. Then, the next iterate $\boldsymbol{x}^{(k)}$ is computed by using the correction formula (10). Except for the matrix D, the scheme does not require to prefix any parameters.
The following theorem shows the convergence of the VPM for the strictly convex QP problem (1).

Theorem 1. Let the problem (1) be feasible with G symmetric positive definite matrix. Let D be a symmetric positive definite matrix. Given an arbitrary vector $\boldsymbol{x}^{(0)}$, the sequence generated by the VPM is convergent to the unique solution $\boldsymbol{x}^*$ of (1).

We sketch the essential ideas underlying the proof; for a detailed proof refer to [21]. Since $\theta_1 = 1$, the point $\boldsymbol{x}^{(1)}$ is feasible; then, the sequence $\{\boldsymbol{x}^{(k)}\}$, $k = 1, 2, \ldots$, is a sequence of feasible points. Furthermore, taking into account that, if $\boldsymbol{y}^{(k)} \neq \boldsymbol{x}^{(k-1)}$, $k = 2, 3, \ldots$, then $\boldsymbol{y}^{(k)} - \boldsymbol{x}^{(k-1)}$ is a descent direction for $f(\boldsymbol{x})$ in $\boldsymbol{x}^{(k-1)}$, it is possible to show the following inequality:

$$f(\boldsymbol{x}^{(k-1)}) - f(\boldsymbol{x}^{(k)}) \geq \frac{1}{2}(\boldsymbol{x}^{(k)} - \boldsymbol{x}^{(k-1)})^T G(\boldsymbol{x}^{(k)} - \boldsymbol{x}^{(k-1)}).$$

Thus, the sequence $f(\boldsymbol{x}^{(k)})$ is monotonically nonincreasing and, since it is bounded below by $f(\boldsymbol{x}^*)$, it converges. Consequently $\{\boldsymbol{x}^{(k)} - \boldsymbol{x}^{(k-1)}\} \to 0$ for $k \to \infty$. From here on, since the sequence $\{\rho_k\}$, $k = 1, 2, \ldots$, is bounded above and below by positive constants, the convergence of $\{\boldsymbol{x}^{(k)}\}$ follows from standard arguments [9].

In the VPM, as well as in all the considered methods, it is required to solve a sequence of strictly convex QP subproblems, having the following form

$$\begin{array}{ll} \textit{minimize} & \frac{1}{2}\boldsymbol{x}^T \Delta \boldsymbol{x} + \boldsymbol{q}^{(k-1)T}\boldsymbol{x} \\ \textit{subject to} & C\boldsymbol{x} = \boldsymbol{d}, \qquad A\boldsymbol{x} \geq \boldsymbol{b} \end{array} \tag{12}$$

where $\boldsymbol{q}^{(k-1)} = (G - \Delta)\boldsymbol{x}^{(k-1)} + \boldsymbol{q}$, $k = 1, 2, ...$, and Δ is an easily solvable matrix, for instance a diagonal or block diagonal matrix ($\Delta = D$ for the SM, $\Delta = \frac{D}{\rho}$ for the PM, $\Delta = I$ for the MPM and $\Delta = \frac{D}{\rho_k}$ for the VPM).
By using the Karush–Kuhn–Tucker optimality conditions, we can derive the solution $\bar{\boldsymbol{x}}$ of (12) in terms of its corresponding Lagrange multipliers $\bar{\boldsymbol{\lambda}}$ and $\bar{\boldsymbol{\mu}}$:

$$\bar{\boldsymbol{x}} = \Delta^{-1}(-\boldsymbol{q}^{(k-1)} + A^T\bar{\boldsymbol{\lambda}} + C^T\bar{\boldsymbol{\mu}})$$

where $\begin{pmatrix} \bar{\boldsymbol{\lambda}} \\ \bar{\boldsymbol{\mu}} \end{pmatrix}$ is the solution of the following mixed Linear Complementarity Problem (LCP):

$$\begin{gathered} \begin{pmatrix} \boldsymbol{u} \\ 0 \end{pmatrix} = \begin{pmatrix} \boldsymbol{z}_1^{(k-1)} \\ \boldsymbol{z}_2^{(k-1)} \end{pmatrix} + M \begin{pmatrix} \boldsymbol{\lambda} \\ \boldsymbol{\mu} \end{pmatrix} \\ \boldsymbol{u} \geq 0, \quad \boldsymbol{\lambda} \geq 0, \quad \boldsymbol{u}^T\boldsymbol{\lambda} = 0 \end{gathered} \tag{13}$$

where $\boldsymbol{z}_1^{(k-1)} = -\boldsymbol{b} - A\Delta^{-1}\boldsymbol{q}^{(k-1)}$, $\boldsymbol{z}_2^{(k-1)} = -\boldsymbol{d} - C\Delta^{-1}\boldsymbol{q}^{(k-1)}$ and the matrix

$$M = \begin{pmatrix} A \\ C \end{pmatrix} \Delta^{-1} \begin{pmatrix} A^T & C^T \end{pmatrix} \tag{14}$$

is a symmetric positive semidefinite matrix of order $\nu = (m_i + m_e)$ with positive diagonal entries. The mixed LCP (13) is solvable because it arises from the strictly convex QP problem (12). In the case of equality constraints only, M is the symmetric positive definite matrix $CD^{-1}C^T$ and (13) is reduced to the following positive definite linear system:

$$M\boldsymbol{\mu} = -\boldsymbol{z}_2^{(k-1)}. \tag{15}$$

Thus, we can determine the solution of (12) by solving an equivalent problem whose size is equal to the number of constraints.

When this number is small, we can solve the problem (13) or (15) by direct methods. In the case of large–scale sparse symmetric monotone LCPs, it appears convenient to use again a splitting iterative scheme, such as the classical Projected SOR scheme of Cryer [2] and the Projected Symmetric SOR scheme of Mangasarian [13]. We can also use an iterative parallel solver, obtained by considering a splitting of the matrix M such that each iteration of the splitting iterative scheme can be decomposed into independent processes. In [10] we show the effectiveness of the splitting method for QP problems combined with some parallel solvers for the inner LCPs, as the Parallel SOR, the Parallel Gradient Projection SOR [14] and the Overlapping Parallel SOR. When the inner subproblem (12) can be reformulated as the linear system (15), we can use as iterative solver the Preconditioned Conjugate Gradient (PCG) Method, that is also well suited for implementation on parallel computers. The absence of a particular structure in the matrix M suggests to use as preconditioner the classical SSOR preconditioner [6] or the Arithmetic Mean preconditioner [7].

4 Computational Experiments

In this section we report the results of a set of numerical experiments that show the computational behaviour of the methods described in the previous sections. All the experiments are carried out on a Digital Alpha 500/333 Mhz workstation, using the double precision (macheps= $2.22 \cdot 10^{-16}$) and a set of programs written in Fortran 77. The considered test problems are randomly generated with assigned features; in particular, we prefix the sizes n, m_i, m_e, the solution $\boldsymbol{x}^*$ and the corresponding Lagrange multipliers, the level of sparsity, the spectral condition number, the euclidean norm and the distribution of singular values of the matrices G and $\begin{pmatrix} A \\ C \end{pmatrix}$. These matrices are generated following a technique similar to that introduced in [23], but using Givens rotations instead of Householder elementary trasformations for obtaining the required level of sparsity. For details see [20].

The values considered for several features of the test problems (e.g. size, condition number, level of sparsity) reflect those of the problems arising in many practical applications. Furthermore, even if these experiments concern test problems without structure, it is well known that the splitting and projection–type methods are suited

to exploit the structures of the Hessian matrix or the constraint matrices that often appear in the real problems.
In all the experiments we use the following stopping rule:

$$\frac{\|\boldsymbol{x}^{(k+1)} - \boldsymbol{x}^{(k)}\|_2}{\|\boldsymbol{x}^{(k+1)}\|_2} \leq tol \tag{16}$$

where $tol = 10^{-6}$ in Table 1–4 and $tol = 10^{-12}$ in Table 5–7. In the tables we denote by *it* and *time* the number of iterations and the time in seconds to obtain the solution and by $er_{\boldsymbol{x}}$ the relative error $\frac{\|\boldsymbol{x}^* - \boldsymbol{x}^{(it)}\|_2}{\|\boldsymbol{x}^*\|_2}$.

Table 1. Behaviour of the PM for different values of ρ

$n = 1000$ $m_e = 600$ $m_i = 0$ spars$(G) = 98\%$ $\|G\|_2 = 1$ spars$(C) = 99\%$ $\|C\|_2 = 10$ $K(C) = 10^2$ sufficient convergence condition: $\rho < 2$				
Projection Method				
$K(G)$	ρ	*it*	*time*	$er_{\boldsymbol{x}}$
10^2	1.43	405	14.2	2.7e-5
	1.67	347	12.3	2.6e-5
	1.94	297	11.0	2.7e-5
	1.98	658	23.3	1.2e-7
	1.995	2742	94.3	1.2e-7
10^3	1.43	2455	84.2	2.7e-4
	1.67	2104	72.4	2.7e-4
	1.94	1806	61.8	2.7e-4
	1.98	1771	60.2	2.7e-4
	1.995	2741	93.7	3.8e-5
10^4	1.43	3800	129.1	5.9e-4
	1.67	3246	111.0	5.9e-4
	1.94	2796	98.9	5.9e-4
	1.98	2741	93.9	5.9e-4
	1.995	2721	93.4	5.9e-4
10^5	1.43	7215	246.3	3.2e-3
	1.67	6178	211.7	3.2e-3
	1.94	5319	181.6	3.2e-3
	1.98	5210	177.5	3.2e-3
	1.995	5171	176.2	3.2e-3

The aim of the first set of experiments is to evaluate the numerical features of the considered methods, removing any possible dependence on the inner iterative solver.

Thus, we use test problems of medium–size ($n = 1000$) with equality constraints only ($m_e = 600$), so that the inner subproblems can be solved by LAPACK routines. In this first set of experiments, the eigenvalues of G have a uniform distribution, the sparsity levels of G and C (denoted by spars(G) and spars(C)) are 98% and 99% respectively, $\|G\|_2 = 1$, $\|C\|_2 = 10$ and $K(C) = 10^2$. The results of these tests are reported in Tables 1 and 2.

With regard to the splitting method (SM), a P–regular splitting of G is obtained by taking $D = \Omega\text{diag}(g_{11}, g_{22}, ..., g_{nn})$, where Ω is a diagonal matrix with entries ω_i, $i = 1, ..., n$, satisfying the condition $\omega_i > \max(1, \frac{\sum_j |g_{ij}|}{2g_{ii}})$; consequently, $2D - G$ is strictly diagonally dominant with positive entries and, then, it is a positive definite matrix. In the projection method (PM) we choose $D = I$. For the MPM, we use $P = I + G$ and we report the results obtained with the empirical optimal value of parameter θ ($\theta = 0.8$). An "a priori" scaling of G and $\boldsymbol{q}$ is performed using the technique suggested in [22]. For the VPM, we use $D = I$ and the rule (11) for updating the parameter ρ_k at each iteration.

Table 1 shows the behaviour of the PM with respect to different values of the parameter ρ. For each prefixed value of the condition number $K(G)$ of G, we try to find the value of ρ that satisfies the sufficient convergence condition $\rho < 2$ and gives the better performance. Nevertheless, in some cases, it is possible to find values of ρ that do not satisfy the sufficient convergence condition and produce a lower number of iterations; for example, when $K(G) = 10^4$ and $\rho = 3.92$, we have $it = 1463$, $time = 51.1$, $er_{\boldsymbol{x}} = 5.1e - 4$ and when $K(G) = 10^5$ and $\rho = 3.98$, we obtain $it = 3199$, $time = 109.8$, $er_{\boldsymbol{x}} = 2.5e - 3$.

Table 2. Comparison among the PM, the SM, the MPM and the VPM on the test problems of Table 1

	PM			SM			MPM			VPM		
$K(G)$	it	$time$	$er_{\boldsymbol{x}}$	it	$time$	$er_{\boldsymbol{x}}$	it	$time$	$er_{\boldsymbol{x}}$	it	$time$	$er_{\boldsymbol{x}}$
10^2	297	11.0	2.7e-5	310	12.2	2.5e-5	167	7.9	2.6e-5	58	3.3	1.1e-5
10^3	1771	60.2	2.7e-4	1993	73.4	2.5e-4	879	42.4	2.7e-4	139	7.0	2.0e-4
10^4	2721	93.4	5.9e-4	2454	90.3	6.0e-4	1507	67.5	4.3e-4	155	7.7	5.9e-4
10^5	5171	176.2	3.2e-3	7519	266.3	2.2e-3	2563	112.6	1.3e-3	181	8.8	2.4e-4

In Table 2 we can observe how the number of iterations of all the methods changes when $K(G)$ increases; for the PM we report the better results obtained with ρ satisfying the sufficient convergence condition. The MPM and VPM are more efficient than the classical SM and PM; in particular, the VPM requires in all cases a very small number of iterations and seems to be weakly dependent on the condition number of G. Furthermore, the results obtained by the VPM do not require an initial scaling of G and $\boldsymbol{q}$, while the efficiency of the MPM is strongly affected by the use of this scaling (for example, for the test problems of Table 2, if we do not use the scaling suggested in [22], it results: $it = 311$ for $K(G) = 10^2$, $it = 1853$ for $K(G) = 10^3$, $it = 2864$ for

$K(G) = 10^4$ and $it = 5437$ for $K(G) = 10^5$). Finally, the rate of convergence of the methods also depends on the distribution of the eigenvalues of G. For example, in the case of the SM, when $K(G) = 10^4$ and we have some hundreds of eigenvalues of G close to the maximum eigenvalue, the number of iterations is 63; on the contrary, if some hundreds of eigenvalues are close to the minimum eigenvalue, $it = 4222$.
When we use a direct solver for solving the inner subproblems, all the considered methods are not affected by the increase of the condition number of the constraint matrix. On the contrary, the choice of an iterative scheme as inner solver is crucial for the effectiveness of the methods.
In Table 3 we report a comparison between two different inner solvers for the VPM: the projected SOR method [2] and the PCG method with the Arithmetic Mean preconditioner [7]. In this case $K(G)$ is fixed and we change $K(C)$; furthermore, in order to avoid the introduction of scaling techniques on the inner solvers, the constraint matrices are generated so that $\|C\|_2 = 1$. For the SOR scheme we report the results obtained with an empirical optimal value of the relaxation parameter ($\omega = 1.5$). We use an inner progressive termination rule, in the sense that the accuracy in the solution of each subproblem depends on the quality of the previous iterate $\boldsymbol{x}^{(k-1)}$: the closer $\boldsymbol{x}^{(k-1)}$ is to satisfying the external stopping criterion for the outer iteration, the more accurately the corresponding subproblem is solved. In this way, unnecessary inner iterations are avoided when $\boldsymbol{x}^{(k-1)}$ is far from the solution. For the PCG scheme the best results are obtained with an inner fixed termination rule.
We point out that the number of outer iterations is about the same in all the cases, while the number of inner iterations (denoted by it_{in}) increases as $K(C)$ increases. Nevertheless, the efficiency of the VPM is preserved if we use the PCG as inner solver.

Table 3. Behaviour of VPM with different iterative inner solvers

$n = 1000$ $m_e = 600$ $m_i = 0$; spars$(G) = 98\%$ $\|G\|_2 = 1$ $K(G) = 10^3$; spars$(C) = 99\%$ $\|C\|_2 = 1$								
	VPM (SOR)				VPM (PCG)			
$K(C)$	it	it_{in}	$time$	$er_{\boldsymbol{x}}$	it	it_{in}	$time$	$er_{\boldsymbol{x}}$
10	29	569	1.6	2.1e-5	30	485	2.4	4.4e-6
30	31	2836	5.4	2.2e-6	31	1082	4.8	4.5e-7
50	30	4323	6.1	2.2e-5	29	1355	4.4	1.5e-6
100	27	8480	11.4	1.5e-5	29	1881	5.9	1.8e-6
300	35	21973	28.0	9.2e-6	29	2385	7.1	3.5e-6

The use of an iterative scheme as inner solver permits to solve large-scale QP problems. In Table 4 we show the results obtained by the SM, the MPM and the VPM for well conditioned test problems with size $n = 8000$. In the following, we use as inner solver the PCG method for the equality constrained problem and the Projected

SOR method for the equality and inequality constrained problem (in this last case, we denote by *nac* the number of constraints that are active in the solution x^*).
We may observe that, also for these cases, the VPM achieves the best performance; furthermore, since the VPM is weakly dependent on the condition number of G (see Table 2) it appears promising also for the solution of not so well conditioned large–scale problems.

Table 4. Well conditioned large–scale QP problems

$n = 8000$

$\text{spars}(G) = 99.8\%$ $\quad \|G\|_2 = 40$ $\quad K(G) = 30$

$\text{spars}\begin{pmatrix} A \\ C \end{pmatrix} = 99.9\%$ $\quad \left\|\begin{pmatrix} A \\ C \end{pmatrix}\right\|_2 = 1$ $\quad K\begin{pmatrix} A \\ C \end{pmatrix} = 10$

	SM				MPM				VPM			
	it	it_{in}	*time*	er_x	*it*	it_{in}	*time*	er_x	*it*	it_{in}	*time*	er_x
$m_e = 3000$ $m_i = 0$	222	2137	128.7	3.5e-6	151	1010	116.2	2.4e-6	46	285	43.6	1.9e-6
$m_e = 2000$ $m_i = 3000$ $nac = 4000$	235	1532	169.9	3.5e-6	144	814	136.3	3.7e-6	48	481	71.9	2.1e-6

In the second set of experiments we are interested to study the behaviour of the VPM in comparison with the active–set method for large and sparse QP problems implemented in the routine E04NKF of the NAG library [17]. This routine is based on parts of the SNOPT and MINOS packages.
In order to improve the coherence in the accuracy of the two approaches, in the subsequent tests the stopping rule (16) for the VPM works with $tol = 10^{-12}$; in this way, we have $|f(x_{it}) - f(x^*)|/|f(x^*)| \leq 10^{-15}$ for both the methods. The two solvers use the same infeasible starting point.

Table 5. Equality constrained QP problems

$n = 5000$ $\quad m_i = 0$

$\text{spars}(G) = 99.90\%$ $\quad \|G\|_2 = 1$ $\quad K(G) = 10^4$

$\text{spars}(C) = 99.95\%$ $\quad \|C\|_2 = 1$ $\quad K(C) = 10^3$

	VPM				E04NKF	
m_e	*it*	it_{in}	*time*	er_x	*time*	er_x
500	208	1861	3.4	6.6(-11)	46.1	5.3(-15)
2000	274	4446	11.6	4.0(-11)	26.1	4.5(-12)
3500	211	1347	10.3	3.1(-11)	15.3	1.1(-13)
4900	206	208	10.8	2.2(-11)	11.0	1.0(-14)

In Table 5 we consider equality constrained QP problems with different number of constraints. When $m_e \ll n$ the VPM appears very convenient (the dimension of the subproblems (15) is m_e) while, as observed in [17], this is the worst case for the NAG routine. The difference between the computational time required by the two solvers decreases as the number of equality constraints increases.

Table 6. QP problems with equality and inequality constraints

$n = 5000$; spars$(G) = 99.90\%$ $\|G\|_2 = 1$ $K(G) = 10^4$; spars$\begin{pmatrix} A \\ C \end{pmatrix} = 99.95\%$ $\left\|\begin{pmatrix} A \\ C \end{pmatrix}\right\|_2 = 1$ $K\begin{pmatrix} A \\ C \end{pmatrix} = 10^3$								
			VPM				E04NKF	
m_e	m_i	nac	it	it_{in}	$time$	er_x	$time$	er_x
4700	1800	4800	166	957	29.3	5.0(-11)	16.0	4.8(-15)
4500	2000	4800	154	920	27.9	4.4(-11)	22.1	3.6(-15)
4000	2500	4800	167	1308	34.9	2.3(-11)	44.4	4.0(-15)
3000	3500	4800	136	1182	32.4	1.9(-11)	134.0	1.1(-14)
3950	550	4000	150	871	8.4	4.8(-11)	18.4	7.7(-15)
3500	1000	4000	182	1326	9.9	3.7(-11)	71.3	1.0(-14)
3000	1500	4000	158	1344	9.5	2.1(-11)	151.8	6.4(-15)
2500	2000	4000	176	1722	10.7	3.1(-11)	246.5	6.1(-15)

Table 7. Different levels of the sparsity of the constraint matrices

$n = 5000$; spars$(G) = 99.90\%$ $\|G\|_2 = 1$ $K(G) = 10^4$								
			VPM				E04NKF	
	spars$\begin{pmatrix} A \\ C \end{pmatrix}$	nz	it	it_{in}	$time$	er_x	$time$	er_x
$m_e = 2500$	99.94	7505	254	4665	13.2	3.2(-11)	20.9	1.1(-14)
$m_i = 0$	99.92	10004	250	12570	39.6	4.6(-11)	22.6	1.6(-14)
$\|C\|_2 = 1$	99.90	12501	375	26977	108.0	6.4(-11)	29.9	8.1(-15)
$K(C) = 10^3$	99.89	13757	431	31612	137.0	7.2(-11)	39.8	3.1(-14)
$m_e = 3500$ $m_i = 1000$	99.93	15752	298	6828	26.6	5.1(-11)	64.5	1.3(-14)
$nac = 4000$	99.91	20250	379	10506	54.5	4.2(-11)	94.4	1.0(-14)
$\left\|(A^T\ C^T)^T\right\|_2 = 1$	99.89	24753	434	16920	133.5	4.4(-11)	116.8	1.6(-14)
$K\left((A^T\ C^T)^T\right) = 10^2$	99.87	29255	481	24450	266.9	4.5(-11)	145.5	1.3(-14)

The numerical results of Table 6 are concerning QP problems with equality and inequality constraints. We analyse the performance of the methods for different values of the active inequality constraints. The routine E04NKF, since it is based on an active-set strategy, benefits by a small value of the number $nac - m_e$ of inequality constraints active in the solution while the behaviour of the VPM is weakly dependent on $nac - m_e$.

Finally, in Table 7 test problems with decreasing value of the sparsity of the constraint matrices are considered. We denote by nz the number of nonzero elements of A and C. The NAG routine appears to be less dependent on the sparsity than the VPM. We conclude that the VPM may be an effective approach for large QP problems with very sparse constraint matrices and, in some cases, a valid alternative to the classical active–set strategy.

Acknowledgment. We are grateful to the referees for their helpful comments and suggestions. We are also indebted to L. Vittori of LASERTEC s.r.l. – La Spezia (Italy) for his support about NAG library.

References

[1] R.W. Cottle, J.S. Pang, R.E. Stone. (1992), *The Linear Complementarity Problem*, Academic Press, San Diego.

[2] C.W. Cryer (1971), "The solution of a quadratic programming problem using systematic overrelaxation," *SIAM J. Control*, 9, 385–392.

[3] S. Dafermos (1980), "Traffic equilibrium and variational inequalities," *Transportation Sci.*, 14, 42–54.

[4] N. Dyn., J. Ferguson (1983), "The numerical solution of equality-constrained quadratic programming problems," *Math. Comp.*, 41, 165–170.

[5] M. Fukushima (1992), "Equivalent differentiable optimization problems and descent methods for asymmetric variational inequality problems," *Math. Programming*, 53, 99–110.

[6] D.J. Evans (1968), "The use of preconditioning in iterative methods for solving linear equations with symmetric positive definite matrices," *J. IMA*, 4, 295–314.

[7] E. Galligani, V. Ruggiero (1992), "The arithmetic mean preconditioner for multivector computers," *Intern. J. Computer Math.*, 44, 207–222.

[8] E. Galligani, V. Ruggiero, L. Zanni (1996), "Splitting methods for constrained quadratic programs in data analysis," *Computers Math. Applic.*, 32, 1–9.

[9] E. Galligani, V. Ruggiero, L. Zanni (1997), "Splitting methods for quadratic optimization in data analysis," *Intern. J. Computer Math.*, 63, 289–307.

[10] E. Galligani, V. Ruggiero, L. Zanni (1998), "Parallel solution of large scale quadratic programs," in *High Performance Algoritms and Software in Nonlinear Optimization*, (R. De Leone, A. Murli, P. Pardalos, G. Toraldo eds.), Kluwer Academic Publishers, Dordrecht, 189–205.

[11] B. He (1994), "Solving a class of linear projection equations," *Numer. Math.*, 68, 71–80.

[12] Y.Y. Lin, J.S. Pang (1987), "Iterative methods for large convex quadratic programs: a survey," *SIAM J. Control Optim.*, 25, 383–411.

[13] O.L. Mangasarian (1977), "Solution of symmetric linear complementarity problems by iterative methods," *J. Optim. Theory Appl.*, 22, 465–485.

[14] O.L. Mangasarian, R. De Leone (1988), "Parallel gradient projection successive overrelaxation for symmetric linear complementarity problems and linear programs," *Ann. Oper. Res.*, 14, 41–59.

[15] P. Marcotte, J.H. Wu (1995), "On the convergence of projection methods: application to the decomposition of affine variational inequalities," *J. Optim. Theory Appl.*, 85, 347–362.

[16] W. Murray (1997), "Sequential quadratic programming methods for large-scale problems," *Computat. Optim. Applic.*, 7, 127–142.

[17] NAG Fortran Library Manual. (1998), Mark 18.

[18] J.M. Ortega. (1972), *Numerical Analysis: A Second Course*, Academic Press, New York.

[19] J.S. Pang, D. Chan (1982), "Iterative methods for variational and complementarity problems," *Math. Programming*, 24, 284–313.

[20] V. Ruggiero, L. Zanni (1999), "On a class of iterative methods for large–scale convex quadratic programs," in Numerical Methods in Optimization (A. Maugeri, E. Galligani eds.), *Rend. Circ. Matem. Palermo*, Serie II, Suppl. 58, 205–212.

[21] V. Ruggiero, L. Zanni, "A modified projection algorithm for large strictly convex quadratic programs," *J. Optim. Theory Appl.*, to appear.

[22] M.V. Solodov, P. Tseng (1996), "Modified projection-type methods for monotone variational inequalities," *SIAM J. Control Optim.*, 34, 1814–1830.

[23] G.W. Stewart (1980), "The efficient generation of random orthogonal matrices with an application to condition estimators," *SIAM J. Numer. Anal.*, 17 , 403–409.

[24] L. Zanni (1992), "On the convergence rate of two projection methods for variational inequalities in R^n," *Calcolo*, 29, 193–212.

A Comparison of Rates of Convergence of Two Inexact Proximal Point Algorithms

Michael V. Solodov (solodov@impa.br)
Instituto de Matemática Pura e Aplicada,
Estrada Dona Castorina 110, Jardim Botânico,
Rio de Janeiro, RJ 22460-320, Brazil

Benar F. Svaiter (benar@impa.br)
Instituto de Matemática Pura e Aplicada,
Estrada Dona Castorina 110, Jardim Botânico,
Rio de Janeiro, RJ 22460-320, Brazil

Abstract

We compare the linear rate of convergence estimates for two inexact proximal point methods. The first one is the classical inexact scheme introduced by Rockafellar, for which we obtain a slightly better estimate than the one given in [16]. The second one is the hybrid inexact proximal point approach introduced in [25, 22]. The advantage of the hybrid methods is that they use more constructive and less restrictive tolerance criteria in inexact solution of subproblems, while preserving all the favorable properties of the classical method, including global convergence and local linear rate of convergence under standard assumptions. In this paper, we obtain a linear convergence estimate for the hybrid algorithm [22], which is better than the one for the classical method [16], even if our improved estimate is used for the latter.

Keywords: Maximal monotone operator, proximal point method, linear convergence.

G. Di Pillo and F. Giannessi (eds.), Nonlinear Optimization and Related Topics, 415–427.

1 Introduction

We consider the classical problem of finding a zero of a maximal monotone operator (or a multifunction) T on a real Hilbert space $\mathcal{H}$:

$$\text{find } x \in \mathcal{H} \text{ such that } 0 \in T(x). \tag{1}$$

This is a well known problem which includes, as special cases, optimization and min-max problems, complementarity problems, and variational inequalities.

One of the fundamental approaches to solving (1) is the proximal point method, which generates the next iterate x^{k+1} by solving the subproblem

$$0 \in c_k T(x) + (x - x^k), \tag{2}$$

where x^k is the current iterate, and $c_k > 0$ is a regularization parameter. The literature on this subject is vast and includes, but is not limited to, [15, 16, 14, 13, 12, 8, 9, 7] (see [11] for a survey).

For developing implementable computational techniques, of particular importance is treating the case when subproblems are solved only approximately. This leads to the following scheme:

$$\begin{aligned}&\text{find } x^{k+1} \in \mathcal{H} \text{ such that}\\ &v^{k+1} \in T(x^{k+1}), \quad 0 = c_k v^{k+1} + (x^{k+1} - x^k) - r^k,\end{aligned} \tag{3}$$

where $r^k \in \mathcal{H}$ is an error associated with inexact solution of subproblem (2). The following tolerance criteria imposed on error terms r^k were first introduced in [16], and they (or their variations) are standard in the literature (see, for example, [1, 26, 4, 6, 5]):

$$\|r^k\| \le \delta_k, \quad \sum_{k=0}^{\infty} \delta_k < \infty, \tag{4}$$

and

$$\|r^k\| \le \delta_k \|x^{k+1} - x^k\|, \quad \sum_{k=0}^{\infty} \delta_k < \infty. \tag{5}$$

The first condition is typically needed to establish global convergence of the proximal point algorithm, while the second is required for local linear rate of convergence result under some additional natural assumptions.

Recently, some new proximal-based algorithms have been proposed [25, 22], which use improved error criteria. The method in [25] is based on projection ideas (related, for example, to [18]). It works as follows. A pair y^k, v^k is considered an admissible approximate solution of proximal point subproblem (2) if

$$v^k \in T(y^k), \quad 0 = c_k v^k + (y^k - x^k) - r^k,$$

and the error r^k satisfies

$$\|r^k\| \le \delta \max\{c_k\|v^k\|, \|y^k - x^k\|\}, \tag{6}$$

where δ is a *fixed* scalar in $[0, 1)$. The next iterate x^{k+1} is obtained by projecting x^k onto the hyperplane

$$H_k = \{z \in \mathcal{H} \mid \langle v^k, z - y^k \rangle = 0\},$$

which can be shown to strictly separate x^k from any zero of T, whenever (6) is satisfied. This projection step is the key for ensuring global convergence of the method, i.e., without it, the inexact proximal method managed by (6) may fail (see an example in [25]). It is important to note that projection onto a hyperplane is explicit, and thus its cost is computationally negligible.

Another method, introduced in [22], is based on the extragradient ideas (related, for example, to [10, 19]). It works as follows:

$$x^{k+1} = x^k - c_k v^k,$$

where

$$v^k \in T^{\varepsilon_k}(y^k), \quad 0 = c_k v^k + (y^k - x^k) - r^k.$$

Here, T^{ε_k} is the ε_k-enlargement (outer approximation) of T introduced by Burachik, Iusem and Svaiter in [2] (see also [3]), and the error tolerance criterion is

$$\|r^k\|^2 + 2c_k\varepsilon_k \le \delta^2\|y^k - x^k\|^2, \tag{7}$$

where again the tolerance parameter $\delta \in [0, 1)$ can be fixed. In this method, inexact solutions of subproblems are treated both, through the explicit error term r^k, and through T^{ε_k}, a current approximation of T. If $v^k \in T(y^k)$ then $\varepsilon_k = 0$ and the above error tolerance criterion reduces to

$$\|r^k\| \le \delta\|y^k - x^k\|,$$

which is somewhat stronger than (6).

Both hybrid methods outlined above preserve all the desirable convergence properties of the exact proximal algorithm (or its inexact version given by (3) with error criteria (4), (5)), including global convergence and local linear rate under standard assumptions. The advantage of the new error criteria (6) or (7) is that they are constructive, and the relaxation parameter δ need not tend to zero to ensure convergence (for example, it can be fixed). Another way of looking at it, is to observe that (6) is equivalent to saying that the bound for the *relative error* in solving the proximal subproblem can be fixed (and need not tend to zero), which is computationally realistic. This compares favorably with the conditions in (4), (5). This hybrid strategy has proved to be useful in many applications (see [20, 21, 17, 23, 24]), among which is the design of truly globally convergent inexact Newton methods.

In the case when the proximal subproblems are solved without error, the hybrid (projection or extragradient) methods coincide with the classical exact proximal point method.

While it has been established in [25, 22] that the hybrid methods possess the linear convergence property under the same assumptions as the classical exact/inexact schemes of [16], the issue of which of the methods converges faster has not been addressed yet. This question has been posed to the authors several times since the methods were first proposed. In this paper, we give the answer by comparing the linear rate of convergence estimates of the method from [16] with a similar estimate of the hybrid extragradient-proximal point method from [22] described above. We note that this comparison is not obvious if one is simply to look at the results in [16] and [22]. The paper is organized as follows. In Section 2, we give a linear convergence rate result for the classical method. In fact, by slightly modifying the argument, we obtain a somewhat improved linear rate estimate than the one given in [16]. In Section 3, we refine the previous analysis in [22] for the hybrid extragradient-proximal point method, and demonstrate that its linear rate of convergence is actually faster than for the classical method, even if the improved result from Section 2 is used for the latter.

2 Linear convergence of the classical method

For the purposes of this section, we assume that the sequence generated by the inexact proximal point method converges (strongly), and that T^{-1} is Lipschitz continuous around zero, i.e., problem (1) has the unique solution, say x^*, and there exist some constants $a > 0$ and $\tau > 0$ such that

$$\|v\| \leq \tau, v \in T(y) \Rightarrow \|y - x^*\| \leq a\|v\|. \tag{8}$$

Specifically, we consider a sequence satisfying the following relations:

$$\begin{aligned} 0 = c_k v^{k+1} + (x^{k+1} - x^k) - r^k, \quad v^{k+1} \in T(x^{k+1}), \\ \|r^k\| \leq \delta_k \|x^{k+1} - x^k\|, \end{aligned} \tag{9}$$

where the regularization parameters sequence $\{c_k\}$ is bounded away from zero. As it has been shown in [16, Theorem 2], the following linear convergence rate estimate holds (for iteration indices k sufficiently large):

$$\|x^{k+1} - x^*\| \leq \theta_k \|x^k - x^*\|, \tag{10}$$

where

$$\theta_k = \frac{\mu_k + \delta_k}{1 - \delta_k}, \tag{11}$$

with

$$1 > \mu_k = \frac{a}{\sqrt{a^2 + c_k^2}}. \tag{12}$$

Clearly, once δ_k is sufficiently small, it holds that $\theta_k < 1$, and so (10) implies linear rate of convergence. Note also that since for the purposes of this study we assume that $\{x^k\}$ converges to the solution, the requirement that $\sum_{k=0}^{\infty} \delta_k < \infty$ is no longer needed, and we can relax it by merely saying that δ_k is small enough to ensure that $\theta_k < 1$.

In what follows, by slightly modifying the analysis in [16], we obtain a better linear convergence estimate, where θ_k in (10) is replaced by

$$\hat{\theta}_k = \mu_k + \frac{\delta_k}{1 - \delta_k} \quad (\leq \theta_k). \tag{13}$$

Observe that $\hat{\theta}_k < \theta_k$ whenever $\delta_k \neq 0$, i.e., the two estimates are the same only for the exact method.

Theorem 1 *Suppose a sequence $\{x^k\}$ defined by (9) converges to x^* satisfying condition (8). Then for all k sufficinetly large, it holds that*

$$\|x^{k+1} - x^*\| \leq \hat{\theta}_k \|x^k - x^*\|,$$

where $\hat{\theta}_k$ is given by (13).

Proof. Let s^k, $w^k \in T(s^k)$ be the exact solution of the k-th proximal point subproblem, that is

$$c_k w_k + s^k - x^k = 0.$$

Recall ([16, Proposition 3]) that the condition $\|r^k\| \leq \delta_k \|x^{k+1} - x^k\|$ implies that

$$\|x^{k+1} - s^k\| \leq \delta_k \|x^{k+1} - x^k\|. \tag{14}$$

Therefore

$$\|s^k - x^k\| \leq \|s^k - x^{k+1}\| + \|x^{k+1} - x^k\| \leq (1 + \delta_k)\|x^{k+1} - x^k\|,$$

which, using the convergence of $\{x^k\}$, implies that $\|x^k - s^k\| \to 0$. Because $\{c_k\}$ is bounded away from zero, we further conclude that $w_k = (1/c_k)(x^k - s^k) \to 0$. Using (8), we have that for indices k sufficiently large

$$\|s^k - x^*\| \leq a\|w^k\| = (a/c_k)\|s^k - x^k\|. \tag{15}$$

We further obtain

$$\begin{aligned} \|x^k - x^*\|^2 &= \|x^k - s^k\|^2 + \|s^k - x^*\|^2 + 2\langle x^k - s^k, s^k - x^* \rangle \\ &= \|x^k - s^k\|^2 + \|s^k - x^*\|^2 + 2c_k \langle w^k, s^k - x^* \rangle \\ &\geq \|x^k - s^k\|^2 + \|s^k - x^*\|^2 \\ &\geq (1 + (c_k/a)^2)\|s^k - x^*\|^2, \end{aligned} \tag{16}$$

where the first inequality follows from the monotonicity of T, and the last follows from (15). Hence,

$$\|s^k - x^*\| \leq \mu_k \|x^k - x^*\|,$$

where μ_k is defined by (12). Using the latter relation and (14), we further obtain

$$\begin{aligned} \|x^{k+1} - x^*\| &\leq \|x^{k+1} - s^k\| + \|s^k - x^*\| \\ &\leq \delta_k \|x^{k+1} - x^k\| + \mu_k \|x^k - x^*\|. \end{aligned} \tag{17}$$

Similarly,

$$\begin{aligned} \|x^{k+1} - x^k\| &\leq \|x^{k+1} - s^k\| + \|s^k - x^k\| \\ &\leq \delta_k \|x^{k+1} - x^k\| + \|x^k - x^*\|, \end{aligned}$$

where also (16) was used in the last inequality. Therefore,

$$\|x^{k+1} - x^k\| \leq (1 - \delta_k)^{-1} \|x^k - x^*\|.$$

Now, combining the latter relation with (17), we obtain the desired result:

$$\|x^{k+1} - x^*\| \leq \left(\mu_k + \frac{\delta_k}{1 - \delta_k}\right) \|x^k - x^*\|.$$

The proof is complete. □

Remark. The difference between the analysis given above and [16, Theorem 2] is basically in the bound for $\|x^{k+1} - x^k\|$ once (17) is obtained. In [16, Theorem 2], the bound

$$\|x^{k+1} - x^k\| \leq \|x^{k+1} - x^*\| + \|x^* - x^k\|$$

is used, which in the end results in a weaker linear rate estimate.

3 Linear convergence of the hybrid method

In this section, we consider the extragradient-proximal point method of [22]. We derive a convergence rate estimate which is strictly better than the one for the classical inexact proximal algorithm.

In [2], Burachik, Iusem and Svaiter defined $T^\varepsilon(x)$, an ε-enlargement of a monotone operator T, as

$$T^\varepsilon(x) := \{v \in \mathcal{H} \mid \langle u - v, y - x\rangle \geq -\varepsilon \quad \forall\, y, u \in T(y)\}, \tag{18}$$

where $\varepsilon \geq 0$. Since T is assumed to be maximal monotone,

$$T^0(x) = T(x)$$

for any x. Furthermore, directly from the definition it follows that

$$0 \le \varepsilon_1 \le \varepsilon_2 \Rightarrow T^{\varepsilon_1}(x) \subseteq T^{\varepsilon_2}(x).$$

Thus T^ε is an *enlargement* of T. The use of elements in T^ε instead of T allows an extra degree of freedom, which is very usefull in various applications (see [2, 3, 24]). On the other hand, setting $\varepsilon = 0$ one retrieves the original operator T, so that the classical methods can be also treated.

We now turn our attention to the hybrid proximal-based method of [22]:

$$\begin{gathered} x^{k+1} = x^k - c_k v^k, \\ v^k \in T^{\varepsilon_k}(y^k), \quad 0 = c_k v^k + (y^k - x^k) - r^k, \\ \sqrt{\|r^k\|^2 + 2c_k\varepsilon_k} \le \delta_k \|y^k - x^k\|. \end{gathered} \tag{19}$$

As in Section 2, we study the case when the sequence $\{x^k\}$ converges to the unique zero of T, with T^{-1} being Lipschitz continuous around zero, i.e., (8) holds. In [22], the following linear rate estimate was given :

$$\|x^{k+1} - x^*\| \le \sqrt{1 - \frac{1-\delta_k}{(1 + a/c_k)^2(1+\delta_k)}}\|x^k - x^*\|.$$

This estimate appears to be not better than the one for the classical method. In this section, by refining the analysis we obtain a new sharper estimate which suggests that the hybrid algorithm attains a faster linear rate of convergence than the classical method. Specifically, we show that the factor of linear convergence for the former is

$$\tilde{\theta}_k := \left(1 + \frac{1-\delta_k^2}{\left(a/c_k + \delta_k\sqrt{1 + (a/c_k)^2}\,\right)^2}\right)^{-1/2}, \tag{20}$$

which is strictly (unless the method is exact) smaller than $\hat{\theta}_k$ in Theorem 1.

Theorem 2 *Suppose a sequence $\{x^k\}$ defined by (19) converges to x^* satisfying condition (8). Then for all k sufficiently large, it holds that*

$$\|x^{k+1} - x^*\| \le \tilde{\theta}_k \|x^k - x^*\|,$$

where $\tilde{\theta}_k$ is given by (20).

Proof. First observe that by the construction of the algorithm,

$$r^k = c_k v^k + y^k - x^k = y^k - x^{k+1}. \tag{21}$$

By direct computation, we obtain

$$\begin{aligned}\|x^k - x^*\|^2 - \|x^{k+1} - x^*\|^2 &= \|y^k - x^*\|^2 + 2\langle y^k - x^*, x^k - y^k\rangle + \|y^k - x^k\|^2 \\ &\quad - \left(\|y^k - x^*\|^2 + 2\langle y^k - x^*, x^{k+1} - y^k\rangle + \|x^{k+1} - y^k\|^2\right) \\ &= 2\langle y^k - x^*, x^k - x^{k+1}\rangle + \|y^k - x^k\|^2 - \|x^{k+1} - y^k\|^2 \\ &= 2c_k\langle v^k, y^k - x^*\rangle + \|y^k - x^k\|^2 - \|r^k\|^2.\end{aligned}$$

Since $v_k \in T^{\varepsilon_k}(y^k)$ and $0 \in T(x^*)$, from (18) it follows that

$$\langle v^k, y^k - x^*\rangle \geq -\varepsilon_k ,$$

so

$$\|x^k - x^*\|^2 - \|x^{k+1} - x^*\|^2 \geq \|y^k - x^k\| - \left(\|r^k\|^2 + 2c_k\varepsilon_k\right) .$$

Using further the error tolerance criterion, we obtain

$$\|x^k - x^*\|^2 \geq \|x^{k+1} - x^*\|^2 + (1 - \delta_k^2)\|y^k - x^k\|^2. \tag{22}$$

Since $\{x^k\}$ converges to x^* (strongly), and $1 > \limsup_{k\to\infty} \delta_k$, it follows that

$$0 = \lim_{k\to\infty} \|y^k - x^k\|. \tag{23}$$

Now from the error tolerance criterion we conclude that

$$0 = \lim_{k\to\infty} \|r^k\| = \lim_{k\to\infty} c_k\varepsilon_k. \tag{24}$$

As before, let s^k, $w^k \in T(s^k)$ be the exact solution of the k-th proximal point subproblem, that is

$$c_k w_k + s^k - x^k = 0.$$

Observe that

$$c_k(v^k - w^k) = s^k - y^k + r^k.$$

Since $v^k \in T^{\varepsilon_k}(y^k)$ and $w^k \in T(s^k)$, using (18) we get

$$\begin{aligned}-c_k\varepsilon_k &\leq c_k\langle v^k - w^k, y^k - s^k\rangle \\ &= \langle s^k - y^k + r^k, y^k - s^k\rangle \\ &= -\|s^k - y^k\|^2 - \langle r^k, s^k - y^k\rangle \\ &= -\|s^k - y^k + r^k/2\|^2 + \|r^k\|^2/4.\end{aligned} \tag{25}$$

Define q^k as the middle point between y^k and x^{k+1}, i.e., $q^k = (y^k + x^{k+1})/2$. In view of (21), we have

$$\begin{aligned}q^k &= y^k - r^k/2 \\ &= x^{k+1} + r^k/2.\end{aligned} \tag{26}$$

Using (25) it now follows that

$$\|s^k - q^k\|^2 \le c_k\varepsilon_k + \|r^k\|^2/4. \tag{27}$$

By the triangle inequality and (26), (27), we have that

$$\begin{aligned}\|s^k - x^k\| &\le \|y^k - x^k\| + \|s^k - q^k\| + \|q^k - y^k\| \\ &\le \|y^k - x^k\| + \sqrt{c_k\varepsilon_k + \|r^k\|^2/4} + \|r^k\|/2.\end{aligned}$$

Using (23) and (24), we conclude that

$$0 = \lim_{k\to\infty} \|s^k - x^k\|.$$

Since c_k is bounded away from zero, it also follows from $w^k = (1/c_k)(x^k - s^k)$ that

$$0 = \lim_{k\to\infty} \|w^k\|.$$

Thus, using (8), for k large enough it holds that

$$\|s^k - x^*\| \le a\|w^k\| = (a/c_k)\|s^k - x^k\| .$$

Using this inequality and the triangle inequality, we obtain

$$\begin{aligned}\|x^{k+1} - x^*\| &\le \|x^{k+1} - s^k\| + \|s^k - x^*\| \\ &\le \|x^{k+1} - s^k\| + (a/c_k)\|s^k - x^k\| \\ &\le \|x^{k+1} - s^k\| + (a/c_k)\|s^k - y^k\| + (a/c_k)\|y^k - x^k\|. \end{aligned} \tag{28}$$

Note that by the Cauchy-Schwarz inequality it holds that

$$\|x^{k+1} - s^k\| + (a/c_k)\|s^k - y^k\| \le \sqrt{\|x^{k+1} - s^k\|^2 + \|s^k - y^k\|^2}\ \sqrt{1 + (a/c_k)^2}\ .$$

By (26),

$$\begin{aligned}s^k - y^k &= (s^k - q^k) - (1/2)r^k, \\ s^k - x^{k+1} &= (s^k - q^k) + (1/2)r^k.\end{aligned}$$

Hence

$$\begin{aligned}\|x^{k+1} - s^k\|^2 + \|s^k - y^k\|^2 &= 2\|s^k - q^k\|^2 + (1/2)\|r^k\|^2 \\ &\le 2c_k\varepsilon_k + \|r^k\|^2 \\ &\le \delta_k^2\|y^k - x^k\|^2,\end{aligned}$$

where (27) was used for deriving the first inequality, and the error criterion was used in the second inequality. Therefore

$$\|x^{k+1} - s^k\| + (a/c_k)\|s^k - y^k\| \le \delta_k\sqrt{1 + (a/c_k)^2}\ \ \|y^k - x^k\|,$$

which combined with (28) gives

$$\|x^{k+1} - x^*\| \le \left[(a/c_k) + \delta_k\sqrt{1 + (a/c_k)^2}\,\right] \|y^k - x^k\|.$$

Using further (22), we get

$$\|x^k - x^*\|^2 \ge \left[1 + \frac{1 - \delta_k^2}{\left((a/c_k) + \delta_k\sqrt{1 + (a/c_k)^2}\,\right)^2}\right] \|x^{k+1} - x^*\|^2.$$

So, for k large enough,

$$\|x^{k+1} - x^*\| \le \tilde{\theta}_k \|x^k - x^*\|,$$

where $\tilde{\theta}_k$ is defined in (20). It remains to compare $\tilde{\theta}_k$ with $\hat{\theta}_k$ given by (13). Denote

$$t_k := c_k/a\ .$$

Then we have

$$\tilde{\theta}_k = \left(1 + \frac{t_k^2(1 - \delta_k^2)}{\left(1 + \delta_k\sqrt{1 + t_k^2}\,\right)^2}\right)^{-1/2}$$

By direct computations,

$$1 + \frac{t_k^2(1 - \delta_k^2)}{(1 + \delta_k\sqrt{1 + t_k^2})^2} = \left(\frac{\delta_k + \sqrt{1 + t_k^2}}{1 + \delta_k\sqrt{1 + t_k^2}}\right)^2$$

Recall that

$$\mu_k = 1/\sqrt{1 + t_k^2} < 1.$$

Hence,

$$\begin{aligned}
\tilde{\theta}_k &= \frac{1 + \delta_k\sqrt{1 + t_k^2}}{\delta_k + \sqrt{1 + t_k^2}} \\
&= \frac{(1/\sqrt{1 + t_k^2}) + \delta_k}{(\delta_k/\sqrt{1 + t_k^2}) + 1} \\
&= \frac{\mu_k + \delta_k}{\mu_k\delta_k + 1} \\
&\le \mu_k + \delta_k \\
&\le \mu_k + \frac{\delta_k}{1 - \delta_k} \\
&= \hat{\theta}_k \quad (\le \theta_k).
\end{aligned}$$

□

Remark. Observe that if $\delta_k = 0$ then $\tilde{\theta}_k = \hat{\theta}_k = \theta_k$ $(= \mu_k)$. This is natural because in that case all methods considered here reduce to the classical exact proximal algorithm, and hence have the same convergence rate. However, all inequalities above are strict, i.e., $\tilde{\theta}_k < \hat{\theta}_k < \theta_k$, whenever $\delta_k \neq 0$, which is the more interesting and important case corresponding to inexact methods.

4 Concluding remarks

We obtained sharper estimates for the linear convergence rates for both the classical inexact proximal point method, and the hybrid extragradient-proximal point algorithm. According to these results, the rate for the latter is faster than for the former. However, it is important to keep in mind that these results are only theoretical estimates. As of now, we have a better estimate for the hybrid algorithm. Since it is not known whether or not the estimates are tight, strictly speaking, our results do not automatically imply that the hybrid methods converge faster locally. However, this theoretical comparison is still useful. And of course, it is possible that one may further improve the rate of convergence estimates for either method through a more intricate analysis.

Acknowledgements. We are grateful to Professor R.T. Rockafellar who posed to us the question of comparison of the rates of convergence discussed in this paper.

Research of the first author is supported by CNPq Grant 300734/95-6 and by PRONEX–Optimization, research of the second author is supported by CNPq Grant 301200/93-9(RN) and by PRONEX–Optimization.

References

[1] A. Auslender. Numerical methods for nondifferentiable convex optimization. *Mathematical Programming Study*, 30:102–126, 1987.

[2] R.S. Burachik, A.N. Iusem, and B.F. Svaiter. Enlargement of monotone operators with applications to variational inequalities. *Set-Valued Analysis*, 5:159–180, 1997.

[3] R.S. Burachik, C.A. Sagastizábal, and B.F. Svaiter. ε-Enlargements of maximal monotone operators: Theory and applications. In M. Fukushima and L. Qi, edi-

tors, *Reformulation - Nonsmooth, Piecewise Smooth, Semismooth and Smoothing Methods*, pages 25–44. Kluwer Academic Publishers, 1998.

[4] J.V. Burke and M. Qian. A variable metric proximal point algorithm for monotone operators. *SIAM Journal on Control and Optimization*, 37:353–375, 1998.

[5] R. Cominetti. Coupling the proximal point algorithm with approximation methods. *Journal of Optimization Theory and Applications*, 95:581–600, 1997.

[6] J. Eckstein. Approximate iterations in Bregman-function-based proximal algorithms. *Mathematical Programming*, 83:113–123, 1998.

[7] J. Eckstein and D.P. Bertsekas. On the Douglas-Rachford splitting method and the proximal point algorithm for maximal monotone operators. *Mathematical Programming*, 55:293–318, 1992.

[8] M.C. Ferris. Finite termination of the proximal point algorithm. *Mathematical Programming*, 50:359–366, 1991.

[9] O. Güler. New proximal point algorithms for convex minimization. *SIAM Journal on Optimization*, 2:649–664, 1992.

[10] G.M. Korpelevich. The extragradient method for finding saddle points and other problems. *Matecon*, 12:747–756, 1976.

[11] B. Lemaire. The proximal algorithm. In J.P. Penot, editor, *New Methods of Optimization and Their Industrial Use. International Series of Numerical Mathematics 87*, pages 73–87. Birkhauser, Basel, 1989.

[12] F.J. Luque. Asymptotic convergence analysis of the proximal point algorithm. *SIAM Journal on Control and Optimization*, 22:277–293, 1984.

[13] B. Martinet. Regularisation d'inequations variationelles par approximations successives. *Revue Française d'Informatique et de Recherche Opérationelle*, 4:154–159, 1970.

[14] G.B. Passty. Weak convergence theorems for nonexpansive mappings in Banach spaces. *Journal of Mathematical Analysis and Applications*, 67:274–276, 1979.

[15] R.T. Rockafellar. Augmented Lagrangians and applications of the proximal point algorithm in convex programming. *Mathematics of Operations Research*, 1:97–116, 1976.

[16] R.T. Rockafellar. Monotone operators and the proximal point algorithm. *SIAM Journal on Control and Optimization*, 14:877–898, 1976.

[17] M. V. Solodov and B. F. Svaiter. A truly globally convergent Newton-type method for the monotone nonlinear complementarity problem, 1998. *SIAM Journal on Optimization*, submitted.

[18] M. V. Solodov and B. F. Svaiter. A new projection method for variational inequality problems. *SIAM Journal on Control and Optimization*, 37:765–776, 1999.

[19] M. V. Solodov and P. Tseng. Modified projection-type methods for monotone variational inequalities. *SIAM Journal on Control and Optimization*, 34:1814–1830, 1996.

[20] M.V. Solodov and B.F. Svaiter. Forcing strong convergence of proximal point iterations in a Hilbert space, 1997. *Mathematical Programming*, submitted.

[21] M.V. Solodov and B.F. Svaiter. A globally convergent inexact Newton method for systems of monotone equations. In M. Fukushima and L. Qi, editors, *Reformulation - Nonsmooth, Piecewise Smooth, Semismooth and Smoothing Methods*, pages 355–369. Kluwer Academic Publishers, 1998.

[22] M.V. Solodov and B.F. Svaiter. A hybrid approximate extragradient–proximal point algorithm using the enlargement of a maximal monotone operator, 1998. *Set-Valued Analysis*, submitted.

[23] M.V. Solodov and B.F. Svaiter. An inexact hybrid generalized proximal point algorithm and some new results on the theory of Bregman functions, 1998. *Mathematics of Operations Research*, submitted.

[24] M.V. Solodov and B.F. Svaiter. Error bounds for proximal point subproblems and associated inexact proximal point algorithms, 1999. *Mathematical Programming*, submitted.

[25] M.V. Solodov and B.F. Svaiter. A hybrid projection – proximal point algorithm. *Journal of Convex Analysis*, 6(1), 1999.

[26] P. Tossings. The perturbed proximal point algorithm and some of its applications. *Applied Mathematics and Optimization*, 29:125–159, 1994.

One way to Construct a Global Search Algorithm for d.c. Minimization Problems

Alexander S. Strekalovsky (strekal@icc.ru)
Institute of System Dynamics and Control Theory, SB of RAS,
Lermontov str., 134, Irkutsk-33, 664033, Russia

Abstract

We give necessary and sufficient Global Optimality Conditions for d.c. (difference of two convex functions) minimization problem and further development of a global search algorithm based on the theory. Finally we test the proposed algorithm on a concrete problem.

Keywords: Global Optimality Conditions, Global Search Algorithm.

1 Introduction

It is well-known [1-4], that a real-valued function $F(\cdot)$ defined on R^n is called to be d.c., if there are two convex functions $g(\cdot)$ and $f(\cdot)$, such that

$$F(x) = g(x) - f(x) \qquad \forall x \in R^n.$$

It turns out that the class DC(R^n) of d.c. functions is the vector space generated by the cone of convex functions, well studied in the literature [1-5].

Futhermore, DC(X) is dense in the set $C(X)$ of continuous functions over compact convex set X, endowed with the topology of uniform convergence over X. In addition, the class DC(X) enjoys a remarkable stability w.r.t. usual operations in Optimization[1-2].

Thus, there are many reasons to study d.c. programming problems. Moreover, some problems of practice directly leads us to d.c. optimization [1-4].

On the other hand, a lot of solving methods may be applied for searching a global solution to a d.c. optimization problem [1-2].

G. Di Pillo and F. Giannessi (eds.), Nonlinear Optimization and Related Topics, 429–443.

But, it seems to us that most of them have no relations with classical approaches of mathematical programming. And this is too natural because standard algorithms of mathematical programming have the ability of being trapped at a local extremum or even a stationary point, if a problem of interest has, for instance, a d.c. structure.

Recently enough, we proposed an approach to Global Optimality Conditions (GOC) for d.c. optimization problem which is connected with the classical extremum theory [5].

Here we will construct global search algorithms based on GOC [6] for d.c. minimization problems. May be, this would be useful in the solving of some application problems.

The aim of the paper is to answer three following questions. Does there exist a possibility to characterize a global solution in the d.c. minimization problem? May one construct a Global Search Algorithm (GSA) based upon GOC? Is it possible to carry out a computational Global Search for any (may-be, test) d.c. problem using this GSA?

On the other hand, we don't touch here the convergence problem for GSA, it will be the goal of a paper in one of Russian Mathematical Journals. Besides, we don't compare our approach with others, this one has been done only for a few real problems (for instance, for well-known combinatorial problem of maximal clique, where the results are rather promising) and will be published in further papers.

The paper consists of 5 sections. In section 2 we will prove the GOC disclosing it's algorithmic features. Note, that the proofs are completely different from these ones presented in [6].

In section 3 we introduce a general conception of the Global Search Algorithm based on GOC. In section 4 we consider a test (Floudas-Pardalos [3]) problem and study some features of the algorithm's realization.

In section 5 we present the first (not always positive) results of computational experiments and try to analyze and discuss some sides of the obtained picture.

2 Global Optimality Conditions

Consider the problem

$$F(x) \triangleq g(x) - f(x) \downarrow \min, \qquad x \in D, \tag{P}$$

where $g(\cdot)$ and $f(\cdot)$ are proper closed convex functions over R^n, s.t.

$$D \subset (int\ dom\ f \cap int\ dom\ g), \tag{1}$$

and D is a convex set from R^n.

In addition, suppose that $f(\cdot)$ is differentiable and there exists $v \in R^n$, s.t.

$$F(v) > F(z) \triangleq \zeta,\ v \in D. \tag{H}$$

It is easy to see that the assumption (H) is too natural in Problem (P), if a point $z \in D$ is suspected to be a global solution to Problem (P).

The following theorem gives a characterization of a global solution to (P).

Theorem 1. *A feasible point z is a global solution to Problem (P) ($z \in Sol(P)$), if and only if for every pair (y, β), such that ($y \in R^n, \beta \in R$)*

$$\left.\begin{array}{l} \beta - f(y) = \zeta \triangleq g(z) - f(z), \\ y \in D, \; g(y) \leq \beta \leq \sup(g, D), \\ \text{one has} \\ g(x) - \beta \geq \langle \nabla f(y), x - y \rangle \;\; \forall x \in D. \end{array}\right\} \qquad (E)$$

Proof. 1) Necessity. Suppose, (E) is broken down with some $(\overline{\beta}, \overline{y}, \overline{x})$:

$$\overline{\beta} - f(\overline{y}) = \zeta, \quad \text{so that} \quad g(\overline{x}) - \overline{\beta} < \langle \nabla f(\overline{y}), \overline{x} - \overline{y} \rangle .$$

Then due to convexity of $f(\cdot)$, one get

$$0 > g(\overline{x}) - \overline{\beta} - f(\overline{x}) + f(\overline{y}) = F(\overline{x}) - \zeta.$$

Thus, $F(\overline{x}) < F(z)$, which contradicts the fact that $z \in Sol(P)$.

2) Sufficiency. Suppose, (E) takes place but there is $u \in D : F(u) < F(z)$. Then, one has

$$g(u) < f(u) + \zeta. \qquad (2)$$

On the other hand, the assumption (H) is equivalent to the inequality

$$g(v) > f(v) + \zeta. \qquad (3)$$

In other words, (2) and (3) mean that

$$(u, g(u)) \notin epi\,[f(\cdot) + \zeta],$$

$$(v, g(v)) \in int\; epi\,[f(\cdot) + \zeta].$$

As a consequence,there exists $\lambda \in]0, 1[$, such that

$$(y, \beta) = \lambda(u, g(u)) + (1 - \lambda)(v, g(v)) \in bd\,[epi\,[f(\cdot) + \zeta]]$$

(where bdA means the boundary of a set A) or

$$\left.\begin{array}{l} f(y) + \zeta = \beta = \lambda g(u) + (1 - \lambda) g(v), \\ y = \lambda u + (1 - \lambda) v. \end{array}\right\} \qquad (4)$$

Then the estimation

$$\inf(g, D) \leq \beta \leq \sup(g, D)$$

becomes obvious, as the inequality by the way

$$g(y) \leq \beta.$$

If follows from (4) that

$$u = \frac{y-(1-\lambda)v}{\lambda}, \quad g(u) = \frac{\beta-(1-\lambda)g(v)}{\lambda}, \tag{5}$$

Since (E) holds, one has due to (H) and convexity of $f(\cdot)$

$$0 \geq \langle \nabla f(y), u-y\rangle + \beta - g(u) =$$

$$\left\langle \nabla f(y), \frac{y-(1-\lambda)v}{\lambda} - y\right\rangle + \beta - \frac{\beta-(1-\lambda)g(v)}{\lambda} =$$

$$= \frac{1-\lambda}{\lambda}\langle \nabla f(y), y-v\rangle + \frac{\lambda-1}{\lambda}[\beta - g(v)] \geq$$

$$\geq \frac{\lambda-1}{\lambda}[f(v) - f(y) + \beta - g(v)] =$$

$$= \frac{\lambda-1}{\lambda}[\zeta - F(v)] > 0.$$

The obtained absurd leads us to the conclusion that the condition (E) is uncompatible with the existence of a feasible point u, which is better than z.#

Remarks. 1) It may seem that the condition (E) has no relations to Classical Extremum Theory. But if we set $y = z$ in (E), we get $\beta = g(z)$ and if $D = R^n$, then one has

$$g(x) - \langle \nabla f(z), x\rangle \geq g(z) - \langle \nabla f(z), z\rangle \quad \forall x \in R^n,$$

so that $z \in Arg\min(\varphi, R^n)$,where

$$\varphi(x) = g(x) - \langle \nabla f(z), x\rangle.$$

Suppose, in addition, $g(\cdot)$ is differentiable. Then we obtain

$$\nabla g(z) = \nabla f(z),$$

what is well-known classical optimality condition for Problem (P), where $D = R^n$.

In the same way, one can obtain the Karush-Kuhn-Tucker conditions, if D is given by, say, equalities and inequalities conditions.

2) From the linearization view-point, one can say, that in order to verify whether a feasible point z is a global solution to (P) $\forall(y,\beta) : \beta - f(y) = \zeta$ we have to solve the linearized problem

$$g(x) - \langle \nabla f(y), x\rangle \downarrow \min, \quad x \in D, \tag{LP}$$

and after that we have to verify the inequality

$$g(x(y,\beta)) - \beta - \langle \nabla f(y), x(y,\beta) - y \rangle \geq 0, \tag{6}$$

where $x(y,\beta)$ is a global(!) solution to Linearized Problem (LP).

It is easy to see that Linearized Problem is simpler to solve than the original problem (P), since the principal non-convexity in (P) generated by the function $f(\cdot)$ does not exist any more in (LP). For instance, if D is convex, (LP) is also the convex problem solvable with usual mathematical programming tools.

So, roughly speaking, Problem (P) is reduced, so to say, to the choice of the parameters $(y,\beta) \in R^n \times R : \beta - f(y) = \zeta$.

3) Let turn our attention to the necessity's proof which express so-called Algorithmic Property of Optimality Condition (E). Actually, if there are the triplets $\left(\overline{\beta}, \overline{y}, \overline{x}\right)$which make the condition (E) broken down, accordingly to the proof, $\overline{x}$ is better than the feasible point z under study. So, in order to find a feasible point, which is better than the point z under study, we have to aspire to break down the inequality in (E) by means of solving Linearized Problems and of choice of $(y,\beta) : \beta - f(y) = \zeta \triangleq F(z)$.

This gives us a chance to propose an algorithm based on Algorithmic Property of GOC.

3 A general conception of Global Search

When searching a solution to a nonconvex problem, the crucial question is that whether a given feasible point z is global solution or not. And if not, how to construct another feasible point which is better than this one of interest?

For d.c. minimization problem (P) Theorem 1 gives a possibility to decide on the subject. Actually, if Condition (E) is not verified, there then exists, according to Remark 3, a better feasible point $\overline{x}$. Therefore, at a given feasible point it would be really constructive to try to break down OC (E). But how to do it? To do so, we have to seek a feasible point $\overline{x}$, such that

$$g(\overline{x}) - \overline{\beta} - \langle \nabla f(\overline{y}), \overline{x} - \overline{y} \rangle < 0,$$

where $\overline{\beta} - f(\overline{y}) = \zeta \triangleq F(z), g(\overline{y}) \leq \overline{\beta} \leq \sup(g, D)$.

Obviously, the simplest way to discover such a triplets $(\overline{x}, \overline{y}, \overline{\beta})$, if one exists, is to solve the following problem:

$$\left.\begin{array}{c} \Phi\,(x,y,\beta) \triangleq g(x) - \beta - \langle \nabla f(y), x - y \rangle \downarrow \min_{(x,y,\beta)}, \\ x \in D, \ (y,\beta) \in R^{n+1} : \beta - f(y) = \zeta \triangleq F(z), \\ g(y) \leq \beta \leq sup(g, D). \end{array}\right\} \tag{7}$$

However, this problem seems to be rather difficult. Nevertheless, one may simplify Problem (7), as suggested by OC (E) and the linearization view-point (see Remark 2), by considering Linearized Problem (LP). But to do so, we need to know some y and β satisfying the constraints in (7).

Let begin by some number β:

$$I \triangleq \inf(g, D) \leq \beta \leq \sup(g, D) \triangleq S.$$

Since in a computational implementation of a Global Search it is impossible to deal with the whole level surface

$$Y(\beta,\zeta) = \{y \in R^n / f(y) = \beta - \zeta\},$$

it would be reasonable to introduce some finite approximation of it

$$R(\beta,\zeta) = \left\{y^1, ..., y^N \mid f\left(y^i\right) = \beta - \zeta,\ y^i = y^i(\beta,\zeta),\ i = 1, ..., N,\ N = N(\beta,\zeta)\right\}.$$

The following operation is really suggested by OC (E) (see Remark 2)

a) For each $y^i \in R(\beta,\zeta)$ solve Linearized Problem:

$$g(x) - \left\langle \nabla f(y^i), x\right\rangle \downarrow \min,\ x \in D. \qquad (PL_i)$$

Recall, that if D is convex, Problem (PL_i) becomes convex and therefore solvable by some well-known convex minimization method suitible in the case.

Hence, the following hypothesis is well-grounded.

(HL). For all $y : f(y) = \alpha - \mu$, where $\alpha \in [I, S]$, $I = inf\,(g, D)$,
$S = sup\,(g, D)$, $\mu \in F(D)$, it is possible to solve (say approximately)
Linearized Problem (LP).

Let u^i be a global (!) solution to (PL_i).

b) Remembering the statement of Problem (7), we see that there exists may be yet one possibility to decrease the obtained value of $\Phi(u^i, y^i, \beta)$ by solving so-called Level Problem

$$\left\langle \nabla f(v), u^i - v\right\rangle \uparrow \max_v,\ f(v) = \beta - \zeta. \qquad (8)$$

Recall, that for the quadatic function $f(\cdot)$. Problem (8) is solvable analyticly [8] (see also Section 4). Therefore, the following hypothesis is also rather reasonable.

(HY). For all $u \in D$ and for all α and μ such as in (HL), one can find an
approximated global solution of Level Problem

$$\langle \nabla f(v), u - v\rangle \uparrow \max_v,\ f(v) = \alpha - \mu.$$

Suppose, w^i be a solution of (8).

c) Now it is possible to form the number

$$\left.\begin{aligned} \eta(\beta) &= \min_{1\leq i\leq N}\{g(u^i) - \beta - \langle \nabla f(w^i), u^i - w^i\rangle\} = \\ &= g(u^j) - \beta - \langle \nabla f(w^j), u^j - w^j\rangle \end{aligned}\right\} \qquad (9)$$

Only having a procedure to construct the number $\eta(\beta)$ for each β one can organize the minimization of the function $\eta(\beta)$ over the interval $[I, S]$. Don't forget that we need this minimization because we aim to minimize the objective function of Problem (7) w.r.t. variables (x, y, β) simultaneously what is generally speaking rather difficult. That is why we minimize $\Phi(x, y, \beta)$ w.r.t. each variable consecutively: first w.r.t. $x \in D((PL_i))$ second w.r.t. y (Problem (8)), finally w.r.t. $\beta \in [I, S]$. This means, that we have actually decomposed Problem (7) into three consecutive problems: (PL_i), (8) and

$$\eta(\beta) \downarrow \min, \quad \beta \in [I, S]$$

We are doing all this in order to break down the OC (E), that is to get $\eta(\beta) < 0$ or what is the same

$$g(u^j) - \beta - \left\langle \nabla f(w^j), u^j - w^j \right\rangle < 0.$$

Then, according to the necessity proof of Theorem 1, one has $F(u^j) < F(z)$, that means we have constructed the better point $u^j \in D$.

It remains to note that the effectiveness of all above constructions depends on a convinient choice of a finite approximation (grid) $R(\beta, \zeta)$ for each pair (β, ζ). One can find the discussion on the matter in [7]-[9].

Now let us put the idea of above in a more algorithmic form. Let be given a feasible point $x^0 \in D$ and a number sequence$\{\varepsilon_k\}$, $\varepsilon_k \downarrow 0$ $(k \to \infty)$. Set $k := 0$ and describe a global search algorithm step by step.

Step 1. Beginning from $x^k \in D$ obtain $z^k \in D$ by one of convenient local search methods for (P), so that z^k is a ε_k-stationary point to (P).

Step 2. Choose $\beta \in [I, S]$. In particular one may begin with $\beta_0 = g(z^k)$.

Step 3. Construct an approximation

$$R_k = \left\{y^1, ..., y^N \mid f\left(y^i\right) = \beta - \zeta_k\right\}, \quad \zeta_k = F\left(z^k\right).$$

Step 4. Introduce the set

$$I_k = I_k(\beta) = \{i \in \{1, ..., N_k\} \mid g(y^i) \leq \beta\}$$

Step 5. For $i \in I_k$ solve Linearized Problem

$$g(x) - \left\langle \nabla f(y^i), x \right\rangle \downarrow \min, \ x \in D. \qquad (PL_i)$$

Let u^i be an ε_k-solution of (PL_i).

Step 6. For $i \in I_k$ solve the level problem, that is, find a point w^i, $f(w^i) = \beta - \zeta_k$, s.t.

$$\left\langle \nabla f(w^i), u^i - w^i \right\rangle + \varepsilon_k \geq \sup_v \left\{ \left\langle \nabla f(v), u^i - v \right\rangle \mid f(v) = \beta - \zeta_k \right\}.$$

Step 7. Set

$$\eta_k(\beta) = g(u^j) - \beta - \left\langle \nabla f(w^j), u^j - w^j \right\rangle =$$

$$= \min_{i \in I_k} \left\{ g(u^i) - \beta - \left\langle \nabla f(w^i), u^i - w^i \right\rangle \right\}.$$

Step 8. If $\eta_k(\beta) < 0$, then set $x^{k+1} := u^j$ and go to step 1.

Step 9. If $\eta_k(\beta) \geq 0$, then set $\beta := \beta + \triangle\beta$ and go to step 3. One can find $\triangle$ by one of well-known one-dimensional minimization methods.

Step 10. Stop, if $\eta_k(\beta) \geq 0 \ \forall\beta \in [I, S]$ (i.e. one-dimensional search of β ove $[I, S]$ is finished) and if $\varepsilon_k \leq \delta$, where δ is a given tolerance.

Remarks. 4) It can be readily seen, that the above method is well-founded give Hypothesises (HL) and (HY).

5) From Step 1 it follows that the method generates the sequence $\{z^k\}$ of ϵ_k stationary point.

6) At Steps 1 and 5 one may use well-known minimization methods of Nonlinea Programming, what can be viewed as beneficial, because of possibility to apply th standard libraries of programmes. Note only, that these methods must be rather fas because one uses them several times at each iteration (!!!).

7) It can be readily seen that the method from above is not , so to say, an algorith in the common sence, because it is not precised exactly what algorithms we have t apply for solving Linearized Problem or Level Problem or what local search metho might be applied at Step 1. Besides, it is not clear, what one-dementional metho may be exploited to minimize $\eta_k(\beta)$ over $[I, S]$. All depends on the nature of Proble of interest. We will see below how this may be done in a concrete case. Therefor this method can be viewed as a principal algorithm [10] or a strategy [7-9].

Nevertheless, this principal method describes quite exactly the strategy of Glob Search founded upon GOC (E) from Theorem 1. Note, that if one began to precis only one unclear point, for instance, an algorithm for solving (PL_i) or one-dimention search on β over $[I, S]$, one would risk to lose the general view of Global Searc Strategy.

That's why we prefered the manner to present Global Search Strategy as above

4 Test problem

Consider the following problem from the famous book of Floudas-Pardalos [3]

$$F\left(x^1, x^2\right) = -0.5 \sum_{s=1}^{10} \lambda_s (x_s - a_s)^2 + 0.5 \sum_{s=11}^{20} \mu_s (x_s - a_s)^2 \qquad (10$$

$$\left.\begin{array}{c} A_1 x^1 + A_2 x^2 \leq b, \\ x^1, x^2 \in R^{10},\ x^1, x^2 \geq 0 \end{array}\right\} \qquad (11$$

All the data can be found in [3] (test-problem 2.10).

The best known solution here is

$$\left.\begin{array}{l} x_6^* = 4.348, \; x_{14}^* = 62.609, \\ x_i^* = 0, \; i \neq 6, 14. \\ F(x^*) = 49318. \end{array}\right\} \quad (12)$$

Now let us describe some particularities of solving the problem (10)-(11).

First, note that it was our first experience of the kind, and some aspects of the general idea's realization can be viewed as not satisfactory. Nevertheless, we would like to estimate them as only first attempts to apply the Theorem 1 for global search in a d.c. minimization problem.

For instance, as a local search method, the conditional gradient method was chosen, because of its simplicity for coding. The slow convergence of the method is well-known and it was confirmed during the computational experiments. The same optimization method was applied for solving the linearized problem, and this influenced a lot on the computational results, as we will see.

4.1 Approximation of the level surface.

Consider the choice of the points y^i on the level surface:

$$f(y^i) = \Theta \triangleq \beta - \zeta_k, \zeta_k = F(z^k), \; where \; f(x) = \frac{1}{2}\sum_{s=1}^{10} \lambda_s (x_s - a_s)^2.$$

Let be given a point z. In the computational experiments $z = z^k$, z^k being a feasible point. As an approximation of the level surface the following set can be taken

$$y^i = (y_s^i, \; s = 1, ..., 20), y_s^i = z_s, \; s \neq i,$$

$$y_i^i = 2a_i - z_i, \; i = 1, ..., 10;$$

$$y^{11} = (y_s^{11}, \; s = 1, ..., 20), y_s^{11} = z_s, \; s > 10,$$

$$y_s^{11} = 2a_s - z_s, \; s = 1, ..., 10;$$

In order to take into account the structure of the feasible set, we introduced into approximation the points which are generated by the rows $A^p \, (p = 1, ..., 10)$ of the matrix A_1

$$y^i = t_i \left(A^{i-11} - a^1\right) + a^1, \; i = 12, ..., 21. \quad (13)$$

It can be readily seen that the parameter t verifies the equality

$$t_i = \left(\frac{\Theta}{f\left(A^{i-11}\right)}\right)^{\frac{1}{2}}, \; i = 12, ..., 21. \quad (14)$$

4.2 The level problem solving.

Remember, the level problem has the form:

$$\left.\begin{aligned} \varphi(v) &\triangleq \langle \nabla f(v), u-v \rangle \uparrow \max, \\ f(v) &= \zeta,\ \zeta > 0,\ v \in R^{10}, \end{aligned}\right\} \tag{15}$$

where

$$f(x) = \frac{1}{2}\sum_{s=1}^{10} \lambda_s (x_s - a_s)^2,\ \nabla f(v) = E_\lambda (x-a) = (\lambda_s [x_s - a_s],\ s = 1, ..., 10)$$

and $u \in R^{10}$ is really only one part u^1 of the vector $u \in R^{20}$.

Let us solve the problem (15) analytically. In order to do it let represent the objective function of (15) as follows

$$\varphi(v) = \langle \nabla f(v), u \rangle - \langle \nabla f(v), v \rangle = \langle E_\lambda (v-a), u \rangle - \langle E_\lambda (v-a), v \rangle =$$
$$= \langle E_\lambda v, u \rangle - \langle E_\lambda a, u \rangle - \langle E_\lambda v, v \rangle + \langle E_\lambda a, v \rangle .$$

Then, it can be readily seen that the gradient of $\varphi(\cdot)$ has the form

$$\nabla\varphi(v) = E_\lambda^T u - 2E_\lambda^T v + E_\lambda^T a = E_\lambda(u+a) - 2E_\lambda v. \tag{16}$$

Now let us apply the Lagrange rule to the problem (15). If $w \in Arg\max(15)$, then there are γ_0 and $\gamma \in R$, such that

$$\gamma_0 \geq 0,\ \gamma_0 + |\gamma| > 0 \qquad \text{and} \qquad -\gamma_0 \nabla\varphi(w) + \gamma\nabla f(w) = 0.$$

Since $\nabla f(w) = E_\lambda(w-a)$, we have due to (16)

$$\gamma_0[2E_\lambda w - E_\lambda(u+a)] + \gamma E_\lambda (w-a) = 0. \tag{17}$$

Suppose, that $\gamma_0 = 0$, then $\gamma E_\lambda (w-a) = 0$.

Since $\gamma \neq 0$ and $E_\lambda = diag\{\lambda_1, ..., \lambda_{10}\}$, $(\lambda_s > 0)$, it follows $w = a$. Then

$$f(w) = \langle E_\lambda (a-a), a-a \rangle = 0,$$

that contradicts the equality in (15).

So, $\gamma_0 > 0$ and one can take $\gamma_0 = 1$.

Then, it follows from (17)

$$E_\lambda(2+\gamma)w = E_\lambda(u+a) + \gamma E_\lambda a,$$

whence

$$w = \frac{1}{2+\gamma}[u + (\gamma+1)a]. \tag{18}$$

Set $t = \frac{1}{2+\gamma}$, then

$$w = t\,[u - a] + a. \tag{18'}$$

In order to find the parameter t let us substitute (18') into the equality in (15)

$$2\zeta = 2f\,(w) = t^2\,\langle E_\lambda(u - a), u - a\rangle\,,$$

whence

$$t_{1,2} = \pm\left(\frac{1}{2\zeta}\,\langle E_\lambda(u - a), u - a\rangle\right)^{\frac{1}{2}} \tag{19}$$

It can be readily seen that

$$\varphi\,(w\,(t)) = \left(t - t^2\right)\langle E_\lambda(u - a), u - a\rangle\,,$$

so that the value of $\varphi\,(w\,(t))$ with $t > 0$ is greater than this one with $t < 0$.

Therefore, in (19) we set the sigh "plus" and w is presented by (18'). The analytical solving the level problem is completed.

5 Computational experiments

We will present below the numeric results of the first attempt to solve the problem (10)-(11) by R-algorithm using a serial PC/AT IBM-486 DX/2/66MHz.

First of all, let us explain how the "worst" initial feasible points were constructed. For instance, one can take this point under the form $(0, x_0^2)$, where $x_0^2 \in R^{10}$ is a solution of the problem ($g(x^2) = \frac{1}{2}\sum_{s=11}^{20}\mu_s\,(x_s - a_s)^2$)

$$g(x^2) \uparrow \max,\; A_2x^2 \leq b,\; x^2 \geq 0. \tag{20}$$

In the same way,we can take the point $(x_0^1, 0)$, where $x_0^1 \in R^{10}$ is a solution of the problem

$$f(x^1) \downarrow \min,\; A_1x^1 \leq b,\; x^1 \geq 0. \tag{21}$$

Further, one can combine two machineries from above by considering the problem

$$f(x^1) \downarrow \min,\; A_1x^1 \leq b - \; A_2x_0^2,\; x^1 \geq 0,$$

after having solved the problem(20).

In the same manner, one can solve the problem

$$g(x^2) \uparrow \max,\; A_2x^2 \leq b - A_1x_0^1,\; x^2 \geq 0.$$

One can imagine a few same way to construct "worst" initial points.

Nevertheless we have to note that R-algorithm turned out to be unfeeling w.r.t. the choice of initial point, as one can see in the table of results.

Now, let $F_0 = F\,(x^0)$ be the initial value of the objective function F;

Table 1

	F_0	F_m	St	PL	T
1	878466	49318	3	531	11:21
2	255043	49318	1	520	11:06
3	547664	49318	3	529	11:39
4	644342	49318	3	948	12:24

F_m- the best obtained value of F;

$St-$ the number of obtained stationary points from which one manage to escape;

$PL-$ the number of solved Linearized Problems, and finally,

T be the solving time (min:sec).

The computational results of solving the d.c. minimization problem (10)-(11) are presented in the Table 1.

Let us do a few remarks about the obtained results.

First, repeat that (it is clear from the table) $R-$algorithm turns out to be unfeeling w.r.t. the choice of initial point.

Second, the best known solution has been reached in all cases, but $R-$algorithm does not ameliorate this point.

Besides, the solving time seems to be too long. One reason to be so is that the linearized problems were solved with the conditional gradient method (CGM) which is well-known to be very slow.

This negative property is confirmed by the table where one can find the number of solved linearized problems equal to almost one thousand. Remembering that every linearized problem was solved by CGM one can get the conclusion that 12 and half minutes of solving time is not so uncredible and unexpected. It became clear that one has to introduce another method instead of CGM, if we rely on the considerable reducing of solving time.

Further, let us pay a bit of attention to the fact that in problem of interest (10)-(11) only three stationary points (including best known solution) were disclosed accordingly to the table of results.

It seems that the complexity of the problem is really of a higher degree than the estimation of only 3 "stationary degrees".

But where is a reason to be so?

To our opinion, one explanation can be given by too simple approximation of the level surface, which did not take into account, for instance, the columnes of the matrix A_1 and the structure of the matrix A_2 As the computational results certify, it suffices to use only first 11 points of approximation $R_1 = (y^1, ..., y^{11})$ for reaching the best known solution (BKS).

On the other hand, we must note that 95% of solving time is devoted to the attempts to escape from the best known solution (which were fruitless), where one used all points of the Approximation.

Finally, let us say a few words about the choice of the number β.

Accordingly to Theorem 1 one has

$$I \triangleq \inf(g, D) \leq \beta \leq \; sup(g, D) \triangleq S$$

Nevertheless, in the computational experiments in order to reach the best known solution it sufficed to use only one value of β, namely $\beta = g(z^k)$.

At BKS we decided to calculate the function

$$\eta(\beta) = \min_{1 \leq i \leq N} \{g(u^i) - \beta - \langle \nabla f(w^i), u^i - w^i \rangle\}$$

over the interval $[I, S]$.

It was observed, that the function $\eta(\beta)$ turned out to be convex w.r.t. the parameter β (which was completely unexpected for us!) and the global minimum of $\eta(\beta)$ is equal to zero! This fact allows us to suspect BKS to be a global solution.

In order to take into account all remarks from above, we desided to replace CGM of the solving the linearized problem (LP) by the Lemke's method for Quadratic Programming [5]. The following tables present the obtained results.

Table 2

	F_0	F_m	St	PL	T
1	876375.7	49318	1	627	0:38.50
2	856442.8	49318	1	627	0:38.50
3	828696.1	49318	1	627	0:38.50
4	73735.5	49318	2	638	0:39.40
5	692703.5	49318	2	638	0:39.40
6	658431.5	49318	2	638	0:39.40
7	592603.5	49318	2	638	0:39.40
8	625671.5	49318	2	638	0:39.40
9	659943.5	49318	2	638	0:39.40
10	547663.5	49318	1	627	0:38.50

Table 3

	F_0	F_m	St	PL	T
1	876376	49318	1	570	1:02.70
2	856443	49318	1	570	1:02.65
3	828696	49318	1	570	1:02.65
4	73735.5	49318	2	580	1:04.80
5	692703.5	49318	2	580	1:04.80
6	658431.5	49318	2	580	1:04.82
7	592603.5	49318	2	580	1:04.80
8	625671.5	49318	2	580	1:04.80
9	659943.5	49318	2	580	1:04.80
10	547663.5	49318	1	570	1:02.70

Table 2 was got using the approximation of level surface $R_1 = \{y^1, ..., y^{11}\}$, while in table 3 we see the results of the computational experiments with the approximation $R_2 = \{y^{12}, ..., y^{21}\}$. First, it can be readily seen that the solving time was considerably reduced what is obviously due to the new (Lemke's) solution method of Linearized Problem.

On the other hand, it is visible that the approximation R_1 is more suitible in the case than the approximation R_2 from the view-point of solution time.

Besides, the number of obtained stationary points is reduced w.r.t. Table 1 what is apparently related to a new local search algorithm, which will be described in one of our later papers. Respectively the number of solved LP is also reduced.

Note, that often even an unfeasible point was choosen as initial one. Nevertheless, in all cases $R-$ algorithm succeed to reach BKS.

Thus, the principal conclusion is positive: $R-$algorithm based on GOC of Theorem 1 has the right to be used for searching a global solution in d.c.-minimization problems.

6 Conclusion

In this paper we proposed the general deterministic approach for solving a d.c.-minimization problem based on Global Optimality Conditions(GOC)

In particular,

a) we presented the GOC for this kind of optimization problems;

b) we have introduced the global search algorithm based on GOC, so called $R-$algorithm (or R- strategy);

c) further, we have studied some features of $R-$algorithm's application for a concrete d.c.-minimization problem;

d) finally, we have demonstrated the possibility of solving the d.c.minimization problem by $R-$algorithm and analyzed the first computational results.

References

[1] Horst R., Tuy H. (1993), *Global Optimization Deterministic Approaches*, 2nd revised edition, Springer Verlag, Heidelberg.

[2] Horst R., Pardalos P.M., Thoai N. (1995), *Introduction to Global Optimization*, Kluwer Academic Publishers, Dordrecht/Boston/London.

[3] Floudas, C.A., Pardalos, P.M. (1990), *A Collection of Test Problems for Constrained Global Optimization Algorithms*, Lecture Notes in Computer Science, 455, Springer Verlag, Berlin.

[4] Hiriart-Urruty J.B. (1989), *From convex optimization to nonconvex optimization. Part I: Necessary and sufficient conditions for optimality*, in Nonsmooth Optimization and related topics, Plenum Press, 219-239.

[5] Bazaraa M.S., Shetty S.M. (1979), *Nonlinear Programming Theory and Algorithms*, J.Wiley and Sons, New York.

[6] Strekalovsky A.S. (1997), *On Global Optimality Conditions for D.C. Programming Problems*, seria "Optimization and Optimal Control", Irkutsk State University Press, Irkutsk.

[7] Strekalovsky A.S., Tsevendorj I. (1998), "Testing the R−strategy for a Reverse Convex Problem," *Journal of Global Optimization*, vol. 13, 61-74.

[8] Strekalovsky A.S. (1993), " On Search of global maximum of a convex function over a feasible set," *Journal of Computational Mathematics and Mathematical Physics*, vol. 33, 349-364.

[9] Kuznetsova A.A., Strekalovsky A.S., Tsevendorj I. (1999), " On an approach to solve integer programming problems," *Journal of Computational Mathematics and Mathematical Physics*, vol. 39, 9-16.

[10] Polak E. (1971), *Computational methods in optimization. An unified approach.* Academic Press, New York/London.

Error Bounds and Superlinear Convergence Analysis of Some Newton-Type Methods in Optimization[1]

Paul Tseng (tseng@math.washington.edu)
Department of Mathematics, Box 354350
University of Washington, Seattle, WA 98195-4350, U.S.A.

Abstract

We show that, for some Newton-type methods such as primal-dual interior-point path following methods and Chen-Mangasarian smoothing methods, local superlinear convergence can be shown without assuming the solutions are isolated. The analysis is based on local error bounds on the distance from the iterates to the solution set.

Keywords: Error bound, superlinear convergence, Newton method, continuation/smoothing method, interior-point method.

1 Introduction

We consider the complementarity problem (CP) of finding an $(x, y) \in \Re^n \times \Re^n$ satisfying

$$x \geq 0, \quad y \geq 0, \quad x^T y = 0, \quad F(x) - y = 0, \tag{1}$$

where $F = (F_1, ..., F_n)^T$ is a continuous mapping from $\Re^n$ (or $\Re^n_+$) to $\Re^n$ that is continuously differentiable on $\Re^n$ (or $\Re^n_{++}$) [10, 12, 26]. We denote by S the set of of solutions of CP, i.e., $S = \{(x, y) \in \Re^n \times \Re^n : (x, y) \text{ satisfy } (1)\}$, which we assume is nonempty. An important special case is when F is affine, giving rise to a linear CP (LCP) [11, 20]. Another important special case is when

$$x = \begin{bmatrix} u \\ v \end{bmatrix}, \quad F(x) = \begin{bmatrix} \nabla f_0(u) + \sum_{i=1}^m v_i \nabla f_i(u) \\ -f_i(u)_{i=1}^m \end{bmatrix},$$

[1]This research is supported by National Science Foundation Grant CCR-9731273.

G. Di Pillo and F. Giannessi (eds.), Nonlinear Optimization and Related Topics, 445–462.

where $f_0, f_1, ..., f_m$ are twice continuously differentiable functions defined on $\Re^l$ (or $\Re^l_{++}$), $m \geq 0, l \geq 1$. In this case, (1) is equivalent to the Karush-Kuhn-Tucker condition associated with the nonlinear program

$$\text{minimize} \ \ f_0(u) \ \ \text{subject to} \ \ u \geq 0, \ f_i(u) \leq 0, \ i = 1, ..., m.$$

Moreover, if $f_0, f_1, ..., f_m$ are convex (respectively, quadratic) on their respective domains, then this F is monotone (respectively, affine) and continuous on $\Re^{l+m}_+$.

Among the various approaches for solving CP, one that has received much attention concerns Newton-type methods based on continuation/smoothing. In this approach, one chooses a continuously differentiable mapping H_μ from $\Re^{2n}$ (or $\Re^{2n}_{++}$) to $\Re^{2n}$, parameterized by $\mu \in \Re_{++}$, having the property that

$$H_\mu(x, y) \to 0, \ (x, y, \mu) \to (\bar{x}, \bar{y}, 0) \qquad \Longrightarrow \qquad (\bar{x}, \bar{y}) \in S.$$

Then, from an initial μ and an initial guess (x, y) of the solution, (x, y) is updated by moving it along a Newton direction for the equation $H_\mu(x, y) = 0$ and afterwards μ is decreased. Termination occurs when μ is sufficiently small. One choice of the Newton direction is the solution $(u, v) \in \Re^{2n}$ of the Newton equation

$$H'_\mu(x, y) \begin{bmatrix} u \\ v \end{bmatrix} + H_\mu(x, y) = 0, \tag{2}$$

which is useful in achieving global convergence. A second choice, useful for practical efficiency as well as achieving local superlinear convergence, is the solution (u, v) of the modified Newton equation

$$H'_\mu(x, y) \begin{bmatrix} u \\ v \end{bmatrix} + H_0(x, y) = 0, \tag{3}$$

where $H_0(x, y) = \lim_{\mu \to 0} H_\mu(x, y)$, assuming this point-wise limit exists. The two directions (2) and (3) may be used in series (i.e., use (2) to update (x, y) and then use (3) to further update (x, y)) or in parallel (i.e., use a convex combination of (2) and (3) to update (x, y)). Primal-dual interior-point path following methods correspond to H_μ given by

$$H_\mu(x, y) = \begin{bmatrix} (x_i y_i - \mu)_{i=1}^n \\ F(x) - y \end{bmatrix}, \qquad x > 0, y > 0. \tag{4}$$

The use of both directions (2) and (3) give rise to the so-called predictor-corrector methods. The literature on such methods is vast and, in the context of CP, include [15, 18, 19, 20, 21, 25, 28, 29, 30, 31, 33, 34, 39, 42]. The smoothing method of Chen and Mangasarian [7, 8] and its extensions [3, 4, 6, 9, 13, 40] correspond to H_μ given by

$$H_\mu(x, y) = \begin{bmatrix} (x_i - \mu g((x_i - y_i)/\mu))_{i=1}^n \\ F(x) - y \end{bmatrix}, \tag{5}$$

where $g : \Re \to \Re$ is any convex continuously differentiable function with the properties that $g(-\tau) \to 0$ and $g(\tau) - \tau \to 0$ as $\tau \to \infty$ and $0 < g'(\tau) < 1$ for all $\tau \in \Re$. We will denote the class of such functions g by $\mathcal{CM}$. [In the case where g is twice continuously differentiable, $\mu g(\tau/\mu)$ may be written as the integral $\int_{-\infty}^{\tau} \int_{-\infty}^{\alpha} \frac{1}{\mu} g''(\xi/\mu) d\xi d\alpha$, where g'' is interpreted as a probability density function. This is the form presented in [8] and used by others. The form (5) is from [40] and is easier to use for our analysis.] Examples of the smoothing function g include one proposed independently by Chen and Harker [5], Kanzow [16], and Smale [32] (see [1, 2, 14, 43] and references therein for methods based on this g):

$$g(\tau) = (\sqrt{\tau^2 + 4} + \tau)/2, \tag{6}$$

and one obtained by integrating the sigmoid function $\tau \mapsto 1/(1 + e^{-\tau})$ used in neural networks [7, 8]:

$$g(\tau) = \ln(e^{\tau} + 1). \tag{7}$$

See [7, 8] for other examples of g. A new example of g that is useful for our analysis is

$$g(\tau) = \begin{cases} (1/2)^{\gamma+2}(1/2 - \tau)^{-\gamma}/\gamma & \text{if } \tau \geq 0 \\ (1/2)^{\gamma+2}(1/2 + \tau)^{-\gamma}/\gamma + \tau & \text{if } \tau < 0 \end{cases}, \tag{8}$$

with $\gamma > 0$. For H_μ given by (5), it is easily seen that $H_0(x, y) = (\min\{x, y\}, F(x) - y)$.

Our focus is on the local superlinear convergence analysis of the preceding Newton-type methods and, for simplicity, we will restrict our attention to H_μ given by (4) and (5). In the typical superlinear convergence analysis, the solutions need to be isolated. An exception to this is the primal-dual interior-point predictor-corrector methods, for which local superlinear convergence can be shown for linear programs and, more generally, monotone LCP having a strictly complementary solution (see [37, 41] and references therein). Further extensions of this analysis to monotone (nonlinear) CP and variational inequalities have been obtained by Sun and Zhao [34], assuming (in addition to existence of strictly complementary solution and some mild assumptions) constant range space and column rank of certain matrices based on $F'(x)$ [34, Assumptions 3-4]. The related analysis of Ralph and Wright [29, 30] allows for nonunique multipliers, but when specialized CP, it still requires uniqueness of solution. Monteiro and Zhou [24] improved the analysis of [29] for the case of linearly constrained convex program, assuming constant null space of certain matrices based on $F'(x)$. This assumption is similar to [34, Assumptions 3-4] and allows for multiple primal solutions. In this paper, we consider a new approach to local superlinear convergence analysis without assuming isolated solutions, based on error bound of the form:

$$\text{dist}(x, X^*) \leq \kappa \| \min\{x, F(x)\} \| \quad \text{whenever } \| \min\{x, F(x)\} \| \leq \delta, \tag{9}$$

where $X^* = \{x : (x, y) \in S \text{ for some } y\}$, $\text{dist}(x, X^*) = \min_{x^* \in X^*} \|x - x^*\|$, and $\kappa > 0, \delta > 0$. In the case of primal-dual interior-point methods for monotone CP, we show that this error bound and strict complementarity, together with some mild assumptions, imply that the Newton direction (u, v) solving (3) is asymptotically in the order of μ (see Lemmas 1, 2), from which a standard argument yields that (x, y) can be moved along (u, v) with a stepsize approaching 1 and with μ decreased at a quadratic rate (see Prop. 1). The same approach applies to the Chen-Mangasarian smoothing method for monotone LCP (see Lemmas 4, 5 and Prop. 2). This gives a new superlinear convergence result for this method without assuming isolated solutions. Although the error bound (9) does not hold for general F, it is known to hold if (i) F is affine or (ii) $F(x) = A^T G(Ax) + b$, with $A \in \Re^{m\times n}$, $b \in \Re^n$, and $G : \Re^m \mapsto \Re^m$ strongly monotone and Lipschitz continuous [22, 36, 27], or (iii) F is Lipschitz continuous and a uniformly P-function on $\Re^n$ [17, Cor. 7.3], [27, p. 320]. In cases (i) and (ii), the elements of S and X^* need not be isolated. In case (iii), S and X^* are singletons. For a good survey of this error bound and its applications, see [27]. It is an open question how this error bound relates to [34, Assumptions 3-4] or [24, Assumption 4].

This paper is organized as follows: in Sec. 2, we describe the Newton-type method based on continuation/smoothing. In Sec. 3, we analyze the local superlinear convergence of this method with H_μ given by (4), assuming that F is monotone, that (9) holds, and that the iterates asymptotically satisfy strict complementarity. In Sec. 4, we do an analogous analysis for H_μ given by (5) with $g \in \mathcal{CM}$. In Sec. 5, we discuss possible extensions of the results.

In our notation, all vectors are column vectors, ^{T} denotes transpose, $\Re^n$ denotes the space of n-dimensional real column vectors, and $\Re^n_+$ (respectively, $\Re^n_{++}$) denotes the nonnegative (respectively, positive) orthant in $\Re^n$. For convenience, we also write $\begin{bmatrix} x \\ y \end{bmatrix}$ as "(x, y)" for any two vectors x and y. We denote $N = \{1, ..., n\}$. For any vector $x \in \Re^n$, we denote by x_i the ith component of x and, for any $I \subset N$, by x_I the vector obtained after removing from x those x_i with $i \notin I$. We denote by $\|x\|$ and $\|x\|_\infty$ the 2- and ∞-norm of x, by X the diagonal matrix with diagonal entries $x_1, ..., x_n$, and by e a vector of 1s. We denote by $\Re^{n\times n}$ the space of $n \times n$ real matrices. For any $M \in \Re^{n\times n}$ and any $I \subset N$ and $J \subset N$, we denote by M_{IJ} the submatrix of M obtained by removing all rows of M with indices outside of I and removing all columns of M with indices outside of J. Also, we denote $\|M\| = \max_{x\in\Re^n:\|x\|=1} \|Mx\|$. For any continuously differentiable function $H = (H_1, ..., H_m)^T : \Re^m \mapsto \Re^m$, we denote its Jacobian by $H' = (\nabla H_1, ..., \nabla H_m)^T$, where ∇H_i denotes the gradient of H_i for $i = 1, ..., m$.

2 A Newton-Type Continuation/Smoothing Method

In order to be amenable to a global convergence analysis, continuation/smoothing methods maintain the iterates (x, y, μ) to lie in a "neighborhood" of the solutions to $H_\mu(x, y) = 0$. There are many choices for the neighborhood. For H_μ given by (4), a reasonable choice is

$$\mathcal{N}_{\beta_1,\beta_2} = \{(x,y,\mu) \in \Re^{2n+1}_{++} : \mu/\beta_1 \le x_i y_i \le \beta_1 \mu \ \forall i, \|F(x) - y\| \le \beta_2 \mu\}, \tag{10}$$

with $\beta_1 > 1, \beta_2 \ge 0$. Alternative choices, discussed in [15, 20, 29, 30, 33, 34, 39, 42], can possibly also be used. For H_μ given by (5), a reasonable choice is

$$\mathcal{N}_{\beta_1,\beta_2} = \{(x,y,\mu) \in \Re^{2n} \times \Re_{++} : |x_i - \mu g((x_i - y_i)/\mu)| \le \beta_1 \mu \ \forall i, \|F(x) - y\| \le \beta_2 \mu\}, \tag{11}$$

with $\beta_1 > g(0), \beta_2 \ge 0$. Alternative choices, discussed in [1, 2, 3, 4, 6, 40, 43], can possibly also be used. We describe the general method, parameterized by H_μ and $\mathcal{N}_{\beta_1,\beta_2}$, formally below.

Algorithm 1 *Choose suitable $\beta^0 \in (0, \beta_1]$ and $(x^0, y^0, \mu^0) \in \mathcal{N}_{\beta^0,\beta_2}$. For $t = 0, 1, \ldots$, generate $(x^{t+1}, y^{t+1}, \mu^{t+1})$ from (x^t, y^t, μ^t) according to the following iteration: let (u^t, v^t) be the solution of* (3) *with $(x, y, \mu) = (x^t, y^t, \mu^t)$. Let*

$$\theta^t \in \arg\min_{\theta \in [0,1]} m(x^t + \theta u^t, y^t + \theta v^t), \tag{12}$$

where we define $m(x, y) = \min\{\mu \in \Re_+ : (x, y, \mu) \in \mathcal{N}_{\beta_1,\beta_2}\}$, and let

$$(\hat{x}^t, \hat{y}^t) = (x^t + \theta^t u^t, y^t + \theta^t v^t), \quad \hat{\mu}^t = m(\hat{x}^t, \hat{y}^t). \tag{13}$$

Generate a $\beta^{t+1} \in (0, \beta_1]$ and $(x^{t+1}, y^{t+1}, \mu^{t+1}) \in \mathcal{N}_{\beta^{t+1},\beta_2}$ satisfying $\mu^{t+1} \le \hat{\mu}^t$.

Note 1. The minimization in (12) and the evaluation of $m(\cdot)$ can be done inexactly (and inexpensively) without affecting our superlinear convergence results, as can be seen from Lemmas 3 and 6.

Note 2. We have left unspecified how $(x^{t+1}, y^{t+1}, \mu^{t+1})$ and β^{t+1} are to be generated so that our results may be applied to different continuation/smoothing methods. One popular choice is

$$(x^{t+1}, y^{t+1}, \mu^{t+1}) = (\hat{x}^t + \hat{\theta}^t \hat{u}^t, \hat{y}^t + \hat{\theta}^t \hat{v}^t, (1 - \sigma \hat{\theta}^t) \hat{\mu}^t), \tag{14}$$

where $(\hat{u}^t, \hat{v}^t)$ is the solution of (2) with $(x, y, \mu) = (\hat{x}^t, \hat{y}^t, \hat{\mu}^t)$, and $\hat{\theta}^t$ is the largest element of $\{1, \psi, \psi^2, \ldots\}$ such that this triple remains in $\mathcal{N}_{\hat{\beta}_1,\beta_2}$, with $\psi \in (0, 1)$ and $\sigma \in (0, 1), \hat{\beta}_1 \in (0, \beta_1)$ suitably chosen (see [1, 2, 3, 6, 29, 39, 40, 42, 43]). Instead of a single Newton step, one can also take multiple Newton steps.

Note 3. In the context of interior-point methods, i.e., H_μ given by (4), Algorithm 1 is reminiscent of the predictor-corrector method of Mizuno, Todd, and Ye for linear programming [23]. Also, Algorithm 1 is an infeasible method in the sense that it does not maintain $F(x) = y$ satisfied at all iterations. If we assume that an initial $(x^0, y^0, \mu^0) \in \mathcal{N}_{\beta^0,\beta_2}$ satisfying $F(x^0) = y^0$ is available, we can alternatively consider a feasible method whereby $y^t + \theta v^t$ in (12) is replaced by $F(x^t + \theta u^t)$ and analogous changes are made in (13) and (14). Such a feasible method has the advantage that "polynomial convergence" can be shown assuming that F satisfies a scaled Lipschitz condition (see [15, 34, 35] and references therein). However, the assumption of a feasible interior solution limits the practicality of such a method.

3 Superlinear convergence analysis of interior-point method

In this section we analyze the local superlinear convergence of Algorithm 1 with H_μ and $\mathcal{N}_{\beta_1,\beta_2}$ given by (4) and (10), respectively. We begin with two technical lemmas that together show that the Newton direction (u,v) given by (3) is in the order of μ, assuming that F is (locally) monotone and Lipschitz continuous, that (9) holds, and that $x+y$ is bounded away from zero componentwise.

Lemma 1 *Assume F is monotone and F' is Lipschitz continuous with constant $\lambda \geq 0$ on $\Re^n_+$. Then, for any $(x,y,\mu) \in \Re^{2n+1}_{++}$ and any $(x^*,y^*) \in S$, we have*

$$\begin{aligned}\|u\| &\leq \|X\|(\lambda\|X\|\|x-x^*\|^2/2 + \|x-x^*\|\|y-y^*\|)/(\min_i x_iy_i) + \|x-x^*\|,\\ \|v\| &\leq \|Y\|(\lambda\|X\|\|x^*-x\|^2/2 + 2\|x-x^*\|\|y-y^*\|)/(\min_i x_iy_i) + \|y-y^*\|,\end{aligned} \tag{15}$$

where (u,v) solves (3) and H_μ is given by (4).

Proof. Using (4), (3) may be written as

$$F'(x)u - v + F(x) - y = 0, \qquad Yu + Xv = -Xy.$$

Using $F(x^*) - y^* = 0$ and $X^*y^* = 0$, this can be rewritten as

$$\begin{aligned}F'(x)(x-x^*+u) - (y-y^*+v) &= F(x^*) - F(x) + F'(x)(x-x^*),\\ Y(x-x^*+u) + X(y-y^*+v) &= (X-X^*)(y-y^*).\end{aligned} \tag{16}$$

Since $\|F(x^*) - F(x) + F'(x)(x-x^*)\| = \|\int_0^1 (F'(x+t(x^*-x)) - F'(x))(x^*-x)dt\| \leq \int_0^1 \lambda t\|x^*-x\|^2 dt = \lambda\|x^*-x\|^2/2$, multiplying the first equation on the left by $(x-x^*+u)^T$ and using the positive semidefiniteness of $F'(x)$ yields

$$\begin{aligned}-(x-x^*+u)^T(y-y^*+v) &\leq (x-x^*+u)^T(F(x^*) - F(x) + F'(x)(x-x^*))\\ &\leq \lambda\|X\|\|X^{-1}(x-x^*+u)\|\|x^*-x\|^2/2.\end{aligned}$$

Then, multiplying the second equation on the left by $(x-x^*+u)^TX^{-1}$ and rearranging terms yield

$$\begin{aligned}&(\min_i x_iy_i)\|X^{-1}(x-x^*+u)\|^2\\ \leq\ & (x-x^*+u)^TYX^{-1}(x-x^*+u)\\ =\ & -(x-x^*+u)^T(y-y^*+v) + (x-x^*+u)^TX^{-1}(X-X^*)(y-y^*)\\ \leq\ & \lambda\|X\|\|X^{-1}(x-x^*+u)\|\|x^*-x\|^2/2 + \|X^{-1}(x-x^*+u)\|\|(X-X^*)(y-y^*)\|.\end{aligned}$$

Thus

$$(\min_i x_iy_i)\|X^{-1}(x-x^*+u)\| \leq \lambda\|X\|\|x^*-x\|^2/2 + \|(X-X^*)(y-y^*)\|. \tag{17}$$

Since $\|X^{-1}(x - x^* + u)\| \geq \|x - x^* + u\|/\|X\| \geq (\|u\| - \|x - x^*\|)/\|X\|$, this yields the first inequality in (15). Finally, multiplying the second equation in (16) by X^{-1} yields

$$\begin{aligned}
\|y - y^* + v\| &= \|X^{-1}(X - X^*)(y - y^*) - YX^{-1}(x - x^* + u)\| \\
&\leq \|(XY)^{-1}\|\|Y\|\|(X - X^*)(y - y^*)\| + \|Y\|\|X^{-1}(x - x^* + u)\| \\
&\leq \|Y\|\|(X - X^*)(y - y^*)\|/(\min_i x_i y_i) + \|Y\|(\lambda\|X\|\|x^* - x\|^2/2 \\
&+ \|(X - X^*)(y - y^*)\|)/(\min_i x_i y_i),
\end{aligned}$$

where the last inequality also uses (17). Rearranging terms yields the second inequality in (15). ∎

Lemma 2 *Assume F is Lipschitz continuous on $\Re^n_+$ with constant $L > 0$ and* (9) *holds for some $\kappa > 0, \delta > 0$. Let H_μ and $\mathcal{N}_{\beta,\beta_2}$ be given by* (4) and (10) *with $\beta > 0, \beta_2 \geq 0$. Then, for any $(x, y, \mu) \in \mathcal{N}_{\beta,\beta_2}$ satisfying $(x_I, y_J) \geq \rho e$ and $\beta\mu \leq \rho^2, \beta_2\sqrt{n}\mu \leq \delta\rho$, where $\rho > 0$ and $I \subset N, J = N\backslash I$, there exists an $(x^*, y^*) \in S$ satisfying*

$$\|x - x^*\| \leq (\kappa\beta\sqrt{n}/\rho)\mu, \qquad \|y - y^*\| \leq (L\kappa\beta\sqrt{n}/\rho + \beta_2)\mu.$$

Proof. Since $(x, y, \mu) \in \mathcal{N}_{\beta,\beta_2}$ so that $Xy \leq \beta\mu e$, then $(x_I, y_J) \geq \rho e$ implies

$$0 < (y_I, x_J) \leq \beta\mu/\rho e. \tag{18}$$

Then, $(x_I, y_J) \geq \rho e$ implies, since $\beta\mu \leq \rho^2$, that $\min\{(x_I, x_J), (y_I, y_J)\} = (y_I, x_J)$, so if in addition $\beta\sqrt{n}\mu \leq \delta\rho$, then (9) and (18) would imply the existence of $(x^*, y^*) \in S$ such that

$$\|x^* - x\| \leq \kappa\|(y_I, x_J)\| \leq \kappa\beta\sqrt{n}\mu/\rho.$$

We also have

$$\|y^* - y\| = \|F(x^*) - F(x) + F(x) - y\| \leq L\|x^* - x\| + \beta_2\mu \leq (L\kappa\beta\sqrt{n}/\rho + \beta_2)\mu.$$

∎

The next lemma shows that, by moving (x, y) in the direction (u, v) given by (3), it is possible to decrease μ by a factor of $\|(u, v)\|$ while remaining in $\mathcal{N}_{\beta_1,\beta_2}$.

Lemma 3 *Assume F' is Lipschitz continuous on $\Re^n_+$ with constant $\lambda \geq 0$. Let H_μ and $\mathcal{N}_{\beta_1,\beta_2}$ be given by* (4) *and* (10), *with $\beta_1 > 0, \beta_2 \geq 0$. Then, for any $(x, y, \mu) \in \mathcal{N}_{\beta,\beta_2}$ with $\beta < \beta_1$ and (u, v) satisfying* (3), *we have $(x + \theta u, y + \theta v, \mu^+) \in \mathcal{N}_{\beta_1,\beta_2}$ whenever*

$$1 - \theta \geq \frac{4/c}{(1 + \sqrt{1 + 4/c})^2}, \tag{19}$$

$$((1-\theta)(\min_i x_i y_i) + \theta^2(\min_i u_i v_i))\beta_1 \geq \mu^+ \geq \max\left\{ \begin{array}{c} ((1-\theta)\|Xy\|_\infty + \theta^2\|Uv\|_\infty)/\beta_1, \\ (1-\theta)\mu + \theta^2\lambda\|u\|^2/(2\beta_2) \end{array} \right\}, \tag{20}$$

where $c = (\beta_1/\beta - 1)\mu/(\|Uv\|_\infty\beta_1 + \max\{\|Uv\|_\infty/\beta_1, \lambda\|u\|^2/(2\beta_2)\})$.

Proof. By (3) and (4), $Yu + Xv + Xy = 0$, so

$$(X + \theta U)(y + \theta v) = Xy + \theta Yu + \theta Xv + \theta^2 Uv = (1-\theta)Xy + \theta^2 Uv.$$

Then, a sufficient condition for $(\mu^+/\beta_1)e \leq (X + \theta U)(y + \theta v) \leq \beta_1\mu^+ e$ is

$$\mu^+/\beta_1 \leq (1-\theta)(\min_i x_i y_i) + \theta^2(\min_i u_i v_i), \quad \beta_1\mu^+ \geq (1-\theta)\|Xy\|_\infty + \theta^2\|Uv\|_\infty.$$

Also, by (3) and (4), $F(x) + F'(x)u - (y+v) = 0$, so

$$\begin{aligned}
\|F(x+\theta u) - (y+\theta v)\| &= \|(1-\theta)(F(x)-y) + F(x+\theta u) - (F(x) + \theta F'(x)u)\| \\
&= \|(1-\theta)(F(x)-y) + \int_0^1 F'(x+t\theta u) - F'(x))\theta u \, dt\| \\
&\leq (1-\theta)\|F(x)-y\| + \lambda \int_0^1 t\|\theta u\|^2 dt \\
&\leq (1-\theta)\beta_2\mu + \lambda\|u\|^2\theta^2/2,
\end{aligned}$$

where the last inequality uses $(x, y, \mu) \in \mathcal{N}_{\beta,\beta_2}$. Thus, $\|F(x+\theta u) - (y+\theta v)\| \leq \beta_2\mu^+$ whenever $\mu^+ \geq (1-\theta)\mu + \lambda\|u\|^2\theta^2/(2\beta_2)$. Thus $(x+\theta u, y+\theta v, \mu^+) \in \mathcal{N}_{\beta_1,\beta_2}$ whenever μ^+ satisfies (20). For such μ^+ to exist, it is sufficient that

$$((1-\theta)(\min_i x_i y_i) + \theta^2(\min_i u_i v_i))\beta_1 \geq \max\left\{ \begin{array}{c} ((1-\theta)\|Xy\|_\infty + \theta^2\|Uv\|_\infty)/\beta_1, \\ (1-\theta)\mu + \theta^2\lambda\|u\|^2/(2\beta_2) \end{array} \right\}.$$

Since $\mu/\beta \leq x_i y_i \leq \beta\mu$ for all i and $\beta < \beta_1$, a sufficient condition for this inequality to hold is

$$((1-\theta)\mu/\beta - \theta^2\|Uv\|_\infty)\beta_1 \geq ((1-\theta)\mu + \theta^2 \max\{\|Uv\|_\infty/\beta_1, \lambda\|u\|^2/(2\beta_2)\}.$$

This is a quadratic inequality of the form $(1-\theta)c - \theta^2 \geq 0$. Solving it gives $\theta \leq 2/(1+\sqrt{1+4/c})$ and (19) follows. ∎

Combining Lemmas 1, 2 and 3, we obtain the following main result of this section.

Proposition 1 *Assume F is monotone and Lipschitz continuous on $\Re^n_+$. Also, assume F' is Lipschitz continuous on $\Re^n_+$ and (9) holds for some $\kappa > 0, \delta > 0$. Let H_μ and $\mathcal{N}_{\beta_1,\beta_2}$ be given by (4) and (10). Let $\{(x^t, y^t, \mu^t, u^t, v^t, \hat{\mu}^t, \beta^t)\}_{t=0,1,\ldots}$ be generated by Algorithm 1. Assume $\{(x^t, y^t)\}$ is bounded, $\mu^t \to 0$, $\lim_{t\to\infty} \inf x^t + y^t > 0$, and $\lim_{t\to\infty} \inf \beta_1/\beta^t > 1$. Then, there exist $\bar{t}$ and $C_1 > 0, C_2 > 0$ such that $\|(u^t, v^t)\| \leq C_1\mu^t$ and $\mu^{t+1} \leq C_2(\mu^t)^2$ for all $t \geq \bar{t}$.*

Proof. Since $\mu^t \to 0$ and $\lim_{t\to\infty} \inf x^t + y^t > 0$, Lemma 2 implies that, for each t exceeding some $\bar{t}$, there is an $(\bar{x}^t, \bar{y}^t) \in S$ with

$$\|x^t - \bar{x}^t\| = O(\mu^t), \qquad \|y^t - \bar{y}^t\| = O(\mu^t).$$

[Here $b = O(a)$ means $b \le Ca$, with C a constant independent of t.] Then Lemma 1, together with the fact $\min_i x_i^t y_i^t \ge \mu^t/\beta^t \ge \mu^t/\beta_1$, yields

$$\|u^t\| = O(\|X^t\|(\|X^t\| + 1)\mu^t + \mu^t), \qquad \|v^t\| = O(\|Y^t\|(\|X^t\| + 1)\mu^t + \mu^t).$$

Since $\{(x^t, y^t)\}$ is bounded, this implies $\|(u^t, v^t)\| \le C_1\mu^t$ for suitable constant $C_1 > 0$. By Lemma 3, we have $(x^t + \bar{\theta}^t u^t, y^t + \bar{\theta}^t v^t, \bar{\mu}^t) \in \mathcal{N}_{\beta_1,\beta_2}$, where

$$\begin{aligned} c^t &= (\beta_1/\beta^t - 1)\mu^t/(\|U^t v^t\|_\infty \beta_1 + \max\{\|U^t v^t\|_\infty/\beta_1, \lambda\|u^t\|^2/(2\beta_2)\}), \\ \bar{\theta}^t &= 1 - \frac{4/c^t}{(1 + \sqrt{1 + 4/c^t})^2}, \\ \bar{\mu}^t &= \max\{\, ((1 - \bar{\theta}^t)\|X^t y^t\|_\infty + (\bar{\theta}^t)^2\|U^t v^t\|_\infty)/\beta_1, (1 - \bar{\theta}^t)\mu^t + (\bar{\theta}^t)^2\lambda\|u^t\|^2/(2\beta_2)\,\}. \end{aligned}$$

Since $\lim_{t\to\infty} \inf \beta_1/\beta^t > 1$ and $\|(u^t, v^t)\| \le C_1\mu^t$, we see that $1/c^t = O(\mu^t)$ for t large. Thus, $1 - \bar{\theta}^t = O(\mu^t)$ and hence

$$\mu^{t+1} \le \hat{\mu}^t \le \bar{\mu}^t = O((\mu^t)^2).$$

∎

We remark that the assumptions that $\mu^t \to 0$, $\lim_{t\to\infty} \inf x^t + y^t > 0$, and $\lim_{t\to\infty} \inf \beta_1/\beta^t > 1$ are reasonable and, in particular, hold when $(x^{t+1}, y^{t+1}, \mu^{t+1})$ is generated by (14), assuming in addition F is affine or uniformly nondegenerate [39, Lemma 2.1(c),(d)]. The same reference gives conditions on F that ensure boundedness of $\{(x^t, y^t)\}$. For the feasible variant, as discussed in Note 3, these assumptions hold if F satisfies a scaled Lipschitz condition [15, 34, 35]. Also, Prop. 1 still holds if the assumptions on F hold only locally around the cluster points of $\{x^t\}$.

4 Superlinear convergence analysis of Chen-Mangasarian smoothing method

In this section we analyze the local superlinear convergence of Algorithm 1 with H_μ and $\mathcal{N}_{\beta_1,\beta_2}$ given by (5) and (11), respectively. We begin with two technical lemmas that are analogous to Lemmas 1 and 2.

Lemma 4 *Assume F is monotone and F' is Lipschitz continuous with constant $\lambda \ge 0$ on $\Re^n$. Let H_μ and $\mathcal{N}_{\beta,\beta_2}$ be given by (5) and (11) with $\beta > 0, \beta_2 \ge 0$, and $g \in \mathcal{CM}$. Then, for any $(x, y, \mu) \in \mathcal{N}_{\beta,\beta_2}$ and any $(x^*, y^*) \in S$ with $(x_I, y_J) \ge \rho e$, $(\beta + g(0))\mu < \rho$ and $(y_I^*, x_J^*) = 0$, where $\rho > 0$ and $I \subset N, J = N \backslash I$, we have*

$$\begin{aligned} \|u\| &\le (\lambda\|x^* - x\|^2/2 + \max\{\|I - D_{II}\|, \|D_{JJ}\|\}(\|x - x^*\| + \|y - y^*\|)) \\ &\quad /\min_i D_{ii}(1 - D_{ii}) \; + \; \|x - x^*\|, \\ \|v\| &\le (\lambda\|x^* - x\|^2/2 + 2\max\{\|I - D_{II}\|, \|D_{JJ}\|\}(\|x - x^*\| + \|y - y^*\|)) \\ &\quad /\min_i D_{ii}(1 - D_{ii}) \; + \; \|y - y^*\|, \end{aligned} \tag{21}$$

where (u, v) *solves* (3), *and* $D \in \Re^{n\times n}$ *is diagonal with diagonal entries* $D_{ii} = g'((x_i - y_i)/\mu)$, $i = 1, ..., n$. *In addition,*

$$\min\{(x_I, x_J), (y_I, y_J)\} = (y_I, x_J), \qquad \|(y_I, x_J)\|_\infty \le (\beta + g(0))\mu. \tag{22}$$

Proof. Using (4), (3) may be written as

$$F'(x)u - v + F(x) - y = 0, \qquad (I - D)u + Dv = -\min\{x, y\}.$$

Using $F(x^*) - y^* = 0$, this can be rewritten as

$$\begin{aligned} F'(x)(x - x^* + u) - (y - y^* + v) &= F(x^*) - F(x) + F'(x)(x - x^*), \\ (I - D)(x - x^* + u) + D(y - y^* + v) &= r, \end{aligned} \tag{23}$$

where we let $r = (I - D)(x - x^*) + D(y - y^*) - \min\{x, y\}$. Since $\|F(x^*) - F(x) + F'(x)(x - x^*)\| = \| \int_0^1 (F'(x + t(x^* - x)) - F'(x))(x^* - x)dt\| \le \int_0^1 \lambda t \|x^* - x\|^2 dt = \lambda\|x^* - x\|^2/2$, multiplying the first equation on the left by $(x - x^* + u)^T$ and using the positive semidefiniteness of $F'(x)$ yields

$$\begin{aligned} -(x - x^* + u)^T(y - y^* + v) &\le (x - x^* + u)^T(F(x^*) - F(x) + F'(x)(x - x^*)) \\ &\le \lambda\|D\|\|D^{-1}(x - x^* + u)\|\|x^* - x\|^2/2. \end{aligned}$$

Then, multiplying the second equation on the left by $(x - x^* + u)^T D^{-1}$ and rearranging terms yield

$$\begin{aligned} &\min_i D_{ii}(1 - D_{ii})\|D^{-1}(x - x^* + u)\|^2 \\ \le\ & (x - x^* + u)^T(I - D)D^{-1}(x - x^* + u) \\ =\ & -(x - x^* + u)^T(y - y^* + v) + (x - x^* + u)^T D^{-1} r \\ \le\ & \lambda\|D\|\|D^{-1}(x - x^* + u)\|\|x^* - x\|^2/2 + \|D^{-1}(x - x^* + u)\|\|r\|. \end{aligned} \tag{24}$$

We have from a consequence of [40, Lemma 3.1] that $0 < \mu g(t/\mu) - \max\{0, t\} \le g(0)\mu$ for all $\mu > 0$ and $t \in \Re$. Thus, for each $i \in N$,

$$\begin{aligned} |\min\{x_i, y_i\}| &= |x_i - \max\{0, x_i - y_i\}| \\ &= |x_i - \mu g((x_i - y_i)/\mu)| + |\mu g((x_i - y_i)/\mu) - \max\{0, x_i - y_i\}| \\ &\le \beta\mu + g(0)\mu, \end{aligned}$$

where the last inequality also uses $(x, y, \mu) \in \mathcal{N}_{\beta,\beta_2}$. Since $(\beta + g(0))\mu < \rho$ and $(x_I, y_J) \ge \rho e$, this yields (22). Rearranging coordinates if necessary, we can assume without loss of generality that $x = (x_I, x_J), y = (y_I, y_J)$. Then (22) and $(y_I^*, x_J^*) = 0$ yield

$$\begin{aligned} r &= (I - D)(x - x^*) + D(y - y^*) - \min\{x, y\} \\ &= (I - D)\begin{bmatrix} x_I - x_I^* \\ x_J \end{bmatrix} + D\begin{bmatrix} y_I \\ y_J - y_J^* \end{bmatrix} - \begin{bmatrix} y_I \\ x_J \end{bmatrix} \\ &= \begin{bmatrix} (I - D_{II})(x_I - x_I^* - y_I + y_I^*) \\ D_{JJ}(y_J - y_J^* - x_J + x_J^*) \end{bmatrix} \end{aligned}$$

This and (24) imply

$$\begin{aligned}&\min_i D_{ii}(1-D_{ii})\|D^{-1}(x-x^*+u)\|\\ \le\ &\lambda\|D\|\|x^*-x\|^2/2+\max\{\|I-D_{II}\|,\|D_{JJ}\|\}(\|x-x^*\|+\|y-y^*\|).\end{aligned} \quad (25)$$

Since $\|D^{-1}(x-x^*+u)\| \ge \|x-x^*+u\|/\|D\| \ge (\|u\|-\|x-x^*\|)/\|D\|$ and $\|D\| \le 1$, this yields

$$\begin{aligned}\|u\| \ \le\ &(\lambda\|x^*-x\|^2/2+\max\{\|I-D_{II}\|,\|D_{JJ}\|\}(\|x-x^*\|+\|y-y^*\|))\\ &/\min_i D_{ii}(1-D_{ii}) \ + \ \|x-x^*\|,\end{aligned}$$

so the first inequality in (21) is shown. Finally, multiplying the second equation in (23) by D^{-1} yields

$$\begin{aligned}&\|y-y^*+v\|\\ &=\|D^{-1}r-(I-D)D^{-1}(x-x^*+u)\|\\ &\le\|(D(I-D))^{-1}\|\|I-D\|\|r\|+\|(I-D)\|\|D^{-1}(x-x^*+u)\|\\ &\le(\max\{\|I-D_{II}\|,\|D_{JJ}\|\}(\|x-x^*\|+\|y-y^*\|))/\min_i D_{ii}(1-D_{ii})\\ &+(\lambda\|x^*-x\|^2/2+\max\{\|I-D_{II}\|,\|D_{JJ}\|\}(\|x-x^*\|+\|y-y^*\|))/\min_i D_{ii}(1-D_{ii}),\end{aligned}$$

where the last inequality also uses (25) and $\|D\| \le 1, \|I-D\| \le 1$. Rearranging terms yields the second inequality in (21). ∎

Lemma 5 *Assume F is Lipschitz continuous on $\Re^n$ with constant $L>0$ and (9) holds for some $\kappa>0, \delta>0$. Let H_μ and $\mathcal{N}_{\beta,\beta_2}$ be given by (5) and (11) with $\beta>0, \beta_2\ge 0$, and $g \in \mathcal{CM}$. Then, for any $(x,y,\mu)\in\mathcal{N}_{\beta,\beta_2}$ with $(x_I,y_J)\ge\rho e$ and $\beta_2\sqrt{n}\mu\le\delta\rho$, $(\max\{1,L\kappa\sqrt{n}\}(\beta+g(0))+\beta)\mu<\rho$, where $\rho>0$ and $I\subset N, J=N\backslash I$, there exists an $(x^*,y^*)\in S$ satisfying $(y_I^*,x_J^*)=0$ and*

$$\|x-x^*\|\le\kappa(\beta+g(0))\sqrt{n}\mu, \quad \|y-y^*\|\le(L\kappa(\beta+g(0))\sqrt{n}+\beta)\mu.$$

Proof. By Lemma 4, we have that (22) holds, so (9) implies the existence of $(x^*,y^*)\in S$ such that

$$\|x^*-x\|\le\kappa\|(y_I,x_J)\|\le\kappa(\beta+g(0))\sqrt{n}\mu.$$

We also have

$$\|y^*-y\|=\|F(x^*)-F(x)+F(x)-y\|\le L\|x^*-x\|+\beta\mu\le(L\kappa(\beta+g(0))\sqrt{n}+\beta)\mu.$$

Since $(x_I,y_J)\ge\rho e$ and $(L\kappa(\beta+g(0))\sqrt{n}+\beta)\mu<\rho$, this implies $(x_I^*,y_J^*)>0$, so $(y_I^*,x_J^*)=0$. ∎

The next lemma is analogous to Lemma 3.

Lemma 6 *Assume F' is Lipschitz continuous on $\Re^n_+$ with constant $\lambda \geq 0$. Let H_μ and $\mathcal{N}_{\beta,\beta_2}$ be given by* (5) *and* (11), *with $\beta > 0, \beta_2 \geq 0$ and $g \in \mathcal{CM}$ satisfying $g'(-\tau) \leq \kappa_1/\tau$ and $1-g'(\tau) \leq \kappa_1/\tau$ for all $\tau > 0$ ($\kappa_1 > 0$). Then, for any $(x, y, \mu) \in \mathcal{N}_{\beta,\beta_2}$ with $(x_I, y_J) \geq \rho e$ and $(\beta+g(0))\mu \leq \rho/2$ and (u, v) satisfying* (3) *with $\|u\|_\infty + \|v\|_\infty < \rho/2$, where $\rho > 0$ and $I \subset N, J = N\backslash I$, we have $(x + u, y + v, \mu^+) \in \mathcal{N}_{\beta_1,\beta_2}$ whenever $\beta_1 > g(0)$ and*

$$\mu^+ \geq \max\{2\kappa_1\mu(\|u\|_\infty + \|v\|_\infty)/(\rho(\beta_1 - g(0))), \lambda\|u\|^2/(2\beta_2)\}. \tag{26}$$

Proof. Since $(x_I, y_J) \geq \rho e$ and $(\beta + g(0))\mu < \rho$, Lemma 4 implies (22) holds. Also, we have from (3) and (5) that

$$(I - D)u + Dv = -\min\{x, y\}, \tag{27}$$

where $D \in \Re^{n\times n}$ is diagonal with diagonal entries $D_{ii} = g'((x_i - y_i)/\mu)$, $i = 1, ..., n$. Since $0 \leq \tau - g(\tau) \leq g(0)$ for all $\tau > 0$ and $\beta_1 > g(0)$, there exists a $\eta > 0$ such that $|\tau - g(\tau - s)| \leq \beta_1$ for all $\tau > 0$ and $|s| \leq \eta$. For each $i \in I$, we have from (22) and (27) that

$$(1 - D_{ii})(u_i - v_i) = -(y_i + v_i).$$

Since $x_i \geq \rho$ and $|y_i| \leq (\beta+g(0))\mu \leq \rho/2$, we have $x_i - y_i \geq \rho/2$ and hence $1 - D_{ii} \leq \kappa_1\mu/(x_i - y_i) \leq 2\kappa_1\mu/\rho$. This implies $|y_i + v_i| = |1 - D_{ii}||u_i - v_i| \leq 2\kappa_1\mu(\|u\|_\infty + \|v\|_\infty)/\rho$. Thus, for $\mu^+ \geq 2\kappa_1\mu(\|u\|_\infty + \|v\|_\infty)/(\rho\eta)$, we have $|y_i + v_i|/\mu^+ \leq \eta$ as well as $x_i + u_i > 0$ (since $\|u\|_\infty < \rho \leq x_i$), implying

$$|(x_i + u_i)/\mu^+ - g(((x_i + u_i) - (y_i + v_i))/\mu^+)| \leq \beta_1.$$

For each $i \in J$, we have from (22) and (27) that

$$D_{ii}(u_i - v_i) = x_i + u_i.$$

Since $y_i \geq \rho$ and $|x_i| \leq (\beta + g(0))\mu \leq \rho/2$, we have $x_i - y_i \leq -\rho/2$ and hence $D_{ii} \leq \kappa_1\mu/|x_i - y_i| \leq 2\kappa_1\mu/\rho$. This implies $|x_i + u_i| = |D_{ii}||u_i - v_i| \leq 2\kappa_1\mu(\|u\|_\infty + \|v\|_\infty)/\rho$. Since $0 \leq g(\tau) \leq g(0)$ for $t < 0$ and $(x_i + u_i) - (y_i + v_i) < 0$ (since $\|u\|_\infty + \|v\|_\infty < \rho/2 \leq y_i - x_i$), this implies

$$|(x_i + u_i)/\mu^+ - g(((x_i + u_i) - (y_i + v_i))/\mu^+)| \leq |(x_i + u_i)/\mu^+| + g(0) \leq \beta_1$$

whenever $2\kappa_1\mu(\|u\|_\infty + \|v\|_\infty)/\rho \leq (\beta_1 - g(0))\mu^+$. Lastly, as in the proof of Lemma 3, we have $\|F(x + u) - (y + v)\| \leq \beta_2\mu^+$ whenever $\mu^+ \geq \lambda\|u\|^2/(2\beta_2)$. Thus, $(x + u, y + v, \mu^+) \in \mathcal{N}_{\beta_1,\beta_2}$ whenever μ^+ satisfies (26). ∎

Combining Lemmas 4, 5 and 6, we obtain the following main result of this section.

Proposition 2 *Assume F is monotone and affine. Let H_μ and $\mathcal{N}_{\beta_1,\beta_2}$ be given by* (5) *and* (11), *with $\beta_1 > 0, \beta_2 \geq 0$, $g \in \mathcal{CM}$ satisfying $g(0) < \beta_1$, $g'(-\tau) \leq \kappa_1/\tau$ and*

$1 - g'(\tau) \le \kappa_1/\tau$ for all $\tau > 0$ ($\kappa_1 > 0$), and, for each $\kappa_2 \ge 1$, there exist $\kappa_3 > 0$ such that $g'(-\tau)/g'(-s) \le \kappa_3$ and $(1 - g'(\tau))/(1 - g'(s)) \le \kappa_3$ whenever $0 < \tau < s \le \kappa_2\tau$. Let $\{(x^t, y^t, \mu^t, u^t, v^t, \hat{\mu}^t, \beta^t)\}_{t=0,1,...}$ be generated by Algorithm 1. Assume $\{(x^t, y^t)\}$ is bounded, $\mu^t \to 0$ and $\lim_{t\to\infty} \inf x^t + y^t > 0$. Then there exist $\bar{t}$ and $C_1 > 0, C_2 > 0$ such that $\|(u^t, v^t)\| \le C_1\mu^t$ and $\mu^{t+1} \le C_2(\mu^t)^2$ for all $t \ge \bar{t}$.

Proof. Since F is affine, then F' is Lipschitz continuous on $\Re^n_+$ with constant $\lambda = 0$ and, by the remark following (9), (9) holds for some $\kappa > 0, \delta > 0$. Since $\mu^t \to 0$ and $\lim_{t\to\infty} \inf x^t + y^t > 0$, Lemma 5 yields there exist $\bar{t}$ and $\rho > 0$ such that $t \ge \bar{t}$ implies $(x^t_{I^t}, y^t_{J^t}) \ge \rho e$ for some $I^t \subset N$, $J^t = N \backslash I^t$, and the existence of $(\bar{x}^t, \bar{y}^t) \in S$ satisfying

$$(\bar{y}^t_{I^t}, \bar{x}^t_{J^t}) = 0, \quad \|x^t - \bar{x}^t\| = O(\mu^t), \quad \|y^t - \bar{y}^t\| = O(\mu^t). \tag{28}$$

[Here $b = O(a)$ means $b \le Ca$, with C a constant independent of t.] Then Lemma 4 with $\lambda = 0$ yields

$$\|(u^t, v^t)\| = O(\mu^t \max\{\|I - D^t_{I^tI^t}\|, \|D^t_{J^tJ^t}\|\}/\min_i D^t_{ii}(1 - D^t_{ii}) + \mu^t),$$

where $D^t \in \Re^{n\times n}$ is the diagonal matrix with diagonal entries $D^t_{ii} = g'((x^t_i - y^t_i)/\mu^t)$, $i = 1, ..., n$. Since $\mu^t \to 0$, (28) implies $(y^t_{I^t}, x^t_{J^t}) \to 0$ so, taking $\bar{t}$ larger if necessary, we can assume $\|(y^t_{I^t}, x^t_{J^t})\|_\infty \le \rho/2$ for all $t \ge \bar{t}$. Then, $(x^t_{I^t}, y^t_{J^t}) \ge \rho e$ implies $x^t_{I^t} - y^t_{I^t} \ge \rho e/2$ and $x^t_{J^t} - y^t_{J^t} \le -\rho e/2$. Since $\{(x^t, y^t)\}$ is bounded, this and our last assumption on g' imply

$$\max\{\|I - D^t_{I^tI^t}\|, \|D^t_{J^tJ^t}\|\}/\min_i D^t_{ii}(1 - D^t_{ii}) \text{ is bounded,} \tag{29}$$

so that $\|(u^t, v^t)\| = O(\mu^t)$, i.e., $\|(u^t, v^t)\| \le C_1\mu^t$ for suitable constant $C_1 > 0$. By Lemma 6 and $\lambda = 0$, we have $(x^t + u^t, y^t + v^t, \tilde{\mu}^t) \in \mathcal{N}_{\beta_1,\beta_2}$, where

$$\tilde{\mu}^t = 2\kappa_1\mu^t(\|u^t\|_\infty + \|v^t\|_\infty)/(\rho(\beta_1 - g(0))).$$

Thus,

$$\mu^{t+1} \le \hat{\mu}^t \le \tilde{\mu}^t = O((\mu^t)^2).$$

■

We remark that the assumption of $\mu^t \to 0$ is reasonable and, in particular, holds when $(x^{t+1}, y^{t+1}, \mu^{t+1})$ is generated by (14) [3, 6, 40, 43]. The same references discuss conditions on F that ensures boundedness of $\{(x^t, y^t)\}$. The reasonableness of the assumption $\lim_{t\to\infty} \inf x^t + y^t > 0$ is unclear and remains to be determined. Notice that g given by (6) or (8) satisfies the assumptions of Prop. 2, whereas g given by (7) does not satisfy these assumptions. Also, like Prop. 1, Prop. 2 still holds if the assumptions on F hold only locally around the cluster points of $\{x^t\}$. Unlike Prop. 1, Prop. 2 does not assume $\lim_{t\to\infty} \inf \beta_1/\beta^t > 1$ but it does assume F is affine. The latter is needed to establish (29). Otherwise we would need to assume instead that

$g'(-\tau) \geq \bar{\kappa}_1/\tau$ and $1 - g'(\tau) \geq \bar{\kappa}_1/\tau$ for some $\bar{\kappa}_1 > 0$, which would imply $g(\tau)$ behaves like $-\ln(-\tau)$ for $\tau < 0$ and like $\tau - \ln(\tau)$ for $\tau > 0$. An example is

$$g(\tau) = \begin{cases} -\ln(1/2 - \tau)/4 & \text{if } \tau \leq 0 \\ \tau - \ln(1/2 + \tau)/4 & \text{if } \tau > 0 \end{cases}.$$

Such a g unfortunately *cannot* belong to $\mathcal{CM}$ due to $g(-\tau) \not\to 0, \tau - g(\tau) \not\to 0$ as $\tau \to \infty$. It is an open question whether Prop. 2 can be suitably extended to such a g.

5 Conclusion and Extensions

We have shown that error bounds can be useful in the superlinear convergence analysis of Newton-type methods, without assuming the solutions are isolated. Although we have restricted our analysis to a specific type of continuation/smoothing methods and to the context of CP, the underlying idea can be applied to other continuation/smoothing methods and to other contexts. For example, both Lemma 1 and Prop. 1 can be extended to monotone semidefinite linear complementarity problem where, instead of $\Re^n$, we work in the space of $n \times n$ symmetric real matrices and componentwise nonnegativity is replaced by nonnegativity of the eigenvalues [38]. It may be possible to extend our analysis to H_μ based on the Kanzow smoothing function $\sqrt{x_i^2 + y_i^2 + 2\mu}$ [16], as well as to nonsmooth Newton methods, but further study would be required.

References

[1] Burke, J. and Xu, S. (1997), "A non-interior predictor-corrector path following algorithm for the monotone linear complementarity problem," Preprint, Department of Mathematics, University of Washington, Seattle.

[2] Burke, J. and Xu, S. (1997), "A non-interior predictor-corrector path-following method for LCP," Preprint, Department of Mathematics, University of Washington, Seattle.

[3] Chen, B. and Chen, X. (1997), "A global linear and local quadratic continuation smoothing method for variational inequalities with box constraints," Report, Department of Management and Systems, Washington State University, Pullman.

[4] Chen, B. and Chen, X. (1997), "A global and local superlinear continuation-smoothing method for $P_0 + R_0$ and monotone NCP," Report, Department of Management and Systems, Washington State University, Pullman; *SIAM J. Optim.*, to appear.

[5] Chen, B. and Harker, P. T. (1993), "A non-interior-point continuation method for linear complementarity problems," *SIAM J. Matrix Anal. Appl.*, 14, 1168-1190.

[6] Chen, B. and Xiu, N. (1997), "A global linear and local quadratic non-interior continuation method for nonlinear complementarity problems based on Chen-Mangasarian smoothing function," Report, Department of Management and Systems, Washington State University, Pullman.

[7] Chen, C. and Mangasarian, O. L. (1995), "Smoothing methods for convex inequalities and linear complementarity problems," *Math. Programming*, 71, 51-69.

[8] Chen, C. and Mangasarian, O. L. (1996), "A class of smoothing functions for nonlinear and mixed complementarity problems," *Comput. Optim. Appl.*, 5, 97-138.

[9] Chen, X., Qi, L., and Sun, D. (1996), "Global and superlinear convergence of the smoothing Newton method and its application to general box constrained variational inequalities," Report, School of Mathematics, University of New South Wales, Sydney; *Math. Comp.*, to appear.

[10] Cottle, R. W., Giannessi, F., and Lions, J.-L., editors. (1980), *Variational Inequalities and Complementarity Problems: Theory and Applications*, John Wiley & Sons, New York.

[11] Cottle, R. W., Pang, J.-S., and Stone, R. E.. (1992), *The Linear Complementarity Problem*, Academic Press, New York.

[12] Ferris, M. C. and Pang, J.-S., editors. (1997), *Complementarity and Variational Problems: State of the Art*, SIAM, Philadelphia.

[13] Gabriel, S. A. and Moré, J. J. (1997), "Smoothing of mixed complementarity problems," in *Complementarity and Variational Problems: State of the Art*, edited by M. C. Ferris and J.-S. Pang, SIAM, Philadelphia, 105-116.

[14] Hotta, K., Inaba, M., and Yoshise, A. (1998), "A complexity analysis of a smoothing method using CHKS-functions for monotone linear complementarity problems," Discussion paper, Institute of Policy and Planning Sciences, University of Tsukuba, Tsukuba.

[15] Jansen, B., Roos, K., Terlaky, T., and Yoshise, A. (1997), "Polynomiality of primal-dual affine scaling algorithms for nonlinear complementarity problems," *Math. Programming*, 78, 315–345.

[16] Kanzow, C. (1996), "Some noninterior continuation methods for linear complementarity problems," *SIAM J. Matrix Anal. Appl.*, 17, 851–868.

[17] Kanzow, C. and Fukushima, M. (1996), "Equivalence of the generalized complementarity problem to differentiable unconstrained minimization," *J. Optim. Theory Appl.*, 90, 581–603.

[18] Kojima, M., Megiddo, N., and Mizuno, S. (1993), "A general framework of continuation methods for complementarity problems," *Math. Oper. Res.*, 18, 945–963.

[19] Kojima, M., Megiddo, N., and Noma T. (1991), "Homotopy continuation methods for nonlinear complementarity problems," *Math. Oper. Res.*, 16, 754–774.

[20] Kojima, M., Megiddo, N., Noma, T. and Yoshise, A. (1991), *A Unified Approach to Interior Point Algorithms for Linear Complementarity Problems*, Lecture Notes in Computer Science 538, Springer-Verlag, Berlin.

[21] Kojima, M., Noma, T., and Yoshise, A. (1994), "Global convergence in infeasible-interior-point algorithms," *Math. Programming*, 65, 43–72.

[22] Luo, Z.-Q. and Tseng, P. (1993), "Error bounds and convergence analysis of feasible descent methods: a general approach," *Ann. Oper. Res.*, 46, 157–178.

[23] Mizuno, S., Todd, M., and Ye, Y. (1993), "On adaptive-step primal-dual interior-point algorithms for linear programming," *Math. Oper. Res.*, 18, 964-981.

[24] Monteiro, R. D. C. and Zhou, F. (1997), "On superlinear convergence of infeasible-interior-point algorithms for linearly constrained convex programs," *Comput. Optim. Appl.*, 8, 245-262.

[25] Nesterov, Y. E. and Nemirovskii, A. S. (1993), *Interior Point Polynomial Methods in Convex Programming: Theory and Applications*, SIAM, Philadelphia.

[26] Pang, J.-S. (1995), "Complementarity problems," in *Handbook on Global Optimization*, edited by R. Horst and P. Pardalos, Kluwer Academic Publishers, Norwell, 271–338.

[27] Pang, J.-S. (1997), "Error bounds in mathematical programming," *Math. Programming*, 79, 299–332.

[28] Potra, F. A. and Ye, Y. (1996), "Interior point methods for nonlinear complementarity problems," *J. Optim. Theory Appl.*, 88, 617–647.

[29] Ralph, D. and Wright, S. J. (1997), "Superlinear convergence of an interior-point method for monotone variational inequalities," in *Complementarity and Variational Problems: State of the Art*, edited by M. C. Ferris and J.-S. Pang, SIAM, Philadelphia, 345–385.

[30] Ralph, D. and Wright, S. J. (1996; revised 1998), "Superlinear convergence of an interior-point method despite dependent constraints," preprint, Mathematics and Computer Science Division, Argonne National Laboratory, Argonne.

[31] Simantiraki, E. M. and Shanno, D. F. (1997), "An infeasible-interior-point algorithm for solving mixed complementarity problems," in *Complementarity and Variational Problems: State of the Art*, edited by M. C. Ferris and J.-S. Pang, SIAM, Philadelphia, 386–404.

[32] Smale, S. (1987), "Algorithms for solving equations," in *Proceedings of the International Congress of Mathematicians*, edited by A. M. Gleason, American Mathematical Society, Providence, 172-195.

[33] Sun, J. and Zhao, G. (1996), "Global linear and local quadratic convergence of a long-step adaptive-mode interior point method for some monotone variational inequality problems," Report, Department of Decision Sciences, National University of Singapore, Singapore; *SIAM J. Optim.*, to appear.

[34] Sun, J. and Zhao, G. (1997), "A long-step quadratically convergent interior point method for nonlinear monotone variational inequality problems," Report, Department of Decision Sciences, National University of Singapore, Singapore.

[35] Tseng, P. (1992), "Global linear convergence of a path-following algorithm for some monotone variational inequality problems," *J. Optim. Theory Appl.*, 75, 265–279.

[36] Tseng, P. (1995), "On linear convergence of iterative methods for the variational inequality problem," *J. Comp. Appl. Math.*, 60, 237–252.

[37] Tseng, P. (1995), "Simplified analysis of an $O(nL)$-iteration infeasible predictor-corrector path following method for monotone LCP," in *Recent Trends in Optimization Theory and Applications*, edited by R. P. Agarwal, World Scientific, Singapore, 423-434.

[38] Tseng, P. (1996), "Analysis of infeasible path-following methods using the Alizadeh-Haeberly-Overton direction for the monotone semi-definite LCP," Report, Department of Mathematics, University of Washington, Seattle.

[39] Tseng, P. (1997), "An infeasible path-following method for monotone complementarity problems," *SIAM J. Optim.*, 7, 386–402.

[40] Tseng, P. (1998), "Analysis of a non-interior continuation method based on Chen-Mangasarian smoothing functions for complementarity problems," in *Reformulation – Nonsmooth, Piecewise Smooth, Semismooth and Smoothing Methods* edited by M. Fukushima and L. Qi, Kluwer Academic Publishers, Boston, 381–404.

[41] Wright, S. J. (1997), *Primal-Dual Interior-Point Methods*, SIAM, Philadelphia.

[42] Wright, S. J. and Ralph, D. (1996), "A superlinear infeasible-interior-point algorithm for monotone complementarity problems," *Math. Oper. Res.*, 21, 815-838.

[43] Xu, S. (1997), "The global linear convergence and complexity of a non-interior path-following algorithm for monotone LCP based on Chen-Harker-Kanzow-Smale smoothing functions," Preprint, Department of Mathematics, University of Washington, Seattle.

A New Derivative-Free Descent Method for the Nonlinear Complementarity Problem

Kenjiro Yamada (ken2ro@jp.ibm.com)
IBM Japan, Ltd., Nihonbashi Hakozaki-cho, Chuo-ku, Tokyo 103-8510 Japan.

Nobuo Yamashita (nobuo@kuamp.kyoto-u.ac.jp)
Department of Applied Mathematics and Physics, Graduate School of Informatics, Kyoto University, Kyoto 606-8501, Japan.

Masao Fukushima (fuku@kuamp.kyoto-u.ac.jp)
Department of Applied Mathematics and Physics, Graduate School of Informatics, Kyoto University, Kyoto 606-8501, Japan.

Abstract

Recently, much effort has been made in solving and analyzing the nonlinear complementarity problem (NCP) by means of a reformulation of the problem as an equivalent unconstrained optimization problem involving a merit function. In this paper, we propose a new merit function for the NCP and show several favorable properties of the proposed function. In particular, we give conditions under which the function provides a global error bound for the NCP and conditions under which its level sets are bounded. Moreover, we propose a new derivative-free descent algorithm for solving the NCP based on this function. We show that any accumulation point generated by the algorithm is a solution of the NCP under the monotonicity assumption on the problem. Also, we prove that the sequence generated by the algorithm converges linearly to the solution under the strong monotonicity assumption.

Keywords: Derivative-free method, nonlinear complementarity problem, merit function, global convergence, linear convergence.

G. Di Pillo and F. Giannessi (eds.), Nonlinear Optimization and Related Topics, 463–487.

1. Introduction

The nonlinear complementarity problem (NCP) [11, 20] is to find a point $x \in R^n$ such that

$$x \geq 0, \quad F(x) \geq 0, \quad x^T F(x) = 0, \tag{1}$$

where $F : R^n \to R^n$ is a given function. Throughout this paper, we assume that F is continuously differentiable. Applications of the NCP can be found in many important fields such as mathematical programming, economics, engineering and mechanics [2, 11].

A variety of methods have been proposed for solving the NCP [9, 11, 20]. One of the powerful approaches that have been studied intensively in recent years is to reformulate the NCP as an equivalent system of nonlinear equations [16, 21, 25] or as an equivalent unconstrained optimization problem [3, 5, 6, 10, 12, 14, 18, 22, 24]. The objective function of such an equivalent unconstrained optimization problem is called a merit function. In other words, a merit function for the NCP is a function whose global minima are coincident with the solutions of the NCP.

To construct a merit function, the class of functions called NCP-functions have played an important role. A function $\phi : R^2 \to R$ is called an NCP-function if it satisfies

$$\phi(a, b) = 0 \iff a \geq 0, \quad b \geq 0, \quad ab = 0.$$

An NCP-function ϕ is said to be nonnegative if

$$\phi(a, b) \geq 0 \quad \text{for all } (a, b)^T \in R^2.$$

It is clear that if ϕ is a nonnegative NCP-function on R^2, then the function $\Psi : R^n \to R$ defined by

$$\Psi(x) = \sum_{i=1}^{n} \phi(x_i, F_i(x)).$$

is a merit function for the NCP. In this paper, we are particularly interested in the following merit functions constructed by some well-known NCP-functions. The natural residual $\Psi_{NR} : R^n \to R^n$ [11] is a merit function defined by

$$\Psi_{NR}(x) = \sum_{i=1}^{n} \phi_{NR}(x_i, F_i(x)), \tag{2}$$

where $\phi_{NR} : R^2 \to R$ is a nonnegative NCP-function given by

$$\phi_{NR}(a, b) = \min\{a, b\}^2.$$

Note that the function ϕ_{NR} is not differentiable everywhere, hence neither is Ψ_{NR}. So, the equivalent unconstrained optimization problem of minimizing the function Ψ_{NR} cannot be solved by using general algorithms that require the differentiability of the objective function. Mangasarian and Solodov [17] proposed the implicit Lagrangian $\Psi_{MS} : R^n \to R$ defined by

$$\Psi_{MS}(x) = \sum_{i=1}^{n} \phi_{MS}(x_i, F_i(x)),$$

where $\phi_{MS} : R^2 \to R$ is a nonnegative NCP-function given by

$$\phi_{MS}(a, b) = ab + \frac{1}{2\alpha}\left\{(-\alpha b + a)_+^2 - a^2 + (-\alpha a + b)_+^2 - b^2\right\},$$

where $\alpha > 1$ is a fixed constant and $(z)_+ = \max\{0, z\}$ for $z \in R$. The function ϕ_{MS} is continuously differentiable, and so is Ψ_{MS}. Any stationary point of Ψ_{MS} is a solution of the NCP whenever F is a P-function [4] (see also [24]). Another well-known merit function is the function $\Psi_{FB} : R^n \to R$ defined by

$$\Psi_{FB}(x) = \sum_{i=1}^{n} \phi_{FB}(x_i, F_i(x)), \tag{3}$$

where $\phi_{FB} : R^2 \to R$ is given by

$$\phi_{FB}(a, b) = \frac{1}{2}(\sqrt{a^2 + b^2} - a - b)^2. \tag{4}$$

The function ϕ_{FB} is a nonnegative NCP-function introduced by Fischer [6] and called the Fischer-Burmeister function. This function has been extensively studied and shown to have a number of favorable properties. For example, ϕ_{FB} is continuously differentiable on R^2, and hence Ψ_{FB} is also continuously differentiable. Moreover, any stationary point of Ψ_{FB} is a solution of the NCP whenever F is a P_0-function [5]. Note that this condition is weaker than the one required for the implicit Lagrangian to have the same property.

For solving an equivalent unconstrained optimization problem based on a merit function, extensions of classical iterative methods such as Newton's method have been proposed. Recently, so-called derivative-free methods have attracted much attention, which do not require computation of derivatives of F and make use of the particular properties of a merit function [8, 10, 12, 18, 24, 27]. Derivative-free methods are particularly suitable for problems where the derivatives of F are not available or are extremely expensive to compute.

For the variational inequality problem (VIP), which contains the NCP as a special case, Fukushima [8] proposed a derivative-free descent method based on the regularized gap function and showed its global convergence under the strong monotonicity assumption on the problem mapping. More recently, Yamashita, Taji and Fukushima [27] proposed a derivative-free descent method for solving the VIP based on the D-gap function and established global convergence of the method under similar assumptions.

For the NCP, Yamashita and Fukushima [24] proposed a derivative-free descent method based on the implicit Lagrangian Ψ_{MS}. They showed that their method is globally convergent under the strong monotonicity assumption on F. Jiang [12] proposed a derivative-free descent method based on the Fischer-Burmeister function. He showed that his method converges globally to a solution of the NCP if F is monotone. Geiger and Kanzow [10] proposed a similar method and proved its global convergence under somewhat stronger assumptions than [12]. Note however that convergence rate of the proposed methods has not been discussed for the above mentioned methods for the VIP and the NCP. Recently, Mangasarian and Solodov [18] slightly modified the derivative-free descent algorithm of [24] and showed that the sequence generated by the modified algorithm converges linearly to the solution of the NCP. However, for their algorithm to have a global convergence property, the strong monotonicity assumption on F is needed. Moreover, in order to ensure the linear rate of convergence, some parameters used in the algorithm have to be specified by using usually unknown constants such as the modulus of strong monotonicity of F.

In this paper, we propose the following merit function $\Psi_\alpha : R^n \to R$ for the NCP:

$$\Psi_\alpha(x) = \sum_{i=1}^{n} \phi_\alpha(x_i, F_i(x)), \tag{5}$$

where $\phi_\alpha : R^2 \to R$ is a nonnegative NCP-function given by

$$\phi_\alpha(a,b) = \frac{\alpha}{2}(ab)_+^2 + \frac{1}{2}(\sqrt{a^2+b^2} - a - b)^2 \tag{6}$$

and $\alpha \geq 0$ is a real parameter. The function Ψ_α is a slight modification of the function proposed in [26] (see also [1]). When $\alpha = 0$, the function ϕ_α reduces to the Fischer-Burmeister function ϕ_{FB}, so ϕ_α is considered an extension of the Fischer-Burmeister function. We show that ϕ_α inherits several favorable properties of the Fischer-Burmeister function. In addition, when $\alpha > 0$, we can show that ϕ_α enjoys an important property that the Fischer-Burmeister function does not have. In particular, we give conditions under which Ψ_α provides a global error bound for the NCP and conditions under which it has bounded level sets. Moreover, we propose a new derivative-free descent algorithm for solving the NCP based on the function Ψ_α. We prove that this algorithm has a global convergence property if F is a monotone function, and that the sequence generated by the algorithm converges linearly to the solution if F is a strongly monotone function. In contrast with the algorithm proposed in [18], proper values of the parameters used in our algorithm do not rely on usually unknown constants such as the modulus of strong monotonicity of F.

This paper is organized as follows. In Section 2, we review some definitions and preliminary results that will be used in the subsequent analyses. In Section 3, we show some important properties of the proposed merit function. In Section 4, we present a derivative-free descent algorithm for solving the NCP and prove global convergence of the algorithm, while in Section 5, we prove its linear convergence. Section 6 makes some concluding remarks.

Throughout the paper, we shall adopt the following notations: For a real-valued matrix A of any dimension, A^T denotes its transpose. For a differentiable vector-valued function $F : R^n \to R^n$, ∇F denotes the transposed $n \times n$ Jacobian matrix whose columns are the gradients of the components of F and the ith component function of F is denoted by F_i. The norm $\|\cdot\|$ denotes the Euclidean norm. For $z \in R$, $\lceil z \rceil$ denotes the smallest integer no less than z. The level set of a function $\Psi : R^n \to R$ is denoted $\mathcal{L}(\Psi, c) := \{x \in R^n \mid \Psi(x) \leq c\}$.

2. Preliminaries

In this section, we recall some definitions and preliminary results that will be used in the subsequent analyses.

The following classes of functions $F : R^n \to R^n$ play an important role in the NCP.

Definition 2.1 *Let S be a convex set in R^n.*

(a) *F is monotone on S if*

$$(x - x')^T(F(x) - F(x')) \geq 0 \quad \text{for all} \ \ x, x' \in S.$$

(b) *F is a P_0-function on S if*

$$\max_{\substack{1 \leq i \leq n \\ x_i \neq x'_i}} (x_i - x'_i)(F_i(x) - F_i(x')) \geq 0 \quad \text{for all} \ \ x, x' \in S, \ \ x \neq x'.$$

(c) *F is a P-function on S if*

$$\max_{1 \leq i \leq n} (x_i - x'_i)(F_i(x) - F_i(x')) > 0 \quad \text{for all} \ \ x, x' \in S, \ \ x \neq x'.$$

(d) *F is strongly monotone with modulus $\mu > 0$ on S if*

$$(x - x')^T(F(x) - F(x')) \geq \mu\|x - x'\|^2 \quad \text{for all} \ \ x, x' \in S.$$

(e) *F is a uniform P-function with modulus $\mu > 0$ on S if*

$$\max_{1 \leq i \leq n} (x_i - x'_i)(F_i(x) - F_i(x')) \geq \mu\|x - x'\|^2 \quad \text{for all} \ \ x, x' \in S.$$

(f) *$\nabla F(x)$ is uniformly positive definite with modulus $\mu > 0$ on S if*

$$d^T \nabla F(x) d \geq \mu\|d\|^2 \quad \text{for all} \ \ x \in S, \ \ d \in R^n.$$

(g) *F is Lipschitz continuous on S if there exists a constant $L > 0$ such that*

$$\|F(x) - F(x')\| \leq L\|x - x'\| \quad \textit{for all} \quad x, x' \in S.$$

□

When the properties mentioned in Definition 2.1 hold with $S = R^n$, we simply omit the words "on S". Note that F is a P_0-function on S if F is monotone on S, and that F is a uniform P-function with modulus $\mu > 0$ on S if F is strongly monotone with modulus $\mu > 0$ on S. Also, when F is continuously differentiable, F is strongly monotone with modulus $\mu > 0$ on S if and only if $\nabla F(x)$ is uniformly positive definite with modulus $\mu > 0$ on S.

We recall some concepts about the NCP which will be used when we discuss boundedness of the solution set of the NCP.

Definition 2.2 *The NCP is said to be strictly feasible if there exists $\hat{x} > 0$ such that $F(\hat{x}) > 0$.* □

The error bound is an important concept that indicates how close an arbitrary point is to the solution set of the NCP. Thus, an error bound may be used to provide stopping rules for an iterative method. The following lemma shows that the natural residual provides a global error bound for the NCP under suitable conditions, which will play an important role in obtaining the linear convergence result of the algorithm to be presented in Section 4.

Lemma 2.1 *([19]) Let Ψ_{NR} be defined by (2). Suppose that F is strongly monotone with modulus $\mu > 0$ and Lipschitz continuous with constant $L > 0$ on some convex set $S \subseteq R^n$. Let x^* be the unique solution of the NCP. Then the following inequality holds:*

$$\|x - x^*\| \leq \frac{L+1}{\mu} \Psi_{NR}(x)^{\frac{1}{2}} \quad \textit{for all} \quad x \in S.$$

□

In the next lemma, we summarize some important properties of the Fischer-Burmeister function ϕ_{FB} defined by (4).

Lemma 2.2 *([10, 13, 23]) The function ϕ_{FB} defined by (4) has the following properties:*

(i) $\phi_{FB}(a, b) = 0 \Longleftrightarrow a \geq 0, b \geq 0, ab = 0$.

(ii) $\phi_{FB}(a, b) \geq 0$ *for all* $(a, b)^T \in R^2$.

(iii) ϕ_{FB} *is continuously differentiable for all* $(a, b)^T \in R^2$.

(iv) $\frac{\partial\phi_{FB}}{\partial a}(a,b)\frac{\partial\phi_{FB}}{\partial b}(a,b) \geq 0$ *at any* $(a,b)^T \in R^2$.

(v) $\frac{\partial\phi_{FB}}{\partial a}(a,b) = 0 \Longleftrightarrow \frac{\partial\phi_{FB}}{\partial b}(a,b) = 0 \Longleftrightarrow \phi_{FB}(a,b) = 0$.

(vi) $\frac{1}{2}\left(\frac{2}{\sqrt{2}+2}\right)^2 \min\{a,b\}^2 \leq \phi_{FB}(a,b) \leq \frac{1}{2}(\sqrt{2}+2)^2 \min\{a,b\}^2$ *for all* $(a,b)^T \in R^2$.

□

Lemma 2.2 (i) and (v) mean that ϕ_{FB}, $\frac{\partial\phi_{FB}}{\partial a}$ and $\frac{\partial\phi_{FB}}{\partial b}$ are all NCP-functions. Moreover, the above lemma indicates that the function Ψ_{FB} defined by (3) is nonnegative on R^n and that $\Psi_{FB}(x) = 0$ if and only if x solves the NCP. Important results about the merit function Ψ_{FB} are summarized in the following theorem.

Theorem 2.1 *([10, 23]) The function Ψ_{FB} defined by (3) has the following properties:*

(i) *If the NCP has at least one solution, then the NCP is equivalent to the unconstrained optimization problem* $\min_{x\in R^n} \Psi_{FB}(x)$.

(ii) *If F is a P_0-function, any stationary point of Ψ_{FB} is a solution of the NCP.*

(iii) *Let Ψ_{NR} be the natural residual defined by (2). Then the following inequalities hold:*

$$\frac{1}{2}\left(\frac{2}{\sqrt{2}+2}\right)^2 \Psi_{NR}(x) \leq \Psi_{FB}(x) \leq \frac{1}{2}(\sqrt{2}+2)^2 \Psi_{NR}(x) \quad \textit{for all } x \in R^n. \tag{7}$$

Moreover, if F is strongly monotone and Lipschitz continuous, then Ψ_{FB} provides a global error bound for the NCP.

(iv) *If F is a uniform P-function, then the level set $\mathcal{L}(\Psi_{FB}, c)$ is bounded for all $c \in R$.* □

3. A New Merit Function and Its Properties

In this section, we propose a new merit function and show that the function has several favorable properties.

Let $\Psi_\alpha : R^n \to R$ be defined by (5) and (6), i.e.,

$$\Psi_\alpha(x) = \sum_{i=1}^{n} \phi_\alpha(x_i, F_i(x)),$$

where $\phi_\alpha : R^2 \to R$ is given by

$$\phi_\alpha(a,b) = \frac{\alpha}{2}(ab)_+^2 + \frac{1}{2}(\sqrt{a^2+b^2} - a - b)^2$$

and $\alpha \geq 0$ is a real parameter. Since the function ϕ_0 reduces to the Fischer-Burmeister function ϕ_{FB}, the function ϕ_α inherits several favorable properties of the Fischer-Burmeister function. Moreover, when $\alpha > 0$, ϕ_α enjoys an important property the Fischer-Burmeister function ϕ_{FB} does not have (see Lemma 3.1 (vi) below).

Lemma 3.1 *Let $\alpha \geq 0$. The function ϕ_α defined by (6) has the following properties:*

(i) $\phi_\alpha(a,b) = 0 \Longleftrightarrow a \geq 0, b \geq 0, ab = 0.$

(ii) $\phi_\alpha(a,b) \geq 0$ *for all* $(a,b)^T \in R^2$.

(iii) ϕ_α *is continuously differentiable at any* $(a,b)^T \in R^2$ *and*

$$\frac{\partial \phi_\alpha}{\partial a}(0,0) = \frac{\partial \phi_\alpha}{\partial b}(0,0) = 0, \tag{8}$$

$$\frac{\partial \phi_\alpha}{\partial a}(a,b) = \alpha b(ab)_+ + \left(\frac{a}{\sqrt{a^2+b^2}} - 1\right)(\sqrt{a^2+b^2} - a - b) \quad \text{for } (a,b) \neq (0,0), \tag{9}$$

$$\frac{\partial \phi_\alpha}{\partial b}(a,b) = \alpha a(ab)_+ + \left(\frac{b}{\sqrt{a^2+b^2}} - 1\right)(\sqrt{a^2+b^2} - a - b) \quad \text{for } (a,b) \neq (0,0). \tag{10}$$

(iv) $\dfrac{\partial \phi_\alpha}{\partial a}(a,b)\dfrac{\partial \phi_\alpha}{\partial b}(a,b) \geq 0$ *for all* $(a,b)^T \in R^2$.

(v) $\dfrac{\partial \phi_\alpha}{\partial a}(a,b) = 0 \Longleftrightarrow \dfrac{\partial \phi_\alpha}{\partial b}(a,b) = 0 \Longleftrightarrow \phi_\alpha(a,b) = 0.$

If $\alpha > 0$, then the function ϕ_α enjoys the following additional property:

(vi) *If $a \to -\infty$ or $b \to -\infty$ or $ab \to \infty$, then $\phi_\alpha(a,b) \to \infty$.*

Proof. Since (i)–(iii) and (vi) can easily be shown, we only show (iv) and (v). We first prove (iv). When $(a,b) = (0,0)$, $\frac{\partial \phi_\alpha}{\partial a}(a,b)\frac{\partial \phi_\alpha}{\partial b}(a,b) \geq 0$ holds obviously. When $(a,b) \neq (0,0)$, it follows from (9) and (10) that

$$\begin{aligned}
\frac{\partial \phi_\alpha}{\partial a}(a,b)\frac{\partial \phi_\alpha}{\partial b}(a,b) \;=\;& \alpha^2 ab(ab)_+^2 + \alpha a(ab)_+ \left(\frac{a}{\sqrt{a^2+b^2}} - 1\right)(\sqrt{a^2+b^2} - a - b) \\
&+ \alpha b(ab)_+ \left(\frac{b}{\sqrt{a^2+b^2}} - 1\right)(\sqrt{a^2+b^2} - a - b) \\
&+ \left(\frac{a}{\sqrt{a^2+b^2}} - 1\right)\left(\frac{b}{\sqrt{a^2+b^2}} - 1\right)(\sqrt{a^2+b^2} - a - b)^2. \quad (11)
\end{aligned}$$

Since

$$ab(ab)_+^2 \geq 0, \quad \frac{a}{\sqrt{a^2+b^2}} - 1 \leq 0, \quad \frac{b}{\sqrt{a^2+b^2}} - 1 \leq 0, \tag{12}$$

it is sufficient to show that the second and third terms of (11) are both nonnegative for all $(a,b)^T \neq (0,0)^T \in R^2$. We first show

$$\alpha a(ab)_+ \left(\frac{a}{\sqrt{a^2+b^2}} - 1\right)(\sqrt{a^2+b^2} - a - b) \geq 0. \tag{13}$$

Consider three cases: $a = 0$, $a < 0$ and $a > 0$. When $a = 0$, (13) holds obviously. When $a < 0$, we have $\sqrt{a^2+b^2} - a - b > 0$, and hence (13) holds. Now we consider the case where $a > 0$. If $b \leq 0$, we have $(ab)_+ = 0$. If $b > 0$, then we have $\sqrt{a^2+b^2} - a - b < 0$. It follows from (12) that (13) holds. Therefore (13) holds for all $(a,b)^T \neq (0,0)^T \in R^2$. In a similar way, we can show that the inequality

$$\alpha b(ab)_+ \left(\frac{b}{\sqrt{a^2+b^2}} - 1\right)(\sqrt{a^2+b^2} - a - b) \geq 0 \tag{14}$$

holds for all $(a,b)^T \neq (0,0)^T \in R^2$. Consequently, it follows from (11)–(14) that

$$\frac{\partial \phi_\alpha}{\partial a}(a,b)\frac{\partial \phi_\alpha}{\partial b}(a,b) \geq 0. \tag{15}$$

We consider (v). Suppose that $\phi_\alpha(a,b) = 0$. Then it follows from the definition of ϕ_α that $ab = 0$ and $\sqrt{a^2+b^2} - a - b = 0$. Thus we obtain $\frac{\partial \phi_\alpha}{\partial a}(a,b) = 0$ and $\frac{\partial \phi_\alpha}{\partial b}(a,b) = 0$ from (9) and (10), respectively. Next we show that $\frac{\partial \phi_\alpha}{\partial a}(a,b) = 0$ implies $\phi_\alpha(a,b) = 0$. Suppose that $\frac{\partial \phi_\alpha}{\partial a}(a,b) = 0$. It then follows from (9) that

$$\alpha b(ab)_+ = -\left(\frac{a}{\sqrt{a^2+b^2}} - 1\right)(\sqrt{a^2+b^2} - a - b). \tag{16}$$

Suppose that both sides of (16) do not vanish. Then we have $b \neq 0$. If $b < 0$, then the left hand side of (16) is not positive and the right hand side of (16) is positive. This is a contradiction. When $b > 0$, $a > 0$ must hold in order to have $(ab)_+ \neq 0$. However, in this case, the left hand side of (16) is positive and the right hand side of (16) is negative. This is a contradiction, too. Therefore both sides of (16) must be equal to zero, and then it can easily be shown that $\phi_\alpha(a,b) = 0$ holds. In a similar way, we can show that $\frac{\partial \phi_\alpha}{\partial b}(a,b) = 0$ implies $\phi_\alpha(a,b) = 0$. The proof is complete. □

As mentioned above, the Fischer-Burmeister function ϕ_{FB} does not have the property stated in Lemma 3.1 (vi); more precisely, $ab \to \infty$ does not necessarily imply $\phi_{FB}(a,b) \to \infty$. Lemma 3.1 (vi) will turn out to be useful in showing the boundedness of level sets of Ψ_α. From the above lemma, we can easily deduce the following theorem.

Theorem 3.1 *Let Ψ_α be defined by (5). Then, $\Psi_\alpha(x) \geq 0$ for all $x \in R^n$ and $\Psi_\alpha(x) = 0$ if and only if x solves the NCP. Moreover, suppose that the NCP has at least one solution. Then, x is a global minimizer of Ψ_α if and only if x solves the NCP.* □

The above theorem says that the NCP can be reformulated as the unconstrained minimization problem:

$$\min_{x \in R^n} \Psi_\alpha(x). \tag{17}$$

In practice, it is not easy to find a global minimum of Ψ_α. Therefore, it is important to know under what conditions any stationary point of Ψ_α is a global minimum.

Theorem 3.2 *Let F be a P_0-function. Then, $x^* \in R^n$ is a global minimum of the unconstrained optimization problem (17) if and only if x^* is a stationary point of Ψ_α.*

Proof. The theorem can be proved by using property (iv) of Lemma 3.1 and the same proof technique as that of Theorem 3.5 in [10]. □

Now we give some preliminary results that will be useful in obtaining the global and linear convergence results for the algorithm to be presented in the next section. We first show that the functions Ψ_α and Ψ_{NR} are of the same order on any bounded set.

Lemma 3.2 *Let Ψ_α and Ψ_{NR} be defined by (5) and (2), respectively, and $\alpha \geq 0$. Then for any bounded set S, the following inequality holds:*

$$\frac{1}{2}\left(\frac{2}{\sqrt{2}+2}\right)^2 \Psi_{NR}(x) \leq \Psi_\alpha(x) \leq \left\{\frac{\alpha}{2}B^2 + \frac{1}{2}(\sqrt{2}+2)^2\right\} \Psi_{NR}(x) \quad \text{for all } x \in S, \tag{18}$$

where B is a constant defined by

$$B = \sup_{x \in S}\left\{\max_{1 \leq i \leq n}\{\max\{|x_i|, |F_i(x)|\}\}\right\} < \infty.$$

Proof. By Theorem 2.1 (iii) and the definitions of Ψ_α and $\Psi_{FB} = \Psi_0$, we have

$$\Psi_\alpha(x) \geq \Psi_{FB}(x) \geq \frac{1}{2}\left(\frac{2}{\sqrt{2}+2}\right)^2 \Psi_{NR}(x) \quad \text{for all } x \in R^n.$$

Therefore the left-hand inequality of (18) holds.

Next we consider the right-hand inequality of (18). Let us show that, for each i, the following inequality holds for all $x \in S$:

$$(x_i F_i(x))_+ \leq B\,|\min\{x_i, F_i(x)\}|. \tag{19}$$

We first consider the case where $F_i(x) \geq x_i$. Then $\min\{x_i, F_i(x)\} = x_i$. When $F_i(x) \geq x_i \geq 0$, it follows from the definition of B that

$$(x_i F_i(x))_+ = x_i F_i(x) = F_i(x)\,|\min\{x_i, F_i(x)\}| \leq B\,|\min\{x_i, F_i(x)\}|.$$

In the case where $F_i(x) \geq 0 \geq x_i$, we have $(x_i F_i(x))_+ = 0$ and hence (19) holds obviously. When $0 \geq F_i(x) \geq x_i$, it follows from the definition of B that

$$(x_i F_i(x))_+ = |x_i F_i(x)| \leq |x_i|^2 \leq B\ |\min\{x_i, F_i(x)\}|.$$

In a similar way, we can show that (19) also holds when $x_i \geq F_i(x)$. Therefore (19) holds for all $x \in S$. Thus it follows from Lemma 2.2 (vi) and (19) that for all $i = 1, \ldots, n$ and $x \in S$

$$\begin{aligned} \phi_\alpha(x_i, F_i(x)) &= \frac{\alpha}{2}(x_i F_i(x))_+^2 + \phi_{FB}(x_i, F_i(x)) \\ &\leq \left\{\frac{\alpha}{2}B^2 + \frac{1}{2}(\sqrt{2}+2)^2\right\} \min\{x_i, F_i(x)\}^2. \end{aligned} \tag{20}$$

Consequently, from the definitions of Ψ_α and Ψ_{NR}, we have the right-hand inequality of (18). The proof is complete. □

Lemma 3.2 indicates that Ψ_α is in the same order of the natural residual Ψ_{NR} on a bounded set. Thus, from Lemma 2.1, Ψ_α provides a global error bound for the NCP if F is strongly monotone and Lipschitz continuous. Moreover, we can show that the function Ψ_α provides a global error bound without assuming Lipschitz continuity of F when $\alpha > 0$. Note that Ψ_{FB} does not have the same property under the strong monotonicity of F only.

Theorem 3.3 *Let Ψ_α be defined by (5) and $\alpha > 0$. Suppose that F is a uniform P-function with modulus μ. Then there exists a positive constant ρ such that*

$$\|x - x^*\| \leq \rho \Psi_\alpha(x)^{\frac{1}{4}} \quad \textit{for all } x \in R^n, \tag{21}$$

where x^* is the unique solution of the NCP.

Proof. Since F is a uniform P-function with modulus μ, we have

$$\begin{aligned} \mu\|x - x^*\|^2 &\leq \max_{1\leq i\leq n}(x_i - x_i^*)(F_i(x) - F_i(x^*)) \\ &= \max_{1\leq i\leq n}\{x_i F_i(x) - x_i^* F_i(x) - F_i(x^*)x_i + x_i^* F_i(x^*)\} \\ &= \max_{1\leq i\leq n}\{x_i F_i(x) - x_i^* F_i(x) - F_i(x^*)x_i\} \\ &\leq \max_{1\leq i\leq n}\tau_i\{(x_i F_i(x))_+ + (-F_i(x))_+ + (-x_i)_+\}, \end{aligned} \tag{22}$$

where

$$\tau_i := \max\{1, x_i^*, F_i(x^*)\}.$$

Now we show that there exists a positive constant τ such that

$$\{(ab)_+ + (-b)_+ + (-a)_+\}^2 \leq \tau\phi_\alpha(a, b) \tag{23}$$

for all $(a,b)^T \in R^2$. First we establish the inequality

$$(-a)_+^2 + (-b)_+^2 \leq \left(\sqrt{a^2+b^2} - a - b\right)^2. \tag{24}$$

Without loss of generality, we assume $a \geq b$. In the case where $a \geq b \geq 0$, (24) holds obviously. In the case where $a \geq 0 \geq b$, we have $\sqrt{a^2+b^2} - a - b \geq -b \geq 0$, which implies that

$$(-a)_+^2 + (-b)_+^2 = b^2 \leq \left(\sqrt{a^2+b^2} - a - b\right)^2.$$

In the case where $0 \geq a \geq b$, since $\sqrt{a^2+b^2} - a - b \geq \sqrt{a^2+b^2}$, we obtain

$$(-a)_+^2 + (-b)_+^2 = a^2 + b^2 \leq \left(\sqrt{a^2+b^2} - a - b\right)^2.$$

We thus have shown that (24) holds for all $(a,b)^T \in R^2$. Therefore, we have for all $(a,b)^T \in R^2$

$$\begin{aligned}
\{(ab)_+ + (-b)_+ + (-a)_+\}^2 &= (ab)_+^2 + (-b)_+^2 + (-a)_+^2 \\
&\quad + 2(ab)_+(-b)_+ + 2(ab)_+(-a)_+ + 2(-a)_+(-b)_+ \\
&\leq (ab)_+^2 + (-b)_+^2 + (-a)_+^2 + ((ab)_+^2 + (-b)_+^2) \\
&\quad + ((ab)_+^2 + (-a)_+^2) + ((-a)_+^2 + (-b)_+^2) \\
&\leq 3\left\{(ab)_+^2 + (-b)_+^2 + (-a)_+^2\right\} \\
&\leq 3\left\{(ab)_+^2 + \left(\sqrt{a^2+b^2} - a - b\right)^2\right\} \\
&\leq \tau\left\{\frac{\alpha}{2}(ab)_+^2 + \frac{1}{2}\left(\sqrt{a^2+b^2} - a - b\right)^2\right\} \\
&= \tau\phi_\alpha(a,b),
\end{aligned}$$

where

$$\tau := \max\left\{\frac{6}{\alpha}, 6\right\} > 0,$$

and the third inequality follows from (24).

Consequently, letting

$$\hat{\tau} := \max_{1\leq i\leq n} \tau_i,$$

and combining (22) and (23), we obtain

$$\begin{aligned}
\mu\|x - x^*\|^2 &\leq \max_{1\leq i\leq n} \tau_i \left\{\tau\phi_\alpha(x_i, F_i(x))\right\}^{\frac{1}{2}} \\
&\leq \hat{\tau}\tau^{\frac{1}{2}} \max_{1\leq i\leq n} \phi_\alpha(x_i, F_i(x))^{\frac{1}{2}} \\
&\leq \hat{\tau}\tau^{\frac{1}{2}} \left\{\sum_{i=1}^n \phi_\alpha(x_i, F_i(x))\right\}^{\frac{1}{2}} \\
&= \hat{\tau}\tau^{\frac{1}{2}} \Psi_\alpha(x, F(x))^{\frac{1}{2}}.
\end{aligned}$$

The proof is complete. □

Next we consider the boundedness of level sets of Ψ_α. The boundedness of level sets of a merit function is important since it ensures that the sequence generated by a descent method has at least one accumulation point. The following theorem gives conditions under which Ψ_α has bounded level sets.

Theorem 3.4 *Suppose that either of the following conditions holds:*

(a) *$\alpha > 0$, F is monotone and the NCP is strictly feasible.*

(b) *$\alpha \geq 0$ and F is strongly monotone.*

Then the level set $\mathcal{L}(\Psi_\alpha, c)$ is bounded for all $c \in R$.

Proof. Under condition (a), the boundedness of the level set $\mathcal{L}(\Psi_\alpha, c)$ is easily shown in a way similar to Theorem 4.1 in [15] by using Lemma 3.1 (vi). Under condition (b), the result can be established by slightly modifying the proof of Theorem 3.8 in [10]. □

Finally we state a preliminary result which is the key to establishing both global and linear convergence of the algorithm. To simplify the notation, we abbreviate the vectors $(\frac{\partial\phi_\alpha}{\partial a}(x_1, F_1(x)), \ldots, \frac{\partial\phi_\alpha}{\partial a}(x_n, F_n(x)))^T$ and $(\frac{\partial\phi_\alpha}{\partial b}(x_1, F_1(x)), \ldots, \frac{\partial\phi_\alpha}{\partial b}(x_n, F_n(x)))^T$ as $\frac{\partial\phi_\alpha}{\partial a}(x, F(x))$ and $\frac{\partial\phi_\alpha}{\partial b}(x, F(x))$, respectively.

Lemma 3.3 *Let $\alpha \geq 0$. Then the following inequality holds for all $x \in R^n$:*

$$\|\frac{\partial\phi_\alpha}{\partial a}(x, F(x)) + \frac{\partial\phi_\alpha}{\partial b}(x, F(x))\|^2 \geq (2 - \sqrt{2})^4 \Psi_{NR}(x).$$

Proof. Suppose $(a, b) \neq (0, 0)$. It then follows from (9) and (10) that

$$\begin{aligned}
&\left(\frac{\partial\phi_\alpha}{\partial a}(a, b) + \frac{\partial\phi_\alpha}{\partial b}(a, b)\right)^2 \\
&= \left\{\alpha(a+b)(ab)_+ + (\sqrt{a^2+b^2} - a - b)\left(\frac{a+b}{\sqrt{a^2+b^2}} - 2\right)\right\}^2 \\
&= \alpha^2(a+b)^2(ab)_+^2 + (\sqrt{a^2+b^2} - a - b)^2\left(\frac{a+b}{\sqrt{a^2+b^2}} - 2\right)^2 \\
&\quad + 2\alpha(a+b)(ab)_+(\sqrt{a^2+b^2} - a - b)\left(\frac{a+b}{\sqrt{a^2+b^2}} - 2\right).
\end{aligned} \tag{25}$$

Now we show that the inequality

$$2\alpha(a+b)(ab)_+(\sqrt{a^2+b^2} - a - b)\left(\frac{a+b}{\sqrt{a^2+b^2}} - 2\right) \geq 0 \tag{26}$$

holds for all $(a,b)^T \neq (0,0)^T \in R^2$. Note that $\frac{a+b}{\sqrt{a^2+b^2}} - 2 \leq 0$ holds for all $(a,b)^T \neq (0,0)^T \in R^2$ When $ab \leq 0$, we have $(ab)_+ = 0$. When $a > 0$ and $b > 0$, we have $a+b > 0$ and $\sqrt{a^2+b^2} - a - b < 0$. When $a < 0$ and $b < 0$, we have $a+b < 0$ and $\sqrt{a^2+b^2} - a - b > 0$. Therefore (26) holds for all $(a,b)^T \neq (0,0)^T \in R^2$. Combining (25) and (26), we obtain for all $(a,b)^T \neq (0,0)^T \in R^2$

$$\begin{aligned}\left(\frac{\partial\phi_\alpha}{\partial a}(a,b) + \frac{\partial\phi_\alpha}{\partial b}(a,b)\right)^2 &\geq (\sqrt{a^2+b^2} - a - b)^2\left(\frac{a+b}{\sqrt{a^2+b^2}} - 2\right)^2 \\ &\geq \left(\frac{2}{\sqrt{2}+2}\right)^2 \min\{a,b\}^2 \left(\frac{a+b}{\sqrt{a^2+b^2}} - 2\right)^2 \\ &\geq \left(\frac{2}{\sqrt{2}+2}\right)^2 (2-\sqrt{2})^2 \min\{a,b\}^2 \\ &= (2-\sqrt{2})^4 \min\{a,b\}^2, \end{aligned} \tag{27}$$

where the second inequality follows from Lemma 2.2 (vi) and the last inequality follows from the fact that

$$2 - \sqrt{2} \leq 2 - \frac{a+b}{\sqrt{a^2+b^2}} \quad \text{for all} \quad (a,b)^T \neq (0,0)^T \in R^2.$$

Note that (27) also holds trivially for $(a,b)^T = (0,0)^T$, and hence it holds for all $(a,b)^T \in R^2$. Consequently, the lemma follows from the definition of Ψ_{NR}. □

4. A Descent Method and Global Convergence

In this section, we propose a derivative-free descent method based on the function ϕ_α. Furthermore, we prove global convergence of this method.

We consider the following search direction:

$$d^k(\beta) := -\frac{\partial\phi_\alpha}{\partial b}(x^k, F(x^k)) - \beta\frac{\partial\phi_\alpha}{\partial a}(x^k, F(x^k)), \tag{28}$$

where β is a parameter such that $\beta \in (0,1)$. Although this search direction is not necessarily a descent direction of Ψ_α at x^k for any $\beta \in (0,1)$, we can choose $\beta > 0$ sufficiently small so that $d^k(\beta)$ is a descent direction of Ψ_α at x^k, provided that the monotonicity assumption on F is fulfilled. We show this result in the following lemma.

Lemma 4.1 *Suppose that F is monotone and $\alpha \geq 0$. If x^k is not a solution of the NCP, then there exists $\bar\beta(x^k) \in (0,1)$ such that, for all $\beta \in [0, \bar\beta(x^k))$ the search direction $d^k(\beta)$ defined by (28) satisfies the descent condition*

$$\nabla\Psi_\alpha(x^k)^T d^k(\beta) < 0. \tag{29}$$

Proof. Since F is continuously differentiable, the function Ψ_α is also continuously differentiable by Lemma 3.1 (iii) and the gradient of Ψ_α at x^k is given by

$$\nabla\Psi_\alpha(x^k) = \frac{\partial\phi_\alpha}{\partial a}(x^k, F(x^k)) + \nabla F(x^k)\frac{\partial\phi_\alpha}{\partial b}(x^k, F(x^k)). \tag{30}$$

From (30) and the definition of $d^k(\beta)$, we have

$$\begin{aligned}
&\nabla\Psi_\alpha(x^k)^T d^k(\beta)\\
&= -\frac{\partial\phi_\alpha}{\partial a}(x^k, F(x^k))^T\frac{\partial\phi_\alpha}{\partial b}(x^k, F(x^k)) - \frac{\partial\phi_\alpha}{\partial b}(x^k, F(x^k))^T\nabla F(x^k)\frac{\partial\phi_\alpha}{\partial b}(x^k, F(x^k))\\
&\quad +\beta\left(-\|\frac{\partial\phi_\alpha}{\partial a}(x^k, F(x^k))\|^2 - \frac{\partial\phi_\alpha}{\partial a}(x^k, F(x^k))^T\nabla F(x^k)\frac{\partial\phi_\alpha}{\partial b}(x^k, F(x^k))\right). \quad (31)
\end{aligned}$$

Now let $p(x^k)$ and $q(x^k)$ be defined by

$$p(x^k) := -\frac{\partial\phi_\alpha}{\partial a}(x^k, F(x^k))^T\frac{\partial\phi_\alpha}{\partial b}(x^k, F(x^k)) - \frac{\partial\phi_\alpha}{\partial b}(x^k, F(x^k))^T\nabla F(x^k)\frac{\partial\phi_\alpha}{\partial b}(x^k, F(x^k))$$

and

$$q(x^k) := -\|\frac{\partial\phi_\alpha}{\partial a}(x^k, F(x^k))\|^2 - \frac{\partial\phi_\alpha}{\partial a}(x^k, F(x^k))^T\nabla F(x^k)\frac{\partial\phi_\alpha}{\partial b}(x^k, F(x^k)),$$

respectively. Then (31) is rewritten as

$$\nabla\Psi_\alpha(x^k)^T d^k(\beta) = p(x^k) + \beta q(x^k).$$

By the monotonicity of F, $\frac{\partial\phi_\alpha}{\partial b}(x^k, F(x^k))^T\nabla F(x^k)\frac{\partial\phi_\alpha}{\partial b}(x^k, F(x^k)) \geq 0$. If x^k is not a solution of the NCP, we have $\frac{\partial\phi_\alpha}{\partial a}(x^k, F(x^k))^T\frac{\partial\phi_\alpha}{\partial b}(x^k, F(x^k)) > 0$ from Lemma 3.1 (iv) and (v). Therefore $p(x^k) < 0$. Let $\bar\beta(x^k)$ be defined by

$$\bar\beta(x^k) := \begin{cases} -p(x^k)/q(x^k) & \text{if } q(x^k) \neq 0 \text{ and } 0 \leq -p(x^k)/q(x^k) \leq 1, \\ 1 & \text{otherwise.} \end{cases}$$

For such $\bar\beta(x^k)$, it is easily seen that, for all $\beta \subset [0, \bar\beta(x^k))$, the search direction $d^k(\beta)$ satisfies the descent condition $\nabla\Psi_\alpha(x^k)^T d^k(\beta) < 0$. □

The above lemma motivates us to propose the following descent algorithm.

Algorithm 4.1

Step1. *Choose* $x^0 \in R^n, \epsilon \geq 0, \sigma \in (0,1), \beta \in (0,1)$ *and* $\gamma \in (0,1)$. *Set* $k := 0$.

Step2. *If* $\Psi_\alpha(x^k) \leq \epsilon$, *stop:* x^k *is an approximate solution of the NCP.*

Step3. *Compute* $x^{k+1} := x^k + \gamma^{l_k} d^k(\beta^{l_k})$, *where*

$$d^k(\beta^{l_k}) := -\frac{\partial \phi_\alpha}{\partial b}(x^k, F(x^k)) - \beta^{l_k} \frac{\partial \phi_\alpha}{\partial a}(x^k, F(x^k))$$

with l_k *being the smallest nonnegative integer* l *satisfying*

$$\Psi_\alpha(x^k + \gamma^l d^k(\beta^l)) - \Psi_\alpha(x^k) \le -\sigma(\gamma^l)^2 \| \frac{\partial \phi_\alpha}{\partial a}(x^k, F(x^k)) + \frac{\partial \phi_\alpha}{\partial b}(x^k, F(x^k)) \|^2. \tag{32}$$

Step4. *Set* $k := k + 1$ *and go to Step 2.*

Note that the above algorithm has no need to compute the gradients of Ψ_α, and therefore there is also no need to compute the Jacobian of F. The most remarkable feature of Algorithm 4.1 is that not only the stepsize but also the search direction itself is adjusted during the backtracking search of Armijo-type, which may be regarded as a sort of curvilinear search.

We are ready to state the global convergence result for Algorithm 4.1 under the monotonicity assumption on F. In the remainder of the paper, we assume that the parameter ϵ used in Algorithm 4.1 is set to be zero and Algorithm 4.1 generates an infinite sequence $\{x^k\}$.

Theorem 4.1 *Suppose that* F *is monotone and* $\alpha \ge 0$. *Then Algorithm 4.1 is well-defined for any initial point* x^0. *Furthermore, if* x^* *is an accumulation point of the sequence* $\{x^k\}$ *generated by Algorithm 4.1, then* x^* *is a solution of the NCP.*

Proof. We first prove that Algorithm 4.1 is well-defined. It suffices to show that Step 3 is well-defined. Assume to the contrary that there is no nonnegative integer l satisfying (32). It then follows that for any integer $l \ge 0$

$$\Psi_\alpha(x^k + \gamma^l d^k(\beta^l)) - \Psi_\alpha(x^k) > -\sigma(\gamma^l)^2 \| \frac{\partial \phi_\alpha}{\partial a}(x^k, F(x^k)) + \frac{\partial \phi_\alpha}{\partial b}(x^k, F(x^k)) \|^2.$$

Dividing the above inequality by γ^l and passing to the limit $l \to \infty$, we have

$$\lim_{l \to \infty} \frac{\Psi_\alpha(x^k + \gamma^l d^k(\beta^l)) - \Psi_\alpha(x^k)}{\gamma^l} \ge 0. \tag{33}$$

Since Ψ_α is continuously differentiable, Ψ_α is locally Lipschitz continuous at x^k, i.e., there exist $L_1 > 0$ and $\delta > 0$ such that

$$\|\Psi_\alpha(x^k) - \Psi_\alpha(x')\| \le L_1 \|x^k - x'\|, \quad \text{for all } x' \text{ such that } \|x^k - x'\| < \delta.$$

Therefore, we have

$$\begin{aligned}
&\lim_{l\to\infty}\frac{\Psi_\alpha(x^k+\gamma^l d^k(\beta^l))-\Psi_\alpha(x^k)}{\gamma^l}\\
&=\lim_{l\to\infty}\left\{\frac{\Psi_\alpha(x^k+\gamma^l d^k(0))-\Psi_\alpha(x^k)}{\gamma^l}+\frac{\Psi_\alpha(x^k+\gamma^l d^k(\beta^l))-\Psi_\alpha(x^k+\gamma^l d^k(0))}{\gamma^l}\right\}\\
&=\nabla\Psi_\alpha(x^k)^T d^k(0)+\lim_{l\to\infty}\frac{\Psi_\alpha(x^k+\gamma^l d^k(\beta^l))-\Psi_\alpha(x^k+\gamma^l d^k(0))}{\gamma^l}\\
&\le \nabla\Psi_\alpha(x^k)^T d^k(0)+\lim_{l\to\infty}L_1\|d^k(\beta^l)-d^k(0)\|\\
&=\nabla\Psi_\alpha(x^k)^T d^k(0).
\end{aligned}\tag{34}$$

It then follows from (33) and (34) that

$$\nabla\Psi_\alpha(x^k)^T d^k(0)\ge 0.$$

This is a contradiction, because from Lemma 4.1, $d^k(0)$ must be a descent direction of Ψ_α at x^k if x^k is not a solution of the NCP. Thus Algorithm 4.1 is well-defined.

Next we prove that any accumulation point x^* of $\{x^k\}$ is a solution of the NCP. Let $\{x^k\}_{k\in K}$ be a subsequence converging to x^*. Then $\{x^k\}_{k\in K}$ is bounded and thus $\{d^k(\beta^{l_k})\}_{k\in K}$ is bounded since ϕ_α is continuously differentiable. Without loss of generality, we assume $d^k(\beta^{l_k})\to d^*$ as $k(\in K)\to\infty$. Since $\Psi_\alpha(x^k)$ decreases at each iteration, the right-hand side of (32) tends to 0. First we consider the case where $\{l_k\}_{k\in K}$ is bounded. Then $\{\gamma^{l_k}\}_{k\in K}$ does not approach 0, so we have

$$\|\frac{\partial\phi_\alpha}{\partial a}(x^*,F(x^*))+\frac{\partial\phi_\alpha}{\partial b}(x^*,F(x^*))\|^2=0.$$

Therefore, it follows from Lemma 3.3 that $\Psi_{NR}(x^*)=0$, i.e., x^* is a solution of the NCP. Next suppose that $\{l_k\}_{k\in K}$ is unbounded, which implies $\{\gamma^{l_k}\}_{k\in K}\to 0$. When $\{l_k\}_{k\in K}\to\infty$, it follows from Step 3 of Algorithm 4.1 that for all $k\in K$

$$\Psi_\alpha(x^k+\gamma^{l_k-1}d^k(\beta^{l_k-1}))-\Psi_\alpha(x^k)>-\sigma(\gamma^{l_k-1})^2\|\frac{\partial\phi_\alpha}{\partial a}(x^k,F(x^k))+\frac{\partial\phi_\alpha}{\partial b}(x^k,F(x^k))\|^2.$$

Dividing the above inequality by γ^{l_k-1} and passing to the limit $k\to\infty$, we have

$$\lim_{k\to\infty}\frac{\Psi_\alpha(x^k+\gamma^{l_k-1}d^k(\beta^{l_k-1}))-\Psi_\alpha(x^k)}{\gamma^{l_k-1}}\ge 0.\tag{35}$$

Since Ψ_α is continuously differentiable, by the Mean-Value theorem there exists some $\tilde\gamma^k\in[0,\gamma^{l_k-1}]$ such that

$$\Psi_\alpha(x^k+\gamma^{l_k-1}d^k(\beta^{l_k-1}))-\Psi_\alpha(x^k)=\gamma^{l_k-1}\nabla\Psi_\alpha(x^k+\tilde\gamma^k d^k(\beta^{l_k-1}))^T d^k(\beta^{l_k-1}).\tag{36}$$

As $k(\in K) \to \infty$, we have $\{\gamma^{l_k-1}\} \to 0$, which in turn implies $\{\tilde{\gamma}^k\}_{k\in K} \to 0$ and hence

$$\lim_{k\to\infty} \tilde{\gamma}^k d^k(\beta^{l_k-1}) = 0. \tag{37}$$

It then follows from (36), (37) and the given assumptions that

$$\begin{aligned}\lim_{k\to\infty} \frac{\Psi_\alpha(x^k + \gamma^{l_k-1} d^k(\beta^{l_k-1})) - \Psi_\alpha(x^k)}{\gamma^{l_k-1}} &= \lim_{k\to\infty} \nabla\Psi_\alpha(x^k + \tilde{\gamma}^k d^k(\beta^{l_k-1}))^T d^k(\beta^{l_k-1}) \\ &= \nabla\Psi_\alpha(x^*)^T d^*, \end{aligned} \tag{38}$$

where $\tilde{\gamma}^k \in [0, \gamma^{l_k-1}]$. Combining (35) and (38), we obtain

$$\nabla\Psi_\alpha(x^*)^T d^* \geq 0.$$

Since $\nabla\Psi_\alpha(x^*)^T d^* \leq 0$ holds obviously, we have $\nabla\Psi_\alpha(x^*)^T d^* = 0$. Moreover, since passing to the limit in (28) and (30) on the subsequence yields

$$d^* = -\frac{\partial\phi_\alpha}{\partial b}(x^*, F(x^*))$$

and

$$\nabla\Psi_\alpha(x^*) = \frac{\partial\phi_\alpha}{\partial a}(x^*, F(x^*)) + \nabla F(x^*)\frac{\partial\phi_\alpha}{\partial b}(x^*, F(x^*)),$$

we obtain

$$\begin{aligned} &\nabla\Psi_\alpha(x^*)^T d^* \\ &= -\frac{\partial\phi_\alpha}{\partial a}(x^*, F(x^*))^T \frac{\partial\phi_\alpha}{\partial b}(x^*, F(x^*)) - \frac{\partial\phi_\alpha}{\partial b}(x^*, F(x^*))^T \nabla F(x^*) \frac{\partial\phi_\alpha}{\partial b}(x^*, F(x^*)) \\ &= 0. \end{aligned}$$

Consequently, by the monotonicity of F and Lemma 3.1 (iv), we have

$$\frac{\partial\phi_\alpha}{\partial a}(x^*, F(x^*))^T \frac{\partial\phi_\alpha}{\partial b}(x^*, F(x^*)) = 0.$$

It then follows from Lemma 3.1 (i) and (v) that x^* is a solution of the NCP. □

Combining the above theorem and Lemma 3.4, we obtain the following global convergence result.

Theorem 4.2 *Suppose that either of the following conditions (a) and (b) holds:*

(a) *$\alpha > 0$, F is monotone and the NCP is strictly feasible.*

(b) *$\alpha \geq 0$ and F is strongly monotone.*

Then Algorithm 4.1 is well-defined for any initial point x^0. Furthermore, the sequence $\{x^k\}$ generated by Algorithm 4.1 has at least one accumulation point and any accumulation point of $\{x^k\}$ is a solution of the NCP. □

5. Linear Convergence

In this section, we show that any sequence generated by Algorithm 4.1 converges linearly to the solution if F is strongly monotone.

Before proving the linear convergence result, we give the following lemma.

Lemma 5.1 *Suppose that F is strongly monotone with modulus $\mu > 0$ and that $\alpha \geq 0$. Let $x^0 \in R^n$ be any given vector. Then there exists an integer $\hat{l} \geq 0$ such that, for each k and for all $l \geq \hat{l}$, the search direction $d^k(\beta^l)$ satisfies*

$$\nabla \Psi_\alpha(x^k)^T d^k(\beta^l) \leq -\frac{\beta^l}{2}\left(\|\frac{\partial \phi_\alpha}{\partial a}(x^k, F(x^k))\| + \|\frac{\partial \phi_\alpha}{\partial b}(x^k, F(x^k))\|\right)^2$$

Proof. It follows from Theorem 3.4 that the level set $\mathcal{L}(\Psi_\alpha, \Psi_\alpha(x^0))$ is bounded. Since $\{x^k\}$ is contained in $\mathcal{L}(\Psi_\alpha, \Psi_\alpha(x^0))$ and ∇F is continuous, there exists a constant $\nu > 0$ such that

$$\|\nabla F(x^k)\| \leq \nu \quad \text{for all} \quad k. \tag{39}$$

In the following, we omit superscript k for simplicity. From (28) and (30), we have

$$\begin{aligned}
&\nabla \Psi_\alpha(x)^T d(\beta^l) \\
&= -\frac{\partial \phi_\alpha}{\partial a}(x, F(x))^T \frac{\partial \phi_\alpha}{\partial b}(x, F(x)) - \frac{\partial \phi_\alpha}{\partial b}(x, F(x))^T \nabla F(x) \frac{\partial \phi_\alpha}{\partial b}(x, F(x)) \\
&\quad - \beta^l \left(\|\frac{\partial \phi_\alpha}{\partial a}(x, F(x))\|^2 + \frac{\partial \phi_\alpha}{\partial a}(x, F(x))^T \nabla F(x) \frac{\partial \phi_\alpha}{\partial b}(x, F(x))\right) \\
&\leq -\mu \|\frac{\partial \phi_\alpha}{\partial b}(x, F(x))\|^2 - \beta^l \|\frac{\partial \phi_\alpha}{\partial a}(x, F(x))\|^2 \\
&\quad - \beta^l \frac{\partial \phi_\alpha}{\partial a}(x, F(x))^T \nabla F(x) \frac{\partial \phi_\alpha}{\partial b}(x, F(x)) \\
&\leq -\mu \|\frac{\partial \phi_\alpha}{\partial b}(x, F(x))\|^2 - \beta^l \|\frac{\partial \phi_\alpha}{\partial a}(x, F(x))\|^2 \\
&\quad + \beta^l \nu \|\frac{\partial \phi_\alpha}{\partial a}(x, F(x))\| \|\frac{\partial \phi_\alpha}{\partial b}(x, F(x))\|,
\end{aligned} \tag{40}$$

where the first inequality follows from Lemma 3.1 (iv) and the strong monotonicity of F and, the second inequality follows from the Cauchy-Schwartz inequality and (39).

Now let $\hat{l}$ be defined by

$$\hat{l} := \left\lceil \log_\beta \frac{2\mu}{\nu^2 + 2\nu + 2} \right\rceil.$$

Then β^l satisfies

$$\beta^l \leq \frac{2\mu}{\nu^2 + 2\nu + 2} \quad \text{for all} \quad l \geq \hat{l}. \tag{41}$$

Since (41) implies $\mu - \frac{\beta^l}{2} \geq 0$, the right-hand side of (40) can be rewritten as

$$\begin{aligned}
&-\mu\|\frac{\partial\phi_\alpha}{\partial b}(x,F(x))\|^2 - \beta^l\|\frac{\partial\phi_\alpha}{\partial a}(x,F(x))\|^2 + \beta^l\nu\|\frac{\partial\phi_\alpha}{\partial a}(x,F(x))\|\|\frac{\partial\phi_\alpha}{\partial b}(x,F(x))\| \\
&= -\frac{\beta^l}{2}\left(\|\frac{\partial\phi_\alpha}{\partial a}(x,F(x))\| + \|\frac{\partial\phi_\alpha}{\partial b}(x,F(x))\|\right)^2 - \left(\mu - \frac{\beta^l}{2}\right)\|\frac{\partial\phi_\alpha}{\partial b}(x,F(x))\|^2 \\
&\quad - \frac{\beta^l}{2}\|\frac{\partial\phi_\alpha}{\partial a}(x,F(x))\|^2 + \beta^l(\nu+1)\|\frac{\partial\phi_\alpha}{\partial a}(x,F(x))\|\|\frac{\partial\phi_\alpha}{\partial b}(x,F(x))\| \\
&= -\frac{\beta^l}{2}\left(\|\frac{\partial\phi_\alpha}{\partial a}(x,F(x))\| + \|\frac{\partial\phi_\alpha}{\partial b}(x,F(x))\|\right)^2 \\
&\quad - \left(\sqrt{\frac{\beta^l}{2}}\|\frac{\partial\phi_\alpha}{\partial a}(x,F(x))\| - \sqrt{\mu - \frac{\beta^l}{2}}\|\frac{\partial\phi_\alpha}{\partial b}(x,F(x))\|\right)^2 \\
&\quad + \left\{\beta^l(\nu+1) - \sqrt{\beta^l(2\mu-\beta^l)}\right\}\|\frac{\partial\phi_\alpha}{\partial a}(x,F(x))\|\|\frac{\partial\phi_\alpha}{\partial b}(x,F(x))\| \qquad (42)
\end{aligned}$$

for all $l \geq \hat{l}$. Since (41) implies

$$\begin{aligned}
\sqrt{\beta^l(2\mu-\beta^l)} &\geq \sqrt{\beta^{2l}(\nu^2+2\nu+2) - \beta^{2l}} \\
&= \sqrt{\beta^{2l}(\nu+1)^2} \\
&= \beta^l(\nu+1),
\end{aligned}$$

the last term of (42) is nonpositive for all $l \geq \hat{l}$. Consequently, it follows from (40) and (42) that for all $l \geq \hat{l}$

$$\nabla\Psi_\alpha(x)^T d(\beta^l) \leq -\frac{\beta^l}{2}\left(\|\frac{\partial\phi_\alpha}{\partial a}(x,F(x))\| + \|\frac{\partial\phi_\alpha}{\partial b}(x,F(x))\|\right)^2.$$

□

Now the linear convergence result for Algorithm 4.1 is presented.

Theorem 5.1 *Suppose that F is strongly monotone and $\alpha \geq 0$. Suppose also that ∇F is Lipschitz continuous on any bounded set. Let the parameters γ and β be chosen to satisfy $\gamma < \beta$. Then the sequence $\{x^k\}$ generated by Algorithm 4.1 converges R-linearly to the solution of the NCP and the sequence $\{\Psi_\alpha(x^k)\}$ converges Q-linearly to zero.*

Proof. We first show that the sequence $\{\Psi_\alpha(x^k)\}$ converges Q-linearly to zero. By Theorem 3.4, the level set $\mathcal{L}(\Psi_\alpha, \Psi_\alpha(x^0))$ is bounded. Then since F, ∇F and $\nabla\Psi_\alpha$ are all Lipschitz continuous on $\mathcal{L}(\Psi_\alpha, \Psi_\alpha(x^0))$, there exists a constant $L_2 > 0$ such that

$$\|\nabla\Psi_\alpha(x) - \nabla\Psi_\alpha(x')\| \leq L_2\|x - x'\| \quad \text{for all } x, x' \in \mathcal{L}(\Psi_\alpha, \Psi_\alpha(x^0)). \qquad (43)$$

Since the sequence $\{\Psi_\alpha(x^k)\}$ is nonincreasing, $\{x^k\} \subseteq \mathcal{L}(\Psi_\alpha, \Psi_\alpha(x^0))$. It then follows from (43) and the Cauchy-Schwartz inequality that for each $t \in [0,1]$

$$\begin{aligned}
\Psi_\alpha(x^k + td^k(\beta^l)) - \Psi_\alpha(x^k) &= \int_0^t \nabla\Psi_\alpha(x^k + sd^k(\beta^l))^T d^k(\beta^l)ds \\
&= t\nabla\Psi_\alpha(x^k)^T d^k(\beta^l) \\
&\quad + \int_0^t \left(\nabla\Psi_\alpha(x^k + sd^k(\beta^l)) - \nabla\Psi_\alpha(x^k)\right)^T d^k(\beta^l)ds \\
&\leq t\nabla\Psi_\alpha(x^k)^T d^k(\beta^l) + L_2\int_0^t s\|d^k(\beta^l)\|^2 ds \\
&= t\left(\nabla\Psi_\alpha(x^k)^T d^k(\beta^l) + \frac{L_2 t}{2}\|d^k(\beta^l)\|^2\right). \qquad (44)
\end{aligned}$$

By Lemma 3.1 (iv) and $\beta^l \in (0,1)$, we have

$$\begin{aligned}
&\|d^k(\beta^l)\|^2 \\
&= \| - \frac{\partial\phi_\alpha}{\partial b}(x^k, F(x^k)) - \beta^l \frac{\partial\phi_\alpha}{\partial a}(x^k, F(x^k))\|^2 \\
&= \|\frac{\partial\phi_\alpha}{\partial b}(x^k, F(x^k))\|^2 + \beta^{2l}\|\frac{\partial\phi_\alpha}{\partial a}(x^k, F(x^k))\|^2 + 2\beta^l\frac{\partial\phi_\alpha}{\partial a}(x^k, F(x^k))^T\frac{\partial\phi_\alpha}{\partial b}(x^k, F(x^k)) \\
&\leq \|\frac{\partial\phi_\alpha}{\partial b}(x^k, F(x^k))\|^2 + \|\frac{\partial\phi_\alpha}{\partial a}(x^k, F(x^k))\|^2 + 2\frac{\partial\phi_\alpha}{\partial a}(x^k, F(x^k))^T\frac{\partial\phi_\alpha}{\partial b}(x^k, F(x^k)) \\
&= \|\frac{\partial\phi_\alpha}{\partial a}(x^k, F(x^k)) + \frac{\partial\phi_\alpha}{\partial b}(x^k, F(x^k))\|^2. \qquad (45)
\end{aligned}$$

Let $\hat{l}$ be as specified in the proof of Lemma 5.1. Then, from (44), (45) and Lemma 5.1, we have for all $l \geq \hat{l}$

$$\begin{aligned}
&\Psi_\alpha(x^k + \gamma^l d^k(\beta^l)) - \Psi_\alpha(x^k) \\
&\leq \gamma^l\left(-\frac{\beta^l}{2}\left(\|\frac{\partial\phi_\alpha}{\partial a}(x^k, F(x^k))\| + \|\frac{\partial\phi_\alpha}{\partial b}(x^k, F(x^k))\|\right)^2 + \frac{L_2\gamma^l}{2}\|d^k(\beta^l)\|^2\right) \\
&\leq \gamma^l\left(-\frac{\beta^l}{2}\|\frac{\partial\phi_\alpha}{\partial a}(x^k, F(x^k)) + \frac{\partial\phi_\alpha}{\partial b}(x^k, F(x^k))\|^2 + \frac{L_2\gamma^l}{2}\|d^k(\beta^l)\|^2\right) \\
&\leq -\frac{\gamma^l}{2}(\beta^l - L_2\gamma^l)\|\frac{\partial\phi_\alpha}{\partial a}(x^k, F(x^k)) + \frac{\partial\phi_\alpha}{\partial b}(x^k, F(x^k))\|^2, \qquad (46)
\end{aligned}$$

where the second inequality follows from the Cauchy-Schwartz inequality and the last inequality follows from (45). By (46), condition (32) is satisfied whenever

$$-\frac{\gamma^l}{2}(\beta^l - L_2\gamma^l) \leq -\sigma(\gamma^l)^2.$$

Since this inequality can be rewritten as

$$\left(\frac{\gamma}{\beta}\right)^l \leq \frac{1}{2\sigma + L_2}, \qquad (47)$$

condition (32) is satisfied for all $l \geq \bar{l}$, where $\bar{l}$ is defined by

$$\bar{l} := \max\left\{\hat{l}, \left\lceil \log_{\frac{\beta}{\gamma}}(2\sigma + L_2)\right\rceil\right\}. \tag{48}$$

Notice that $\bar{l}$ does not depend on k. Thus, we have $l_k \leq \bar{l}$ for all k since l_k is the smallest nonnegative integer l satisfying (32). It then follows from (32) and Lemma 3.3 that

$$\begin{aligned}
\Psi_\alpha(x^{k+1}) - \Psi_\alpha(x^k) &\leq -\sigma(\gamma^{l_k})^2 \|\frac{\partial \phi_\alpha}{\partial a}(x^k, F(x^k)) + \frac{\partial \phi_\alpha}{\partial b}(x^k, F(x^k))\|^2 \\
&\leq -\sigma(\gamma^{\bar{l}})^2 \|\frac{\partial \phi_\alpha}{\partial a}(x^k, F(x^k)) + \frac{\partial \phi_\alpha}{\partial b}(x^k, F(x^k))\|^2 \\
&\leq -(2-\sqrt{2})^4 \sigma \gamma^{2\bar{l}} \Psi_{NR}(x^k).
\end{aligned} \tag{49}$$

Since the level set $\mathcal{L}(\Psi_\alpha, \Psi_\alpha(x^0))$ is bounded, it follows from Lemma 3.2 that

$$\Psi_\alpha(x^k) \leq \left\{\frac{\alpha}{2}B^2 + \frac{1}{2}(\sqrt{2}+2)^2\right\} \Psi_{NR}(x^k) \quad \text{for all } x^k \in \mathcal{L}(\Psi_\alpha, \Psi_\alpha(x^0)), \tag{50}$$

where B is given by

$$B = \sup\left\{ \max_{1\leq i\leq n} \left\{\max\{|x_i^k|, |F_i(x^k)|\}\right\} \,\middle|\, x \in \mathcal{L}(\Psi_\alpha, \Psi_\alpha(x^0))\right\}.$$

Therefore from (49) and (50) we have

$$\Psi_\alpha(x^{k+1}) - \Psi_\alpha(x^k) \leq -\frac{2(2-\sqrt{2})^4}{\alpha B^2 + (\sqrt{2}+2)^2} \sigma \gamma^{2\bar{l}} \Psi_\alpha(x^k). \tag{51}$$

Consequently, we obtain

$$\left(1 - \frac{2(2-\sqrt{2})^4 \sigma \gamma^{2\bar{l}}}{\alpha B^2 + (\sqrt{2}+2)^2}\right) \Psi_\alpha(x^k) \geq \Psi_\alpha(x^{k+1}) \geq 0.$$

On the other hand, since $(\sqrt{2}+2)^2 > 2(2-\sqrt{2})^4$, $\sigma\gamma^{2\bar{l}} \in (0,1)$ and $\alpha B^2 \geq 0$, we have

$$0 < 1 - \frac{2(2-\sqrt{2})^4 \sigma \gamma^{2\bar{l}}}{\alpha B^2 + (\sqrt{2}+2)^2} < 1.$$

Therefore the sequence $\{\Psi_\alpha(x^k)\}$ converges Q-linearly to zero.

Next we prove that the sequence $\{x^k\}$ converges R-linearly to the unique solution x^* of the NCP. From Lemma 2.1 and Lemma 3.2, there exists a positive constant κ such that

$$\|x - x^*\| \leq \kappa \Psi_\alpha(x)^{\frac{1}{2}} \quad \text{for all } x \in \mathcal{L}(\Psi_\alpha, \Psi_\alpha(x^0)).$$

Since the sequence $\{\Psi_\alpha(x^k)\}$ converges Q-linearly to zero, the sequence $\{x^k\}$ converges R-linearly to the solution of the NCP. □

The conditions in the above theorem are weaker than those in [18]. For the linear convergence, we only need the condition $\gamma < \beta$, which does not involve unknown parameters. Also, note that some parameters used in Algorithm 4.1 do not rely on usually unknown constants such as the modulus of strong monotonicity of F.

6. Concluding Remarks

In this paper, we have proposed a merit function for the NCP and shown its properties. We have also proposed a new derivative-free descent algorithm based on this function for solving the NCP and proven global and linear convergence theorems under appropriate assumptions.

The most interesting part of Algorithm 4.1 is Step 3, where the search direction and the stepsize are adjusted simultaneously, whereas search directions in the algorithms proposed in [10, 12, 18, 24] are fixed during a linesearch process. This particular feature enables us to establish the linear convergence of the algorithm under more practical assumptions compared with the algorithm of [18].

Derivative-free descent algorithms have also been proposed for solving the VIP [8, 27]. However, the convergence rate results for those algorithms have been scarce. Also, those algorithms generally use parameters that have to be specified by using unknown constants. It is an interesting future topic to extend the results of this paper to the more general class of VIPs.

References

[1] B. Chen, X. Chen and C. Kanzow (1997), A penalized Fischer-Burmeister NCP-function: Theoretical investigation and numerical results, Working paper, School of Mathematics, University of New South Wales.

[2] R.W. Cottle, J.-S. Pang and R.E. Stone (1992), *The Linear Complementarity Problem*, Academic Press, New York.

[3] S. Dafermos (1983), An iterative scheme for variational inequalities, *Mathematical Programming*, vol. 26, pp. 40-47.

[4] F. Facchinei and C. Kanzow (1997), On unconstrained and constrained stationary points of the implicit Lagrangian, *Journal of Optimization Theory and Applications*, vol. 92, pp. 99-115.

[5] F. Facchinei and J. Soares (1997), A new merit function for nonlinear complementarity problems and a related algorithm, *SIAM Journal on Optimization*, vol. 7, pp. 225-247.

[6] A. Fischer (1992), A special Newton-type optimization method, *Optimization*, vol. 24, pp. 269-284.

[7] A. Fischer (1995), An NCP-Function and its use for the solution of complementarity problems, *Recent Advances in Nonsmooth Optimization*, D.Z. Du, L. Qi, and R.S. Womersley (eds.), World Scientific Publishers, Singapore, pp. 88-105.

[8] M. Fukushima (1992), Equivalent differentiable optimization problems and descent methods for asymmetric variational inequality problems, *Mathematical Programming*, vol. 53, pp. 99-110.

[9] M. Fukushima (1996), Merit functions for variational inequality and complementarity problems, *Nonlinear Optimization and Applications*, G. Di Pillo and F. Giannessi (eds.), Plenum Press, New York, pp. 155-170.

[10] C. Geiger and C. Kanzow (1996), On the resolution of monotone complementarity problems, *Computational Optimization and Applications*, vol. 5, pp. 155-173.

[11] P.T. Harker and J.-S. Pang (1990), Finite dimensional variational inequality and nonlinear complementarity problems: A survey of theory, algorithms and applications, *Mathematical Programming*, vol. 48, pp. 161-220.

[12] H. Jiang (1996), Unconstrained minimization approaches to nonlinear complementarity problems, *Journal of Global Optimization*, vol. 9, pp. 169-181.

[13] C. Kanzow (1994), An unconstrained optimization technique for large-scale linearly constrained convex minimization problems, *Computing*, vol. 53, pp. 101-117.

[14] C. Kanzow (1996), Nonlinear complementarity as unconstrained optimization, *Journal of Optimization Theory and Applications*, vol. 88, pp. 139-155.

[15] C. Kanzow, N. Yamashita and M. Fukushima (1997), New NCP-functions and their properties, *Journal of Optimization Theory and Applications*, vol. 94, pp. 115-135.

[16] O.L. Mangasarian (1976), Equivalence of the complementarity problem to a system of nonlinear equations, *SIAM Journal on Applied Mathematics*, vol. 31, pp. 89-92.

[17] O.L. Mangasarian and M.V. Solodov (1993), Nonlinear complementarity as unconstrained and constrained minimization, *Mathematical Programming*, vol. 62, pp. 277-297.

[18] O.L. Mangasarian and M.V. Solodov (forthcoming), A linearly convergent descent method for strongly monotone complementarity problems, *Computational Optimization and Applications*.

[19] J.-S. Pang (1987), A posteriori error bounds for the linearly-constrained variational inequality problem, *Mathematics of Operations Research*, vol. 12, pp. 474-484.

[20] J.-S. Pang (1994), Complementarity problems, *Handbook of Global Optimization*, R. Horst and P. Pardalos (eds.), Kluwer Academic Publishers, Boston, Massachusetts, pp. 271-338.

[21] J.-S. Pang (1990), Newton's method for B-differentiable equations, *Mathematics of Operations Research*, vol. 15, pp. 311-341.

[22] J.-S. Pang and D. Chan (1982), Iterative methods for variational and complementarity problems, *Mathematics Programming*, vol. 24, pp. 284-313.

[23] P. Tseng (1996), Growth behaviour of a class of merit functions for the nonlinear complementarity problem, *Journal of Optimization Theory and Applications*, vol. 89, pp. 17-37.

[24] N. Yamashita and M. Fukushima (1995), On stationary points of the implicit Lagrangian for nonlinear complementarity problems, *Journal of Optimization Theory and Applications*, vol. 84, pp. 653-663.

[25] N. Yamashita and M. Fukushima (1997), Modified Newton methods for solving a semismooth reformulation of monotone complementarity problems, *Mathematical Programming*, vol. 76, pp. 469-491.

[26] N. Yamashita and M. Fukushima (1998), A new merit function and a descent method for semidefinite complementarity problems, *Reformulation: Nonsmooth, Piecewise Smooth, Semismooth and Smoothing Methods*, M. Fukushima and L. Qi (eds.), Kluwer Academic Publishers, Boston, Massachusetts, pp. 401-416.

[27] N. Yamashita, K. Taji and M. Fukushima (1997), Unconstrained optimization reformulations of variational inequality problems, *Journal of Optimization Theory and Applications*, vol. 92, pp. 439-456.

Applied Optimization

1. D.-Z. Du and D.F. Hsu (eds.): *Combinatorial Network Theory.* 1996
ISBN 0-7923-3777-8
2. M.J. Panik: *Linear Programming: Mathematics, Theory and Algorithms.* 1996
ISBN 0-7923-3782-4
3. R.B. Kearfott and V. Kreinovich (eds.): *Applications of Interval Computations.* 1996 ISBN 0-7923-3847-2
4. N. Hritonenko and Y. Yatsenko: *Modeling and Optimimization of the Lifetime of Technology.* 1996 ISBN 0-7923-4014-0
5. T. Terlaky (ed.): *Interior Point Methods of Mathematical Programming.* 1996
ISBN 0-7923-4201-1
6. B. Jansen: *Interior Point Techniques in Optimization.* Complementarity, Sensitivity and Algorithms. 1997 ISBN 0-7923-4430-8
7. A. Migdalas, P.M. Pardalos and S. Story (eds.): *Parallel Computing in Optimization.* 1997 ISBN 0-7923-4583-5
8. F.A. Lootsma: *Fuzzy Logic for Planning and Decision Making.* 1997
ISBN 0-7923-4681-5
9. J.A. dos Santos Gromicho: *Quasiconvex Optimization and Location Theory.* 1998
ISBN 0-7923-4694-7
10. V. Kreinovich, A. Lakeyev, J. Rohn and P. Kahl: *Computational Complexity and Feasibility of Data Processing and Interval Computations.* 1998
ISBN 0-7923-4865-6
11. J. Gil-Aluja: *The Interactive Management of Human Resources in Uncertainty.* 1998
ISBN 0-7923-4886-9
12. C. Zopounidis and A.I. Dimitras: *Multicriteria Decision Aid Methods for the Prediction of Business Failure.* 1998 ISBN 0-7923-4900-8
13. F. Giannessi, S. Komlósi and T. Rapcsák (eds.): *New Trends in Mathematical Programming.* Homage to Steven Vajda. 1998 ISBN 0-7923-5036-7
14. Ya-xiang Yuan (ed.): *Advances in Nonlinear Programming.* Proceedings of the '96 International Conference on Nonlinear Programming. 1998 ISBN 0-7923-5053-7
15. W.W. Hager and P.M. Pardalos: *Optimal Control.* Theory, Algorithms, and Applications. 1998 ISBN 0-7923-5067-7
16. Gang Yu (ed.): *Industrial Applications of Combinatorial Optimization.* 1998
ISBN 0-7923-5073-1
17. D. Braha and O. Maimon (eds.): *A Mathematical Theory of Design: Foundations, Algorithms and Applications.* 1998 ISBN 0-7923-5079-0

Applied Optimization

18. O. Maimon, E. Khmelnitsky and K. Kogan: *Optimal Flow Control in Manufacturing.* Production Planning and Scheduling. 1998 ISBN 0-7923-5106-1
19. C. Zopounidis and P.M. Pardalos (eds.): *Managing in Uncertainty: Theory and Practice.* 1998 ISBN 0-7923-5110-X
20. A.S. Belenky: *Operations Research in Transportation Systems:* Ideas and Schemes of Optimization Methods for Strategic Planning and Operations Management. 1998 ISBN 0-7923-5157-6
21. J. Gil-Aluja: *Investment in Uncertainty.* 1999 ISBN 0-7923-5296-3
22. M. Fukushima and L. Qi (eds.): *Reformulation: Nonsmooth, Piecewise Smooth, Semismooth and Smooting Methods.* 1999 ISBN 0-7923-5320-X
23. M. Patriksson: *Nonlinear Programming and Variational Inequality Problems.* A Unified Approach. 1999 ISBN 0-7923-5455-9
24. R. De Leone, A. Murli, P.M. Pardalos and G. Toraldo (eds.): *High Performance Algorithms and Software in Nonlinear Optimization.* 1999 ISBN 0-7923-5483-4
25. A. Schöbel: *Locating Lines and Hyperplanes.* Theory and Algorithms. 1999 ISBN 0-7923-5559-8
26. R.B. Statnikov: *Multicriteria Design.* Optimization and Identification. 1999 ISBN 0-7923-5560-1
27. V. Tsurkov and A. Mironov: *Minimax under Transportation Constrains.* 1999 ISBN 0-7923-5609-8
28. V.I. Ivanov: *Model Development and Optimization.* 1999 ISBN 0-7923-5610-1
29. F.A. Lootsma: *Multi-Criteria Decision Analysis via Ratio and Difference Judgement.* 1999 ISBN 0-7923-5669-1
30. A. Eberhard, R. Hill, D. Ralph and B.M. Glover (eds.): *Progress in Optimization.* Contributions from Australasia. 1999 ISBN 0-7923-5733-7
31. T. Hürlimann: *Mathematical Modeling and Optimization.* An Essay for the Design of Computer-Based Modeling Tools. 1999 ISBN 0-7923-5927-5
32. J. Gil-Aluja: *Elements for a Theory of Decision in Uncertainty.* 1999 ISBN 0-7923-5987-9
33. H. Frenk, K. Roos, T. Terlaky and S. Zhang (eds.): *High Performance Optimization.* 1999 ISBN 0-7923-6013-3
34. N. Hritonenko and Y. Yatsenko: *Mathematical Modeling in Economics, Ecology and the Environment.* 1999 ISBN 0-7923-6015-X
35. J. Virant: *Design Considerations of Time in Fuzzy Systems.* 2000 ISBN 0-7923-6100-8

Applied Optimization

36. G. Di Pillo and F. Giannessi (eds.): *Nonlinear Optimization and Related Topics*. 2000
ISBN 0-7923-6109-1

KLUWER ACADEMIC PUBLISHERS – DORDRECHT / BOSTON / LONDON

GPSR Compliance

The European Union's (EU) General Product Safety Regulation (GPSR) is a set of rules that requires consumer products to be safe and our obligations to ensure this.

If you have any concerns about our products, you can contact us on ProductSafety@springernature.com

In case Publisher is established outside the EU, the EU authorized representative is:

Springer Nature Customer Service Center GmbH
Europaplatz 3
69115 Heidelberg, Germany

Batch number: 10817789

Printed by Printforce, the Netherlands